Plant Galls

Denis Coelho de Oliveira
Rosy Mary dos Santos Isaias
Editors

Plant Galls

Structure and Functions

Springer

Editors
Denis Coelho de Oliveira
Institute of Biology
Universidade Federal de Uberlândia (UFU)
Uberlândia, Minas Gerais, Brazil

Rosy Mary dos Santos Isaias
Department of Botany
Universidade Federal de Minas
Gerais (UFMG)
Belo Horizonte, Minas Gerais, Brazil

ISBN 978-3-031-80066-5 ISBN 978-3-031-80064-1 (eBook)
https://doi.org/10.1007/978-3-031-80064-1

This Springer imprint is published by the registered company Springer Nature Switzerland AG
The registered company address is: Gewerbestrasse 11, 6330 Cham, Switzerland

If disposing of this product, please recycle the paper.

Preface

This book gathers essential information on galls, comprising scientists' interest in these fascinating structures worldwide. It also reflects the scientific maturity of our research group, "Structure, Chemistry, and Physiology of Neotropical Galls," registered at the Brazilian National Council for Research (Conselho Nacional de Desenvolvimento Científico e Tecnológico—CNPq). Over the years, we have interacted with researchers from Brazil and some other countries, contributing to the production of knowledge on plant-insect interactions. Various inspirations led us to propose the construction of this book and invite our colleagues. The most important was our desire to provide relevant information on gall development, structure, and functions to enthusiasts, students, and young researchers in plant biology, ecology, zoology, and other scientific fields.

The study of plant galls bloomed in the mid-1980s, with research primarily focused on plant gall inventories and testing widely held ecological and evolutionary hypotheses, especially regarding the distribution of galls and gall-inducing organisms across different biomes on Earth. By the early 2000s, studies on gall structure and metabolism became prominent, leading to a deeper understanding of the complexity of galls. Recently, molecular analyses based on transcriptome annotation have corroborated this structural and metabolic complexity, highlighting the need to understand the functionality of gall tissues and how it provides adaptive significance for gall-inducing organisms. The studies presented in this book synthesize current knowledge and open new perspectives for a deeper understanding of gall structure and functions, with contributions from students and researchers worldwide.

Gall development goes beyond the commonly reported cell hypertrophy and tissue hyperplasia. The peculiar cytological and histological diagnostic features demand chemical and physiological host plant traits that, under the control of the gall-inducing organism, result in the high diversity of gall shapes found in nature. Among gall-inducing organisms, insects stand out for their diversity and ability to manipulate host plant tissues into complex structures, and their diversity is also peculiarly reflected in gall structures and functions. In this book, the readers will find recent information about how host plant tissues respond to gall-inducing organisms, especially insects, at the molecular, cellular, tissue, and metabolic levels. Despite the control that gall-inducing organisms exert over gall induction and development, our focus is on the structure and function of plant gall tissues. A gall is, after all, plant tissue.

With this focus on structural and functional perspectives, we have divided this book into four main sections. The first section, comprising six chapters, discusses the molecular and host tissue changes during gall development. The second section, with four chapters, covers important physiological and chemical properties of galls. The third section, with seven chapters, focuses on the structure and function of galls induced by different taxa of gall-inducing organisms. The final section, with three chapters, provides relevant information on how interactions with other trophic levels can alter gall structure and metabolism and determine ecological and evolutionary trends.

Through this book, we aim to bridge the gap between researchers from different fields and draw attention to the importance of understanding host plant cell and tissue transformations during gall development, as well as their adaptive significance for gall-inducing organisms and associated fauna. Editing this book has been a challenge for us, but we have succeeded. We extend our gratitude to all the collaborators who joined us on this journey.

Uberlândia, Minas Gerais, Brazil — Denis Coelho de Oliveira
Belo Horizonte, Minas Gerais, Brazil — Rosy Mary dos Santos Isaias

Contents

Meristematic Features Influence on Gall Development

Rosy Mary dos Santos Isaias, Ígor Abba Arriola, and Bruno Garcia Ferreira

1 Introduction

The interaction between host plants and the gall inducers demands the choice of adequate sites for oviposition and gall induction. Even though inducers can stimulate mature cells and tissues, ovipositing at responsive sites, such as meristems and developing tissue sites, is a common preference among galling organisms (Rohfritsch 1992). Gall induction in meristems or developing tissues is an adequate condition to trigger gall development and the success of the life cycle of inducers. Meristems are responsible for determining plant cell fates and constructing the plant body, and as such, they must maintain a balance between the self-renewal of a reservoir of central stem cells and organ initiation from peripheral cells (Ha et al. 2010). On the other hand, gall induction involves reprogramming the original plant meristematic cells, changing the rates and patterns of division, expansion, and differentiation, which culminates in gall morphogenesis. This morphogenesis is so peculiar to each host plant-gall inducer system that such reprogramming is species-specific and results in various gall morphotypes (Fig. 1) (Isaias et al. 2013, 2014). By the time of gall maturation, some morphotypes may maintain their meristems active, or the parenchyma cells may redifferentiate into cambium-like cells, adding new cell layers to the outer parenchyma or the nutritive tissues (Bragança et al. 2021a, b; Ferreira et al. 2022). A true vascular cambium may occur in stem, petiolar, or midrib galls, producing secondary xylem and phloem cells, which improve the flux of water,

R. M. dos Santos Isaias (✉) · Í. A. Arriola
Department of Botany, Universidade Federal de Minas Gerais (UFMG), Belo Horizonte, Minas Gerais, Brazil
e-mail: rosy-isaias@ufmg.br

B. G. Ferreira
Department of Botany, Universidade Federal do Rio de Janeiro (UFRJ), Rio de Janeiro, Brazil

D. C. de Oliveira, R. M. dos Santos Isaias (eds.), *Plant Galls*,
https://doi.org/10.1007/978-3-031-80064-1_1

Fig. 1 Bud galls on *Celtis iguanaea* (Cannabaceae) (**a**), *Myrcia splendens* (Myrtaceae) (**b**), and bud and leaf galls on Sapindaceae (**c**)

photoassimilates, and minerals crucial for gall development, sometimes in distinct proportions, which is also indicative of peculiar metabolic demands (Bragança et al. 2021a; Jorge et al. 2022; Arriola et al. 2024). The meristematic property of parenchyma, which renews gall cortical cells, supports the longevity of the gall cycle and guides the diversity of shapes. In this process, procambium neoformation is also important for vascularizing new gall tissues connected to the original vascular bundles of the host organs (Yamaguchi et al. 2017). The new origins and fates of the cells at gall developmental sites involve the signaling and trafficking of molecules from the original or the ectopic meristems to the gall tissue compartments (Isaias et al. 2024).

The oviposition site is a determinant for gall development and the establishment of structural and functional traits, making meristems special because their cells can follow every possible fate in plant development. The number of fourth-level shoots, which indicates the plant branching complexity and the availability of apical and axillary meristems, is sometimes indicated as a trait influencing not only cell fates but also gall richness at the species level (Espírito-Santo et al. 2007). Accordingly, patterns of meristem abundance may explain the discrepancies in gall richness among monocots and eudicots (Silvia and Connor 2012). The influence of apical and axillary shoots on oviposition determines gall abundance (Craig et al. 1986; Larson and Whitham 1997; Price 2003), which is predominantly related to shoot growth rates, meristem activity, and the ability to respond to gall-inducing stimuli. However, the direct association of taller and more complex plants with higher gall abundance is conflicting, especially for stem-like galls like those induced on the axillary meristems of *Macairea radula* (Melastomataceae) by *Palaeomystella oligophaga* (Lepidoptera: Momphidae), characterized as stems with shortened internodes (Rezende et al. 2023), so, high gall abundance impact host plant height and ramification. Despite that, the high occurrence of oviposition in young and growing shoots is evident (Price 2003), so we herein assume the importance of the first level of shoots as sites of oviposition and the development of galls in their different sizes and shapes (Fig. 2).

Plant meristems are essential for growth and morphogenesis in all plant organs and, therefore, are essentially modified in host plants under the influence of gall

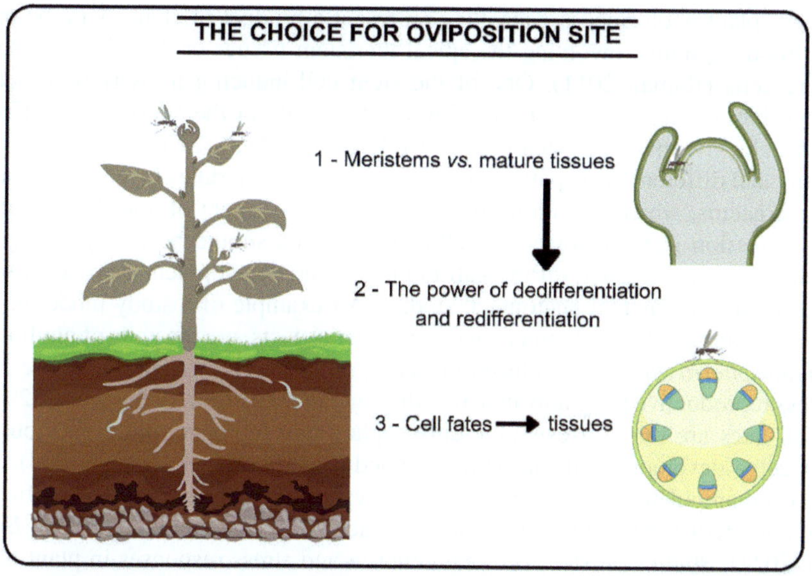

Fig. 2 Hypothetical host plant with three levels of modules of growth. The availability of meristematic tissues pushes the preference for buds as oviposition sites. Gall development alters ordinary cell fates toward specialized cells and tissues in galls. Additionally, the processes of dedifferentiation and redifferentiation can occur and result in galls with peculiar anatomical profiles

inducers. In addition, responsive plant cells are distributed along all plant bodies, and gall inducers can change their developmental status during gall growth and development. The changes in cell divisions, cell growth, and elongation patterns, and redifferentiation of ectopic meristems are key factors for creating the neoformed organs, the galls (Shorthouse et al. 2005). The key changes in plant development that lead to gall formation, involving the plant cell reactions during gall induction, modifications of apical and lateral meristems, and dedifferentiation of mature cells are discussed herein, as future perspectives to developmental studies in plants that lead to changes in cell differentiation patterns.

1.1 Meristems in Plants and Cell Reactions During Gall Formation

The ordinary pattern of meristem development and the remaining stem cells, the promeristem, leads to the formation of three apical meristem tissues (Evert 2006). The protodermal cells differentiate into ordinary epidermal cells and specializations; the ground meristem differentiates into parenchyma, collenchyma, and/or sclerenchyma; and the procambium initials originate the variety of xylem and phloem cells, with small groups of cells maintaining meristem dormancy. As the galls are plant neoformed organs, they represent a product of changes in the three basic tissue systems, involving the apical meristem tissues or dedifferentiation of mature cells (Raman 2011). One of the stem cell-inducing transcription factors responsible for the identity and destiny of stem cells in the shoot is *WUSCHEL* (*WUS*), along with the secreted peptide *CLAVATA3* (*CLV3*) (Zeng et al. 2024). The identity and differentiation pattern of the apical growth modules is repeated in axillary meristems, which remain dormant until the impairment of apical dominance. The expression genes involved in cell differentiation should be changed in galls, which can be investigated during gall induction and developmental phases associated with developmental anatomical studies. An example of a study model in this line, associating gall development and molecular aspects, are the galls of phylloxera (*Daktulosphaira vitifoliae*) induced on wild grapevine (*Vitis riparia*), whose cells express reproductive programs in leaves during gall formation (Schultz et al. 2019).

Cell fates are rarely irreversible during plant development, and even though a plant cell's developmental fate is determined by its position (Fosket 1994), gall inducers manipulate such fates in peculiar ways, triggering processes of dedifferentiation and redifferentiation (Oliveira and Isaias 2010; Ferreira et al. 2019a; Isaias et al. 2024). Studies on stem cell maintenance and stress responses in plant meristems are at the forefront of plant biology, however, our current understanding of how environmental cues and stress inputs are integrated into signaling for plant stem cell homeostasis is limited (Zeng et al. 2024). As shoot apical meristem (SAM) activity is influenced by various environmental factors that modulate phytohormone signaling and the *WUS*-*CLV3* feedback loop, the biotic influence of gall-inducing herbivores is expected to occur. Host-plant morphogenetic constraints, including the synthesis of growth promotors, mediate the attacks of gall-inducing insects. Auxins and cytokinins differential accumulation may be stimulated by remobilization and synthesis in gall tissues (Tooker and Helms 2014), and they are also reported in the saliva of gall inducers (Bartlett and Connor 2014). The mapping of hormone accumulation in gall tissues is particularly interesting for gall studies, as those performed via histochemical techniques (Bedetti et al. 2014, 2017; Carneiro et al. 2017), which demonstrated that the detected levels of auxins and cytokinins are much more evident in gall tissues than in ordinary meristems of the model plants (Bedetti et al. 2018).

Additionally, the symptoms of the activity of these growth regulators are fundamental for gall development. Accordingly, the effects of insect feeding and stimuli at gall developmental sites are continuous but for a specific period and synchronized with the gall inducer's life cycle, which dictates gall morphogenesis (Raman 2011). Cell hypertrophy, one of the common aspects of gall development, demands alterations in the osmotic status of the cell (Oliveira et al. 2017; Arriola et al. 2024). This type of stress affects the electrical properties of the plasma membrane and impacts IAA activity, which in turn alters H^+-transport systems (Raman 2011). These physical/physiological responses occurring at the cell level during insect feeding disrupt host plant cell walls, liberating degenerated wall materials that act as elicitors.

Cell dedifferentiation in leaves affected by gall-inducing stimuli leads to mesophyll homogenization and local hyperplasia, which are sometimes responsible for leaf blade invagination (Jorge et al. 2018; Marquesine et al. 2024) and sometimes for local thickening of the host organ (Ferreira and Isaias 2013; Isaias et al. 2018; Bragança et al. 2021b). Cellular degeneration has been detected at the cytological level, supporting Raman's (2011) proposal of the eliciting effect of degenerated cell wall materials on gall development.

1.2 Galls as Products of Apical Meristem Changes or Neoformation

As the primary shoot apical meristem (SAM) determines the main axis of the plant body, its modification leads to bud gall formation (Ferreira and Isaias 2013; Fleury et al. 2015; Reale et al. 2016). The activity of the additional subtended meristems modifies the plant architecture, and the interaction among the SAM and the axillary meristems during plant development is responsible for the structural complexity of the shoot, as it determines the branching patterns. These patterns depend on the active or dormant status of the lateral buds. Although such patterns are plastic, the developmental program that specifies branching complexity in different plant species is fundamentally important for generating species-specific plant forms (Shimizu-Sato and Mori 2001). The forms generated by the apical and axillary buds can be altered by the action of galling organisms, resulting in globose, pyriform, flask-shaped, clavate, or other singular shapes (Mani 1964).

Galls induced on stem buds involve changes in preexisting meristems induced by external or internal inducers and are distinguished from non-galled buds by their enormous size (Mani 1964). Typically, in some plants, the axillary meristems may undergo immediate development. In contrast, in other plants, the axillary meristems might initiate a few leaves and then become developmentally dormant due to inhibition determined by the terminal bud (Shimizu-Sato and Mori 2001). Nevertheless, gall-inducing herbivores alter their developmental pattern by breaking the dormancy of the axillary buds, inducing cell proliferation, and suppressing internode elongation, which implies a deep alteration in the node–internode unity (Fig. 3). The resulting morphotype may be a rosette gall with several diminutive, congested leaves. This phenomenon occurred in *Guapira opposita* (Nyctaginaceae), whose galls are induced by *Pisphondylia brasiliensis* (Cecidomyiidae) (Maia et al. 2010), with two morphogenetic pathways occurring concomitantly. The congested leaves on these rosette galls develop via the shortening of internodes and by the over-differentiation of axillary buds (Fleury et al. 2015). The galled bud has numerous leafy projections arranged side-by-side without significantly elongating the internodes, which were also observed in axillary bud galls induced on *M. radula* (Melastomataceae) by *P. oligophaga* (Lepidoptera) (Rezende et al. 2019, 2023). Peculiarly, in *G. opposita*—*P. brasiliensis* galls, at the base of each leaf projection, a group of cells remains in meristematic conditions and repeats the production of new leaves. The rosette galls on *G. opposita* host several larvae internal to bud

Fig. 3 Rosette galls on axillary buds of *Neomitrantes obscura* (Myrtaceae) (**a**) and *Davilla* cf. *elliptica* (Dilleniaceae) (**b**). The longitudinal section shows the larval chamber and gall tissues surrounded by leaves in shortened internodes

tissues that share the resources drained to the gall developmental site, as is common for other gall systems (Inbar and Wool 1995; Fleury et al. 2015). Another interesting phenomenon occurs in larger rosette galls on *G. opposita*, which originate from the coalescence of several oviposition sites and the differentiation of a master cambium, which is typical of Nyctaginaceae stems. The term "master cambium" was proposed by Carlquist (1977) and refers to the rings of cambial cells developed from cortical parenchyma cells, which originate fibers and strands of typical vascular cambia and cork cambium (Tamaio et al. 2009; Fleury et al. 2015). The master cambium is a lateral meristem with complex meristematic activity that generates specific secondary vascular tissues in Nyctaginaceae species. Therefore, *P. brasiliensis* oviposit several eggs in an axillary bud, and larval activity is responsible for manipulating the primary and secondary meristematic activities of *G. opposita*, generating a modified shortened shoot—the gall—with modified unusual secondary vascular tissues. In *P. brasiliensis—G. opposita* galls, the master cambium differentiates and is modified by the coalescence of several galls induced on the axillary buds. Therefore, such changes in shoot development generate not only the overproduction of axillary buds but also of secondary tissues, which indicates high energy demand (sink strength) for polytalamous (multichambered) galls. The activity of the master cambium occurs at a site with a cluster of primary meristems, resulting in major vascularization and ensuring the survival of gall inducers (Fleury et al. 2015). The internode shortening is common in bud galls, changing the patterns of a lateral branch and causing profound changes in leaf development patterns (Rezende et al. 2023).

The polymorphism in bud jar-shaped galls induced by *Rhopalomyia* spp. (Diptera: Cecidomyiidae) on the axillary buds of *Artemisia montana* (Asteraceae) in Japan and *A. princeps* in Japan and Korea reveals an interesting but controversial case (Ganaha et al. 2007), the conclusion that the *Rhopalomyia* gall midges associated with different gall morphotypes belong to the same species. The arguments used to affirm that the three supposed species are, in fact, conspecific are based on the morphological features and molecular sequencing data of the gall inducers (Ganaha et al. 2007). In this case, the differences in size and shape and the disposition and number of leafy-like projections, which were anciently used to distinguish the three species, do not support the identification of galls by their associated morphotypes (Ganaha et al. 2007). Nevertheless, we must consider that changes in buds induced by gall inducers may be influenced by the site of oviposition, e.g., in apical or axillary, vegetative or reproductive buds, as well as the constraints of the host plant tissues, which could explain the small variations in bud gall morphology in *Rhopalomyia–Artemisia* spp. systems.

In some cases, instead of stimulating the abnormal growth of buds, the induction of bud galls may involve bud dormancy and consequently the dormancy of gall-inducing insects, as described for the *Castanea sativa–Dryocosmus kuriphilus* system. The infested buds remain asymptomatic from oviposition until the bud bursts in the following spring. Later, the chestnut tissues are altered toward green- or rose-colored galls 5–20 mm in diameter (Reale et al. 2016). Previous studies have suggested that cynipids need meristematic or otherwise totipotent cells to initiate gall growth (Stone et al. 2002). However, Reale et al. (2016) reported that the differentiation of the larval chamber starts with changes in epidermal cells, which ensures the first phase of larval chamber formation. The mitotic activity continues in September and produces many cell layers, which die and cause the larval chamber to increase in size, probably providing the larvae with nutrients. From March to May, the larval chamber increased greatly in size, the layers of parenchyma cells were reduced, and the vascular bundles became closer to the larval chamber (Jara-Chiquito et al. 2021). These vascular bundles connect to those of the host stem and, as is true for many other galls, supply gall tissues with energetic metabolites (Reale et al. 2016). The presence of nutrients in infected buds is also controlled by gall inducers, and the soluble sugars commonly detected in gall nutritive and storage tissues can originate from starch stored in cells close to the larval chamber, which was reported to other gall systems (Meyer and Maresquele 1983; Oliveira et al. 2010). In *C. sativa* galls induced by *D. kuriphilus*, starch grains are abundant during the "dormant" period in the bud and are not detected in gall tissues near the young larval chamber (Reale et al. 2016), indicating their mobilization to gall and gall inducer development.

There are distinct developmental patterns in bud galls, in which the overproduction of leaves may not occur. Instead, complete suppression of leaf development occurs, and the 'soft' trichomes of the normal bud turn into stiff, woody spines, as occurs with the vegetative meristem of the Indian species *Hopea ponga* (Dipterocarpaceae) stimulated by *Mangalorea hopeae* (Hemiptera). The development of lignified cell walls at gall developmental sites has been associated with the generation of reactive oxygen species (ROS), which are responsible for several metabolic and structural-related responses (Isaias et al. 2015). Another example is the complex globoid gall induced by *Eucecidoses minutanus* (Cecidosidae) on

Schinus engleri (Anacardiaceae), which is induced on axillary buds, changing procambial patterns and impairing leaf primordia development (Ferreira et al. 2022).

The cecidial shoots described by Rohfritsch and Anthony (1992) for *Shorea curtisii*, Dipterocarpaceae-*Gallacoccus anthonyae*, and Hemiptera: Beesoniidae are constructed of scale leaves with a meristem identical to the SAM but with no axillary buds, which culminate in determinate growth (Raman 2011). In contrast, galls induced by *Ditylenchus gallaeformans* on the leaves of *Miconia* spp. have ectopic apical-like meristems responsible for the indeterminate growth of the galls (Fig. 4). This gall inducer, a nematode, infects several species of *Miconia* and induces galls on young leaves (Fig. 4a, b), stems, and inflorescences (Ferreira et al. 2017a). The colony-feeding activity on the surface of the host organs induces the maintenance of promeristematic features of the cells in contact with the nematodes, which are called nutritive cells (Fig. 4c–f). Maintaining promeristematic features is linked to the accumulation of cytokinins, which stimulate cell division, and the activity of invertases, leading to the accumulation of reducing sugars in this region (Ferreira et al. 2017b, 2019b). The maintenance of promeristematic features in the nutritive cells of *D. gallaeformans* galls is important for increasing nematode nutrition and indeterminate growth. Promeristematic nutritive tissues, such as ectopic apical-like meristems, can originate from covering tissues surrounding gall inducer colonies (Fig. 4d–f). In addition to the increase in the size of colonies, promeristematic nutritive tissues increase the inner surface area available for feeding and lead to new cover emergences, promoting the indeterminate growth of these galls, which may be present on the host organs for several months (Ferreira et al. 2017a, b, 2019b; Arriola et al. 2020; Arriola and Isaias 2021). Some regions enter senescence in the same gall with indeterminate growth, while others originate novel feeding sites. An ectopic phellogen is observed forming from these older nutritive tissue regions (Fig. 4g), originating phellem-like cells that accumulate phenolics (Fig. 4h).

Ditylenchus gallaeformans—*Miconia* gall systems are good models for studying meristem cell development control by exogenous factors, such as gall inducers, from molecular, physiological, biochemical, and anatomical perspectives. For example, small adventitious neoformed leaves are frequently observed in such galls (Fig. 5), most likely because of changes in hormonal balance caused by the nematode colony (Arriola and Isaias 2021).

Fig. 4 (continued) surfaces. (**b**) Mature leaves with older and larger galls (arrows). (**c**–**h**) Transverse section of a leaf with a gall induced on the abaxial surface. Asterisks indicate covering emergences formed in response to nematode colony infection; arrows indicate the nematodes. Young emergences originated from nutritive cells in (**e**) and (**f**). In older regions, nutritive cells originate ectopic phellogen (Pg) and a phellem-like tissue accumulating phenolics. Fungi (Fu) are present in these older regions. Abbreviations: Dr druses, Fu fungi, MR midrib, NGL non-galled leaf portion, NT nutritive tissue, Ph phenolic compounds, ST storage tissue, Tr trichomes, VB neoformed vascular bundles, YE young emergences

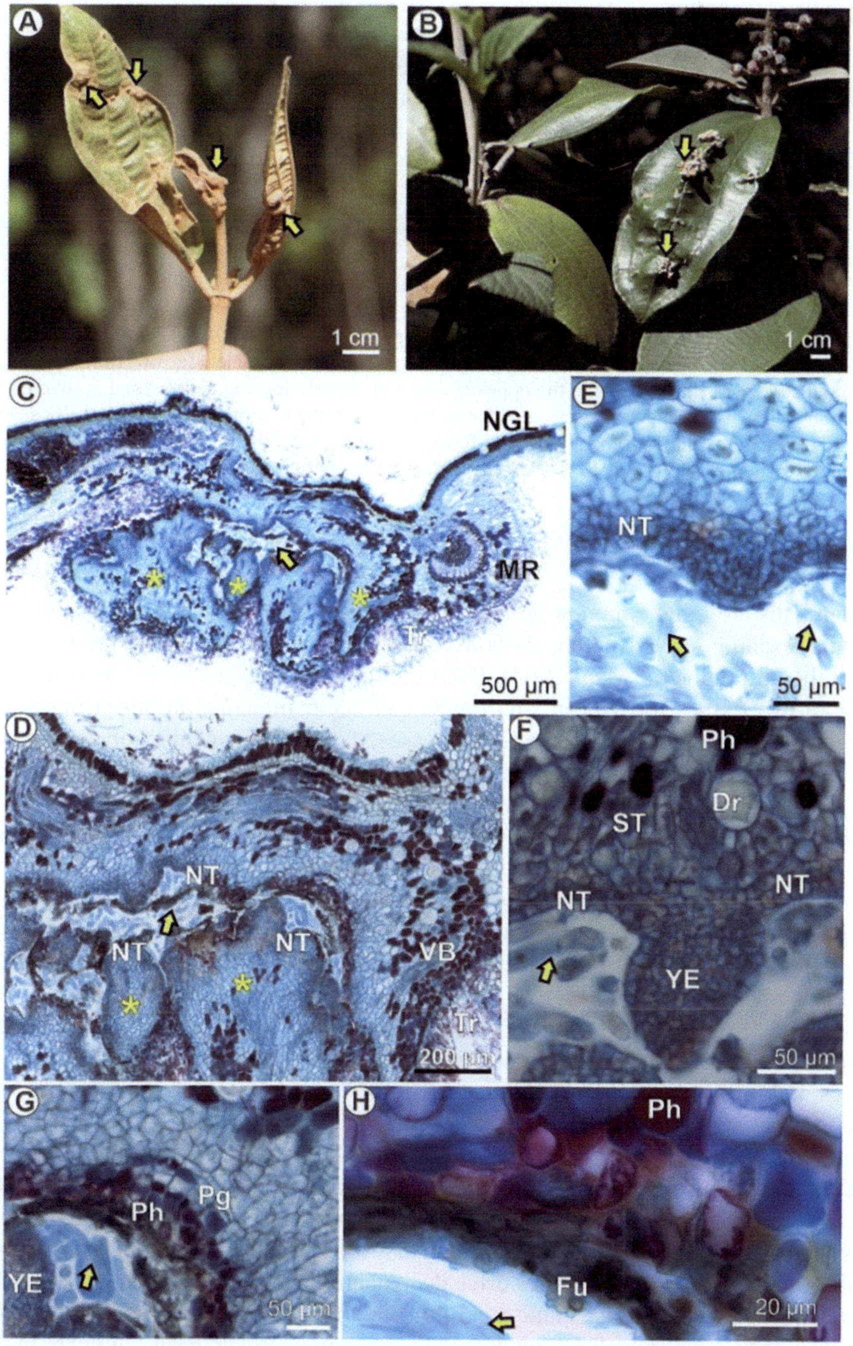

Fig. 4 Galls induced by *Ditylenchus gallaeformans* (Nematoda) on *Miconia* spp. young organs with indeterminate growth. (**a, b**) *M. albicans*. (**a**) Young leaves with galls (arrows) on both leaf

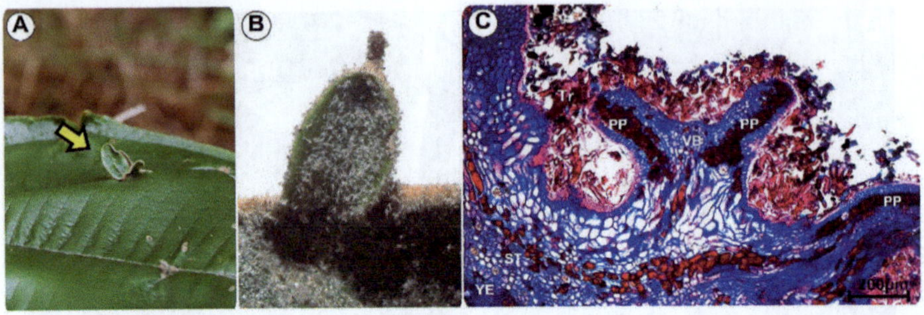

Fig. 5 Adventitious neoformed leaves over the galls induced by *Ditylenchus gallaeformans* (Nematoda) on *Miconia corallina* (Melastomataceae). The neoformed leaf (arrow) develops on the adaxial surface of the host leaf in the opposite direction of gall development (**a**) and is covered by trichomes (**b**). The transverse section of the connection of the neoformed leaf with the host leaf demonstrates vascularization and tissue differentiation (**c**). Abbreviations: Tr trichomes, PP palisade parenchyma, ST storage tissue, VB neoformed vascular bundles, YE young emergences

1.3 Galls Affecting Meristems on Stems and Leaves

Stem galls are quite common in nature, just behind leaf galls in frequency. Owing to the essential features of the shoot axis, galls induced on stems may grow in both length and thickness (Mani 1964). Even though gall inducers prefer meristematic responsive tissues to oviposit, stem galls can occur in stems during primary and secondary growth. Nevertheless, in both cases, their development involves conspicuous alterations in the cortical, vascular, or pith parenchyma (Fig. 6). Stem galls can be frequently induced by Diptera and Cecidomyiidae, but other associated organisms, such as inquilines, may occupy their chambers, which may also be responsible for alterations in gall structure and functional profiles.

While galls induced on stem buds involve changes in existing meristems, those induced on other plant parts usually trigger responses in parenchymatic cells and also the differentiation of ectopic meristems, as is the case for the horn-shaped galls on *Copaifera langsdorffi* (Fig. 7). This gall shifts from a dome-like structure to the elevation of the lateral two horns from the growth and development phase toward maturation, which involves the reacquisition of cell meristematic competence at two sites within the gall body (Carneiro et al. 2017). The shift in shape demands cell dedifferentiation and redifferentiation at the center of the gall, forming a group of cells where auxins are histochemically detected by the Erlich reagent (Carneiro et al. 2017). The intense cell divisions at the central ectopic meristem site cause the gall body to emerge from inside a central volcano-like structure. In the second step, two new meristematic centers at the flanks are responsible for the parenchymatic horns. These meristematic zones drive deep anatomical reprogramming toward the reacquisition of the meristematic condition and the establishment of new functional traits of the tissues, which are not deeply involved in photosynthesis but rather in the maintenance of the associated gall inducer.

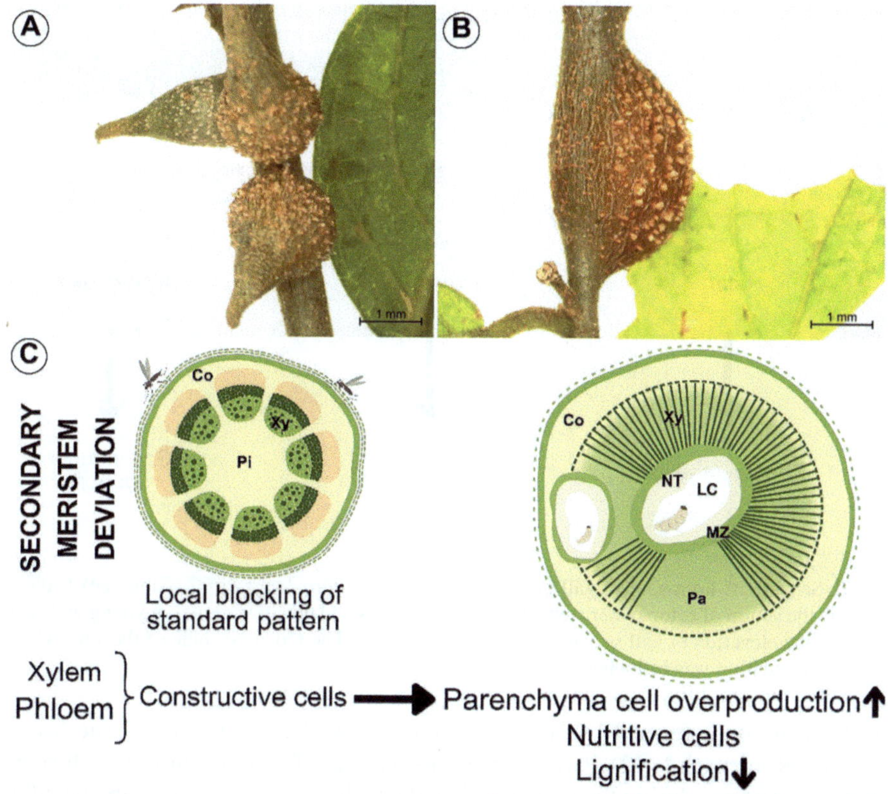

Fig. 6 Gall development affects secondary meristems on *Celtis iguanae* (Cannabaceae). Conical galls (**a**); and globoid galls (**b**). Representation of the secondary meristem deviation toward gall development (**c**). Abbreviations: Co cortical parenchyma, LC larval chamber, MZ mechanical zone, NT nutritive tissue, Pa parenchyma, Pi pith, Xy Xylem

Galls induced on stems during primary growth usually exhibit alterations in eustele structure, as the overproduction of parenchyma sets the vascular bundles apart (Fig. 6). Neovascularization may occur because the procambium strands redifferentiate from the totipotent parenchyma cells, resulting in the differentiation of xylem and phloem in various proportions. When the stem is in secondary growth, the larval chamber may be in the cortical or pith parenchyma, and in both situations, the medullary rays overdifferentiate, and the vascular tissues may dissociate (Mani 1964). The galls induced on *Malus domestica* (Rosaceae) by *Eriosoma lanigerum* (Aphididae) are a good example of the impairment of the ordinary pattern of vascular cambium toward a completely altered anatomical profile (Nogueira et al. 2024). These galls support a colony of insects that stay on the stem or root surface but cause deep alterations in vascular tissues with the overproduction of parenchyma cells. In the case of multiple oviposition routes inside the stem of leaf tissues, several gall chambers develop, each one involving its nutritive tissue, one or two

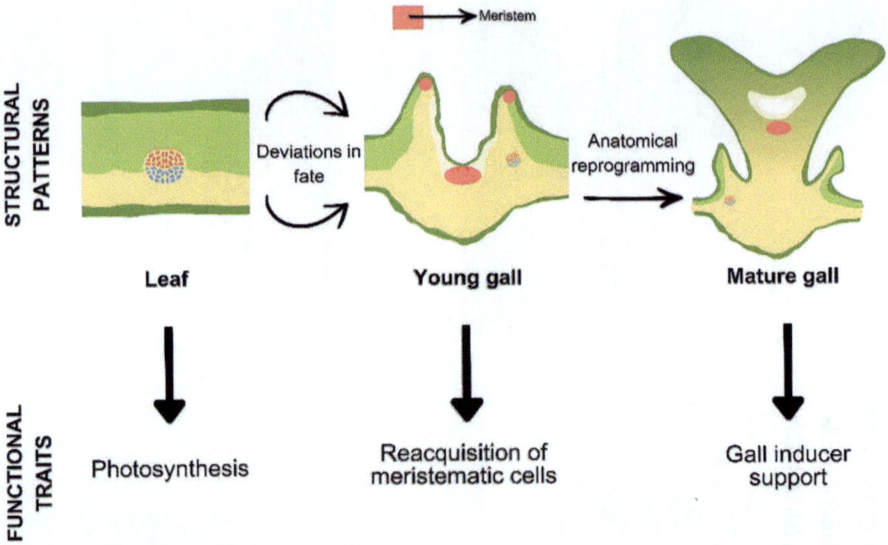

Fig. 7 Scheme of horn-shaped gall development on *Copaifera langsdorffii* (Fabaceae) leaflets, showing the action of ectopic meristems forming the valves and nutritive tissues in young galls and acting in the elevation of gall structure in the mature phase. The red areas indicate the sites where auxin accumulation was histochemically detected

mechanical zones formed by cells with lignified walls, and storage tissue layers outward (Rohfritsch 1992). Parenchyma cells may redifferentiate into newly formed ectopic meristems during the gall growth and development phases in some peculiar host plant-galling organism systems. The development of ectopic meristems demands that cells respond to external stimuli, as is commonly the case for epidermal and parenchyma cells, which retain the capacity to alter their fates toward several other structures and functions. These ectopic meristems may produce valves, as is the case for galls induced by *Lopesia* spp. on *Mimosa gemmulata* (Costa et al. 2021) and by *Euphalerus ostreoides* on *Lonchocarpus muehlbergianus* (Oliveira et al. 2006; Isaias et al. 2011). In addition to morphogenetic changes due to ectopic meristems, cell redifferentiation in mature tissues may occur, generating groups of new stem cells to establish specialized tissue compartments and their associated functional traits.

1.4 Phellogen in Galls

The phellogen or cork cambium is a secondary meristem whose derivatives differentiate as phellem outward and phelloderm inward. The activation of phellogen at sites where it does not occur, such as in leaves, is common in the dermal system of galls and the covering of the gall chamber. As this secondary meristem substitutes the epidermis to accompany the development of the tissues underneath, at gall developmental sites, the products of phellogen, i.e., the phellem and phelloderm, are important in buffering the inner microenvironment and conferring more parenchymatic cells while the gall is moving toward the end of its life cycle. Stem and leaf brown mature galls are usually covered externally by phellem, given the natural color of the stem, or neoformed in leaf galls (Isaias et al. 2014), as occurs in *C. langsdorffii* (Fabaceae) globoid galls (Oliveira et al. 2008) and *Struthanthus vulgaris* (Loranthaceae) leaf galls (Kraus et al. 2002).

In the *Neolasioptera* sp.-*Eremanthus erythropappus* system, an ectopic phellogen differentiates, covering the gall chamber and the escape channel of the inducer when the gall enters senescence. This anatomical feature was associated with the movement of ants in and out of the galls (Jorge et al. 2022). In erinea leaf galls induced by Eriophyidae (Acari) on *Avicennia schauerianna* (Acanthaceae), a phellogen is formed from spongy parenchyma in the final phase of gall maturation, isolating the outer epidermis with neoformed hairs—used by the mites for nutrition—from the chlorophyllian mesophyll tissues (Nobrega et al. 2023).

The impairment of phellogen differentiation was observed in *E. minutanus*-*S. engleri* galls induced on axillary buds. The gall surface, which is distinct from the non-galled stems of *S. engleri*, is covered by a multiseriate epidermis, which is not observed in stems during primary growth or in the leaves of this host plant species (Ferreira et al. 2022).

1.5 Oviposition in Mature Tissues

Even though gall inducers prefer young tissues with nondifferentiated cells for oviposition, they may oviposit in mature tissues, implying structural and relative functional differences in galls. This differential ability to manipulate young and mature tissues from the same insect on the same host plant was reported for the midrib galls induced by an undescribed species of Cecidomyiidae on *C. langsdorffi*. The galls induced in both young and mature tissues are similar. Nevertheless, in terms of the anatomical profile, those induced in young tissues have greater investment in nutritive reserves. In comparison, those induced in mature tissues invest more in lignified cells (Oliveira and Isaias 2009). These differences indicate not only a deviation in plant manipulation, but also divergences in functional aspects related to alleviating oxidative stress in cells and tissues with meristematic potential and mature tissues. This model is an important example of the incredible ability of plant cells to redifferentiate, even if they are already differentiated and occur in mature organs, such as expanded leaves. The ability to manipulate plant cells, returning them to the meristematic stage, is due to the oxidative stress generated by the gall inducers and signals present on their feeding or oviposition secretions and reveals the potential of plant development to be modulated under various external factors.

2 Final Remarks

The galls are elegant models of study of plant cell development, as they result from alterations in meristematic features both in bud galls and secondary growth organs. Still, ectopic meristems may be induced in gall induction sites. Alterations in signals are essential in determining cell fates, and the galling organisms may induce dedifferentiation of ground, dermal, and vascular tissue cells or changes in meristem patterns. Anatomical studies associated with histochemical and physiological analyses are important steps to understanding plant cell versatility, and integrating molecular studies will be an important step to understanding not only gall but also plant cell development.

References

Arriola ÍA, Isaias RMS (2021) Extending the knowledge on the histological patterns of leaf galls induced by *Ditylenchus gallaeformans* (Nematoda) on *Miconia* (Melastomataceae) hosts. Flora 274:151753

Arriola ÍA, Figueiredo MA, Boanares D, França MGC, Isaias RMS (2020) Apoplast-symplast compartmentalization and functional traits of iron and aluminum in promeristematic tissues of nematode induced galls on *Miconia* spp. Plant Physiol Biochem 154:360–368

Arriola ÍA, Costa EC, Oliveira DC, Isaias RMS (2024) Soil–plant–gall relationships: from gall development to ecological patterns. Biol Rev 99:1948–1975. https://doi.org/10.1111/brv.13106

Bartlett L, Connor EF (2014) Exogenous phytohormones and the induction of plant galls by insects. Arthropod Plant Interact 8:339–348

Bedetti CS, Modolo LV, Isaias RMS (2014) The role of phenolics in the control of auxin in galls of *Piptadenia gonoacantha* (Mart.) MacBr (Fabaceae: Mimosoideae). Biochem Syst Ecol:5553–5559

Bedetti CS, Bragança GPP, Isaias RMS (2017) Influence of auxin and phenolic accumulation on the patterns of cell differentiation in distinct gall morphotypes on *Piptadenia gonoacantha* (Fabaceae). Aust J Bot 65:411–420

Bedetti CS, Jorge NC, Trigueiro FCG, Bragança GP, Modolo LV, Isaias RMS (2018) Detection of cytokinins and auxin in plant tissues using histochemistry and immunocytochemistry. Biotech Histochem 93:149–154

Bragança GPP, Ferreira BG, Isaias RMS (2021a) Distinct cytological mechanisms for food availability in three *Inga ingoides* (Fabaceae)—Cecidomyiidae gall systems. Protoplasma 259:155–162. https://doi.org/10.1007/s00709-021-01646-w

Bragança GPP, Freitas MSC, Isaias RMS (2021b) The influence of gall position over xylem features in leaflets of *Inga ingoides* (Rich.) Willd. (Fabaceae: Caesalpinioideae). Trees 35:199–209

Carlquist S (1977) Ecological factors in wood evolution: a floristic approach. American Journal of Botany 64:887–896

Carneiro RGS, Isaias RMS, Moreira ASFP, Oliveira DC (2017) Reacquisition of new meristematic sites determines the development of a new organ, the Cecidomyiidae gall on *Copaifera langsdorffii* Desf. (Fabaceae). Front Plant Sci 8:1622. https://doi.org/10.3389/fpls.2017.01622

Costa EC, Oliveira DC, Ferreira DKL, Isaias RMS (2021) Structural and nutritional peculiarities related to lifespan differences on four *Lopesia* induced bivalve-shaped galls on the single super-host *Mimosa gemmulata*. Front Plant Sci 12:660557. https://doi.org/10.3389/fpls.2021.660557

Craig TP, Price PW, Itami JK (1986) Resource regulation by a stem-galling sawfly, *Euura lasiolepis*, on the arroyo willow, *Salix lasiolepis*. Ecology 67:419-42

Espírito-Santo MM, Neves FS, Andrade-Neto FR, Fernandes GW (2007) Plant architecture and meristem dynamics as the mechanisms determining the diversity of gall-inducing insects. Oecologia 153:353–364. https://doi.org/10.1007/s00442-007-0737-8

Evert RF (2006) Esau's plant anatomy: meristems, cells, and tissues of the plant body: their structure, function, and development. Wiley, Hoboken

Ferreira BG, Isaias RMS (2013) Developmental stem anatomy and tissue redifferentiation induced by a galling Lepidoptera on *Marcetia taxifolia* (Melastomataceae). Botany 91:752–760. https://doi.org/10.1139/cjb-2013-0125

Ferreira BG, Avritzer SC, Isaias RMS (2017a) Totipotent nutritive cells and indeterminate growth in galls of *Ditylenchus gallaeformans* (Nematoda) on reproductive apices of *Miconia*. Flora 227:36–45

Ferreira BG, Álvarez R, Avritzer SC, Isaias RMS (2017b) Revisiting the histological patterns of storage tissues: beyond the limits of gall-inducing taxa. Botany 95:173–184

Ferreira BG, Álvarez R, Bragança GP, Alvarenga DR, Pérez-Hidalgo N, Isaias RMS (2019a) Feeding and other gall facets: patterns and determinants in gall structure. Bot Rev 85:78–106

Ferreira BG, Freitas MSC, Bragança GP, Moreira ASFP, Carneiro RGS, Isaias RMS (2019b) Enzyme-mediated metabolism in nutritive tissues of galls induced by *Ditylenchus gallaeformans* (Nematoda: Anguinidae). Plant Biol 21:1052–1062

Ferreira BG, Moreira GRP, Carneiro RGS, Isaias RMS (2022) Complex meristematic activity induced by *Eucecidoses minutanus* on *Schinus engleri* turns shoots into galls. Am J Bot 109:209–225. https://doi.org/10.1002/ajb2.1798

Fleury G, Ferreira BGF, Soares GLG, Isaias RMS (2015) Elucidating the determination of the rosette galls induced by *Pisphondylia brasiliensis* Couri and Maia 1992 (Cecidomyiidae) on *Guapira opposita* (Nyctaginaceae). Aust J Bot 63:608–617

Fosket DE (1994) Plant growth and development: A molecular approach. Elsevier Inc., New York

Ganaha T, Nohara M, Sato S, Uechi N, Yamagishi K, Yamauchi S, Yukawa J (2007) Polymorphism of axillary bud galls induced by *Rhopalomyia longitubifex* (Diptera: Cecidomyiidae) on *Artemisia princeps* and *A. montana* (Asteraceae) in Japan and Korea, with designation of new synonyms. Entomol Sci 10:157–169

Ha CM, Jun JH, Fletcher JC (2010) Shoot apical meristem form and function. In: Timmerman MCP (ed) Plant development, vol 91, pp 103–140. https://doi.org/10.1016/S0070-2153(10)91004-1

Inbar M, Wool D (1995) Phloem-feeding specialists sharing a host tree: resource partitioning minimizes interference competition among galling aphid species. Oikos 73:109–119

Isaias RMS, Oliveira DC, Carneiro RGS (2011) Role of *Euphalerus ostreoides* (Hemiptera: Psylloidea) in manipulating leaflet ontogenesis of *Lonchocarpus muehlbergianus* (Fabaceae). Botany 89:581–592

Isaias RMS, Carneiro RGS, Oliveira DC, Santos JC (2013) Illustrated and annotated checklist of Brazilian gall morphotypes. Neotrop Entomol 42:230–239

Isaias RMS, Carneiro RGS, Santos JC, Oliveira DC (2014) Gall morphotypes in the Neotropics and the needs to standardize them. In: Santos JC, Fernandes GW (eds) Neotropical insect galls. Springer, New York

Isaias RMS, Oliveira DC, Moreira ASFP, Soares GLG, Carneiro RGS (2015) The imbalance of redox homeostasis in arthropod-induced plant galls: mechanisms of stress generation and dissipation. Biochim Biophys Acta 1850:1509–1517

Isaias RMS, Ferreira BG, Alvarenga DR, Barbosa LR, Salminen JP, Steinbauer MJ (2018) Functional compartmentalization of nutrients and phenolics in the tissues of galls induced by *Leptocybe invasa* (Hymenoptera: Eulophidae) on *Eucalyptus camaldulensis* (Myrtaceae). Austral Entomol 57:238–246

Isaias RMS, Kraus JE, Costa EC, Carneiro RGS (2024) The anatomy of neotropical galls and the untold lessons about the morphogenetical potentialities of plants. Rodriguésia 75:e01542023. https://doi.org/10.1590/2175-7860202475007

Jara-Chiquito JL, Pujade-Villar J, Ferreira BG, Álvarez R (2021) Histological changes induced by the cynipid wasp *Dryocosmus kuriphilus* (Hymenoptera: Cynipidae) in leaves of the chestnut

Castanea sativa (Fagaceae): mechanisms of galling impact on host vigor. Arthropod Plant Interact 15:223–233

Jorge NC, Alvarenga DR, Cavalleri A, Isaias RMS (2018) Anatomical and histometric explanations for leaf folding on *Holopothrips striatus* galls in *Myrcia retorta*. Flora 244:24–28

Jorge NC, Freitas MSC, Caffaro RM, Vale FHA, Lemos-Filho JP, Isaias RMS (2022) Vascular traits of stem galls: cell increment versus morphogenetic constraints in wood anatomy. Plant Biol 24:450–457

Kraus JE, Arduin M, Venturelli M (2002) Anatomy and ontogenesis of hymenopteran leaf galls of *Struthanthus vulgaris* Mart. (Loranthaceae). Rev Bras Bot 25:449–458

Larson KC, Whitham T (1997) Competition between gall aphids and natural plant sinks: plant architecture affects resistance to galling. Oecologia 109:575–582

Maia VC, Fleury G, Soares GLS, Isaias RMS (2010) Description of the female, pupa and gall of *Pisphondylia brasiliensis* Couri and Maia, 1992 (Diptera: Cecidomyiidae, Schizomyiina) with new records. Braz J Biol 70:1059–1063

Mani MS (1964) Ecology of plant galls. Monographiae biologicae. Springer, Dordrecht

Marquesine RR, Canaveze Y, Ferreira BG (2024) Ontogenetic differences in sun and shade galls of *Clinodiplosis profusa* on *Eugenia uniflora* leaves and the cytological antioxidant mechanisms in gall cells. Protoplasma. https://doi.org/10.1007/s00709-024-01973-8

Meyer J, Maresquele HJ (1983) Anatomie des Galles. Gebruder Borntraeger, Berlin

Nobrega LP, Sá-Haiad B, Ferreira BG (2023) Epidermal and subepidermal changes during the formation of hairy galls induced by Eriophyidae on *Avicennia schaueriana* leaves. Sci Nat 110(5):49. https://doi.org/10.1007/s00114-023-01876-3

Nogueira RM, Freitas MSC, Picoli EAT, Isaias RMS (2024) Implications of cell wall immunocytochemical profiles on the structural and functional traits of root and stem galls induced by Eriosoma lanigerum on Malus domestica. Protoplasma 261:911–926. https://doi.org/10.1007/s00709-024-01939-w

Oliveira DC, Isaias RMS (2009) Influence of leaflet age in anatomy and possible adaptive values of the midrib gall of *Copaifera langsdorffii* (Fabaceae: Caesalpinioideae). Rev Biol Trop 57:293–302

Oliveira DC, Isaias RMS (2010) Redifferentiation of leaflet tissues during midrib gall development in *Copaifera langsdorffii* (Fabaceae). S Afr J Bot 76:239–248

Oliveira DC, Christiano JCS, Soares GLG, Isaias RMS (2006) Reações de defesas químicas e estruturais de *Lonchocarpus muehlbergianus* Hassl. (Fabaceae) à ação do galhador *Euphalerus ostreoides* Crwf. (Hemiptera, Psyllidae). Braz J Bot 29:657–667

Oliveira DC, Drummond MM, Moreira ASFP, Soares GLG, Isaias RMS (2008) Potencialidades morfogênicas de *Copaifera langsdorffii* Desf. (Fabaceae): superhospedeira de herbívoros galhadores. Rev Biol Neotrop 5:31–39

Oliveira DC, Magalhães TA, Carneiro RGS, Alvim MN, Isaias RMS (2010) Do Cecidomyiidae galls of *Aspidosperma spruceanum* (Apocynaceae) fit the pre-established cytological and histochemical patterns? Protoplasma 242:81–93

Oliveira DC, Moreira AFSP, Isaias RMS, Martini V, Rezende UC (2017) Sink status and photosynthetic rate of the leaflet galls induced by *Bystracoccus mataybae* (Eriococcidae) on *Matayba guianensis* (Sapindaceae). Front Plant Sci 8. https://doi.org/10.3389/fpls.2017.01249

Price PW (2003) Macroevolutionary theory on macroecological patterns. Cambridge University Press, Cambridge

Raman A (2011) Morphogenesis of insect-induced plant galls: facts and questions. Flora 206(6):517–533

Reale L, Tedeschini E, Rondoni G, Ricci C, Bin F, Frenguelli G, Ferranti F (2016) Histological investigation on gall development induced by a worldwide invasive pest, *Dryocosmus kuriphilus*, on *Castanea sativa*. Plant Biosyst 150:35–42

Rezende UC, Cardoso JCF, Kuster VC, Gonçalves LA, Oliveira DC (2019) How the activity of natural enemies changes the structure and metabolism of the nutritive tissue in galls? Evidence

from the *Palaeomystella oligophaga* (Lepidoptera)-*Macairea radula* (Melastomataceae) system. Protoplasma 256:669–677
Rezende UC, Gonçalves PHP, Kuster VC, Carneiro RGS, Oliveira DC (2023) Every end is a new beginning: histological features of galls induced on *Macairea radula* (Melastomataceae) allow a post-senescence colonization. Flora 306:152369
Rohfritsch O (1992) Patterns in gall development. In Shorthouse J, Rohfritsch O (eds) Biology of insect-induced galls (pp. 60–86). Oxford University Press, New York
Rohfritsch O, Anthony M (1992) Strategies on gall induction by two groups of homopterans. In Shorthouse J, Rohfritsch O (eds) Biology of insect-induced galls (pp. 102–117). Oxford University Press, New York
Schultz JC, Edger PP, Body MJ, Appel HM (2019) A galling insect activates plant reproductive programs during gall development. Sci Rep 9:1833
Shimizu-Sato S, Mori H (2001) Control of outgrowth and dormancy in axillary buds. Plant Physiol 127:1405–1413
Shorthouse JD, Wool D, Raman A (2005) Gall-inducing insects – nature's most sophisticated herbivores. Basic Appl Ecol 6:407–411
Silvia MS, Connor EF (2012) Differences in meristems between monocots and dicots and susceptibility to attack by gall-inducing insects. Arthropod Plant Interact 11:485–494
Stone GN, Schönrogge K, Atkinson RJ, Bellido D, Pujade-Villar J (2002) The population biology of oak gall wasps (Hymenoptera: Cynipidae). Annu Rev Entomol 47:633–668
Tamaio N, Vieira RC, Angyalossy VA (2009) Origin of successive cambia on stem in three species of Menispermaceae. Revta Brasil Bot 32(4):839–848
Tooker JF, Helms AM (2014) Phytohormone dynamics associated with gall insects, and their potential role in the evolution of the gall-inducing habit. J Chem Ecol 40:742–753
Yamaguchi YL, Suzuki R, Cabrera J, Nakagami S, Sagara T, Ejima C, Sano R, Aoki Y, Olmo R, Kurata T, Obayashi T, Demura T, Ishida T, Escobar C, Sawa S (2017) Root-knot and cyst nematodes activate procambium-associated genes in Arabidopsis roots. Front Plant Sci 8:1195. https://doi.org/10.3389/fpls.2017.01195
Zeng J, Gneg X, Zhao Z, Zhou W (2024) Tipping the balance: The dynamics of stem cell maintenance and stress responses in plant meristems. Curr Op Plant Biol 78:102510

What Transcriptomics Can and Cannot Tell Us About Insect Gall Development

Jack C. Schultz, Heidi M. Appel, Patrick P. Edger, and Melanie J. A. Body

1 Introduction

Plant galls are highly specialized structures arising from atypical development of plant tissue induced by another organism. Many kinds of organisms can induce galls on plants, including viruses, fungi, bacteria, nematodes, mites, and insects (Redfern 2011). However, insects make galls that are more structurally complex and diverse than those made by other gall-inducing organisms (Takeda et al. 2021). Galls provide enhanced nutrition and protection from natural enemies and environmental stresses (Redfern 2011). Insects induce galls by eliciting ectopic expression of plant development genes, sometimes referred to as "molecular hijacking" (Takeda et al. 2019).

Because altered plant gene expression underlies formation of insect galls, investigators are increasingly studying plant gene expression, often using RNA sequencing (RNAseq) to identify genes and pathways involved in insect gall formation (Table 1). This approach has provided many important insights on insect gall development. However, its usefulness is often limited by shortcomings inherent in the approach, including lack of galls on transformable model plant species, a lack of sequenced hostplant genomes and functional annotations, incomplete and multiple overlapping functional annotations, dilution effects arising from sampling entire

J. C. Schultz (✉) · H. M. Appel
Division of Plant Science & Technology, University of Missouri, Columbia, MO, USA

Department of Biology and Biochemistry, University of Houston, Houston, TX, USA
e-mail: jcschul3@central.uh.edu; happel@central.uh.edu

P. P. Edger · M. J. A. Body
Division of Biological Sciences, University of Missouri, Columbia, MO, USA

Department of Horticulture, Michigan State University, East Lansing, MI, USA
e-mail: edgerpat@msu.edu

D. C. de Oliveira, R. M. dos Santos Isaias (eds.), *Plant Galls*,
https://doi.org/10.1007/978-3-031-80064-1_2

Table 1 Recent studies of gene expression in insect or mite galls

Galler taxa/Authors	Date	URL
Cynipid wasps		
Acquadro A, Torello Marinoni D, Sartor C, et al.	2020	https://doi.org/10.1007/s00438-019-01607-2
Betancourt EK, Soto PH, Cortés NC, et al.	2020	https://doi.org/10.1007/978-3-030-46012-9_17
Hearn J, Blaxter M, Schönrogge K, et al.	2019	https://doi.org/10.1371/journal.pgen.1008398
Kariñho-Betancourt E, Hernández-Soto P, Rendón-Anaya M, et al.	2019	https://doi.org/10.1080/17429145.2019.1603404
Martinson EO, Werren JH, Egan SP	2022	https://doi.org/10.1111/mec.16159
Pinsupa S, Tongmark K, Aesomnuk W, et al.	2023	https://doi.org/10.3390/horticulturae9020127
Zhu C, Shi F, Chen Y, et al.	2019	https://doi.org/10.3390/ijms20040855
Zhu C, Wang W, Chen Y, et al.	2022	https://doi.org/10.3389/fpls.2022.874434
Chalcid wasp		
Oates CN, Denby KJ, Myburg AA, et al.	2021	https://doi.org/10.1111/pce.13930
Oates CN, Külheim C, Myburg AA, et al.	2015	https://doi.org/10.1093/pcp/pcv064
Aphids		
Davidovich-Rikanati R, Bar E, Hivert G, et al.	2022	https://doi.org/10.1093/jxb/erab289
Hirano T, Kimura S, Sakamoto T, et al.	2020	https://doi.org/10.3389/fpls.2020.00471
Hirano T, Okamoto A, Oda Y, et al.	2023	https://doi.org/10.1038/s41598-023-29302-8
Hua J, Liu J, Zhou W, et al.	2023	https://doi.org/10.1016/j.plaphy.2023.03.007
Lu Q, Chen X, Yang Z, et al.	2021	https://doi.org/10.3390/ijms22105166
Ruan Z-Y, Chen X-M, Yang P, Wang B-Y	2017	https://doi.org/10.1071/FP16436
Takei M, Ito S, Tanaka K, et al.	2017	https://doi.org/10.1080/09168451.2017.1285685
Wang H, Cui K, Shao S, et al.	2017	https://doi.org/10.1080/17429145.2017.1392627
Zhou W, Jia M, Zhang G, et al.	2021	https://doi.org/10.1093/treephys/tpaa132
Phylloxera		
Nabity PD, Haus MJ, Berenbaum MR, DeLucia EH	2013	https://doi.org/10.1073/pnas.1220219110
Schultz JC, Edger PP, Body MJA, Appel HM	2019	https://doi.org/10.1038/s41598-018-38475-6
Psyllid		
Bailey S, Percy DM, Hefer CA, Cronk QCB	2015	https://doi.org/10.1186/s12864-015-2109-9
Cecidomyiidae midges		
Liu X, Bai J, Huang L, et al.	2007	https://doi.org/10.1007/s10886-007-9382-2

(continued)

Table 1 (continued)

Galler taxa/Authors	Date	URL
Khajuria C, Wang H, Liu, X, et al.	2013	https://doi.org/10.1186/1471-2164-14-423
Shih T-H, Lin S-H, Huang M-Y, et al.	2018	https://doi.org/10.1371/journal.pone.0205265
Rawat N, Neeraja CN, Nair S, et al.	2012	https://doi.org/10.1186/1939-8433-5-8
Mites		
Paponova SS, Chetverikov PE, Pautov AA, et al.	2018	https://doi.org/10.1111/aab.12399
Yang M, Li H, Qiao H, et al.	2023	https://doi.org/10.3390/ijms24129839
Mealybug (Pseudococcidae)		
Takano S, Utsumi Y, Nagano A, et al.	2023	https://doi.org/10.1007/s13355-023-00832-x
Multiple taxa		
Giron D, Huguet E, Stone GN, Body M	2016	https://doi.org/10.1016/j.jinsphys.2015.12.009
Takeda S, Hirano T, Ohshima I, Sato MH	2021	IJMS 22:9424. https://doi.org/10.3390/ijms22179424
Takeda S, Yoza M, Amano T, et al.	2019	https://doi.org/10.1371/journal.pone.0223686

galls, and the correlative nature of transcriptomics. Consequently, caution is needed in drawing conclusions using this approach. In this chapter, we point out some of these challenges and their impact on conclusions and illustrate the problem using a study of auxin metabolism and signaling in early development of the phylloxera (*Daktulosphaira vitifoliae*, Phylloxeridae) gall formed on grape (*Vitis* spp.) leaves.

2 An RNAseq Study of Gall Formation

We studied galls formed on *Vitis riparia* Michx leaves by the grape phylloxera, *Daktulosphaira vitifoliae* (Fitch, 1855) (Hemiptera: Phylloxeridae: Aphidoidea) (Fig. 1). Phylloxera elicits formation of complex galls on abaxial leaf surfaces. Details of the gall development have been described by Sterling (1952). Females emerging from eggs in the spring feed on parenchyma from the upper surfaces of the youngest leaves. Within 24–48 hours a disk-shaped depression forms under and around the feeding insect. Cell mitosis decreases and cell expansion increases in this area and soon a bulge or ridge surrounds the feeding insect (Sterling 1952; Fig. 1). Within 48–72 hours the abaxial depression containing the insect deepens due to differential cell division and expansion the adaxial wall closes over it, leaving a narrow opening protected by dense trichomes. Development takes 4–5 days, at which point the insect has matured and begins producing eggs. "Crawlers" hatch

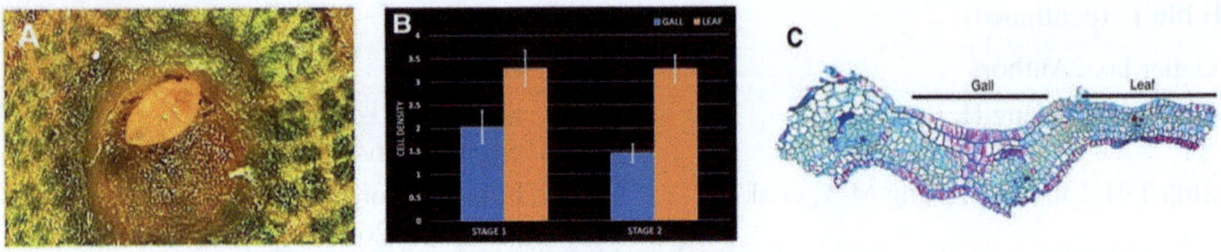

Fig. 1 (**a**) Early leaf gall formation, showing expanding disk and formation of surrounding ridge. (**b**) Density of cells in the disk (blue bars) is lower than in the leaf (orange bars) as cells in the disk expand. (**c**) Expanded cells in the feeding disk are larger than in a comparative non-galled leaf section

from eggs in the gall, exit through the abaxial opening, and proceed to feed and induce galls on younger leaves.

We have used RNAseq to identify differentially expressed genes (DEGs) in the early stages of gall development as compared with non-galled leaf, tried to determine their functions, and characterized the molecular processes underlying gall development (Schultz et al. 2019). We have focused on auxin production and signaling during the earliest gall development stages for 3 reasons. First, early gall development exhibits cell expansion (Fig. 1) which is regulated by auxin (Majda and Robert 2018), so we limited our study to the first 2 gall stages, which take place during the first 72 hours of feeding (Sterling 1952; Schultz et al. 2019). Second, elevated auxin concentrations are often found in insect galls (Tooker and Helms 2014). And third, many galling insects, including phylloxera, are known or thought to produce auxin and may inject it during gall initiation (Schaller 1968; Tooker and Helms 2014).

Most insect gall hostplants lack a sequenced genome, and most model plants (e.g., *Arabidopsis*) lack a tractable gall system. We resolved this issue using genome-wide syntenic analysis to identify orthologous genes in the sequenced *V. vinifera* genome (Schultz et al. 2019). Few *V. vinifera* genes are functionally annotated, so we also used syntenic analysis to identify *Arabidopsis thaliana*—*Vitis vinifera* orthologs because of the much better functional annotation of *Arabidopsis* genes (Schultz et al. 2019). We were able to find annotations for more than 86% of all coding sequences in the *V. vinifera* genome this way. Most insect gall hostplants lack a sequenced genome, and most model plants (e.g., *Arabidopsis*) lack a tractable gall system. To resolve the lack of sequenced genome, we use the very close relatedness of the *V. riparia* and *V. vinifera* genomes. That did not, however, allow inferences about gene functions, since there is little functional annotation of *V. vinifera* genes (especially for leaves). Our approach was then to identify *Arabidopsis* orthologs for which there is functional annotation. As in most studies of this type, we sought information about gene function using Gene Ontology Biological Process categories, which group genes based on similarities in what is known about their biological functions.

We determined which of the DEGs we found in galls occurred in the genes listed at The Arabidopsis Information Resource (TAIR: https://www.arabidopsis.org) in

GO:0009850, "Auxin metabolism" and GO:0009733, "Response to Auxin". Twenty-two (22) genes involved in auxin metabolism were differentially expressed in galls; expression of 12 was upregulated and 10 downregulated compared with ungalled leaf (Table 2). Expression of 26 genes known to respond to auxin was greater in the gall than the leaf, indicating likely responses to auxin. But because some upregulated genes may suppress biological functions and some downregulated ones may promote them, merely finding a DEG in a GO collection says little or nothing about its role in that GO's biological process. We needed to know whether those auxin-related DEGs indicated that auxin was being synthesized and auxin signaling was active.

We sought detailed functional information available for many *Arabidopsis* genes at TAIR. But because the TAIR information lags the published literature, gene ontologies can be incomplete. So, we also did primary literature searches to obtain the most recent and accurate functional information for each DEG. We also found that many genes are classified in more than one GO category, lending uncertainty to conclusions about their role in gall formation.

Based on their functions, 14 of the 22 auxin-metabolism DEGs would increase auxin production (Table 2). Several auxin synthesis pathways have been described but only the tryptophan-dependent pathway is thought to be the main source of free auxin in plant tissues (Zhao 2012). Expression of the genes underlying the tryptophan-dependent pathway, from tryptophan synthesis to *YUCCA 6* (*YUC6*), which catalyzes the rate-limiting irreversible oxidative decarboxylation of

Table 2 Log_2 (-fold difference between gall and leaf) in expression of GO:0009850, "Auxin-metabolism" DEGs

V. vinifera locus	A. thaliana ortholog	Gall 1	Gall 2	Protein ID	Protein name
GSVIVT01018990001	AT5G25620	4.19	5.54	YUCCA6	YUCCA6
GSVIVT01006656001	AT1G70560	2.29	2.55	TAA1	TRYPTOPHAN AMINOTRANSFERASE 1
GSVIVT01002586001	AT5G14420	1.78	0.00	RGLG2	RING DOMAIN LIGASE 2
GSVIVT01022111001	AT4G36930	1.29	1.24	SPT	SPATULA
GSVIVT01036477001	AT5G38530	1.45	1.29	TSBTYPE2	TRYPTOPHAN SYNTHASE BETA SUBUNIT TYPE 2
GSVIVT01001997001	AT4G27070	0.94	1.29	TSB2	TRYPTOPHAN SYNTHASE B2
GSVIVT01000716001	AT3G54640	0.91	1.33	TSA1	TRYPTOPHAN SYNTHASE A1
GSVIVT01027971001	AT2G47750	0.63	3.15	GH3.9	GRETCHEN HAGEN 3.9
GSVIVT01009854001	AT4G31500	0.00	3.73	SUR2	SUPPERROOT
GSVIVT01001997001	AT4G27070	0.00	1.29	TSB2/TRP2	TRYPTOPHAN SYNTHASE BETA SUBUNIT 2
GSVIVT01030905001	AT1G71090	0.00	0.95	PILS2	PIN-LIKES 2
GSVIVT01016288001	AT1G48850	0.00	0.78	CS	CHORISMATE SYNTHASE
GSVIVT01035717001	AT1G80360	0.00	-0.72	VAS1	REVERSAL OF SAV3 PHENOTYPE 1
GSVIVT01011155001	AT3G08510	-1.47	0.00	PLC2	PHOSPHOLIPASE C 2
GSVIVT01014545001	AT5G54510	-1.27	-2.36	GH3.6	GRETCHEN HAGEN 3.6
GSVIVT01028471001	AT4G28720	0.00	-3.43	YUCCA8	YUCCA8
GSVIVT01000176001	AT2G14960	0.00	-2.59	GH3.1	GRETCHEN HAGEN3.1
GSVIVT01037730001	AT5G55250	0.00	-1.67	IAMT1	IAA CARBOXYLMETHYLTRANSFERASE 1
GSVIVT01008852001	AT1G44350	0.00	-1.45	ILL6	ILR1-LIKE 6
GSVIVT01011006001	AT1G48910	0.00	-1.33	YUCCA10	YUCCA10
GSVIVT01037267001	AT5G12330	0.00	-0.90	LRP1	LATERAL ROOT PRIMORDIUM 1
GSVIVT01035469001	AT4G33210	0.00	-0.65	SLOMO	SLOW MOTION

Values in orange boxes indicate gene expression significantly greater in gall than in leaf; blue indicates lower expression in gall than in leaf. Protein IDs in yellow would increase auxin concentrations; those in blue would decrease concentrations. Protein names in orange boxes identify members of the tryptophan-dependent auxin biosynthesis pathway

Table 3 Log_2 of significant increase in expression of GO:0009733 ("Response to Auxin") DEGs in gall compared with leaf

V. vinifera locus	A. thaliana ortholog	Gall 1	Gall 2	Protein ID	Protein name
GSVIVT01016698001	AT1G72430	2.91	2.32	SAUR78	SMALL AUXIN UPREGULATED RNA 78
GSVIVT01028342001	AT1G30690	2.23	1.32	PATL4	Sec14p-like phosphatidylinositol transfer family
GSVIVT01031314001	AT3G53480	1.83	0.00	PIS1	POLAR AUXIN TRANSPORT INHIBITOR SENSITIVE 1
GSVIVT01032619001	AT3G55120	1.79	3.43	CFI	CHALCONE FLAVANONE ISOMERASE
GSVIVT01031659001	AT1G10840	0.92	0.80	TIF3H1	TRANSLATION INITIATION FACTOR 3 SUBUNIT H1
GSVIVT01008907001	AT5G08640	0.00	3.44	FLS1	FLAVONOL SYNTHASE 1
GSVIVT01027971001	AT2G47750	0.00	3.15	GH3.9	GRETCHEN HAGEN 3.9
GSVIVT01016739001	AT5G07990	0.00	2.30	TT7	CYTOCHROME P450 75B1
GSVIVT01007877001	AT5G07990	0.00	1.53	TT7	CYTOCHROME P450 75B1
GSVIVT01007878001	AT5G07990	0.00	1.38	TT7	CYTOCHROME P450 75B1
GSVIVT01030905001	AT1G71090	0.00	0.95	PILS2	PIN-LIKES 2
GSVIVT01035713001	AT4G32690	0.00	0.78	GLB3	HEMOGLOBIN 3
GSVIVT01038651001	AT5G24520	0.00	0.78	TTG1	TRANSPARENT TESTA GLABRA 1
GSVIVT01029853001	AT5G01990	0.00	0.75	PILS6	PIN-LIKES 6
GSVIVT01026077001	AT4G09160	0.00	0.68	PATL5	Sec14p-like phosphatidylinositol transfer family protein

indole-3-pyruvate acid (IPyA) to form indole-3-acetic acid (IAA) (Zhao 2012), was significantly increased in the developing gall (Table 2). This makes it very likely, if not certain, that auxin was being synthesized by the plant in the earliest gall tissues. Additional free auxin may be retained by downregulated expression of *Gretchen Hagen 3.6* and *3.1* (*GH3.6*, *GH3.1*) which conjugate auxin to amino acids (Korasick et al. 2013) and *IAA CARBOXYL METHYLTRANSFERASE 1* (*IAMT1*), which methylates auxin (Korasick et al. 2013). In addition, elevated expression of *PIN-LIKES 2* (*PILS2*) would slow auxin efflux, potentially increasing local concentrations (Feraru et al. 2022). Overall, the results strongly suggest that auxin synthesis was elevated in early gall stages and that auxin was likely to accumulate there.

Evidence of auxin accumulation in the gall comes from the expression of DEGs found in GO:0009733, "Response to Auxin". Expression of 26 genes putatively responsive to auxin were upregulated in the gall as compared with leaves. However, 11 of these genes also respond to other phytohormones. After removing them from consideration, we were left with 15 DEGs thought to respond only to auxin (Table 3). Finding that some auxin-responsive genes whose expression was upregulated in the gall also respond to other phytohormones does not, however, mean that they were not responding to auxin in the gall. We just can't be sure without further study. Clearly expression of some genes that respond only to auxin was upregulated in the gall.

Taken together, our results strongly suggest that auxin is synthesized in the early developing phylloxera leaf gall and likely influences development there. There would be no need for the insect to provide auxin, although it must be stimulating auxin synthesis.

3 Challenges with the Approach and Possible Solutions

There are several challenges inherent in studies using RNA sequencing to study gall development. These include a lack of genomic and functional annotation for the non-model host species with galling parasites, the incompleteness and correlative nature of relationships in gene ontologies, complex interactions among gene expression networks, highly localized gene expression and developmental events, the possibility that functions involve post-transcriptional regulation, and the correlative nature of transcriptomics.

Most insect galls form on 'non-model' plant species, whose genomes may not have been sequenced. We approached this by identifying orthologs from *Arabidopsis*, which are relatively well-characterized. Genome sequencing has now become accessible to more gall investigators, and some are beginning to sequence the parasite's host plant (Martinson et al. 2022; Hearn et al. 2019). This still leaves the investigator with very limited information about the functions of genes they find. Comparison with better-studied orthologs may still be needed, although inferring function across taxa ignores the possibility of refunctionalization; genes and proteins with similar sequences may not have precisely the same function in two unrelated taxa.

While functional information may be found in Gene Ontologies, even relatively complete gene ontology lists are incomplete, and new genes are added frequently. Frequent literature searching is required to find functions of all the genes that belong in a GO category and new functional discoveries. Ontology gene lists indicate only that expression of those genes is somehow associated with a function. Those associations are often only correlative, based on co-expression patterns. Because some genes can be promoting or inhibiting a function depending on whether their expression is up- or down-regulated, finding a gene in a functional category says little or nothing about its specific role. It is necessary to survey the literature to find what is known about each gene's function. Merely listing significantly enriched GO categories tells us little or nothing about how transcription relates to function.

Plant gene expression networks and pathways are complex and interact. Expression of many, perhaps most, genes is found in more than one network or pathway, indicating more than one likely function. Similarly, signaling pathways interact so that the expression of many genes is influenced by more than one signal. This leaves the investigator to determine which of the functions a gene may participate in is the one occurring in the developing gall. The ability to draw a correct conclusion depends on how much is known about the genes in question. But it simply may not be possible to be certain about the role(s) of some genes being differentially expressed in the gall. Being certain about auxin signaling activity was made more difficult for us by interacting hormone signaling pathways. We were fortunate in our study that expression of 3 genes encoding proteins that are only responsible for synthesizing auxin were upregulated, allowing a relatively clear conclusion.

Insect galls are complex, heterogenous collections of tissues (Martinson et al. 2022). Sampling multiple tissues (e.g., whole galls) makes it difficult to capture an accurate transcriptional picture of such events (Martinson et al. 2022). For example, auxin's impact on plant development often involves local biosynthesis and very localized activity (Brumos et al. 2018). Our ability to detect this is hindered by dilution unless we can sample only the responding cells. For example, the nutritive cells that are characteristic of most galls including the phylloxera leaf gall are often unique metabolically and structurally. And other cells of different tissues within a gall frequently differ in size, shape, organelle composition, and cell wall structure. These traits must be governed by differential gene expression and their development is known or likely to be regulated by auxin in normal plant development. Because auxin is not the only signal influencing cellular development, it is difficult or impossible to infer this from histology alone. Confirming auxin's local accumulation and impact on development will require analytical chemistry and gene expression studies done on those specific cells. Ultimately, transformation with a reporter construct or gene editing will be required.

The meaning of transcriptome results is weakened by the fact that not all plant developmental events are transcriptionally regulated. Focusing on gene transcription using RNAseq ignores post transcriptional regulation of protein activity. For example, activity of the key auxin biosynthesis protein TAA1 is regulated by post-transcriptional phosphorylation (Wang et al. 2020). RNAseq also fails to account for the contribution of epigenetic regulation of gene expression, as has been suggested for some gall systems (e.g., Martinson et al. 2022). Protein-protein interactions and degradation may also play important roles. Only expanded protein and epigenetic studies can overcome these limitations.

Finally, it is important to keep in mind that all RNAseq results are strictly correlative. We cannot conclude, for example, that localized auxin synthesis is responsible for the structure of the early gall. Doing so would require experimentally manipulating the expression of one or more of the key genes and assessing the consequences for development. This has been hampered by inability to genetically transform non-model host plants. That kind of manipulation is becoming tractable in more plants with the advent of CRISPR-CAS gene editing but was not available to us for grapevine.

4 Conclusion

We used RNAseq successfully to gather transcriptional evidence that auxin is synthesized by the plant during initiation and early development of the gall elicited by phylloxera on wild grape leaves. We also found evidence that auxin-responsive genes are activated in the developing gall. RNAseq can be a useful approach in studying gall formation. However, since its results are correlative, it cannot provide definitive conclusions about the causal impact of differential gene expression on gall formation. Only gene expression manipulation can identify genes required for gall formation. Development of new gene editing techniques may make this possible. Investigators need to be aware of this limitation of the method to avoid overstating conclusions.

5 Methods: RNAseq Experiment

Methods for sampling tissue, extracting RNA, constructing the Illumina library, processing Illumina reads, quantifying expression, and validating RNAseq results by digital droplet PCR are also described in Schultz et al. (2019).

Tissue sampling—Galled and ungalled leaves were collected between 09:00 am and 10:00 am from, April to August 2014 from wild *Vitis riparia* Michx. vines near Rocheport, Missouri, USA (38° 58′ 16.424″ N, 92° 32′ 54.118″ W). Galls from three different vines were separated by size into four developmental categories and dissected on ice; midribs were removed from ungalled control leaves. Because the two earliest gall stages developed on the same leaves, there were only three control leaf size classes matched to the four gall stages. To obtain enough RNA, samples were pooled from three individual vines, producing three biological replicates (N = 3) for each of the four gall developmental stages and three control leaf sizes (i.e., originating from 12 independent grapevines for galled tissue, and nine for control leaves). All tissues were immediately frozen in liquid nitrogen and stored at −80 °C.

RNA Extraction—RNA was extracted and DNase1-treated, on column, using the Spectrum Plant Total RNA Kit (Sigma #STRN50-1KT; protocol A and Appendix). The resulting RNA was further purified and concentrated with the RNeasy MinElute Cleanup Kit (Qiagen #74204) and eluted with water. The quality of the resulting RNA was assessed using the Agilent 2100 BioAnalyzer (Agilent, Santa Clara, CA, USA), and all RNA integrity number values were found to be above 8.

Illumina Library and Construction—The Illumina libraries (three biological replicates for each of the four gall developmental stages and three control leaf sizes, for a total of 21 libraries) were constructed using the RNA TruSeq Kit (Illumina, Inc., San Diego, CA, USA), barcoded (TACT ungalled; GTAT galled), and sequenced single-end with 100 bp reads on the Illumina HiSeq-2000 platform at the University of Missouri DNA Core (http://dnacore.missouri.edu/; University of Missouri, Columbia, MO, USA).

Illumina read processing and expression quantification—A custom Perl script was used to parse the libraries and remove barcode sequences resulting in approximately 40.9 million reads per biological replicate for the ungalled library and 40.3 million reads per biological replicate for the galled library. NextGENe V2.3.3.1 (SoftGenetics, LLC., State College, PA, USA) was used to quality filter the fastq data, remove reads with a median quality score of less than 22, trim reads at positions that had three consecutive bases with a quality score of less than 20, and remove any trimmed reads with a total length less than 40 bp. The reads were aligned against Vitis vinifera V2 genome (DOE-JGI; ftp://ftp.jgi-psf.org/pub/compgen/phytozome/v9.0/Vvinifera/), thus eliminating transcripts originating from phylloxera insects. Gene expression was quantified using TopHat/Cufflinks software (Ghosh and Chan 2016).

Differential expression between galled and ungalled leaf tissue was analyzed for each mapping, using two discrete probability distribution-based methods, DESeq and edgeR (https://bioconductor.org) and the annotated *Vitis vinifera* V2 genome (DOE-JGI; ftp://ftp.jgipsf.org/pub/compgen/phytozome/v9.0/Vvinifera/). Read counts and RPKM values (reads per kilobase per million) were calculated for each library. An RPKM cutoff of 0.1 per gene model was applied for comparing expression values. Functional analyses were limited to genes with a differential expression significance <0.05 and >1.5-fold difference. Fold-change between galls and their respective ungalled control leaves was calculated for each gene by subtracting the base-2 logarithm of the RPKM value of galls to the base-2 logarithm of the RPKM value of ungalled control leaves.

Genome-wide syntenic analyses were performed to identify Arabidopsis thaliana—*Vitis vinifera* orthologs using CoGe (http://genomevolution.org/CoGe/). In addition, Arabidopsis—Vitis orthologs were identified using reciprocal BLASTp analyses (protein databases) with a 0.00001 p-value cutoff resulting in the annotation of ~86.7% of all coding sequences in the Vitis vinifera V2 genome. Because the Vitis genome is not yet fully annotated, we used orthology with *A. thaliana* genes to infer gene function to our *V. riparia* differentially expressed genes (DEGs). Gene Ontology (GO) enrichment analyses were performed for each of the gall and leaf gene expression sets using the PANTHER classification system (Mi et al. 2017). Statistical significance for enrichment scores was set at <0.005. RNAseq results were validated successfully using digital droplet PCR. The results are reported in Schultz et al. (2019). Data availability—RNAseq data that were generated for this study are available at NCBI Gene Expression Omnibus (https://www.ncbi.nlm.nih.gov/geo/) under study accession GSE118569.

6 Tissue Preparation for Microscopy (Fig. 1c)

Thin section slides were dried on a heat block at 58 °C for 10–15 min. Samples were deparaffinized using xylene for 5 min (repeat twice), then subjected to an ethanol series from 100% to 50%. Sections were then stained using the Wagner 3A triple stain (Wacker 2006). Samples were mounted on slides using xylene-based Permount and a coverslip.

Acknowledgements This work was supported by US National Science Foundation grant NSF-IOS-1757358 (JCS HMA) and by JCS and HMA's personal funds. We thank Wade T. Dismukes and the University of Missouri Informatics Research Core Facility for help analyzing and organizing data and Dean E. Bergstrom for fieldwork and RNA extraction.

References

Brumos J, Robles LM, Yun J et al (2018) Local auxin biosynthesis is a key regulator of plant development. Dev Cell 47:306–318.e5. https://doi.org/10.1016/j.devcel.2018.09.022

Feraru E, Feraru MI, Moulinier-Anzola J et al (2022) PILS proteins provide a homeostatic feedback on auxin signaling output. Development 149:dev200929. https://doi.org/10.1242/dev.200929

Ghosh S, Chan CKK (2016) Analysis of RNA-seq data using TopHat and Cufflinks. In: Edwards D (ed) Plant bioinformatics. Methods in molecular biology, vol 1374. Humana Press, New York. https://doi.org/10.1007/978-1-4939-3167-5_18

Hearn J, Blaxter M, Schönrogge K et al (2019) Genomic dissection of an extended phenotype: oak galling by a cynipid gall wasp. PLoS Genet 15:e1008398. https://doi.org/10.1371/journal.pgen.1008398

Korasick DA, Enders TA, Strader LC (2013) Auxin biosynthesis and storage forms. J Exp Bot 64:2541–2555. https://doi.org/10.1093/jxb/ert080

Majda M, Robert S (2018) The role of auxin in cell wall expansion. Int J Mol Sci 19:951. https://doi.org/10.3390/ijms19040951

Martinson EO, Werren JH, Egan SP (2022) Tissue-specific gene expression shows a cynipid wasp repurposes oak host gene networks to create a complex and novel parasite-specific organ. Mol Ecol 31:3228–3240. https://doi.org/10.1111/mec.16159

Mi H, Huang X, Muruganujan A et al (2017) PANTHER version 11: expanded annotation data from Gene Ontology and Reactome pathways, and data analysis tool enhancements. Nucleic Acids Res 45:D183–D189. https://doi.org/10.1093/nar/gkw1138

Nabity PD, Haus MJ, Berenbaum MR, DeLucia EH (2013) Leaf-galling phylloxera on grapes reprograms host metabolism and morphology. Proc Natl Acad Sci USA 110:16663–16668. https://doi.org/10.1073/pnas.1220219110

Oates CN, Denby KJ, Myburg AA et al (2021) Insect egg-induced physiological changes and transcriptional reprogramming leading to gall formation. Plant Cell Environ 44:535–547. https://doi.org/10.1111/pce.13930

Redfern M (2011) Plant galls. HarperCollins Publishers Limited. https://books.google.com/books?id=_D0mzgEACAAJ

Schaller G (1968) Untersuchen zur Erzeugung kunstlicher Pflanzengallen. Marcellia 25:131–153

Schultz JC, Edger PP, Body MJA, Appel HM (2019) A galling insect activates plant reproductive programs during gall development. Sci Rep 9:1833. https://doi.org/10.1038/s41598-018-38475-6

Sterling C (1952) Ontogeny of the phylloxera gall of grape leaf. Am J of Bot 39:6–15. https://doi.org/10.1002/j.1537-2197.1952.tb13038.x

Tooker JF, Helms AM (2014) Phytohormone dynamics associated with gall insects, and their potential role in the evolution of the gall-inducing habit. J Chem Ecol 40:742–753. https://doi.org/10.1007/s10886-014-0457-6

Takeda S, Yoza M, Amano T, Ohshima I, Hirano T, Sato MH, Sakamoto T, Kimura S (2019) Comparative transcriptome analysis of galls from four different host plants suggests the molecular mechanism of gall development. PloS One, 14:e0223686. https://doi.org/10.1371/journal.pone.0223686

Wacker R (2006) Eine neue und einfache methode zur polychromatischen anfärbung von paraffinschnitten pfanzlicher gewebe für durlicht- and fluoreszenzmikroskopie. Mikrokosmos 95:210–212

Wang Q, Qin G, Cao M et al (2020) A phosphorylation-based switch controls TAA1-mediated auxin biosynthesis in plants. Nat Commun 11:679. https://doi.org/10.1038/s41467-020-14395-w

Zhao Y (2012) Auxin biosynthesis: a simple two-step pathway converts tryptophan to indole-3-acetic acid in plants. Mol Plant 5:334–338. https://doi.org/10.1093/mp/ssr104

Epidermis: The First Line of Contact Between Plants and Gall-Inducing Herbivores

Rosy Mary dos Santos Isaias, Bruno Garcia Ferreira, and Renê Gonçalves da Silva Carneiro

1 Epidermal Cells: Origin and Fates

The morphological complexity of galls is a result of various developmental and structural patterns that involve the deviation of plant organogenesis by a series of changes in the number of divisions and in the degree and polarity of cell growth in the tissues manipulated by gall inducers (*cf.* Mani 1964; Meyer and Maresquelle 1983). The key processes in most gall growth and development phases are hyperplasia, cell hypertrophy, dedifferentiation, and cell death triggered by different mechanisms, either autolysis or injuries caused by insect feeding (Rey 1992). These processes lead to the formation of specialized tissues derived from the three tissue systems of plants: the dermal, ground, and vascular systems. Epidermal cells are the first line of contact between the host plant and the gall inducer.

The epidermis is a complex tissue composed of various cell types with different degrees of morphological and functional specializations, which depend both on the fates of the cell lineage to which they belong and on interaction-based mechanisms (Glover 2000). These two factors explain the standard epidermis organization in plants, which is composed of a mosaic of common pavement cells, stomatal guard cells, and trichomes (if present) that differentiate under conservative patterns. Even undifferentiated protodermal cells are individually marked as meristemoid mother

---

R. M. dos Santos Isaias (✉)
Department of Botany, Universidade Federal de Minas Gerais (UFMG), Belo Horizonte, Minas Gerais, Brazil
e-mail: rosy-isaias@ufmg.br

B. G. Ferreira
Department of Botany, Universidade Federal do Rio de Janeiro (UFRJ), Rio de Janeiro, Brazil

R. G. S. Carneiro
Department of Botany, Universidade Federal de Goiás (UFG), Goiânia, Goiás, Brazil

D. C. de Oliveira, R. M. dos Santos Isaias (eds.), *Plant Galls*,
https://doi.org/10.1007/978-3-031-80064-1_3

cells that originate stomata, trichome precursors, or pavement cells, which will follow their fates during the cycles of cell division (Glover 2000; Liu et al. 2021). The epidermal patterning in aerial plant parts follows genetic programs and has crucial functional implications for the survival of plants in the face of environmental challenges, both biotic and abiotic. Thus, despite being conserved within species, patterns of stomata and trichome differentiation may be variable when analyzed from evolutionary and ecological perspectives (Martin and Glover 2007). So, in the face of environmental stressors, it is plausible to assume that plant epidermis might differentiate into different patterns, which is the case when galls start to develop in the plant body.

Gall development and the modifications occurring in epidermal cells depend on the developmental phase in which the tissue is and on the competence of the cells for further division and differentiation. These are crucial steps for the establishment of the epidermis of galls, which has new structural and functional patterns compared to nongalled host-plant organs (Isaias et al. 2014, 2024; Ferreira et al. 2019). Under the influence of a gall inducer, epidermal and adjacent cells respond to the stimuli promoted either by oviposition or larval feeding, depending on the gall-inducing taxon (Rohfritsch 1992). Epidermal cells may have multiple functions, sometimes protective and sometimes attractive (Evert 2006). As these cells are the first line of contact with external factors, they must respond to environmental changes during their differentiation or even during the additional surface growth of an organ. This makes them good target cells for recognizing induction sites by gall-inducing herbivores. In galls, epidermal cells also play multiple functions. Still, the patterns of cell differentiation by asymmetric divisions that result in pavement cells, stomatal guard cells, and trichomes are often altered, both structurally and functionally. The simple cell hypertrophy and hyperplasia of the subepidermal layers, which occur in galls, change the expansion and division rates of these cells, leading to changes in their shapes and sizes, which may occur simply to compensate for subepidermal growth or to generate novel structural patterns (Moura et al. 2008; Ferreira and Isaias 2013; Amorim et al. 2017; Nobrega et al. 2023) (Fig. 1).

Changes in epidermal traits begin very early during gall induction, when chemical and mechanical stimuli generated by gall-inducing organisms lead to an imbalance in redox homeostasis in plant cells, as hypothesized by Isaias et al. (2015). This imbalance most commonly translates into developmental signals for gall ontogenesis, but some plant species may have sensitive epidermal cells that respond differently. In this case, a strong reaction that causes epidermal and adjacent cells to die rapidly after the first contact with galling organisms impairs the establishment of the gall inducer. This process, which was first known to occur only in response to infection by pathogens, was later observed in insect–plant interactions and recognized as a hypersensitive response (HR) (Fernandes et al. 2000). In this chapter, we discuss the first responses of epidermal cells to gall induction stimuli and the morphophysiological impacts of the changes induced by galling organisms during gall development in the outer epidermis, the layer covering the galls externally and acting as a barrier and a contact of the gall tissues with the external environment, and the inner epidermis, which is in contact with the gall inducers.

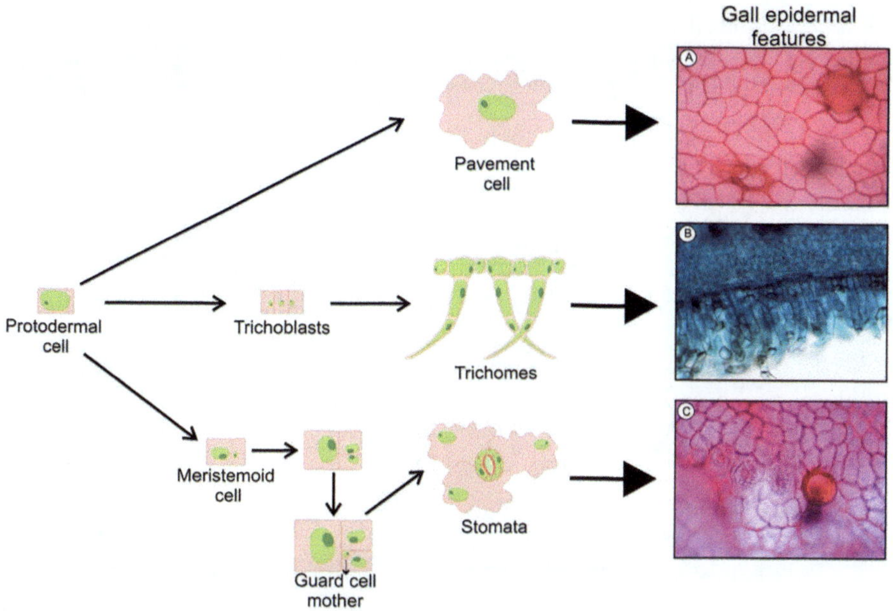

Fig. 1 Ordinary developmental fates of epidermal cells, and deviations toward the loss of sinuosity in pavement cells (**a**), overdifferentiation of trichomes (**b**), and stomatal malformations (**c**) due to gall induction by Eriophyidae on *Avicennia schaueriana* (Acanthaceae). (Photos by Leticia P. Nobrega)

2 Hypersensitive Response

The hypersensitive response (HR), as a resistance mechanism of plants against the attack of gall-inducing herbivores (Fernandes 1998), may be recognized by biochemical, cytological, histological, and morphological events that lead to local cell death and, consequently, to the elimination of the gall inducer (Isaias and Oliveira 2012). Cell death involves necrosis caused by extrinsic factors that trigger the accumulation of phytotoxic molecules after a traumatic stress event (Breusegem and Dat 2006), such as the attempt to induce galls either by oviposition or larval feeding. The events that characterize necrotic plant cell death are passive and indiscriminate and are characterized by a progressive loss of membrane integrity that results in cytoplasmic swelling and the release of cellular constituents (Breusegem and Dat 2006), together with a burst in the production of reactive oxygen species (ROS) (Isaias and Oliveira 2012; Isaias et al. 2015).

In many cases, HR is visualized in a single cell, or sometimes in a region involving tens or hundreds of cells, visible to the naked eye (Balint-Kurti 2019). Morphologically, the HR is observed as brown spots on leaves that indicate the sites of necrotic cell death (Fig. 2a–d) (Fernandes et al. 2000). The symptoms of HR are commonly observed as rounded areas in which the tissues surrounding the site of

Fig. 2 Hypersensitive responses (HR) in leaves of *Aspidosperma macrocarpon* (Apocynaceae) due to unsuccessful oviposition. (**a–d**) Development of necrotic tissues forming the typical HR halo. Bars = 1 cm

gall induction become necrotic, with obliteration of the cell lumen. In *Bauhinia brevipes-Contarinia* sp. (Diptera, Cecidomyiidae), the HR symptoms are spread through the palisade and spongy parenchyma, reaching the cells of the abaxial epidermal surface, on the other side where galls are tentatively induced (Fernandes et al. 2000). Surprisingly, the cells of the adaxial epidermal surface, closest to the site of gall induction, are free of death symptoms, which are likely blocked by the accumulation of mucilage. The HR site that develops around sedentary larvae impairs their nutrition (Fernandes 1998; Fernandes et al. 2000) and kills them by starvation. These reactions are responsible for more than 90% of gall failure initiation, indicating that this might be a mechanism that controls the population of gall inducers. In such a sense, the hypersensitive response may be considered an adaptive strategy of the host plants since gall development is usually detrimental to the host plant's fitness, as galls are physiological sinks of carbohydrates and other important metabolites (Oliveira et al. 2017).

Even though HR negatively affects gall inducers, some species can take advantage of this phenomenon. For example, in Thysanoptera-induced galls on *Myrcia splendens*, which are leaf rolls, the histological alterations are simple, with areas of homogeneous tissue altering with areas with necrotic cells (Jorge et al. 2018). In this particular case, the HR that causes necrosis also acts as a morphogenetic event that leads the flattened leaves to develop into leaf-rolling galls, thus benefiting the gall inducer, not the plant. In fact, when the HR is controlled by a gall inducer, the accumulation of ROS is followed by the accumulation of phenolics, which impairs the death process by interacting with IAA oxidases, a crucial step for gall induction and development (Isaias et al. 2015).

3 Alterations in Epidermal Differentiation Patterns

If galling organisms successfully control the HR, tissue homeostasis is maintained (Isaias et al. 2015), and the gall initiates its development by the stimulation of epidermal and adjacent cells to divide and enclose the larva (e.g., Isaias et al. 2011; Marquesine et al. 2024). In such a situation, epidermal cells display increased cell divisions, usually anticlinal, and hypertrophy, compensating for the increase in surface area caused by the intumescence of the host organ during gall growth (Álvarez et al. 2009; Oliveira and Isaias 2010; Isaias et al. 2011; Ferreira and Isaias 2013, 2014; Nobrega et al. 2021; Castro et al. 2023). The increase in volume of the ground and vascular tissues underneath is followed by an increase in epidermal surface area, a phenomenon observed in the development of virtually all gall morphotypes.

The structural complexity of galls varies according to the degree of tissue alteration; the more altered the tissues are, the more complex a gall may be. Alterations in the epidermis often involve the suppression or malformation of stomata, overdifferentiation of trichomes, stratification of the epidermis or its substitution by a periderm coating the outer gall surface (Ferreira et al. 2019). In the inner gall epidermis, in contact with the gall chamber, the epidermis may be part of a typical nutritive tissue in Cecidomyiidae galls (Marquesine et al. 2024), may be formed by vacuolated sugar-rich cells or differentiated hairs in Eriophyidae galls (Moura et al. 2008; Ferreira et al. 2017; Nobrega et al. 2023), or may lose the typical compactness of epidermal cells via the development of holes or channels in aphid-induced galls (Álvarez 2011, 2012; Kutsukake et al. 2012; Álvarez et al. 2014). In complex aphid galls, the holes or channels in the inner epidermis effectively reabsorb the excess honeydew produced by aphids during their phloem-sucking feeding activity (Kutsukake et al. 2012). This process prevents aphids from drowning and enables plant tissues to recover and reuse part of the energy lost to aphid feeding. The differentiation of all cell types present on the epidermis of nongalled organs may occur in galls, sometimes changing their shape or dimensions, or may be impaired. A clearly defined gall epidermis is generally characterized by hypertrophied cells that are larger than those of nongalled host organs on the external surface (Meyer and Maresquelle 1983; Carneiro et al. 2014a; Oliveira et al. 2020; Castro et al. 2023). Additionally, such cells may display thickened cuticles and cell walls, sometimes lignified (Mani 1964). In the frontal view, changes in the outline of epidermal cells are commonly observed, as the anticlinal cell walls change from sinuous to straight or curved, as a sign of hyperplasia or hypertrophy (Amorim et al. 2017); in the case of the predominance of hyperplasia in the epidermis, hypotrophy of such cells is also common (Meyer and Maresquelle 1983).

3.1 The Pavement Cells

Compared with the nongalled portion of the host organ, the pavement cells covering the galls are often hypotrophied, which is a consequence of hyperplasia without cell enlargement but may also hypertrophy adjusting to the size of the new structure (Magalhães et al. 2014; Bedetti et al. 2017; Castro et al. 2023). Tissue hyperplasia or cell hypertrophy is evident on the side of gall development and is sometimes inconspicuous on the opposite surface of the host organ. The jigsaw puzzle shape of epidermal cells on leaves is commonly related to the expansion of the leaf lamina in all directions, and it is also important along with the expansion of the gall morphotypes. Nevertheless, not all the epidermal cells in galls maintain such complex shapes, and cell elongation at gall developmental sites may be responsible for straight or slightly curved anticlinal cell walls (Amorim et al. 2017; Castro et al. 2023). In the case of stem and root galls, the axis of cell elongation may be maintained, and the cells may hypertrophy while maintaining the original elongation axis.

In the outer covering of galls, the cell wall strata and cuticle may be thicker than those of nongalled portions of the host organ, enhancing protection against the loss of water and excess light and against fungal attack (Oliveira and Isaias 2009; Kurzfeld-Zexer et al. 2015; Martinez et al. 2018). This phenomenon is observed in the galls induced by *Nothotrioza* spp. (Hemiptera: Triozidae) in *Psidium* spp. (Myrtaceae), in which the outer epidermis displays a thicker cuticle and, sometimes, thicker secondary cell walls (Carneiro and Isaias 2015; Carneiro et al. 2015) than nongalled leaves. Interestingly, the galls induced by *N. myrtoidis* on *P. myrtoides*, which occur in the Brazilian Cerrado under a seasonally dry climate (Carneiro and Isaias 2015), have thicker cuticles and cell walls in the outer epidermis (RGS Carneiro pers. obs.) than the galls induced by *N. cattleiani* on *P. cattleyanum*, which occur in the constantly humid Atlantic Forest (Carneiro et al. 2015). This illustrates the important role that the environment plays in determining the epidermal characteristics that protect galls and gall inducers against water loss. Other aspects, especially those related to gall development, may influence the phenotype of the pavement epidermal cells; in the association between *Copaifera langsdorffii* (Fabaceae) and an undescribed species of Diptera, Cecidomyiidae, a midrib gall is induced both in young and mature leaflets (Oliveira and Isaias 2009). In this case, leaflet age affects epidermal traits, as galls induced in young leaves display pavement epidermal cells with thinner walls. In contrast, those on mature leaves have thick external periclinal walls, especially along the ostiole on the adaxial surface of the gall, which forms as swelled invagination of the leaflet lamina along the midrib (Oliveira and Isaias 2009). In fact, the ostiole of galls is a place where deep epidermal alterations are often observed. Dead lignified cells (Carneiro et al. 2014a) and crystal inclusions in the cell walls (Dias et al. 2013) are characteristics that reinforce the gall structure near the ostiole and secure an adequate environment for the inducers to grow. Other alterations, such as the overdifferentiation of trichomes, may also occur near the gall ostiole and will be addressed further ahead (see the Sect. 3.3).

While the outer epidermis of galls maintains its protective role, similar to ungalled organs, in some galls, one of the epidermal surfaces originates from the inner covering of the larval chamber (Fig. 3), where the protective role is almost completely lost. The structural-functional characteristics of such cells shift toward a nutritional role (Ferreira et al. 2017, 2019). The inner epidermal cells that differentiate into nutritive cells are usually small, thin-walled, and phenotypically different from the outer epidermal cells, as observed in the midrib gall on *Copaifera langsdorffii* (Oliveira and Isaias 2009, 2010). Such nutritive cells are provided with granulose cytoplasm and hypertrophied nuclei, as in galls induced by *Clinodiplosis profusa* on *Eugenia uniflora* (Marquesine et al. 2024), or with dense accumulation of reducing sugars, as in the case of the *filzgalls* on *Miconia ibaguensis* (Ferreira et al. 2017). An example of the distinction between outer and inner epidermal cells is observed in galls induced on axillary buds of *Marcetia taxifolia* by an unidentified Cecidomyiidae (Ferreira and Isaias 2014). In these galls, the outer epidermal cells are larger than the inner epidermal cells are, and the outer epidermis has trichomes and stomata, which are absent on the inner gall surface (Fig. 3). In this case, trichomes on the inner surface are present only in the scape channel, where they seal the gall aperture, protecting the larva during gall maturation against desiccation of the internal microenvironment and preventing the invasion of the gall by opportunistic organisms or natural enemies (Fig. 3f).

In galls induced by insects with piercing-sucking mouthparts and that feed from subepidermal tissues (mainly vascular and perivascular parenchyma), the inner epidermal cells are poorly developed, small, thinly walled. They may even constitute a discontinuous layer, as we may observe in *N. myrtoidis* (Psylloidea) galls on *P. myrtoides* (Myrtaceae) (Fig. 4a, b) (Carneiro and Isaias 2015; Carneiro et al. 2015). This further illustrates how gall development alters epidermal fates structurally and functionally. This facilitates the access of the gall inducer to its feeding sites.

During gall development or at the end of the gall life cycle, epidermal cells may be suberized or metacutinized, as observed in Hemiptera galls on *Aspidosperma spruceanum* (Formiga et al. 2011). Suberization occurs due to the substitution of the epidermis for a periderm, which sometimes represents a divergence of the ordinary cell differentiation observed on leaves, in which the epidermis forms the common dermal system throughout leaf life (Isaias et al. 2024). The hymenoptera-induced globoid leaf galls on *Struthanthus vulgaris* (Loranthaceae) have a periderm formed from a one-layered phellogen from which a phellem composed of large and slightly suberized loosely packed cells differentiates (Kraus et al. 2002). This newly formed tissue indicates the ability of the inducer to trigger a program normally present on the host plant organs other than the leaves and to take advantage of the impermeability of the new dermal system. In other cases, however, the periderm is formed in response to the increase in volume caused by the proliferation of vascular and ground tissues, as observed in root-knot nematode galls (Cabrera et al. 2023). A similar phenotype is observed for nongalled roots, in which the epidermis is naturally replaced by a periderm in response to increased volume caused by cambial activity during secondary growth (Barros et al. 2024). Even though the causes are different (e.g., induction of root-knot galls *vs.* secondary root growth), the resulting

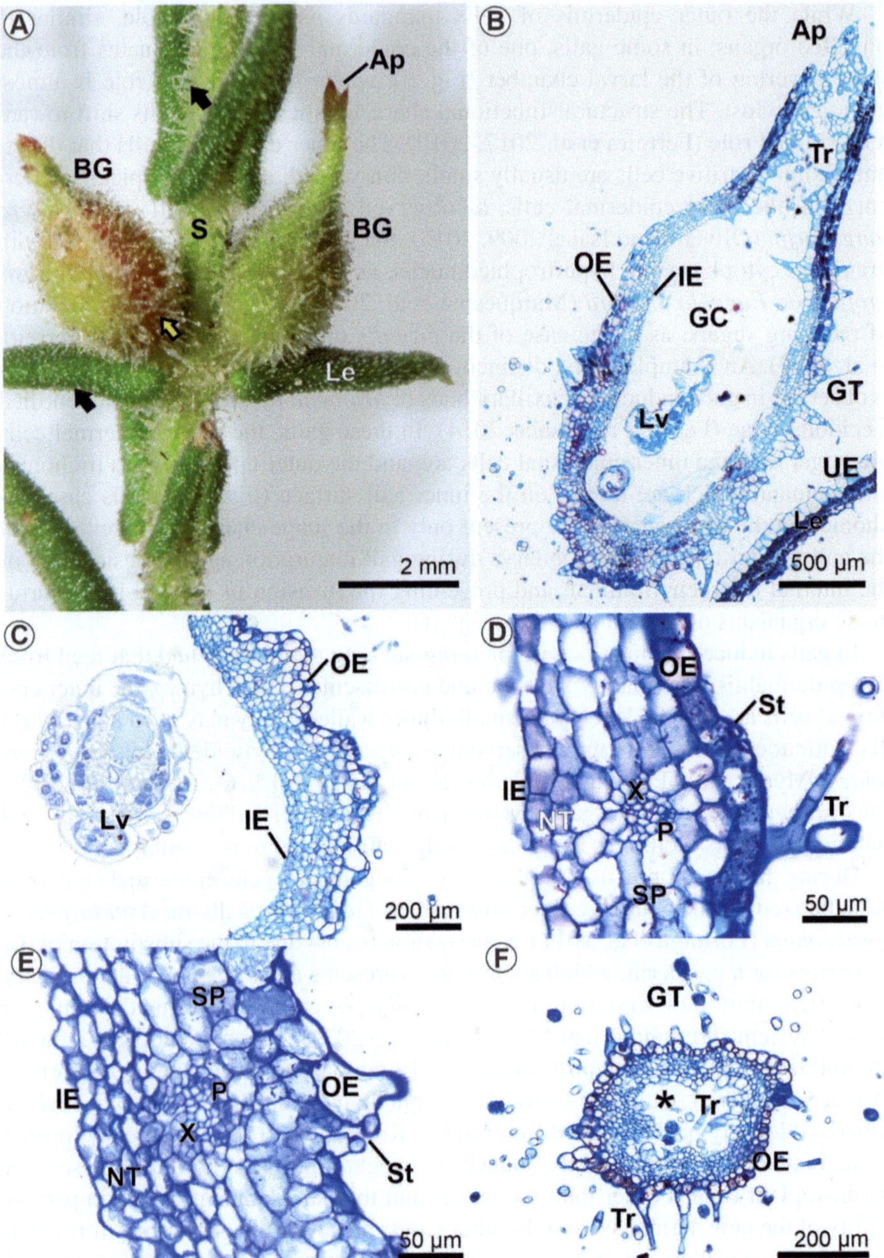

Fig. 3 Differences in the inner and outer gall surfaces of galls induced by a Cecidomyiidae on the axillary buds of *Marcetia taxifolia*. (**a**) Pistil-shaped bud galls observed via stereomicroscopy. Note the greater trichomes on the galls (yellow arrow) than on the stems and leaves (black arrows). (**b**) Longitudinal section of the gall and part of a nongalled leaf. (**c–f**) Transverse sections. (**c**) Differences in the outer and inner epidermis. (**d**) Trichome and closed stomata on the outer surface.

phenotypes during gall and nongalled organ development may be similar, indicating that gall inducers do not always affect the developmental programs of plants; sometimes, they simply hijack them for their benefit. For example, root-knot nematodes are often attracted to previously established galls and ultimately infect roots at the same sites (Dutta and Akhil 2024). In this case, the periderm not only prevents water loss but also may act as a barrier against reinfection by more nematodes, possibly contributing to avoiding intraspecific competition.

3.2 The Stomata

One of the striking characteristics of the gall epidermis, in general, is that the stomata are present in reduced numbers compared with the normal pattern and, in many cases, are completely absent, as in two leaf Cecidomyiidae gall morphotypes on *Mikania glomerata* (Asteraceae). However, another Cecidomyiidae gall morphotype on the same plants apparently has normal stomata (Amorim et al. 2017). In other cases, even when stomata are present, they may be ineffective because of the collapse of one or both guard cells (Mani 1964; Meyer and Maresquelle 1983). Concerning the stomata, there are a few possible alterations in terms of developmental patterns. In intralaminar galls, whose cells and tissues are on a *continuum* with the host organ, the stomata are generally similar to those of nongalled galls, as in Cecidomyiidae galls on *Avicennia schaueriana* (Acanthaceae) (Nobrega et al. 2021). Nevertheless, in extralaminar galls, the differentiation of one or two guard cells may be impaired, as observed in the bivalve-shaped galls induced by *Euphalerus ostreoides* on *Lonchocarpus muehlbergianus* (Oliveira et al. 2006). In such cases, gall tissues may be subjected to hypercarbia, as gas diffusion in and out of the gall structure is difficult (Pincebourde and Casas 2016; Oliveira et al. 2017). This seems to be the case for the globoid leaf gall of *N. myrtoidis* on *P. myrtoides*, which has a compact cortex and fewer stomata than the nongalled leaves do (Fig. 4c–h) and is photosynthesis deficient (Carneiro et al. 2014b), which means that photosynthetic rates are insufficient to consume excess carbon dioxide, possibly leading to hypercarbia. In the horn-shaped galls induced by Cecidomyiidae on *Copaifera langsdorfii*, however, the galls were found to photosynthesize at adequate levels (Carneiro et al. 2017), thus compensating for tissue compactness and the lack of functional stomata to avoid the negative effects of hypercarbia. In other cases, such as those of the extralaminar Cecidomyiidae galls induced on *Eugenia uniflora*,

Fig. 3 (continued) (**d**) Open stomata on the outer surface. (**e**) Inner epidermis, part of the nutritive tissue, and outer epidermis with an open stomata. (**f**) Apical edge of the gall, showing an aperture (asterisk) with glandular and nonglandular trichomes on the inner and outer surfaces. Staining: **b–d, f**: Astra blue and safranin. **e**: Toluidine O blue. Abbreviations: Ap gall aperture, BG bud gall, GC gall chamber, GT glandular trichome, IE inner epidermis, Le leaf, Lv larva, NT nutritive tissue, OE outer epidermis, P phloem, S stem, SP storage parenchyma, St stoma, Tr trichomes, UE upper leaf epidermis, X xylem

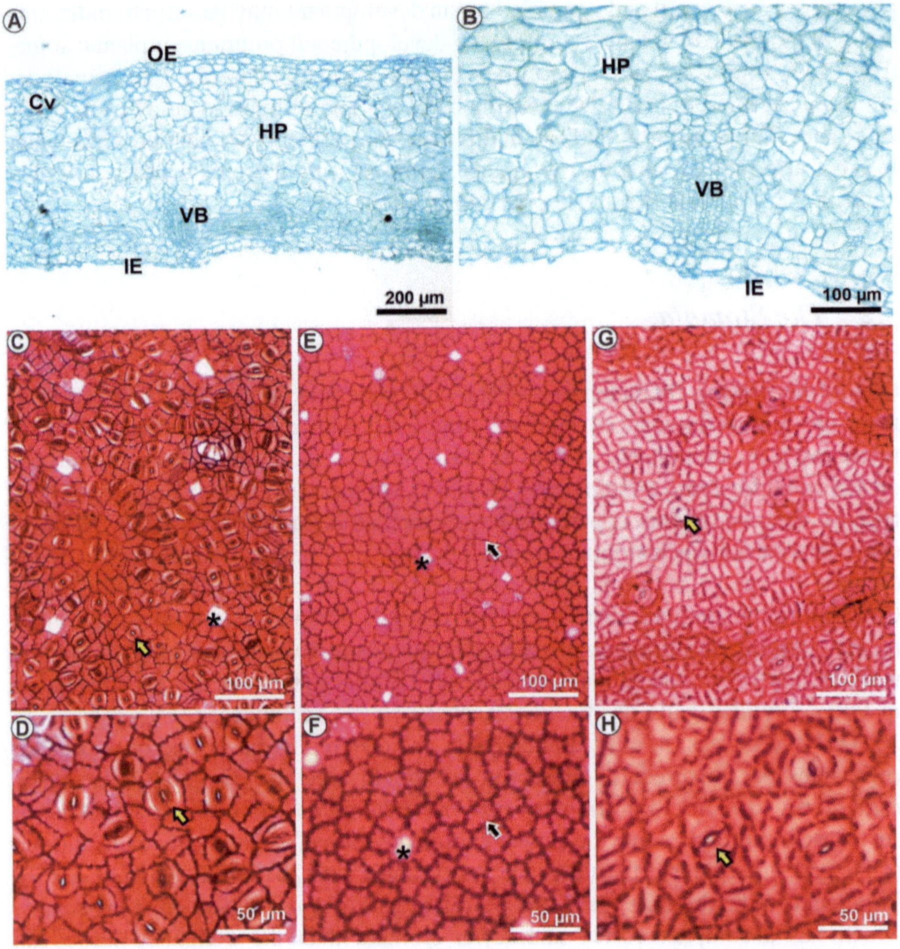

Fig. 4 Epidermal features of galls induced by *Nothotrioza myrtoidis* (Hemiptera: Psylloidea) on *Psidium myrtoides* (Myrtaceae) leaves. (**a, b**) Transverse section of the galls. Note the differences between the outer and inner epidermis, the latter with a discontinuous layer of cells. The vascular bundles near the inner epidermis facilitate the feeding of the galling hemiptera. (**c–h**) Dissociated epidermis in frontal view. (**c, d**) Abaxial surface of a nongalled leaf. (**e, f**) Adaxial surface of a nongalled leaf. (**g, h**) Outer epidermis of the gall. The asterisks indicate the tops of the Myrtaceae secretory cavities. Yellow arrows indicate stomata, and black arrows indicate puzzle-shaped ordinary epidermal cells (common pavement cells) with sinuous anticlinal cell walls. Staining: **a, b**: Astra blue and safranin. **c–h**: Safranin. Abbreviations: Cv secretory cavity, HP homogeneous parenchyma, IE inner epidermis near the gall chamber, OE outer epidermis, VB vascular bundle

leaves have hypertrophied stomata that are apparently normal, and if they are functional, their increase could compensate for the smaller stomatal index and stomatal density compared with those of nongalled leaves (Castro et al. 2023).

The type of stomata may differ in the nongalled portions of the host leaves and galls, from diacytic to paracytic, or the guard cells may not expand equally, being smaller, as demonstrated on mite galls induced by an Eriophyidae on *Avicennia schaueriana* (Acanthaceae) (Nobrega et al. 2023). In particular, pavement cells may cover guard cells completely or partially, causing their dysfunction (Mani 1964). Further investigations on stomatal differentiation and patterning in galls are needed to provide more insights into the functional roles played by their redifferentiated epidermis.

3.3 Trichomes

Trichomes are commonly the first epidermal cell type to differentiate on the ordinary pattern of host leaves; their initiation ends before the end of epidermal cell division, so their density is affected by the number of divisions of the pavement cells. Due to these developmental patterns, the distribution of trichomes in mature organs may not reflect their original distribution in young organs (Larkin et al. 1997). Trichomes are commonly related to mechanical protection against natural enemies but also guarantee the microclimate inside the gall, as proposed by Moura et al. (2008) and Dias et al. (2013) for arthropod-induced galls. Occasionally, galling parasites may induce the formation of new trichomes that can be protective, glandular, or nutritious. Among the nonglandular trichomes, those formed in the ostiole are functionally important, as they obliterate the gall aperture, as observed in *Calophya duvauae* galls on *Schinus polygama* (Dias et al. 2013). By the end of the gall life cycle, such trichomes may senesce, allowing the escape of the adult inducer (Dias et al. 2013). The overdifferentiation of trichomes is also observed during the initial developmental steps of the horn-shaped gall induced by a Cecidomyiidae on *Copaifera langsdorffii*, in which a dense indumentum protects the gall inducer, while the gall chamber is not fully closed (Carneiro et al. 2017). Some galls remain open during the entire morphogenesis and display a greater aperture in which trichomes may not be as tightly arranged, as in pistil-shaped galls induced by Cecidomyiidae on axillary buds of *Marcetia taxifolia* (Fig. 2) (Ferreira and Isaias 2014). The trichomes formed on the surface of galls may be similar to those of non-galled leaves or may be modified in some way. In the first case, the indumentum is normally denser than it is on a normal plant organ; in the second case, the gall trichomes vary in size (Fig. 2), shape, and general structure (Mani 1964; Nobrega et al. 2023).

The overdifferentiation of trichomes and of emergences may be the only process involved in gall formation and is the diagnostic feature of erineas (Ferreira et al. 2019; Guedes et al. 2024), which are galls typically induced by mites belonging to the Eriophyidae. These galls are open and involve the overexpression of trichoblasts, with the indumentum constituting the whole dermal system. Otherwise, trichoblast expression may not occur, and the gall covering, even in a hairy leaf, may be completely glabrous. The first case guarantees that the microenvironment is buffered from abiotic alterations, supporting the microenvironment hypothesis (Price et al. 1987).

In Eriophyidae-induced galls, the neoformation of elongated filiform trichomes occurs on the abaxial surface, involving colonies of mites. These trichomes accumulate reducing sugars and proteins, which add to the functions of protection and nutrition of gall-inducing agents (Nobrega et al. 2023).

3.4 Epidermal Cell Wall Functional Clues

The epidermis, as the first line of contact with the environment, is affected by abiotic and biotic factors, such as gall inducers. In addition to the diagnostic features of macro- and microscopic alterations, cell wall compounds have been mapped via immunocytochemical labeling with monoclonal antibodies (MAbs), revealing functional traits at the cytological level.

Plant cell walls are composed of multiple biopolymers and have a complex structural network. The cellulose microfibrils are interconnected by hemicelluloses and interspersed by pectins and proteins; each of these components performs specific functions alone or in connection. However, plant cell walls are the least understood cellular structures (Zhang et al. 2021). Immunocytochemical approaches reveal alterations in several epitopes of pectins, hemicelluloses, and proteins in the epidermal cell walls covering galls. The immunocytochemical profiles have demonstrated the asymmetrical deposition of wall compounds involved in epidermal physiological functions, such as determining cell shape, conferring porosity and rigidity, and maintaining juvenile features (Formiga et al. 2013; Carneiro et al. 2014a; Bragança et al. 2020; Santos et al. 2024). For example, in mature galls on *Baccharis reticularia* (Asteraceae), the MAbs LM2, LM5, LM6, JIM5, and JIM7 intensely labeled the epidermis of the rolling and pocket galls but not in the kidney-shaped galls. The epitopes of arabinogalactan proteins (AGPs) labeled by LM2 are related to the prevention of programmed cell death (PCD). In contrast, the epitopes of arabinans and galactans labeled by LM5 and LM6 indicate the control of cell expansion in the new functional design of the rolling and pocket gall morphotypes. Concerning the three gall morphotypes, the homogalacturonans (HGs) labeled by JIM5 and JIM7 express the equilibrium between cell expansion and rigidity among the three gall morphotypes (Formiga et al. 2013).

Advancing in the relationship of the labeling of epitopes on cell wall compounds but on a developmental basis, Carneiro et al. (2014a) detected the epitopes of AGPs, arabinans, and methyl-esterified homogalacturonans (HGs) on epidermal cell walls—with the MAbs LM2, LM6, and JIM7—in galls of *N. myrtoidis* on *Psidium myrtoides* (Myrtaceae). The association of the arabinans labeled by LM6 and the HGs labeled by JIM7 in epidermal cell walls enhances cell wall porosity and flexibility due to the depletion of calcium-mediated cross bonds between the side chains of HGs (Verhertbruggen et al. 2009; Carneiro et al. 2014a). The epitopes of HGs are labeled during gall induction and the growth and development phase in the inner epidermis, i.e., when the cell layer is in contact with the inducer. The outer epidermis of this gall exhibited increased detection of cell wall epitopes from the induction phase to maturation. All the epitopes labeled by JIM5, JIM7, LM2, LM5, and LM6 were detected during gall development, except for the epitopes of galactans, which were undetectable by gall senescence (Carneiro et al. 2014a). Additionally, the labeling of the epitopes of AGPs by LM2 and arabinans by LM6 on the epidermal cell walls of the globoid galls induced by *N. myrtoidis* on *P. myrtoides* is involved in the control of cell expansion and the prevention of premature cell death, which is reversed by gall senescence when the AGPs are absent. Accordingly, the epitopes of xyloglucans, hemicelluloses involved in cell wall rigidity, were labeled by the MAb LM15 in the young and mature stages of the globoid galls on *Inga ingoides* (Fabaceae), which was not observed for the other two gall morphotypes associated with this species. In lenticular galls, the epitopes of these hemicelluloses are detected only in the young phase, and in fusiform galls, xyloglucans are detected only in the senescent stage (Bragança et al. 2020).

The labeling of the epitopes of arabinans with LM6 seems to reduce in the epidermal cell walls from gall maturation to senescence on galls induced by *Baccharopelma dracunculifoliae* on *Baccharis dracunculifolia* (Oliveira et al. 2014). The dynamics of the other investigated epitopes were similar. The cell wall composition, ultrastructure, and thickness vary locally and influence cell expansion and shape (Riglet et al. 2021). In the globoid galls induced on *Macairea radula* (Melastomataceae) by the gall-inducing *Palaeomystella oligophaga* (Lepidoptera), a clear tissue compartmentalization was observed, with the nutritive tissue showing intense labeling, especially for epitopes of HGs (Santos et al. 2024). Studies on the composition of cell walls at different developmental stages suggest that cell wall changes are dynamic throughout the development of both nongalled organs and galls. These changes in the functional aspects of cell walls may enhance the reinforcement of outer epidermal cells and support multiple functions in inner epidermal cells, such as flexibility, reinforcement, porosity, and the regulation of programmed cell death (PCD) (Formiga et al. 2013; Carneiro et al. 2014a; Oliveira et al. 2014; Teixeira et al. 2018; Martini et al. 2019; Bragança et al. 2020). These findings highlight the critical role of cell walls in determining the structural-functional traits of plant tissues.

4 Final Considerations

The maintenance of a typical epidermis such as that of the nongalled host organ may be observed in several galls. Nevertheless, changes in the thickening of the cuticle, cell dimensions, sinuosity of the anticlinal walls of common epidermal cells, and density of trichomes and stomata are recorded. Cecidogenetic activity affects the differentiation of sinuosity, growth polarity, and size of epidermal cells. Changes in size occur because of cellular hypertrophy, hypotrophy (cell division without cell growth), and hyperplasia. Trichomes may vary in quantity and, sometimes, in morphological type, which indicates greater genetic control of the morphogenic potential of the host organs by the inducing insects. The increase in trichome density observed in galls is associated with mechanical defense against herbivory and thermal and desiccation control, but it also involves considerable alterations that are yet to be discussed in terms of the metabolic perspective of plants. The peculiar traits of epidermal cell walls guarantee the maintenance of cell juvenility, form a mosaic of cell wall rigidity and flexibility, accompanying variations in size during gall development and allowing the trafficking of molecules due to cell wall porosity. All the chemical dynamics of the cell wall in the gall dermal system seem to facilitate gall inducer establishment and the success of the gall cycle.

References

Álvarez R (2011) Initial stages in the formation of galls induced by *Geoica utricularia* in *Pistacia terebinthus* leaflets: origin of the two vascular bundles which characterize the wall of the galls. Am J Plant Sci 2:175–179. https://doi.org/10.4236/ajps.2011.22019

Álvarez R (2012) Microscopic study of the walls of galls induced by *Geoica utricularia* and *Baizongia pistaciae* in *Pistacia terebinthus*: a contribution to the phylogeny of Fordini. Arthropod Plant Interact 6:137–145. https://doi.org/10.1007/s11829-011-9154-3

Álvarez R, Encina A, Pérez-Hidalgo N (2009) Histological aspects of three *Pistacia terebinthus* galls induced by three different aphids: *Paracletus cimiformis*, *Forda marginata* and *Forda formicaria*. Plant Sci 176:303–314. https://doi.org/10.1016/j.plantsci.2008.11.006

Álvarez R, Molist P, González-Sierra S, Martinez JJI, Nieto-Nafría JM (2014) The histo structure of galls induced by aphids as a useful taxonomic character: the case of *Rectinasus* (Hemiptera, Aphididae, Eriosomatinae). Zootaxa 3861:487–492. https://doi.org/10.11646/zootaxa.3861.5.6

Amorim DO, Ferreira BG, Fleury G (2017) Plant potentialities determine anatomical and histochemical diversity in *Mikania glomerata* Spreng. Galls. Braz J Bot 40:517–527

Balint-Kurti P (2019) The plant hypersensitive response: concepts, control and consequences. Mol Plant Pathol 20(8):1163–1178

Barros PM, Sapeta H, Belo J, Leal AR, Oliveira MM (2024) 6 The root periderm. In: Plant roots: the hidden half. CRC Press, p 104

Bedetti CS, Bragança GP, Isaias RMS (2017) Influence of auxin and phenolic accumulation on the patterns of cell differentiation in distinct gall morphotypes on *Piptadenia gonoacantha* (Fabaceae). Aust J Bot 65(5):411–420

Bragança GPP, Alencar CF, Freitas MSC, Isaias RMS (2020) Hemicelluloses and associated compounds determine gall functional traits. Plant Biol 22:981–991

Breusegem FV, Dat JF (2006) Reactive oxygen species in plant cell death. Plant Physiol 141:384–390

Cabrera VA, Doucet ME, Lax P (2023) Histopathology of the root-knot nematode, *Meloidogyne incognita*, on ornamental plants (Crassulaceae). J Plant Dis Prot 130(4):891–897

Carneiro RGS, Isaias RMS (2015) Cytological cycles and fates in *Psidium myrtoides* are altered toward new cell metabolism and functionalities by the galling activity of *Nothotrioza myrtoidis*. Protoplasma 252:637–646

Carneiro RGS, Oliveira DC, Isaias RMS (2014a) Developmental anatomy and immunocytochemistry reveal the neo-ontogenesis of the leaf tissues of *Psidium myrtoides* (Myrtaceae) towards the globoid galls of *Nothotrioza myrtoidis* (*Triozidae*). Plant Cell Rep 33:2093–2106

Carneiro RGS, Castro AC, Isaias RMS (2014b) Unique histochemical gradients in a photosynthesis-deficient plant gall. S Afr J Bot 92:97–104

Carneiro RGS, Pacheco P, Isaias RMS (2015) Could the extended phenotype extend to the cellular and subcellular levels in insect-induced galls? PLoS One 10(6):e0129331

Carneiro RG, Isaias RM, Moreira AS, Oliveira DC (2017) Reacquisition of new meristematic sites determines the development of a new organ, the Cecidomyiidae gall on *Copaifera langsdorffii* Desf. (Fabaceae). Front Plant Sci 8:1622

Castro RRM, Barbosa PEF, Sant'Anna LG, Pereira CM, Ferreira BG (2023) Sun and shade galls of *Clinodiplosis profusa* (Cecidomyiidae) on *Eugenia uniflora* (Myrtaceae): are there differences in their establishment and growth? Flora 303:15228

Dias GG, Ferreira BG, Moreira GRP, Isaias RMS (2013) Developmental pathway from leaves to galls induced by a sap-feeding insect on *Schinus polygamus* (Cav.) Cabrera (Anacardiaceae). An Acad Bras Cienc 85:187–200

Dutta TK, Akhil VS (2024) The relative infection potential of *Meloidogyne incognita* and *M. graminicola* in the basmati rice cultivar PB1121. Plant Pathol 73(2):262–271

Evert RF (2006) Esau's plant anatomy: meristems, cells, and tissues of the plant body: their structure, function, and development. Wiley, Hoboken

Fernandes GW (1998) Hypersensitivity as a phenotypic basis of plant induced resistance against a galling insect (Diptera: Cecidomyiidae). Environ Entomol 27:260–267

Fernandes GW, Cornelissen TG, Isaias RMS, Lara AF (2000) Plants fight gall formation: hypersensitivity. Ciênc Cult 52:49–54

Ferreira BG, Isaias RMS (2013) Developmental stem anatomy and tissue redifferentiation induced by a galling Lepidoptera on *Marcetia taxifolia* (Melastomataceae). Botany 91:752–760

Ferreira BG, Isaias RMS (2014) Floral-like destiny induced by a galling Cecidomyiidae on the axillary buds of *Marcetia taxifolia* (Melastomataceae). Flora 209:391–400

Ferreira BG, Álvarez R, Avritzer SC, Isaias RMS (2017) Revisiting the histological patterns of storage tissues: beyond the limits of gall-inducing taxa. Botany 95:173–184

Ferreira BG, Álvarez R, Bragança GP, Alvarenga DR, Pérez-Hidalgo N, Isaias RMS (2019) Feeding and other gall facets: patterns and determinants in gall structure. Bot Rev 85:78–106

Formiga AT, Soares GLG, Isaias RMS (2011) Responses of the host plant tissues to gall induction in *Aspidosperma spruceanum* Müell. Arg. (Apocynaceae). Am J Plant Sci 2:823–834

Formiga AT, Oliveira DC, Ferreira BG, Magalhães TA, Castro AC, Fernandes GW, Isaias RMS (2013) The role of pectic composition of cell walls in the determination of the new shape-functional design in galls of *Baccharis reticularia* (Asteraceae). Protoplasma 250:899–908

Glover BJ (2000) Differentiation in plant cell epidermis. J Exp Bot 51:497–505

Guedes LM, Henríquez IA, Sanhueza C, Rodríguez-Cerda L, Figueroa C, Gavilán E, Aguilera N (2024) Alterations induced by Colomerus vitis on the structural and physiological leaf features of two grape cultivars. Exp Appl Acarol 92(2):183–201

Isaias RMS, Oliveira DC (2012) Gall phenotypes – product of plant cells defensive responses to the inducers attack. In: Mérillon JM, Ramawat KG (eds) Plant defense: biological control. Progress in biological control 12. Springer Science+Business Media B.V, pp 273–290

Isaias RMS, Oliveira DC, Carneiro RGS (2011) Role of *Euphalerus ostreoides* (Hemiptera: Psylloidea) in manipulating the leaflet ontogenesis of *Lonchocarpus muehlbergianus* (Fabaceae). Botany 89:581–592

Isaias RMS, Oliveira DC, Carneiro RGS, Kraus JE (2014) Developmental anatomy of galls in the neotropics: arthropods stimuli versus host plant constraints. In: Fernandes GW, Santos JC (eds) Neotropical insect galls. Springer, New York, pp 15–34

Isaias RMS, Oliveira DC, Moreira ASFP, Soares GLG, Carneiro RGS (2015) The imbalance of redox homeostasis in arthropod-induced plant galls: mechanisms of stress generation and dissipation. Biochim Biophys Acta 1850:1509–1517

Isaias RMDS, Kraus JE, Costa EC, Carneiro RGS (2024) The anatomy of neotropical galls and the untold lessons about the morphogenetical potentialities of plants. Rodriguésia 75:e01542023

Jorge NC, Souza-Silva ES, Alvarenga DR, Saboia G, Soares GLG, Zini CA, Cavalleri A, Isaias RMS (2018) Structural and chemical profiles of *Myrcia splendens* (Myrtaceae) leaves under the influence of the galling *Nexothrips* sp. (Thysanoptera). Front Plant Sci 9:1521

Kraus JE, Arduin M, Venturelli M (2002) Anatomy and ontogenesis of hymenopteran leaf galls of *Struthanthus vulgaris* Mart. (Loranthaceae). Rev Bras Bot 25:449–458

Kurzfeld-Zexer L, Lev-Yadun S, Inbar M (2015) One aphid species induces three gall types on a single plant: comparative histology of one genotype and multiple extended phenotypes. Flora 210:19–30

Kutsukake M, Meng X-Y, Katayama N, Nikoh N, Shibao H, Fukatsu T (2012) An insect-induced novel plant phenotype for sustaining social life in a closed system. Nat Commun 3:1187–1194

Larkin JC, Marks MD, Nadeau J, Sack F (1997) Epidermal cell fate and patterning in leaves. Plant Cell 9:1109–1120

Liu S, Jobert F, Rahneshan Z, Doyle SM, Robert S (2021) Solving the puzzle of shape regulation in plant epidermal pavement cells. Annu Rev Plant Biol 17(72):515–550

Magalhães TA, Oliveira DC, Suzuki AYM, Isaias RMS (2014) Patterns of cell elongation in the determination of the final shape in galls of *Baccharopelma dracunculifoliae* (Psyllidae) on *Baccharis dracunculifolia* DC (Asteraceae). Protoplasma 251:747–753

Mani MS (1964) Ecology of plant galls. W. Junk Publishers, The Hague

Marquesine RR, Canaveze Y, Ferreira BG (2024) Ontogenetic differences in sun and shade galls of Clinodiplosis profusa on *Eugenia uniflora* leaves and the cytological antioxidant mechanisms in gall cells. Protoplasma:1–20

Martin C, Glover BJ (2007) Functional aspects of cell patterning in aerial epidermis. Curr Opin Plant Biol 10(1):70–82

Martinez JJ, Moreno-González V, Jonas-Levi A, Álvarez R (2018) Quantitative differences detected in the histology of galls induced by the same aphid species in different varieties of the same host. Plant Biol 20(3):516–524

Martini VC, Moreira ASFP, Kuster VC, Oliveira DC (2019) Galling insects as phenotype manipulators of cell wall composition during the development of galls induced on leaves of *Aspidosperma tomentosum* (Apocynaceae). S Afr J Bot 127:226–233

Meyer J, Maresquelle HJ (1983) Anatomie des galles. Gerbrüder Borntraeger, Berlin, p 662

Moura MZD, Soares GLG, Isaias RMS (2008) Species-specific changes in tissue morphogenesis induced by two arthropod leaf gallers in *Lantana camara* L. (Verbenaceae). Aust J Bot 56:153–160

Nobrega LP, Silva JB, Luna BN, Ferreira BG (2021) Modulation of anatomical adaptations of leaves of *Avicennia schaueriana* (Acanthaceae) by a galling Meunieriella (Cecidomyiidae). Flora 274:151750. https://doi.org/10.1016/j.flora.2020.151750

Nobrega LP, Haiad BS, Ferreira BG (2023) Epidermal and subepidermal changes during the formation of hairy galls induced by Eriophyidae on *Avicennia schaueriana* leaves. Sci Nat 110(49):1–15

Oliveira DC, Isaias RMS (2009) Influence of leaflet age in anatomy and possible adaptive values of the midrib gall of *Copaifera langsdorffii* (Fabaceae: Caesalpinioideae). Rev Biol Trop 57(1–2):293–302

Oliveira DC, Isaias RMS (2010) Redifferentiation of leaflet tissues during midrib gall development in *Copaifera langsdorffii* (Fabaceae). S Afr J Bot 76:239–248

Oliveira DC, Christiano JCS, Soares GLG, Isaias RMS (2006) Reações de defesas químicas e estruturais de Lonchocarpus muehlbergianus Hassl. (Fabaceae) à ação do galhador Euphalerus ostreoides Crawf. (Hemiptera: Psyllidae). Braz J Bot 29:657–667

Oliveira DC, Magalhães TA, Ferreira BG, Teixeira CT, Formiga AT, Fernandes GW, Isaias RMS (2014) Variation in the degree of pectin methylesterification during the development of *Baccharis dracunculifolia* kidney-shaped gall. PLoS One 9(4):e94588

Oliveira DC, Moreira ASFP, Isaias RMS, Martini V, Rezende UC (2017) Sink status and photosynthetic rate of the leaflet galls induced by *Bystracoccus mataybae* (Eriococcidae) on *Matayba guianensis* (Sapindaceae). Front Plant Sci 8:1249

Oliveira DC, Martini VC, Moreira ASFP, Fuzaro L, Gonçalves LA (2020) *Pseudophacopteron longicaudatum* (Hemiptera) induces intralaminar leaf galls on *Aspidosperma tomentosum* (Apocynaceae): a qualitative and quantitative structural overview. An Acad Bras Cienc 92:e20181002

Pincebourde S, Casas J (2016) Hypoxia and hypercarbia in endophagous insects: larval position in the plant gas exchange network is key. J Insect Physiol 84:137–153
Price PW, Fernandes GW, Waring GL (1987) Adaptative nature of insect galls. Environ Entomol 16:15–24
Rey LA (1992) Developmental morphology of two types of hymenopterous galls. In: Shorthouse JD, Rohfritsch O (eds) Biology of insect-induced galls. Oxford University Press, New York, pp 87–101
Riglet L, Gatti S, Moyrouind E (2021) Sculpting the surface: structural patterning of plant epidermis. iScience 24:103346
Rohfritsch O (1992) Patterns in gall development. In: Shorthouse JD, Rohfritsch O (eds) Biology of insect-induced galls. Oxford University Press, New York, pp 60–86
Santos PD, Bragança GPP, Kuster VC, Rocha DI, Oliveira DC (2024) Cell wall epitope distribution in the functional compartments of galls induced by *Palaeomystella oligophaga* (Lepidoptera) in *Macairea radula* (Melastomataceae). Aust J Bot 72:BT23064
Teixeira CT, Oliveira DC, Kuster VC, Isaias RMS (2018) Immunocytochemical demonstration of cell wall components related to tissue compartments in the globoid galls induced by *Clinodiplosis* sp. (Cecidomyiidae) on *Croton floribundus* Spreng. (Euphorbiaceae). Botany 96(1):9–18
Verhertbruggen Y, Marcus SE, Haeger A, Ordaz-Ortiz JJ, Knox JP (2009) An extended set of monoclonal antibodies to pectic homogalacturonan. Carbohydr Res 344:1858–1862
Zhang B, Gao Y, Zhand L, Zhou Y (2021) The plant cell wall: biosynthesis, construction, and functions. J Integr Plant Biol 63:251–272

Patterns of Xylem (Re)differentiation and Distribution in Galls

Gracielle Pereira Pimenta Bragança, Ana Maria Abreu Santos, Mariana de Sousa Costa Freitas, and Ravena Malheiros Nogueira

1 Introduction

The evolution of an intricate vascular system has facilitated the extensive dispersion of vascular plants throughout diverse terrestrial habitats (Lucas et al. 2013). This vascular system comprises two tissues, xylem and phloem, whose ontogenetic development, cellular composition, and organization within the plant are crucial for the hydraulic maintenance of the organisms and the distribuition of resources (Evert et al. 2006). Thus, it reveals an excellent model for studying plant-pathogen interactions.

The primary function of the xylem is to transport water and mineral salts throughout the plant via tracheal elements (Evert et al. 2006). However, this tissue also comprises parenchyma and fibers that significantly influence its physical and chemical properties, ensuring functions under high pressure and enabling both local and long-distance hydraulic supply (Słupianek et al. 2021). The perception that xylem is mostly composed of rigid or even "dead at maturity" cells, such as vessel elements, might suggest that it is not a dynamic tissue capable of responding to various stimuli. However, this notion is contradicted by diverse studies (Fonti et al. 2010; Pandey 2021) including plant-pathogen interactions (Jorge et al. 2022; Nobrega et al. 2023). Herein, we aim to discuss the symplastic and apoplastic responses of xylem cells on galls induced by different gall-inducing taxa. Furthermore, we intend to explore the hormonal factors influencing the patterns of xylem cell differentiation and shed light on existing knowledge gaps that may inspire new research perspectives in the field of xylem function in galls and tumoral plant development.

G. P. P. Bragança (✉) · A. M. A. Santos · M. de Sousa Costa Freitas · R. M. Nogueira
Department of Botany, Universidade Federal de Minas Gerais (UFMG), Belo Horizonte, MG, Brazil

D. C. de Oliveira, R. M. dos Santos Isaias (eds.), *Plant Galls*,
https://doi.org/10.1007/978-3-031-80064-1_4

2 Hormones and the Differentiation of Xylem in Galls

The redifferentiation of vascular cells in galls promotes a connection with the vascular cells of the host plant, whose characteristics are related to the origin of the new vascular tissues (Kraus et al. 2002; Ferreira and Isaias 2013; Aranda-Rickert et al. 2017; Dolzblasz et al. 2018; Bragança et al. 2021). Once the plant hormones regulate vascular differentiation (Fukuda et al. 1998; Aloni 2021), they are essential for gall development due to the hydraulic demand. Activating vascular cell redifferentiation requires the signaling of auxins, cytokinins (Aloni 2013; Matsumoto-Kitano et al. 2008; Schuetz et al. 2013), and ethylene (Aloni 2021).

The main hormones related to gall development are auxins and cytokinins, which have been extensively investigated (Yamaguchi et al. 2012; Tooker and Helms 2014; Hearn et al. 2019; Ferreira et al. 2019; Bragança et al. 2021). The histolocalization of auxins is related to cellular redifferentiation mediated by phenolic compounds, which blocks the activity of IAA oxidases, consequently increasing the concentration of auxins at the site of gall development (Hori 1992; Bedetti et al. 2014, 2017). The flow of auxins contributes to cell hypertrophy and the formation of new vascular bundles connecting the gall to the host organ (Aloni 2013, 2015). In addition to auxins, the cytokinins are essential for the differentiation of tracheal elements (Fukuda 1997; Schuetz et al. 2013). The accumulation of cytokinins positively regulates cambial activity in fusiform galls of *Inga ingoides* induced by Cecidomyiidae, for example, increasing auxin-sensitive cambium cells (Bragança et al. 2021). Cytokinins can stimulate slow auxin transport in the cambium, which results in slow differentiation of vessel elements and their large expansion before secondary cell wall deposition (Aloni 2001, 2015, 2021). Cytokinins can also be produced and secreted by galling herbivores, such as nematodes and insects (Siddique et al. 2015; Takei et al. 2015; Andreas et al. 2020), during gall development, stimulating vascular cell formation.

Ethylene, a hormone generated in stressful conditions (Suttle 1988; Junghans et al. 2004), is produced during the gall induction process and can disrupt the polar auxin flow, resulting in a localized accumulation of auxins (Aloni 2013, 2015). This hormone is also important for the differentiation of vascular tissues, with peaks found during the maturation of tracheal elements correlated with the activity of the biosynthetic oxidase of ethylene 1-aminocyclopropane-1-carboxylic acid (Pesquet and Tuominen 2011). Moreover, during secondary growth, ethylene diffuses in the centrifugal direction, inducing cell divisions in the cambial initials. This process promotes the formation of rays and the expansion of the existing ones. (Aloni 2015, 2021). Ethylene also triggers enzymatic activity for xylem lignification (Hennion et al. 1992; Kuriyama and Fukuda 2001) and modifies cell wall polysaccharide composition for xylogenesis (Eklund 1991; Ingemarsson et al. 1991; Kuriyama and Fukuda 2001). The over-differentiation of axial parenchyma, and parenchymatic rays in the xylem may be related to the effects of ethylene, as described in *Populus* sp. (Junghans et al. 2004) and seen in galls on *Eremanthus erytropappus* induced by

Neolasioptera sp. (Jorge et al. 2022), and in galls on *Malus domestica* induced by *Eriosoma lanigerum* (Freitas 2022).

3 Xylem Differentiation in Galls Induced by Distinct Groups of Gall-Inducing Organisms

3.1 Bacteria Galls

There are many known taxa of bacteria, both gram-positive and gram-negative, that can induce galls. These bacteria are largely distributed among the plant families (Thapa et al. 2019), and the most common galls they induce are the root and stem crown galls and nodule galls induced in roots.

The crown galls are primarily induced by *Agrobacterium tumefaciens* and are commonly known as plant tumors. They are considered a mass of tissue equipped with specific adaptive mechanisms for water and nutrient acquisition (Aloni and Ullrich 2008). These bacteria induce galls by transferring the bacterial T-DNA into the host plant's genome (Zupan et al. 2000; Venturi and Fuqua 2013), increasing cytokinin, auxin, and ethylene levels, which leads to accelerated growth, limited only by the host plant's water and nutrient reserves (Chalupowicz et al. 2006; Aloni and Ullrich 2008). Increased cytokinin levels are the main factor responsible for stimulating tissue and vessel differentiation within the gall (Aloni et al. 1998). Additionally, cytokinins aid in the regeneration of vessel elements and sieve tubes in wounded plant tissue in response to the presence of auxins (Aloni et al. 1995; Spaepen and Vanderleyden 2011, Jackson 2008). The increase in cytokinin levels is associated with the rapid differentiation of vessels and the phloem anastomoses in the gall (Chalupowicz et al. 2006; Jackson 2008). The increase in auxin levels creates an electrochemical proton gradient, which enhances the nutrient sink condition of the gall (Spaepen and Vanderleyden 2011). Furthermore, these levels stimulate ethylene production within gall tissues, which affects the gall morphogenesis and the differentiation of the adjacent host plant tissues (Aloni et al. 1998; Chalupowicz et al. 2006). Ethylene is also known to reduce vessel width (Aloni 1991; Nonaka and Ezura 2014) and, as such, plays a role in gall constriction, as the reduction of vessel caliber on the adjacent tissue restricts water flow to the vessels at the base of the tumor, as well as increase water uptake to the gall. Alternatively, it has been shown that reducing ethylene levels at the moment of induction improves gene transfer efficiency from the galling bacteria to the host plant (Nonaka and Ezura 2014).

In contrast to crown galls, root nodule galls are mainly induced by a variety of *Rhizobia* species (Venturi and Fuqua 2013) and play a significant supporting role in the metabolism of the host plant, notably through their ability to fix nitrogen. The induction site is marked by swelling on the root, varying between a singular emergence and several nodules at once (Selami et al. 2014). The infecting bacteria release several nod factors (Harris and Pitzschke 2020), inducing physiological responses from the host plant, such as plasma membrane depolarization, rhizosphere alkalinization, and calcium ion oscillations in the nucleus (Popp and Ott 2011). The nodule formation and development are dictated mainly by the pathogen-triggered immunity (PTI) mechanism, which uses a combination of gene and hormone signaling pathways to facilitate the establishment of the bacterial colony in the root cortex (Kende 2000; Harris and Pitzschke 2020). This generates a stress response in the host plant and must be regulated by both parts to maintain the host plant metabolism and prevent the overgrowth of the bacterial colony (Venturi and Fuqua 2013; Berrabah et al. 2018; Kende 2000). The connection and nutrient exchange between the host plant and the gall is done by a few peripheral vascular bundles which converge toward the apex of the nodule (Guan et al. 2013).

3.2 *Nematoda Galls*

Nematode galls are typically identified by the swelling of plant organs due to cell hypertrophy and tissue hyperplasia. These galls are generally simpler than those induced by other gall-inducing organisms, such as insects (Mani 1964). Nematode galls can be induced on vegetative and reproductive organs of plants (Mani 1964; Ferreira et al. 2017a, b; Arriola and Isaias 2021; Garita et al. 2021), causing serious damage to economically important crops (Palomares-Rius et al. 2017; Garita et al. 2021; Vilela et al. 2019, 2021, 2023). The diversity of anatomical and morphological alterations in the host organs may be associated with the feeding sites and the lifestyle of the nematode. The nematode either inserts its stylets or produces fluids that modify the cytoplasm (Palomares-Rius et al. 2017). These alterations can be localized or extend around the feeding site of the nematode and may even have long-distance effects (Mani 1964; Ferreira et al. 2017a, b; Garita et al. 2021).

Nematode galls may occur on roots (Mani 1964), especially induced by a sedentary endoparasitic nematode such as root-knot nematodes (*Meloidogyne* spp.) (Vilela et al. 2019, 2021, 2023) and cysts (*Heterodera* spp., *Globodera* spp.) (Bartlem et al. 2014). Root-knot nematodes migrate intercellularly to vascular cell precursors in the root elongation region and, upon establishing themselves, induce the formation of a group of nutritive cells called giant cells (Bartlem et al. 2014; Escobar et al. 2015; Palomares-Rius et al. 2017). Nematode cysts directly target differentiated vascular cells and induce the formation of syncytia, which are nutritive cells that increase in size through wall dissolution and fusion of adjacent cells (Bartlem et al. 2014; Palomares-Rius et al. 2017). The redifferentiation of parenchyma cells or the differentiation of meristematic cells into nutritive cells or giant cells impacts the cell organization of the vascular system and the components of cell wall (Bartlem et al. 2014; Ferreira et al. 2019; Meidani et al. 2019; Vilela et al. 2021, 2023), as observed in *Meloidogyne incognita* galls induced on *Abelmoschus esculentus* (Vilela et al. 2021) and *Petroselinum crispum* (Vilela et al. 2023). In *A. esculentus* galls, the dynamics of the cell wall change due to the de-methylesterification process, and the presence of low methyl-esterified HG epitopes contributes to increased porosity, favoring the exchange of nutrients between the host tissues and the galls (Vilela et al. 2021). In *P. crispum* galls, the HG epitopes observed in giant cells reveal the continuous elongation of giant cells (Vilela et al. 2023) (Fig. 1).

The formation of giant cells by root-knot nematodes involves the differentiation of meristematic cells that disrupt the vascular system (Haegeman et al. 2012; Bartlem et al. 2014; Vilela et al. 2021, 2023; Tapia-Vázquez et al. 2022). However, xylem and phloem cells differentiate around giant cells, ensuring the continuity of the vascular tissue and enabling access to water and solutes (Jones and Goto 2011; Bartlem et al. 2014). Giant cells can also induce distortion, crushing (Sasanelli et al.

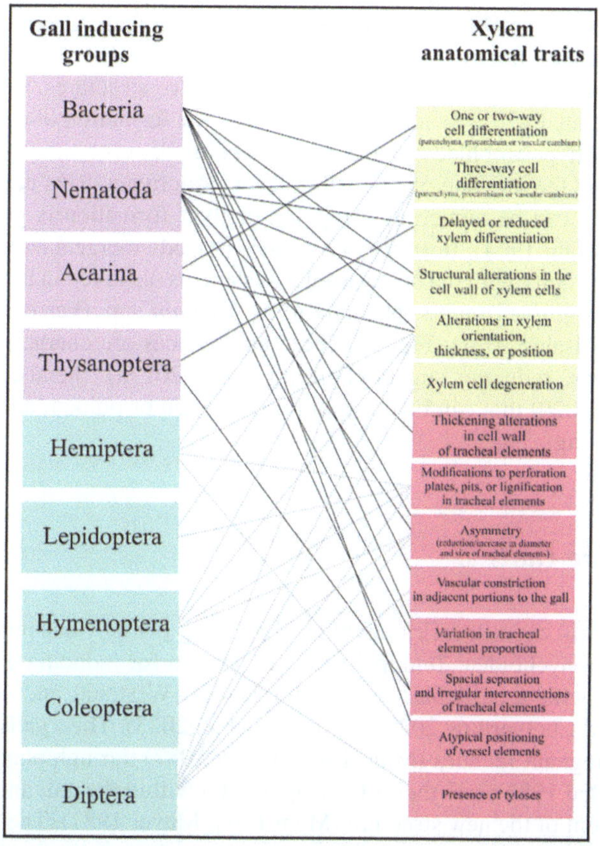

Fig. 1 Network of gall-inducing groups and xylem anatomical traits. For gall-inducing groups, purple squares indicate simple galls and green squares indicate intermediate or complex galls (*sensu* Ferreira et al. 2019). For xylem anatomical traits, yellow squares represent tissue changes, and red squares represent changes in tracheal elements. Uninterrupted lines connect simple galls to xylem anatomical attributes, and dotted lines connect intermediate and complex galls to their respective xylem anatomical attributes

2015), compression, and obliteration of vascular bundles (Asmus 2001). The neoformed tracheal elements exhibit characteristics of "wound" type, displaying asymmetry, shorter length, irregular interconnections, and a distinctive thickening of the cell walls (Jones and Goto 2011; Bartlem et al. 2014). The differentiation mechanism of these new tracheal elements and their connection with existing vascular tissues result from cambial activation or re-differentiation of parenchyma cells, as observed in other gall systems (Bartlem et al. 2014; Bragança et al. 2021; Anand and Ramani 2021) (Fig. 1).

The establishment of the syncytium induced by cysts occurs in a differentiated vascular system (Bartlem et al. 2014); therefore, its formation involves cells with the potential for cellular differentiation or redifferentiation, such as cambial, pericycle, and parenchyma cells (Sobczak et al. 1997; Palomares-Rius et al. 2017; Gómez et al. 2017). The main changes observed are displacement or rupture of the conducting tissues, causing an asymmetry of the stele and the obliteration of the cortex (Hernández-López et al. 2006; Gómez et al. 2017) (Fig. 1).

In galls induced on stems, leaves, and reproductive structures, the main alterations are associated with vascular tissues, such as the formation of cells with dense cytoplasm and hypertrophied nucleus and nucleolus between the vascular bundles and close to the feeding site of the gall inducer (Palomares-Rius et al. 2017). The vessel elements can be numerous, irregular, or their formation is greatly retarded and reduced (Mani 1964). In some cases, vascularized emergences are formed and can ensure continuity with the host vascular system, as observed in the galls induced by *Ditylenchus gallaeformans* (Nematoda) on *Miconia* spp. (Ferreira et al. 2017a, b; Arriola and Isaias 2021) (Fig. 1). These emergences are characterized by cells rich in proteins, reducing sugars (Ferreira et al. 2017a), metals, and minerals (Arriola et al. 2020) that can be transported via xylem or phloem and redirected to nematode feeding.

3.3 *Acarina Galls*

Within the acariform mites, Eriophyoidea, including Eriophyidae and Phytoptidae, as well as Tetranychoidea: Tenuipalpidae, are known for inducing galls in various plant organs, with a particular affinity for leaves (Westphal and Manson 1986; Desnitskiy and Chetverikov 2022; Desnitskiy et al. 2023). Their galls are generally induced by pregnant females (Mani 1964; Meyer 1987) through plant tissue perforation (Westphal 1983), and their offspring remain within the gall and contribute to the development of the new structure (Mani 1964; Meyer 1987). The first alteration observed in Eriophyidae galls is the formation of nutritive cells that differentiate in the perforation site (Westphal et al. 1981; Westphal 1983), which are usually epidermal cells (Mani 1964). When piercing the cell, the mite injects saliva possibly conducted to the adjacent area of the host organ through plasmodesmata and the vascular system (Desnitskiy et al. 2023). These cells initiate hyperplasia and hypertrophy processes, giving rise to emergences that accumulate especially reducing sugars (Ferreira et al. 2017a). In some cases, invagination of the plant tissue may occur (Anand and Ramani 2021), or the mites may enter the host plant tissues, forming a gall chamber (Ferreira et al. 2017a, 2019). Regardless of gall morphology, the vascular system remains similar to the host plant's. However, in some cases, the organization of the vascular system can be altered (Mani 1964; Moura et al. 2008) (Fig. 1) as new vascular cells can redifferentiate from parenchyma cells (Anand and Ramani 2021) or meristematic cells as new vessel elements (protoxylem and metaxylem) observed in the root-tip region of Arabidopsis seedlings by different doses of *Schlechtendalia chinensis* extract (Hirano et al. 2023).

3.4 *Insecta Galls*

The impacts caused by insects on three plant tissue systems vary according to the type of mouthparts of each insect order (Mani 1964; Rohfritsch 1992; Oliveira et al. 2016). These impacts are classified as simple in galls induced by Thysanoptera (Mani 1964), and as intermediate or complex in galls induced by Hemiptera, Lepidoptera, Hymenoptera, Coleoptera, and Diptera (Ferreira et al. 2019). Groups with intermediate and complex impacts typically feature the formation of tissue compartments containing specialized cells that perform specific functions within the galls, along with the presence or absence of vascular tissues (Rohfritsch 1992; Bragança et al. 2016; Ferreira et al. 2019). The xylem and phloem play crucial roles in transporting water and photoassimilates, thereby fulfilling the energy requirements of the galls and supporting the nutrition of the gall inducers. The variety of feeding habits among the gall-inducing insects leads to a broad spectrum of responses in the xylem, which exhibits distinct anatomical features. These responses have been extensively researched across all host organs and within all the mentioned orders, except for Thysanoptera and Coleoptera, where reports are relatively limited (Fig. 1).

Thysanoptera-induced galls are characterized by leading to the partial or complete folding of the leaf or floral bud (Ananthakrishnan 1992; Ferreira et al. 2019). However, "pocket" galls, horns, rosette, tubular, and stem galls can also occur (Ananthakrishnan 1992). During the feeding process of one or more adult individuals that pierce the cells and suck out the cellular contents (Mound 2020; Panizzi et al. 2021), the Thysanoptera induce cellular redifferentiation that culminates in gall formation (Ananthakrishnan and Raman 1989; Ananthakrishnan 1992; Mound 2020). In these galls, the main changes observed include an increase in mesophyll layers, homogenization of the parenchyma, a hypersensitive reactive area, and epidermal changes (Ananthakrishnan and Raman 1989; Ananthakrishnan 1992; Souza et al. 2000; Mound 2020; Jorge et al. 2018a, b). In many studies, the characteristics of the xylem are either not described, or no significant changes are observed. The main alterations associated with the vascular system typically entail vascular proliferation, resulting in the separation of vessel elements (Ananthakrishnan and Raman 1989; Ananthakrishnan 1992). In galls induced by *Gynaikothrips ficorum* on *Ficus microcarpa* (Moraceae), there was some inhibition of vascular tissue development at the early stage of gall formation. Additionally, the parenchymatic tissues surrounding the vascular tissues displayed hypertrophy (Souza et al. 2000).

Gall-inducing insects within Hemiptera are characterized by phloem-sucking behavior, inserting their mouthparts directly into the phloem (Dreger-Jauffret and Shorthouse 1992; Carneiro et al. 2014, 2017; Álvarez et al. 2020) or parenchyma cells (Dreger-Jauffret and Shorthouse 1992; Álvarez et al. 2021). Their galls are formed through the hypertrophy and hyperplasia of parenchyma cells, have numerous vascular bundles, and, in some cases, lignified cells (Meyer 1987; Isaias et al. 2011; Muñoz-Viveros et al. 2014; Oliveira et al. 2020; Beltrame et al. 2022; Silva et al. 2023; Rosa et al. 2024). Regardless of where the stylets are inserted for feeding, the main changes described in the vascular system are the neoformation and hypertrophy of the vascular bundles (Álvarez et al. 2021; Rosa et al. 2024; Oliveira et al. 2020). This hypertrophy is most evident in phloem cells, but impacts on the xylem are associated with positioning concerning the nymphal chamber. The xylem can be oriented towards either the nymphal chamber or outer cortex and epidermis (Álvarez et al. 2013, 2020), thereby potentially affecting the acquisition of nutrients. In galls with the xylem facing the epidermis of the larval chamber, the mouth apparatus of the galling herbivore must bypass the xylem to access the phloem (Álvarez et al. 2020). In some galls, the xylem and bundle sheath may be absent, facilitating the penetration of the stylets directly into the phloem, as observed in galls induced by *Populus angustifolia* on *Pemphigus betae* (Richardson et al. 2017). Tracheal elements can differentiate from the parenchyma cells, as observed in the galls of *Diaspis echinocacti* on *Cereus* sp. In these galls, tracheids with helical thickening differentiate between the water-storage parenchyma cells and may be associated with the stress induced during gall development (Beltrame et al. 2022).

Lepidoptera galls result from oviposition and the voracious feeding of larvae throughout their development (Miller 2005; Ferreira et al. 2019). Recently, Guedes et al. (2023a) revisited 13 host-plant systems, investigating galls induced on ferns and angiosperms across temperate and neotropical regions. They observed that vascular reorganization is a common anatomical characteristic across all gall morphotypes. Additionally, conservative traits, including those associated with the host plants' vascular system, are preserved within these galls and contribute to their categorization. Apart from vascular reorganization, the position of the vascular system varies among galls, occurring either in the outer or inner tissue compartments (Guedes et al. 2023a). Gall-inducing Lepidoptera can stimulate cells of the procambium toward hyperplasia and also the hypertrophy of pericycle cells, as observed in galls induced on *Microgramma vacciniifolia* (Martins et al. 2023) and *Microgramma squamulosa* (Kraus et al. 1993). In cases of secondary growth, the cambium activity may be altered or the formation of a meristem resembling the vascular cambium may occur, as in galls on *Schinus engleri* (Anacardiaceae) (Ferreira et al. 2022). The differentiation of specific cells in the radial parenchyma in galls induced by *Parthenium hysterophorus* (Anacardiaceae) also occurs (Raman and Dhileepan 1999), or the impairment of cambium activity, maintaining the primary and conspicuous condition of vascular cells in *Schinus polygama* (Anacardiaceae) (Guedes et al. 2023a). Distinctively, galls induced on *Schinus polygama* exhibit the differentiation of rosette-shaped procambial strands around the typical nutritive tissue, associated with improved water supply in these galls (Guedes et al. 2023a).

In Hymenoptera galls, the outer compartment comprises the dermal system, parenchymatous cells, and the mechanical zone composed of sclerenchyma cells (Rohfritsch 1992; Aguilera et al. 2022). The internal compartment is formed by nutritive cells surrounding the larval chamber (Rohfritsch 1992; Bomfim et al. 2021; Aguilera et al. 2022). The vascular system can originate from the procambium, vascular cambium, or parenchymatic cells (Taft and Bissing 1988; Aguilera et al. 2022). In Hymenoptera galls, the absence of vascular neoformation can occur, as observed in *Ophelimus migdanorum* (Hymenoptera: Eulophidae) galls on *Eucalyptus globulus* (Myrtaceae) (Guedes et al. 2023d). This absence is attributed to the association between the leaves' vascular bundles and the stem's secondary vascular cylinder. However, in the majority of Hymenoptera galls, alterations in vascularization related to various cell responses (Kraus et al. 2002; Jara-Chiquito et al. 2021). In the galls induced by *Dryocosmus kuriphilus* (Hymenoptera: Cynipidae) on *Castanea sativa* (Fagaceae) leaves (Jara-Chiquito et al. 2021), the hyperplasia and hypertrophy of meristematic tissues led to the separation of vascular arches, forming vascular bundle units located around the larval chambers. In addition, in these galls, the gall-inducer alters the differentiation pattern of the procambium, generating collateral or amphicrival bundles of different sizes. This alteration, coupled with the apparent reduction in the diameter of the vessel elements, may impact hydraulic conductivity in the gall tissues (Jara-Chiquito et al. 2021).

Concerning tracheal elements, intriguing modifications are observed in galls induced by *Callirhytis cornigera* (Hymenoptera: Cynipidae) on *Quercus palustris* (Fagaceae) (Taft and Bissing 1988), *Aulacidea hieracii* on *Hieracium umbellatum* (Asteraceae) and *Diplolepis spinosa* on *Rosa blanda* and *R. rugosa* (Rosaceae) (Silva et al. 2006). In *Q. palustris* galls, the activity of the vascular cambium changed, influencing the orientation and pattern of the xylem cells. Numerous tracheids with helical or irregularly thickened cell walls were observed in the basal region of the gall, a pattern not observed in the non-galled xylem. These tracheids may be associated with the movement of water and nutrients to the upper regions, especially in the projections that characterize the gall as horned-shaped (Taft and Bissing 1988). Interestingly, vessel elements in the non-galled regions adjacent to these galls contained numerous tyloses. The presence of these tyloses can disrupt the continuous flow of water to the regions above the galls, leading to the death of those regions, as observed in the death of young branches above the galls (Taft and Bissing 1988).

The analysis of gall development on *A. hieracii* and *D. spinosa*, a basal and a derived gall-inducers, respectively, revealed a difference in the initiation of vascular tissue differentiation (Silva et al. 2006). In the *A. hieracii* galls, the mechanical zone seems to be formed primarily by lignified tracheal elements (Silva et al. 2006). A notable aspect of *D. spinosa* behavior is oviposition in tissues close to the procambium. This precision in locating its eggs implies a differentiation pattern of tracheal elements that, compared to *A. hieracii*, is not as pronounced. Although some tracheal elements in the gall of *D. spinosa* develop spiral thickenings, their walls are thinner than the tracheal elements of the non-galled stem. This suggests a suppression of lignification in the vascular tissue, allowing rapid water and nutrient conduction to the larval chambers and reducing mechanical support. The mechanical support is ensured by the lignification of the interfascicular parenchyma around the gall vascular system (Silva et al. 2006).

The chewing mouthparts of both larvae and adults of Coleoptera cause significant structural damage and create multiple larval chambers (Fernandes et al. 2012). However, in most cases, the galls exhibit a resemblance to their host organs, characterized by the formation of mines or tunnels within the organs, hyperplasia, hypertrophy of the cambium, and abnormal development of vascular bundles (Dreger-Jauffret and Shorthouse 1992; Isaias et al. 2014). The analysis of the xylem in Coleoptera galls is little explored, and some records indicate the secondary growth is similar between non-galled stems and galls, featuring thick-walled tracheal elements and a narrow lumen (Singh et al. 2017). Commonly, Coleoptera galls can exhibit nectaries with a similar arrangement of vascular bundles, featuring tracheal elements that extend to the proximal region of the secretory epidermis (Fernandes et al. 2021).

The feeding impact of the Diptera: Cecidomyiidae is characterized by cell disruption caused by the prothoracic spatula puncturing superficial cells (Rohfritsch 1992; Harris 1994), inducing the formation of specialized tissues (Rohfritsch 1992; Bronner 1992) that are compartmentalized into two distinct regions: the outer tissue compartment and the inner tissue compartment (Bragança et al. 2016; Ferreira et al. 2019). Among the peculiar characteristics of these compartments is the presence of a vascular system, which can occur both in the outer and inner regions of the gall (Dorchin et al. 2002; Bragança et al. 2016; Costa et al. 2022). Studies related to the differentiation of the vascular system in the galls of Cecidomyiidae reveal that the xylem and phloem can differentiate from parenchyma cells, procambium activity, and vascular cambium (Bragança et al. 2021; Jorge et al. 2022; Nobrega et al. 2023; Guedes et al. 2023b, c). In galls induced on *Pongamia glabra* flowers, the xylem has a mixed origin, arising from the activity of the vascular cambium (normal xylem) and the differentiation of ground and xylem parenchyma (reaction xylem), which are arranged in various patterns (Govindarajalu and Lourdusamy 1980). Furthermore, the vessel elements of normal and reaction xylem differ in size, perforations' location, and vessel pitting alterations (Govindarajalu and Lourdusamy 1980). The xylem thickness can also increase, as observed in galls induced by *Dasineura* sp. on *Peumus boldus* (Monimiaceae), functioning as a mechanical tissue that protects and supports the gall (Guedes et al. 2023b). However, early in the development of these galls, the inducer induces delignification of xylem cells, which is an intriguing factor for galls in general (Guedes et al. 2023b, c). Additionally, the vessel elements are shorter and wider, implying a high-water flow for the development of these galls, even with a lower number of vessel elements per unit area. The analysis of the vascular system during the senescence of galls in *P. boldus* also revealed that after the galled leaves, the vascular connection with the host tissue is broken, and some branches die (Guedes et al. 2023b, c). This emphasizes the importance of maintaining these vascular connections between the gall and its host plant to maintain the water status of the portions above the gall.

Other families of Diptera commonly induce galls in Asteraceae (Tephritidae) (Freidberg 1998) and Poaceae (Chloropidae) (Fernandes et al. 2012) and can also induce galls in Bryophytes (Agromyzidae) (Ohgue et al. 2018). Among the observed changes in galls induced on vascular plants, Tephritidae and Chloropidae gall-inducers stimulate the redifferentiation of vascular cells, as seen by the formation of concentric vascular bundles in galls induced by *Lipara lucens* (Chloropidae) on *Phragmites australis* (Poaceae) (Vandevyvere and Bruyn 1998). Additionally, the redifferentiation of vascular system cells occurs via parenchyma cells that anastomose with the vascular system of the organ adjacent to the gall, as observed in *Urophora cardui* (Tephritidae) galls (Shorthouse and Lalonde 1988). In some cases, such as in the galls of *Vernonia potyanthes* (Asteraceae), xylem cells may also degenerate (Andrade et al. 1995).

4 Vascular Constriction Hypothesis

Gall-inducing organisms can potentially manipulate the host plant tissues, inducing galls with different degrees of structural complexity, implying different functional traits toward the metabolic maintenance of the gall (Mani 1964; Oliveira et al. 2016; Ferreira et al. 2019). Changes involving the vascular system and their functional implications have recently been investigated (Aloni et al. 1995; Aloni 2021; Bragança et al. 2021; Jorge et al. 2022; Nobrega et al. 2023), revealing new patterns of xylem differentiation.

The study conducted by Aloni et al. (1995) demonstrated that the secondary xylem formed in the crown gall could potentially diminish the water supply to the young organs situated above it. This reduction occurs due to decreased vessel diameter caused by ethylene production from the crown gall tissue. Consequently, the water and nutrient supply is compromised in favor of the developing gall. However, as the vessel size below the gall was normal, it was well supplied with water, which led to the description of the "gall constriction hypothesis". This hypothesis predicts that the regions of the stem below the gall exhibit more pronounced xylem differentiation, characterized by regular-sized rays and vessel elements. Conversely, the region above the gall is expected to feature narrower vessel elements, enlarged rays, and the absence of fibers. These structural changes may restrict water transport to the plant's aerial part (Aloni et al. 1995, 1998; Aloni 2021).

The constriction hypothesis has been histologically tested in different gall systems (Fig. 2). In *Inga ingoides* (Fabaceae), the hypothesis was corroborated for the intralaminar fusiform galls, due to alterations in cambium activity (Bragança et al. 2021). The galls induced by Cecidomyiidae (Diptera) on *I. ingoides* disrupt the pattern of vascular development by stimulating the production of large vessels in gall portions proximal to the petiole. The alterations in the width of vessel elements seem to be associated with the continuity between the galled tissue and the host

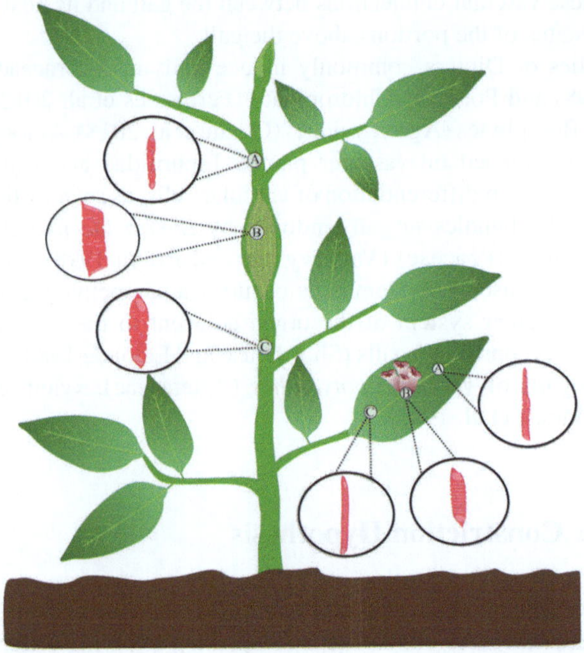

Fig. 2 Hypothetical representation of vascular constriction observed in Neotropical galls. In the gall induced in the stem, which maintains continuity with the host organ, the hypothesis of vascular constriction (*sensu* Aloni et al. 1995) is corroborated by tracheal elements above the gall region (**a**) compared to the gall (**b**) and below the gall (**c**). In the extralaminar gall induced in the leaf, where morphological continuity with the host organ does not occur, the hypothesis of vascular constriction is not corroborated, with tracheal elements varying in size between the regions above the gall, within the gall, and below the gall

leaflet veins (Bragança et al. 2021). However, in the leaves of *Avicennia schaueriana* (Acanthaceae) induced by *Meunieriella* sp. (Cecidomyiidae), the vessel elements in the regions distal and proximal to the *Meunieriella* sp. galls had larger sizes than the vessels inside the galls, which does not support the constriction hypothesis. Fewer and smaller vessel elements were observed inside the galls, which implies reduced hydraulic conductivity in the xylem. As a compensatory mechanism, water storage and spongy parenchyma cells expand in the galls, explaining the higher growth in more vascularized regions. Also, in *E. erythropappus*, the changes in the patterns of secondary vascular tissues do not corroborate the constriction hypothesis. In the galls induced by *Neolasioptera* sp., the vessels are wider both in the galls and in the portions of the stem above the galls (Jorge et al. 2022). The increase in the overproduction of parenchyma and the dimensions of vessel elements indicate the priority of water supply to the galls and stem portions above the galls, ensuring the vitality and maintenance of the growth and development phase of stem galls (Jorge et al. 2022).

In *Malus domestica* (Rosaceae), the stem galls are induced by a colony of the exophytophagous *Eriosoma lanigerum* (Hemiptera), whose stimuli cause the differentiation of anomalous vascular elements, divergent radial and axial orientation. The similarity of the dimensions of the vessel elements in the portions below and above the stem galls indicates the limitation of xylem cells in increasing the water supply to the galls on *M. domestica.* The vessel elements redifferentiated from parenchyma cells potentially increase water storage in the gall developmental site, a compensatory mechanism ensuring the water status in the stem galls.

5 Conclusions and Remarks

Despite being complex and presenting conservative tissue features, the xylem cells can be significantly altered due to galling stimuli due to hormone signaling. The main hormones related to xylem responses are auxins, cytokinins, and ethylene. However, other classes of hormones and/or the synergistic action of several growth regulators can impact the pattern of the vascular system, revealing a knowledge gap yet to be filled. The main changes observed in the xylem include cell differentiation or redifferentiation, allowing vascular connection between the gall and the non-galled host organ portions, favoring water supply. Elucidating the mechanisms that trigger vascular cell origins and differentiation pathways from procambium, cambium, and parenchyma cells implies advancing knowledge of cellular morphogenetic potential using galls as morphogenetic study models. Furthermore, galls induced by insects, especially Diptera, Lepidoptera, Hymenoptera, and Coleoptera, proved to be elegant models of study to understand the mechanisms of changes in the xylem and their relation with compensatory mechanisms toward guaranteeing hydraulic conductivity along the gall and adjacent tissues.

References

Aguilera N, Isaias RMS, Jorge NC, Conejeros MJ, Becerra J, Nieves-Aldrey JL, Guedes LM (2022) Distinctive anatomical and histochemical responses of *Nothofagus obliqua* (Mirb.) Oerst (Nothofagaceae) to two galling Pteromalidae (Hymenoptera: Chalcidoidea) in Chile. Flora 290:152056

Aloni R (1991) Wood formation in deciduous hardwood trees. In: Raghavendra AS (ed) Physiology of Trees. Wiley, New York, pp 175–197

Aloni R (2001) Foliar and axial aspects of vascular differentiation: hypotheses and evidence. J Plant Growth Regul 20:22–34

Aloni R (2013) Role of hormones in controlling vascular differentiation and the mechanism of lateral root initiation. Planta 238:819–830

Aloni R (2015) Ecophysiological implications of vascular differentiation and plant evolution. Trees 29(1):1–16

Aloni R (2021) Vascular differentiation and plant hormones. ISBN:978-3-030-53201-7

Aloni R, Ullrich CI (2008) Biology of crown gall tumors. In: Tzfira T, Citovsky V (eds) *Agrobacterium*: from biology to biotechnology. Springer, New York, pp 565–591

Aloni R, Pradel KS, Ullrich CI (1995) The three-dimensional structure of vascular tissues in *Agrobacterium tumefaciens*-induced crown galls and in the host stems of *Ricinus communis* L. Planta 196:597–605

Aloni R, Wolf A, Feigenbaum P, Avni A, Klee HJ (1998) The never ripe mutant provides evidence that tumor-induced ethylene controls the morphogenesis of *Agrobacterium tumefaciens*-induced crown galls on tomato stems. Plant Physiol 117:841–849

Álvarez R, González-Sierra S, Candelas A, Martinez JJI (2013) Histological study of galls induced by aphids on leaves of *Ulmus minor*: Tetraneura ulmi induces globose galls and *Eriosoma ulmi* induces pseudogalls. Arthropod Plant Interact 7:643–650

Álvarez R, Moreno-González V, Martinez JJI, Ferreira BG, Hidalgo NP (2020) Microscopic study of nine galls induced in *Populus nigra* by aphids of the Iberian Peninsula. Arthropod Plant Interact 14:799–809

Álvarez R, García-Ferreira B, López B, Martinez JJI, Boso S, Martínez MC (2021) Histological study of leaf galls induced by phylloxera in *Vitis* (Vitaceae) leaves. Proc Natl Acad Sci India Sect B Biol Sci 91:117–122

Anand PP, Ramani N (2021) Dynamics of limited neoplastic growth on *Pongamia pinnata* (L.) (Fabaceae) leaf, induced by *Aceria pongamiae* (Acari: Eriophyidae). BMC Plant Biol 21:1

Ananthakrishnan TN (1992) Unique aspects in the biology of thrips-induced galls. In: J D Sorthouse e O R, pp 185–195

Ananthakrishnan TN, Raman A (1989) Morphology of galls. In: Ananthakrishnan TN, Raman A (eds) Thrips and gall dynamics. Oxford and IBH Publishing, New Delhi, pp 67–73

Andrade GI, Silva IM, Fernandes GW, Scatena VL (1995) Aspectos biológicos das galhas de *Tomoplagia rudolphi* (Diptera: Tephritidae) em *Vernonia polyanthes* (Asteraceae). Rev Bras Biol 55:819–829

Andreas P, Kisiala A, Emery RJN, De Clerck-Floate R, Tooker JF, Price PW, Miller DG III, Chen MS, Connor EF (2020) Cytokinins are abundant and widespread among insect species. Plan Theory 9(2):208

Aranda-Rickert A, Rothen C, Diez P, González AM, Marazzi B (2017) Sugary secretions of wasp galls: a want-to-be extrafloral nectar? Ann Bot 120:765–774

Arriola IA, Isaias RMS (2021) Extending the knowledge on the histological patterns of leaf galls induced by *Ditylenchus gallaeformans* (Nematoda) on *Miconia* (Melastomataceae) hosts. Flora 274:151753

Arriola IA, Figueiredo MF, Boanres D, França MGC, Isaias RMS (2020) Apoplast-symplast compartmentalization and functional traits of iron and aluminum in promeristematic tissues of nematode induced galls on *Miconia* spp. Plant Physiol Biochem 154:360–368

Asmus GL (2001) Danos causados à cultura da soja por nematoides do gênero *Meloidogyne*. In: Ferraz LCCB, Asmus GL, Carneiro RG, Mazaferra P, Silva JFV (eds) Relações parasito-hopedeiro nas meloidoginoses da soja. Embrapa Soja, Londrina, pp 39–56

Bartlem DG, Jones MGK, Hammes UZ (2014) Vascularization and nutrient delivery at root-knot nematode feeding sites in host roots. J Exp Bot 65:1789–1798

Bedetti CS, Modolo LV, Isaias RMS (2014) The role of phenolics in the control of auxin in galls of *Piptadenia gonoacantha* (Mart.) Mc.Br (Fabaceae: Mimosoideae). Biochem Syst Ecol 55:53–99

Bedetti CS, Bragança GP, Isaias RMS (2017) Influence of auxin and phenolic accumulation on the patterns of cell differentiation in distinct gall morphotypes on *Piptadenia gonoacantha* (Fabaceae). Aust J Bot 65(5):411

Beltrame WS, Daquila BV, Caleffe RRT, Wolff VRS, Santos ES, Machado MFPS, Conte H, Ruvolo-Takasusuki MCCR (2022) Gall formation in *Cereus* sp. infected with *Diaspis echinocacti* Bouché, 1833 (Hemiptera: Diaspididae). Flora 290:152042

Berrabah F, Balliau T, Aït-Salem EH, George J, Zivy M, Ratet P, Gourion B (2018) Control of the ethylene signaling pathway prevents plant defenses during intracellular accommodation of the rhizobia. New Phytol 219:310–323

Bomfim PMS, Kuster VC, Caetano APS, Gonçalves LA, Oliveira DC (2021) Ovule-gall stimulating a large fake fruit on *Miconia chamissois* Naudin (Melastomataceae): a structural overview. Botany 99:115–126

Bragança GP, Oliveira DC, Isaias RMS (2016) Compartmentalization of Metabolites and Enzymatic Mediation in Nutritive Cells of Cecidomyiidae Galls on Piper Arboreum Aubl. (Piperaceae) J. Plant Stud 6(1):11

Bragança GPP, Freitas MSC, Isaias RMS (2021) The influence of gall position over xylem features in leaflets of *Inga ingoides* (Rich.) Willd. (Fabaceae: Caesalpinioideae). Trees 35:199–209

Bronner R (1992) The role of nutritive cells in the nutrition of cynipids and cecidomyiids. In: Shorthouse JD, Rohfritsch O (eds) Biology of insect-induced galls. Oxford University Press, New York, pp 118–140

Carneiro RGS, Oliveira DC, Isaias RMS (2014) Developmental anatomy and immunocytochemistry reveal the neo-ontogenesis of the leaf tissues of *Psidium myrtoides* (Myrtaceae) towards the globoid galls of *Nothotrioza myrtoidis* (Triozidae). Plant Cell Rep 33:2093–2106

Carneiro RGS, Isaias RMS, Moreira ASFP, Oliveira DC (2017) Reacquisition of new meristematic sites determines the development of a New Organ, the Cecidomyiidae Gall on Copaifera langsdorffii Desf. (Fabaceae). Front Plant Sci 8:1622

Chalupowicz L, Barash I, Schwartz M, Aloni R, Manulis S (2006) Comparative anatomy of gall development on *Gypsophila paniculata* induced by bacteria with different mechanisms of pathogenicity. Planta 224:429–437

Costa EC, Freitas MSC, Carneiro RGS, Oliveira DC, Isaias RMS (2022) The ontogenesis of four *Lopesia* Rübsaamen (Cecidomyiidae) galls on the super-host *Mimosa gemmulata* Barneby (Fabaceae) reveals peculiar anatomical traits. Flora 287:151996

Desnitskiy AG, Chetverikov PE (2022) Induction of leaf galls by four-legged mites (Eriophyoidea) as a problem of developmental biology. Russ J Dev Biol 53:6–14

Desnitskiy AG, Chetverikov PE, Ivanova LA, Kuzmin IV, Ozman-Sullivan SK, Sukhareva SI (2023) Molecular aspects of gall formation induced by mites and insects. Life (Basel) 13(6):1347

Dolzblasz A, Myskow E, Gola EM (2018) Meristems of seedless vascular plants: the state of the art. In: Fernandez H (ed) Current Advances in Fern Research. Springer, Berlin, pp 47–73

Dorchin N, Freidberg A, Aloni R (2002) Morphogenesis of stem gall tissues induced by larvae of two cecidomyiid species (Diptera: Cecidomyiidae) on *Suaeda monoica* (Chenopodiaceae). Can J Bot 80(11):1141–1150

Dreger-Jauffret F, Shorthouse JD (1992) Diversity of gall-inducing insects and their galls. In: Shorthouse JD, Rohfritsch O (eds) Biology of insect-induced galls. Oxford University Press, New York, pp 8–33

Eklund L (1991) Relations between indoleacetic acid, calcium ions and ethylene in the regulation of growth and cell wall composition in Picea abies. J Exp Bot 42:785–789

Escobar C, Barcala M, Cabrera J, Fenoll C (2015) Chapter One – Overview of root-knot nematodes and giant cells. In: Escobar C, Fenoll C (eds) Advances in botanical research, vol 73. Academic, pp 1–32

Evert RF, Esau K, Esau K, Esau (2006) Plant anatomy: meristems, cells, and tissues of the plant body: their structure, function, and development, 3rd edn. Wiley-Interscience, Hoboken. ISBN:978-0-471-73843-5

Fernandes GW, Carneiro MAA, Isaias RMS (2012) Gall-inducing insects: from anatomy to biodiversity. In: Panizzi AR, Parra JRP (eds) Insect bioecology and nutrition for integrated pest management. CRC Press, Boca Raton, pp 369–395

Fernandes GW, Aguirre-Jaimes A, Contreras-Varela X, Cocoletzi E, Sousa WO, Araujo L, Nunes B, Angeles G, Quesada M (2021) Morphological and histological characterization of extrafloral nectaries induced by *Coelocephalapion* galls on *Croton antisyphiliticus* in the Brazilian Cerrado. Flora 274:151758

Ferreira BG, Isaias RMS (2013) Developmental stem anatomy and tissue redifferentiation induced by a galling Lepidoptera on *Marcetia taxifolia* (Melastomataceae). Botany 91:752–760

Ferreira BG, Álvarez R, Avritzer SC, Isaias RMS (2017a) Revisiting histological patterns of storage tissues: beyond the limits of gall-inducing taxa. Botany 95:173–184

Ferreira BG, Avritzer SC, Isaias RMS (2017b) Totipotent nutritive cells and indeterminate growth in galls of *Ditylenchus gallaeformans* (Nematoda) on reproductive apices of *Miconia*. Flora 227:36–45

Ferreira BG, Álvarez R, Bragança GP, Alvarenga DR, Pérez-Hidalgo N, Isaias RMS (2019) Feeding and other gall facets: patterns and determinants in gall structure. Bot Rev 85:78–106

Ferreira BG, Moreira GRP, Isaias RMS, Carneiro RGS (2022) Complex meristematic activity induced by *Eucecidoses minutanus* on *Schinus engleri* turns shoots into galls. Am J Bot 109(2):209–225

Fonti P, Arx VG, García-González I, Eilmann B, Sass-Klaassen U, Gärtner H, Eckstein D (2010) Studying global change through investigation of the plastic responses of xylem anatomy in tree rings. New Phytol 185(1):42–53

Freidberg A (1998) Tephritid galls and gall Tephritidae revisited, with special emphasis on myopitine galls. In: Csóka G, Mattson WJ, Stone GN, Price PW (eds) The biology of gall-inducing arthropods. USDA Forest Service, North Central Research Station, General Technical Report NC-199, pp 36–43

Freitas MS (2022) Galhas induzidas por *Eriosoma lanigerum* Hausmann (Hemiptera: Aphididae) em *Malus domestica* Borkh. (Rosaceae). Dissertação de Mestrado. Universidade Federal de 525 Minas Gerais- UFMG. Belo Horizonte

Fukuda H (1997) Tracheary element differentiation. Plant Cell 9(7):1147–1156

Fukuda H, Watanabe Y, Kuriyama H, Aoyagi S, Sugiyama M, Yamamoto R, Demura T, Minami A (1998) Programming of cell death during xylogenesis. J Plant Res 111:253–256

Garita AS, Bernardo VF, Gonzales M, Ripodas JI, Arango MC, Rusciti M (2021) The false root-knot nematode: modification of the root anatomy and alteration of the physiological performance in tomato plants. Rhizosphere 20:100424

Gómez YIH, Vera CDP, Jiménez PY, García EA (2017) Comparative histology of feeding sites induced by *Cactodera rosae, Globodera mexicana,* and *Meloidodera astonei* (Nematoda: Heteroderidae). Nematropica 47:114–119

Govindarajalu E, Lourdusamy TA (1980) Morphological and anatomical astudies of the floral gall of *Pongamia glabra* Vent. Fabaceae (Papilionaceae). Proc Indian Acad Sci 89:331–345

Guan D, Stacey N, Liu C, Wen J, Mysore KS, Torres-Jerez I, Vernié T, Tadege M, Zhou C, Wang Z, Udvardi MK, Oldroyd GED, Murray JD (2013) Rhizobial infection is associated with the development of peripheral vasculature in nodules of *Medicago truncatula*. Plant Physiol 162(1):107–115

Guedes LM, Costa EC, Isaias RMS, Sáez Carillo K, Aguilera N (2023a) Structural patterns of Lepidoptera galls and the case of *Andescecidium parrai* (Cecidosidae) galls on *Schinus polygama* (Anacardiaceae). J Plant Res 136:715–728

Guedes LM, Gavilán E, Pérez C, Laurans F, Lopez- Vaamond C, Giron D (2023b) Impact of the galling *Dasineura* sp. on the structural and chemical profile of *Peumus boldus* stems. Trees 37:545–553

Guedes LM, Aguilera N, Gavilán E, Péndola JA, Villagrán NE (2023c) Vascular implications of *Dasineura* sp. galls' establishment on *Peumus boldus* stems. Plant Biol 25(6):965–972

Guedes LM, Gavilán E, Aguilera N (2023d) Do different gall induction organs and leaf microsites determine different structural profiles?: the case of *Ophelimus migdanorum* (Hymenoptera: Eulophidae) galls on *Eucalyptus globulus* (Myrtaceae). S Afr J Bot 152:11–18

Haegeman A, Mantelin S, Jones JT, Gheysen G (2012) Functional roles of efectors of plant-parasitic nematodes. Gene 492:19–31

Harris KM (1994) Gall midges (Cecidomyiidae): classification and biology. In: Williams MAJ (ed) Plant galls: organisms, interactions, populations. The Systematics Association Special, vol 49, pp 201–211
Harris MO, Pitzschke A (2020) Plants make galls to accommodate foreigners: some are friends, most are foes. New Phytol 225:1852–1872
Hearn J, Blaxter M, Schönrogge K, Nieves-Aldrey JL, Pujadé-Villar J, Huguet E, Drezen JM, Shorthouse JD, Stone GN (2019) Genomic dissection of an extended phenotype: Oak galling by a cynipid gall wasp. PLoS Genet 15(11):e1008398–e1008398
Hennion S, Little CHA, Hartmann C (1992) Activities of enzymes involved in lignification during the post-harvest storage of etiolated asparagus spears. Physiol Plant 86:474–478
Hernandez YG, Vera ICDP, Jiménez PY, Esteva AG (2017) Comparative histology of feeding sites induced by *Cactodera rosae*, *Globodera mexicana*, and *Meloidodera astonei* (Nematoda: Heteroderidae). Nematropica 47:114–119
Hernández-López J, Tovar-Soto A, Carvajal-Sandoval A, Coronel-Torres R (2006) Alterações anatómicas inducidas por *Cactodera galinsogae* (Nemata: Heteroderinae) en miembros de Asteraceae. Polibotânica 22:1–8
Hirano T, Okamoto A, Oda Y, Sakamoto T, Takeda S, Matsuura T, Ikeda Y, Higaki T, Kimura S, Sato MH (2023) Ab-GALFA, A bioassay for insect gall formation using the model plant *Arabidopsis thaliana*. Sci Rep 13:2554
Hori K (1992) Insect secretions and their effect on plant growth, with special reference to hemipterans. In: Shorthouse JD, Rohfritsch O (eds) Biology of insect-induced galls. Oxford University Press, New York, pp 157–170
Ingemarsson BSM, Eklund L, Eliasson L (1991) Ethylene effects on cambial activity and cell wall formation in hypocotyls of *Picea abies* seedlings. Physiol Plant 82:219–224
Isaias RMS, Oliveira DC, Carneiro RGS (2011) Role of *Euphalerus ostreoides* (Hemiptera: Psylloidea) in manipulating leaflet ontogenesis of *Lonchocarpus muehlbergianus* (Fabaceae). Botany 89:581–592
Isaias RMS, Oliveira DC, Carneiro RGS, Kraus JE (2014) Developmental anatomy of galls in the neotropics, arthropods stimuli versus host plant constraints. In: Fernandes GW, Santos JC (eds) Neotropical insect galls. Springer, pp 15–34
Jackson RS (2008) Vineyard practice. In: Jackson RS (ed) Food science and technology. Wine science. Academic, pp 108–238
Jara-Chiquito JL, Pujade-Villar J, Ferreira BG, Álvarez R (2021) Histological changes induced by the cynipid wasp *Dryocosmus kuriphilus* (Hymenoptera: Cynipidae) in leaves of the chestnut *Castanea sativa* (Fagaceae): Mechanisms of galling impact on host vigor. Arthropod Plant Interact 15:223–233
Jones MGK, Goto DB (2011) Root-knot nematodes and giant cells. In: Jones J, Gheysen G, Fenoll C (eds) Genomics and molecula genetics of plant–nematode interactions. Springer, Dordrecht, pp 83–100
Jorge NC, Alvarenga DR, Cavalleri A, Isaias RMS (2018a) Anatomical and histometric explanations for leaf folding on *Holopothrips striatus* galls in *Myrcia retorta*. Flora 244–245:24–28
Jorge NC, Souza-Silva ÉA, Alvarenga DR, Saboia G, Soares GLG, Zini CA, Cavalleri A, Isaias RMS (2018b) Structural and chemical profiles of Myrcia splendens (Myrtaceae) leaves under the influence of the galling Nexothrips sp. (Thysanoptera). Front Plant Sci 9:1521
Jorge NC, Freitas MSC, Caffaro RM, Vale FHA, Lemos-Filho JP, Isaias RMS (2022) Vascular traits of stem galls: cell increment versus morphogenetic constraints in wood anatomy. Plant Biol 24:450–457
Junghans U, Langenfeld-Heyser R, Polle A, Teichemann T (2004) Effect of auxin transport inhibitors and ethylene on the wood anatomy of poplar. Plant Biol 6:22–29
Kende H (2000) Biochemistry and molecular biology of plants. Science 290(5492):719
Kraus JE, Montenegro G, Kim AJ (1993) Morphological studies on entomogenous stem galls of *Microgramma squamulosa* (Kauf.) Sota (Polypodiaceae). Am Fern J 83:120–128

Kraus JE, Arduin M, Venturelli M (2002) Anatomy and ontogenesis of hymenopteran leaf galls of *Struthanthus vulgaris* Mart. (Loranthaceae). Braz J Bot 25(4):449–458

Kuriyama H, Fukuda H (2001) Regulation of tracheary element differentiation. J Plant Growth Regul 20:35–51

Lucas WJ, Groover A, Lichtenberger R, Furuta K, Yadav SR, Helariutta Y, He XQ, Fukuda H, Kang J, Brady SM, Patrick JW, Sperry J, Yoshida A, López-Millán AF, Grusak MA, Kachroo P (2013) The plant vascular system: evolution, development and functions. J Integr Plant Biol 55:294–388

Mani MS (1964) Ecology of plant galls. Dr. W. Junk Publishers, The Hague, p 434

Martins GS, Santos MG, Reis A, Ferreira BG (2023) Alterations induced by Tortrimosaica polypodivora on the stems of *Microgramma vacciniifolia*: simple or complex galls? Rodriguésia 74:e00532023

Matsumoto-Kitano M, Kusumoto T, Tarkowski P, Kinoshita-Tsujimura K, Vaclavikova K, Miyawaki K, Kakimoto T (2008) Cytokinins are central regulators of cambial activity. Proc Natl Acad Sci 105(50):20027–20031

Meidani C, Ntalli NG, Giannoutsou E, Adamakis IS (2019) Cell wall modifications in giant cells induced by the plant parasitic nematode *Meloidogyne incognita* in wild-type (Col-0) and the fra2 *Arabidopsis thaliana* Katanin Mutant. Int J mol sci 20(21):5465

Meyer J (1987) Plant galls and gall inducers. Gebrüder Borntraeger, Berlin

Miller WE (2005) Gall-inducing Lepidoptera. In: Raman A, Schaefer CW, Withers TM (eds) Biology, ecology and evolution of gall-inducing arthropods. Science Publishers, Enfield, pp 431–465

Mound LA (2020) Host-plant relationships of gall-inducing Thysanoptera – the lack of patterns. J Insect Biodivers Syst 6(4):475–485

Muñoz-Viveros AL, Martinez JJI, Molist P, González-Sierra S, González Julián P, Álvarez R (2014) Microscopic study of galls induced by three species of *Geopemphigus* (Hemiptera: Aphididae: Eriosomatinae) on *Pistacia mexicana*. Arthropod Plant Interact 8:531–538

Moura MZD, Soares GLG, Isaias RMS (2008) Species-specific changes in tissue morphogenesis induced by two arthropod leaf gallers in Lantana camara L. (Verbenaceae) Aust J Bot 56:153–160

Nobrega LP, Sa-Haiad B, Ferreira B (2023) Mechanisms of hydraulic conductivity in the leaf galls of *Meunieriella* sp. (Cecidomyiidae) in *Avicennia schaueriana* (Acanthaceae): does vascularization explain the preferred sites of induction? Plant Biol 25:198–207

Nonaka S, Ezura H (2014) Plant-Agrobacterium interaction mediated by ethylene and super-Agrobacterium conferring efficient gene transfer. Front Plant Sci 5:681

Ohgue T, Imada Y, Sato W, Armando A, Salazar JRL, Makoto K (2018) The first insect-induced galls in bryophytes. Bryophyte Divers Evol 40(1):1

Oliveira DC, Isaias RMS, Fernandes GW, Ferreira BG, Carneiro RGS, Fuzaro L (2016) Manipulation of host plant cells and tissues by gall-inducing insects and adaptive strategies used by different feeding guilds. J Insect Physiol 84:103–113

Oliveira DC, Martini VC, Moreira ASFP, Fuzaro L, Gonçalves LA (2020) *Pseudophacopteron longicaudatum* (Hemiptera) induces intralaminar leaf galls on *Aspidosperma tomentosum* (Apocynaceae): a qualitative and quantitative structural overview. An Acad Bras Cien 92:e20181002

Palomares-Rius JE, Escobar C, Cabrera J, Vovlas A, Castillo P (2017) Anatomical alterations in plant tissues induced by plant-parasitic nematodes. Front Plant 16(8):1987

Pandey S (2021) Climatic influence on tree wood anatomy: a review. J Wood Sci 67:24

Panizzi AR, Lucini T, Mitchell PL (2021) Electronic monitoring of feeding behavior of phytophagous true bugs (Heteroptera). Entomology in Focus

Pesquet E, Tuominen H (2011) Ethylene stimulates tracheary element differentiation in *Zinnia elegans* cell cultures. New Phytol 190:138–149

Popp C, Ott T (2011) Regulation of signal transduction and bacterial infection during root nodule symbiosis. Curr Opin Cell Biol 14(4):458–467

Raman A, Dhileepan K (1999) Qualitative evaluation of damage by *Epiblema strenuana* (Lepidoptera: Tortricidae) to the weed *Parthenium hysterophorus* (Asteraceae). Ann Entomol Soc Am 92:717–723

Richardson RA, Body M, Warmund MR, Schultz JC, Appel HM (2017) Morphometric analysis of young petiole galls on the narrow-leaf cottonwood, *Populus angustifolia,* by the sugarbeet root aphid, *Pemphigus betae*. Protoplasma 254:203–216

Rohfritsch O (1992) Patterns in gall development. In: Shorthouse JD, Rohfritsch O (eds) Biology of insect induced galls. Oxford University Press, New York, pp 60–86

Rosa LMP, Silva MS, Carneiro RGS (2024) Hemiptera-induced galls of *Sapium glandulosum* have histological and cytological compartmentalization created with a large amount of carbohydrate. Protoplasma 261(3):593–606

Sasanelli N, Vovlas N, Cantalapiedra-Navarrete C, Lucarelli G, Palomares-Rius JEC (2015) Parasitism and pathogenicity of curly-leaf parsley with the root-knot nematode *Meloidogyne javanica* in Southern Italy. Helminthologia 52:348–354

Schuetz M, Smith R, Ellis B (2013) Xylem tissue specification, patterning, and differentiation mechanisms. J Exp Bot 64:11–31

Selami N, Auriac MC, Catrice O, Capela D, Kaid-Arche M, Timmers T (2014) Morphology and anatomy of root nodules of *Retama monosperma* (L.) Boiss. Plant Soil 379:109–119

Shorthouse JD, Lalonde RG (1988) Role of *Urophora cardui* (L.) (Diptera, Tephritidae) in growth and development of its gall on stems of *Canada thistle*. Can Entomol 120:639–646

Siddique S, Radakovic ZS, De La Torre CM, Chronis D, Novák O, Ramireddy E, Holbein J, Matera C, Hütten M, Gutbrod P, Anjam MS, Rozanska E, Habash S, Elashry A, Sobczak M, Kakimoto T, Strnad M, Schmülling T, Mitchum MG, Grundler FMW (2015) A parasitic nematode releases cytokinin that controls cell division and orchestrates feeding site formation in host plants. PNAS 112(41):12669–12674

Silva MD, Joseph D, Shorthouse JD (2006) Comparison of the development of stem galls induced by *Aulacidea hieracii* (Hymenoptera: Cynipidae) on hawkweed and by *Diplolepis spinosa* (Hymenoptera: Cynipidae) on rose. Can J Bot 84(7):1052–1074

Silva AFM, Oliveira DC, Isaias RMS (2023) Influence of the sex of galling insects on gall metabolism: the case of eriococcid galls. Biochem Syst Ecol 108:104643

Singh SM, Patni V, Arora DK (2017) Histopathological studies on stem gall of *Acacia leucophloea* Willd. induced by insect. Indian J Plant Physiol 6(3):45–49

Słupianek A, Dolzblasz A, Sokołowska K (2021) Xylem parenchyma – role and relevance in wood functioning in trees. Plan Theory 2(10):1247

Sobczak M, Golinowski W, Grundler FM (1997) Changes in the structure of *Arabidopsis thaliana* roots induced during development of males of the plant parasitic nematode *Heterodera schachtii*. Eur J Plant Pathol 103:113–124

Souza SCPM, Kraus JE, Isaias RMS, Neves LJ (2000) Anatomical and ultrastructural aspects of leaf galls in *Ficus microcarpa* L.F. (Moraceae) induced by *Gynaikothrips ficorum* Marchal (Thysanoptera). Acta Bot Bras 14:57–69

Spaepen S, Vanderleyden J (2011) Auxin and plant-microbe interactions. Cold Spring Harb Perspect Biol 3(4):a001438

Suttle JC (1988) Effect of ethylene treatment on polar IAA transport, net IAA uptake and specific binding of N-1-naphthylphthalamic acid in tissues and microsomes isolated from etiolated pea epicotyls. Plant Physiol 88:795–799

Taft JB, Bissing DR (1988) Developmental anatomy of the Horned Oak Gall induced by *Callirhytis cornigera* on *Quercus palustris* (Pin Oak). Am J Bot 75(1):26–36

Takei M, Yoshida S, Kawai T, Hasegawa M, Suzuki Y (2015) Adaptive significance of gall formation for a gall-inducing aphids on Japanese elm trees. J Insect Physiol 72:43–51

Tapia-Vázquez I, Montoya-Martínez AC, Santos-Villalobos S, EkRamos MJ, Montesinos-Matías R, Martínez-Anaya C (2022) Root-knot nematodes (*Meloidogyne* spp.) a threat to agriculture in Mexico: biology, current control strategies, and perspectives. World J Microbiol Biotechnol 38(6):26

Thapa SP, Davis EW, Lyu Q, Weisberg AJ, Stevens DM, Clarke CR, Coaker G, Chang JH (2019) The evolution, ecology, and mechanisms of infection by Gram-positive, plant-associated bacteria. Annu Rev Phytopathol 57:341–365

Tooker JF, Helms AM (2014) Phytohormone dynamics associated with gall insects, and their potential role in the evolution of the gall-inducing habit. J Chem Ecol 40:742–753

Vandevyvere I, Bruyn L (1998) Morphological and histochemical analysis of galls of *Lipara lucens* (Diptera, Chloropidae) on *Phragmites australis* (Poaceae). Can J Bot 76:1374–1384

Venturi V, Fuqua C (2013) Chemical signaling between plants and plant-pathogenic bacteria. Annu Rev Phytopathol 51:17–37

Vilela RMIF, Martini VC, Gonçalves LA, Kuster VC, Oliveira DC (2019) Structure and development of root gall induced by *Meloidogyne javanica* in *Glycine max* L. Semina: Série botânica 40:1033–1048

Vilela RMIF, Kuster VC, Magalhães TA, Moraes CA, Paula FAC, Oliveira DC (2021) Impact of *Meloidogyne incognita* (nematode) infection on root tissues and cell wall composition of okra (*Abelmoschus esculentus* L. Moench, Malvaceae). Protoplasma 258(5):979–990

Vilela RMIF, Kuster VC, Magalhães TA, Martini VCM, Oliveira RM, Oliveira DC (2023) Galls induced by a root-knot nematode in *Petroselinum crispum* (Mill.): impacts on host development, histology, and cell wall dynamics. Protoplasma 260:1287–1302

Westphal E (1983) Adaptation of gall mites (Acari, Eriophyoidea) to live in galls. In: Margaris NS (ed) Adaptations to terrestrial environments. Plenum, New York, pp 69–75

Westphal E, Manson DCM (1986) Feeding effects on host plants: gall formation and other distortions. In: Lindquist EE, SabelisMW, Bruin J (eds) Eriophyoid mites – their biology, natural enemies and control. Elsevier, Amsterdam, pp 231–242

Westphal E, Bronner R, Le Ret M (1981) Changes in leaves of susceptible and resistant Solanum dulcamara infested by the gallmite *Eriophyes cladophthirus* (Acarina, Eriophyoidea). Can J Bot 59:875–882

Yamaguchi H, Tanaka H, Hasegawa M, Tokuda M, Asami T, Suzuki Y (2012) Phytohormones and willow gall induction by a gall-inducing sawfly. New Phytol 196(2):586–595

Zupan J, Muth TR, Draper O, Zambryski P (2000) The transfer of DNA from *Agrobacterium tumefaciens* into plants: a feast of fundamental insights. Plant J 23:11–28

The Protagonism of Phloem in Gall Development

Ravena Malheiros Nogueira, Gracielle Pimenta Pereira Bragança, Elaine Cotrim Costa, Edgard Augusto de Toledo Picoli, and Rosy Mary dos Santos Isaias

1 Introduction

Galls result from the relationship of specific parasites capable of manipulating plant tissues (Mani 1964) toward changes in cell metabolism, and accumulation of phenols, auxins, enzymes, carbohydrates, proteins, and nucleic acids, for instance (Hori 1992; Albert et al. 2011; Bedetti et al. 2017). Such contents influence cell functional traits, as key processes that are essential for the development and maintenance of gall structure (Ullrich and Aloni 2000). All the influences and peculiar gall tissue organization have been related to the taxa of the galling herbivores and their feeding habits (Bronner 1992; Rohfritsch and Anthony 1992; Ferreira et al. 2017, 2019a), which has been diagnosed by anatomical analyses.

Hemiptera-induced galls have a great richness of shapes and sizes and can be induced on different organs of different host plants (Chen and Qiao 2012). The mouthparts of the Hemiptera consist of a proboscis containing two pairs of stylets: one injects saliva and the other sucks fluids from the plant (Gullan and Martin 2003). Since the Hemiptera have piercing-sucking mouthparts, which are inserted into plant tissues, their nutrition occurs, as a rule, directly from phloem (Rohfritsch and Anthony 1992; Prado and Tjallingii 1994) but can also occur associated with

---

R. M. Nogueira · G. P. P. Bragança · R. M. dos Santos Isaias (✉)
Department of Botany, Universidade Federal de Minas Gerais (UFMG), Belo Horizonte, Minas Gerais, Brazil
e-mail: rosy@icb.ufmg.br

E. C. Costa
Instituto de Ciências Biológicas, Universidade Federal do Rio Grande (FURG), Rio Grande, Rio Grande do Sul, Brazil

E. A. de T. Picoli
Department of Plant Biology, Universidade Federal de Viçosa (UFV), Viçosa, Minas Gerais, Brazil

D. C. de Oliveira, R. M. dos Santos Isaias (eds.), *Plant Galls*,
https://doi.org/10.1007/978-3-031-80064-1_5

xylem (Sandanayaka and Hale 2003; Sandanayaka et al. 2003; Hao et al. 2020) or parenchyma cells (Meyer 1987). In contrast, for example, rasping-sucking (Diptera), and chewing (Lepidoptera) galling insects actively feed on typical nutritive cells around the larval chamber of their galls (Ferreira et al. 2019a; Guedes et al. 2023).

The complexity of gall structures can differ, sometimes including the differentiation of new vessels facilitating the transport of nutrients (Meyer and Maresquelle 1983; Bartlem et al. 2014), sometimes maintaining the original bundles of host organ in intralaminar galls (Bedetti et al. 2017). The feeding behavior of gall inducers is one of the determinant factors for the neoformations induced in the host plant tissues (Mani 1964; Meyer and Maresquelle 1983). The neoformation of vascular bundles may indicate the transition of vascular parenchyma cells toward re-differentiation of procambium, followed by the differentiation of xylem and phloem elements (Mani 1964; Stone and Schönrogge 2003; Lev-Yadun 2003; Ullrich and Aloni 2000; Oliveira and Isaias 2010; Costa et al. 2018). Many researches on structural and functional traits of the vascular system in galls have extensively focused on the xylem (*e.g.* Aloni 2021; Bragança et al. 2020a; Jorge et al. 2022). Nevertheless, the approaches specifically focusing on phloem in galls are still in the early stages, and are revisited here toward elucidating the state of the art. We point out gaps in the understanding of structural and functional alterations in phloem cells in gall developmental sites, with a principal focus on galls induced by Hemiptera that may feed directly from the sieve tube elements, or alternatively from xylem or vascular parenchyma cells.

2 Methodology

A survey of scientific papers on phloem in galls was conducted, without publication period criterion. The data were accessed through Portal de Periódicos da Coordenação de Aperfeiçoamento de Pessoal de Nível Superior (CAPES), Web of Science, and Google Scholar, using the keywords: "phloem in gall", "phloem cell function galls", "phloem anatomy galls", and "sieve tube element galls", with articles from 1976 to 2023 being covered. Only studies ($n = 52$ publications) that examined changes in phloem in galls were selected, which was verified by reading the title, abstract and/or keywords of the papers. A table based on the separation criteria of galling organism taxa, botanical area, and the year of publication from the oldest to the most recent was organized (Table 1).

3 Results

The 52 recovered publications were divided among the study areas: anatomy (51.9%), histochemistry (34.6%), immunocytochemistry (7.6%), article revisions (5.7%), genetics (5.7%), and cytology (1.9%) (Figs. 1 and 2). The altered parameters in the galls in relation to the non-galled organ for each study area. The

Table 1 Phloem characteristics evidenced in galls induced by different groups of inducers

Gall-inducers	Botanical areas	Host plants	Organs	Phloem traits	References
Eriosoma lanigerum (Insects:Hemiptera-Aphididae)	Anatomy	*Malus domestica* Borkh. (Rosaceae)	Root	Cell proliferation	Brown et al. (1991)
Rectinasus buxtoni (Insects:Hemiptera-Aphididae)	Anatomy	*Pistacia palaestina* Boiss. (Anacardiaceae)	Leaflet	Schizogenous ducts	Álvarez et al. (2014)
Geopemphigus moral, G. torsus and *G. blackmani* (Insects:Hemiptera-Aphididae)	Anatomy	*Pistacia mexicana* Kunth (Anacardiaceae)	Leaflet	Schizogenous ducts and druses	Muñoz-Viveros et al. (2014)
Geopemphigus blackmani and *G. torsus* (Insects:Hemiptera-Aphididae)	Anatomy	*Pistacia mexicana* Kunth (Anacardiaceae)	Leaflet	Schizogenous ducts and cell hypertrophy	Álvarez et al. (2016)
Pemphigus betae (Insects: Hemiptera-Aphididae)	Anatomy	*Populus angustifolia* E. James. (Salicaceae)	Leaf	Numerous and larger cells	Richardson et al. (2016)
Eriosoma ulmi and *Geoica utricularia* (Insects:Hemiptera-Aphididae)	Anatomy	*Pistacia terebinthus* L. (Anacardiaceae)	Leaf	Increase in area, and secretory ducts	Ferreira et al. (2017)
Eriosoma ulmi (Insects:Hemiptera-Aphididae)	Revision of anatomy	*Ulmus minor* Mill. (Ulmaceae)	Leaf	Cell hypertrophy	Ferreira et al. (2019a)
Pemphigus vesicarius (Insects: Hemiptera-Aphididae)	Anatomy	*Populus nigra* L. (Salicaceae)	Leaf	Cell hypertrophy	Álvarez et al. (2020)
Tetraneura spp. (Insects: Hemiptera-Aphididae)	Histochemistry	*Ulmus davidiana* var. *japonica* (Ulmaceae)	Leaf	Amino acid	Suzuki et al. (2009)
Euphalerus ostreoides Crawf. (Insects: Hemiptera-Psyllidae)	Anatomy	*Lonchocarpus muehlbergianus* Hassl. (Fabaceae)	Leaflets	Hyperplasia	Oliveira et al. (2006)

(continued)

Table 1 (continued)

Gall-inducers	Botanical areas	Host plants	Organs	Phloem traits	References
Pauropsylla tuberculata (Insects:Hemiptera-Psyllidae)	Anatomy	*Alstonia scholaris* (L.) R.Br. (Apocynaceae)	Leaf	Cell differentiation	Albert et al. (2011)
	Histochemistry			Proteins	
Pauropsylla tuberculata(Insects: Hemiptera-Psyllidae)	Anatomy	*Alstonia scholaris* (L.) R.Br. (Apocynaceae)	Leaf	Size reduction of the cells	Kumar (2016)
Pauropsylla tuberculata (Insects: Hemiptera-Psyllidae)	Anatomy	*Alstonia scholaris* (L.) R. Br	Fruit	Cell hypertrophy	Chauhan et al. (2020)
Pseudophacopteron longicaudatum (Insects: Hemiptera-Psyllidae)	Anatomy	*Aspidosperma tomentosum* Mart. (Apocynaceae)	Leaf	Cell hypertrophy	Martini et al. (2020)
	Histochemistry			Proteins	
Calophyidae mammifex (Insects: Hemiptera-Calophyidae)	Anatomy	*Schinus polygama* (Cav.) Cabrera (Anacardiaceae)	Leaf	Secretory ducts	Guedes et al. (2018a)
	Histochemistry			Lipids and reducing sugars	
Calophya rubra Blanchard (Insects: Hemiptera-Calophyidae)	Histochemistry		Stem	Lipids	Guedes et al. (2018b)
Bystracoccus mataybae (Insects: Hemiptera-Eriococcidae)	Histochemistry	*Matayba guianensis* Aubl. (Sapindaceae)	Stem	Proteins, phenolics, and reactive oxygen species	Silva et al. (2019)
Nothotrioza cattleiani (Insects: Hemiptera-Triozidae)	Immunocytochemistry	*Psidium cattleianum* Sabine (Myrtaceae)	Leaf	Galactans and arabinans, low methylesterified Homogalacturo-nans	Carneiro et al. (2015)
Ceropsylla pouteriae Burckhardt sp. Nov. (Insects: Hemiptera-Triozidae)	Histochemistry	*Pouteria ramiflora* (Mart.) Radlk. (Sapotaceae)	Leaf	Proteins, starch, lipids and reactive oxygen species	Oliveira et al. (2019)
Izeniola obesula (Insects: Diptera-Cecidomyiidae)	Anatomy	*Suaeda monoica* Forssk. Ex J. F. Gmel. (Chenopodiaceae)	Stem	Cell differentiation	Dorchin et al. (2002)

Lopesia sp. (Insects: Diptera-Cecidomyiidae)	Histochemistry	*Lonchocarpus cultratus* (Vell.) A.M.G.Azevedo & H.C.Lima (Fabaceae)	Leaf	Reactive oxygen species	Suzuki et al. (2015)
Liodiplosis cylindrica (Insects: Diptera-Cecidomyiidae)	Histochemistry	*Mikania glomerata* Spreng. (Asteraceae)	Leaf	Secretory ducts	Amorim et al. (2017)
Not identified (Insects: Diptera-Cecidomyiidae)	Histochemistry	*Inga ingoides* (Rich.) Willd. (Fabaceae)	Leaf	Auxins and reactive oxygen species	Bragança et al. (2020a)
Clinodiplosis sp. (Insects: Diptera-Cecidomyiidae)	Immunocytochemistry	*Croton floribundus* Spreng. (Euphorbiaceae)	Leaf	Galactans, and homogalacturo-nans	Teixeira et al. (2017)
Not identified (Insects: Diptera-Cecidomyiidae)	Immunocytochemistry	*Inga ingoides* (Rich.) Willd. (Fabaceae)	Leaf	Xyloglucans	Bragança et al. (2020b)
Andescecidium parrai (Lepidoptera)	Anatomy	*Shinus dependens* Ortega, *S. polygama* (Cav.) Cabrera, and *S. engleri* Barkley (Anacardiaceae)	Bud	Secretory ducts	Guedes et al. (2023)
	Histoche-mistry	*Shinus dependens*		Phenolics	
Not identified (Lepidoptera)	Histochemistry	*Marcetia taxifolia* (Melastomataceae)	Stem	Reactive oxygen species	Ferreira et al. (2015)
Meloidogyne arenaria (Nematode)	Anatomy	*Coleus blumei* Benth. (Lamiaceae)	Root	Non-functional phloem	Jones and Northcote (1972)
Melendogyne spp. (Nematode)	Histochemistry	*Daucus carota* L. (Apiaceae)	Root	Auxins	Knypl et al. (1975)
Meloidogyne hapla (Nematode)	Histochemistry	*Daucus caroia* L. cv. Perfekcja (Apiaceae)	Root	Phenolics and chlorogenic acid	Knypl and Janas (1976)
Meloidogyne incognita (Nematode)	Anatomy	*Glycine max* (L.) Merrill (Fabaceae)	Root	Atypical phloem	Robab et al. (2010)

(continued)

Table 1 (continued)

Gall-inducers	Botanical areas	Host plants	Organs	Phloem traits	References
Meloidogyne incognita (Nematode)	Anatomy	*Lycopersicon esculentum* Mill. (Solanaceae)	Root	Atypical phloem	Azam et al. (2011)
Meloidogyne incognita (Nematode)	Anatomy	*Luffa Cylindrica* (L.) M. Roem. (Cucurbitaceae)	Root	Atypical phloem	Singh and Saxena (2012)
Heterodera schachtii (Nematode)	Anatomy	*Arabidopsis thaliana* (L.) Heynh. (Brassicaceae)	Root	Syncytia-associated phloem; proliferation of companion cells and sieve elements (metaphloem)	Absmanner et al. (2013)
Meloidogyne incognita (Nematode)	Genetics			The RS6 sieve element marker	
Meloidogyne incognita (Nematode)	Anatomy	*Arabidopsis thaliana* (L.) Heynh. (Brassicaceae)	Root	Cell proliferation	Bartlem et al. (2014)
Nacobbus aberrans (Nematode)	Anatomy	*Solanum tuberosum* (L.) (Solanacea)	Root	Size reduction	Tordable et al. (2018)
Meloidogyne incognita (Nematode)	Anatomy	*Trachyspermum ammi* (L.) (Apiaceae)	Root	Atypical phloem	Danish et al. (2021)
Meloidogyne spp. (Nematode)	Revision of Anatomy	*Arabidopsis thaliana* (L.) Heynh. (Brassicaceae)	Root	Cell proliferation	Engler et al. (2016)
Ditylenchus gallaeformans (Nematode)	Histochemistry	*Miconia albicans* (Sw.) Steud. (Melastomataceae)	Leaf	Auxins	Ferreira et al. (2019b)
				Cytokinins	
Ditylenchus gallaeformans (Nematode)	Histochemistry	*Miconia lacunosa* (Cogn.) R. Goldenb. (Melastomataceae)	Leaf	Aluminum	Arriola et al. (2020)
Ditylenchus gallaeformans (Nematode)	Immunocy-tochemistry	*Miconia albicans* (Sw.) (Melastomataceae)	Leaf	Galactans, xyloglucans, and homogalacturo-nans	Ferreira et al. (2020)
Heterodera schachtii (Nematode)	Genetics	*Arabidopsis thaliana* (L.) Heynh. (Brassicaceae)	Root	Companion cell-specific AtSUC2 promoter	Hoth et al. (2005)

Meloidogyne incognita (Nematode)	Genetics	*Populus tremula* (L.) (Salicaceae)	Root	KANADI 1 and KANADI 2, encoding positive regulators of phloem formation; LBD1, involved in secondary phloem	Baldacci-Cresp et al. (2020)
Agrobacterium tumefaciens (Bacteria)	Histochemistry	*Ricinus communis* L. var. *gibsonii* cv. Carmencita (Euphorbiaceae)	Stem	Auxins and cytokinins	Pradel et al. (1999)
Pantoea agglomerans (Bacteria)	Anatomy	*Vaccinium macrocarpon* (Cranberry) (Ericaceae)	Stem	Cell differentiation, cell hypertrophy, and hyperplasia	Aloni and Ullrich (2004)
Agrobacterium tumefaciens (Bacteria)	Revision of anatomy	*Ricinus communis* L. (Euphorbiaceae)	Stem	Anastomoses	Aloni et al. (2008)
Agrobacterium tumefaciens (Bacteria)	Anatomy	*Ricinus communis* L. (Euphorbiaceae)	Stem	Anastomoses	Aloni et al. (1995)
Agrobacterium tumefaciens (Bacteria)	Anatomy	*Launaea capitata* Spreng. (Asteraceae)	Stem	Increase in area	Hussein et al. (2014)
Fiji disease virus	Anatomy	*Saccharum oficinarum* L. cv. (Poaceae)	Leaf	Atypical phloem and cell proliferation	Hatta and Francki (1976)
Virus RBSDV, genus *Fijivirus*	Citology	*Zea mays* L. (Poaceae)	Leaf	Presence of plastids in absence of vacuoles sieve element	Shen et al. (2016)
Cronartizrm ribicola Fischer (Fungi)	Anatomy	*Ribes nigrum* L. (Grossulariaceae)	Stem	Cell hypertrophy and hyperplasia	Boyer (1967)
	Histochemistry			Auxins	
Plasmodiophora brassicae (Fungi)	Histochemistry	*Arabidopsis thaliana* (L.) Heynh. (Brassicaceae)	Hypocotyl	Accumulation of sugars	Walerowski et al. (2018)

Fig. 1 Representative models of different study areas of the phloem in *Malus domestica* Borkh. (Rosaceae)-*Eriosoma lanigerum* (Hemiptera) system, evidencing anatomical (longitudinal plane, blue square), histochemical, immunocytochemical (transverse plane, green square), and cytological (longitudinal plane, blue square) phloem traits

anatomical responses (51.9%) included: cell hypertrophy (13.5%), atypical phloem development (9.6%), schizogenous duct formation (11.5%), cell proliferation (7.7%), hyperplasia (5.7%), increase in area, more numerous and larger sieve tube elements and companion cells, and formation of anastomoses (3.8%, each), and occurrence of druses, differentiating phloem cells and non-functional phloem (1.9%, each). For histochemistry (34.6%), the responses included the detection of proteins (5.6%), lipids (4.2%), reducing sugars (2.8%), starch grains (1.3%), and phenolics (4.2%). The detection of reactive oxygen species (5.5%), amino acids (1.3%), aluminum (1.3%), auxins (5.6%), and cytokinins (2.8%) was also recovered. For the immunocytochemistry of cell wall components (7.6%), (1 → 5) α-L-arabinans (0.8%), (1 → 4) β-D-galactans (2.5%), homogalacturonans (2.5%), and xyloglucans (1.6%) were labeled in phloem cells. In addition, genetic ($n = 1$) and cytological ($n = 1$) aspects of the phloem were little investigated (Fig. 3 and Table 1).

Regarding the galls induced by insects ($n = 28$; 51.8%), nematodes ($n = 17$; 31.4%), bacteria ($n = 5$; 9.3%), fungi ($n = 2$; 3.7%), and viruses ($n = 2$; 3.7%), there

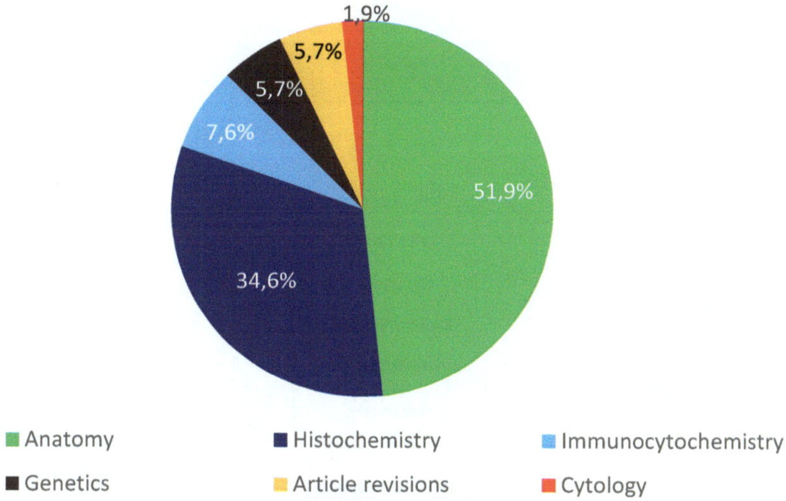

Fig. 2 Comparative proportional approaches of the studies on phloem traits in galls

is a varied investment in the analysis of phloem alterations. Among the insects, studies on phloem were carried out in galls induced by Hemiptera (n = 19; 35.1%), Diptera (n = 6; 11.1%), and Lepidoptera (n = 2; 3.7%). The phloem characteristics for galls of Hemiptera have been recorded with a higher frequency in leaves, and species distributed among five families: Aphididae (n = 9), Psyllidae (n = 5), Calophyidae, Triozidae (2 each), and Eriococcidae (1) (Table 1). These galls occur in 19 plant families, with higher occurrence in Apocynaceae (n = 5) and Anacardiaceae (n = 5) (Table 1). The phloem aspects were reported in 7 morphotypes of Diptera-induced galls, with a higher frequency in leaves (n = 5) of 5 plant families, with higher occurrence in Fabaceae (n = 4) (Table 1). Regarding the Lepidoptera galls, phloem aspects were reported just in 1 morphotype of a stem gall on Melastomataceae (Table 1).

3.1 Phloem Patterns in Galls Induced by Insects

The Aphididae (Hemiptera), which feed on the phloem sap, comprises the major group of gall inducers whose phloem aspects have been studied (Table 1). They suck the phloem sap of the host plant (Bartlem et al. 2014; Ferreira et al. 2019a), and their galls are characterized by cell hypertrophy and tissue hyperplasia in the vascular bundles, specifically of the phloem cells (Table 1), which directly respond to their stimuli (Boyer 1967; Oliveira et al. 2006; Oliveira and Isaias 2010; Álvarez et al. 2020). In Hemiptera galls, the phloem exhibits cell proliferation and differentiation (Brown et al. 1991; Albert et al. 2011; Oliveira et al. 2006), which can be considered facilitating factors by increasing the availability of feeding sites. In the

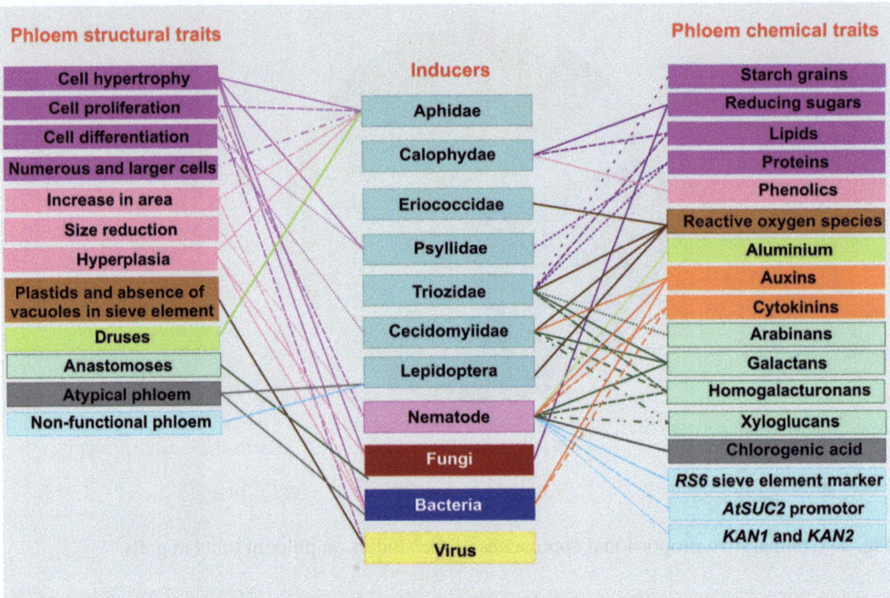

Fig. 3 Network of structural and chemical traits associated with phloem cells in galls already studied. Each phloem trait is linked by lines to the taxa of the gall inducers. For each trait a different line type is used (straight line to dashed lines). The groups of inducers were also separated by color for insects, nematodes, fungi, bacteria, and viruses. In structural traits, the purple box is related to cell features, the pink box indicates tissue features, and the other color boxes relate to different anatomical features occurring in galls. For phloem chemical traits, the purple boxes are related to the primary metabolites, the pink box is related to phenolics, the brown box are related to reactive oxygen species, the light green lemon box is related to aluminum, the orange boxes are related to hormones, the gray box is related to chlorogenic acid, the green boxes are related to cell wall composition, and the blue boxes are related to genetic features

case of the leaflet galls on *Pistacia* spp. (Anacardiaceae), cell hypertrophy is accompanied by the formation of schizogenous ducts and druses associated to the phloem (Álvarez et al. 2009, 2014; Muñoz-Viveros et al. 2014; Álvarez et al. 2016), which are specific traits of the vascular bundles of the Anacardiaceae (Álvarez et al. 2009, 2016; Guedes et al. 2023). The increase of the vascular bundles, particularly of the phloem cells, has been observed in leaf galls on *Populus angustifolia* E. James (Salicaceae) induced by *Pemphigus betae* (Hemiptera: Aphididae) (Richardson et al. 2017), and *Aspidosperma tomentosum* (Apocynaceae) induced by *Pseudophacopteron longicaudatum* (Hemiptera:Psyllidae) (Martini et al. 2020). In contrast, in leaf galls on *Alstonia scholaris* (L.) R.Br. (Apocynaceae) induced by *Pauropsylla tuberculata* (Hemiptera: Psyllidae), there is a reduction in the size of phloem cells (Kumar 2016; Tordable et al. 2018). These variations in the anatomical traits of phloem reveal diverse morphogenetic potential of the host plants under the stimuli of galling insects with the same feeding habit.

In addition to the changes of the phloem structural traits, the accumulation of lipids, reducing sugars, starch grains, and proteins (Guedes et al. 2018a; Martini et al. 2020), as well as of phenolics, and reactive oxygen species have been observed (Oliveira et al. 2019; Silva et al. 2019) (Table 1). The accumulation of the primary metabolites in the phloem cells of galls indicates metabolic activity related to the maintenance of the gall structure, and flow of nutritional resources that favors the feeding of the gall-inducing Hemiptera (Ferreira et al. 2017; Guedes et al. 2018a, b). In galls of *Calophya rubra* (Hemiptera: Calophyidae) on *Schinus polygama* (Cav.) Cabrera (Anacardiaceae), the detection of lipids in phloem is also indirectly linked to the nutrition of the gall inducer (Guedes et al. 2018b). As the hemipterans suck directly on vascular bundles, their galls do not have a typical nutritive tissue (Bronner 1992). The accumulation of starch grains has been detected in the vascular parenchyma of *Ceropsylla pouteriae* (Hemiptera: Triozidae) galls on *Pouteria ramiflora* (Mart.) Radlk. (Sapotaceae) (Oliveira et al. 2019), and also in other tissues of Hemiptera galls induced by Psyllidae (Oliveira et al. 2006) and Aphididae (Álvarez et al. 2009). Nevertheless, such cells are not the feeding source for the inducers (Oliveira et al. 2006, 2019; Ferreira et al. 2017, 2019a). The accumulation of starch grains in galls indicates active sinks of carbohydrates translocated from the photosynthetic tissue sites toward the galls, where this reserve substance can be metabolized for gall maintenance (Oliveira et al. 2006).

The detection of reactive oxygen species and proteins in phloem cells of *C. pouteriae* galls was associated with oxidative stress and high cellular metabolism, respectively (Oliveira et al. 2019). In view of the oxidative stress generated in the gall developmental site, the production of phenolic compounds (Silva et al. 2019) promotes the dissipation of the oxidative stress molecules (Isaias et al. 2015; Bedetti et al. 2014, 2017; Suzuki et al. 2015). Such processes are involved in the maintenance of tissue homeostasis in gall developmental sites, being crucial for the establishment and development of the gall inducers (Oliveira et al. 2016).

Some aphid species reproduce parthenogenetically, resulting in numerous nymphs, which demand a high nutritive support, involving amino acids as nitrogen sources (Suzuki et al. 2009). An example of such relationship was reported for galls on *Ulmus davidiana* var. *japonica* (Ulmaceae), where a high concentration of amino acids in phloem is related to the support of the large number of aphid nymphs. Also, these amino acids accumulated in gall tissues may be translocated into the phloem toward the feeding site of the inducers (Suzuki et al. 2009).

By phloem cells, not only nutrients flow, but also some molecules responsible for gall development. In the Cecidomyiidae galls (Diptera) on *Inga ingoides* (Rich.) Willd. (Fabaceae), the 3-indolacetic acid (IAA) was histochemically detected in specific phloem cells (Bragança et al. 2020a). Also, the association of reactive oxygen species, phenolics and IAA reported in the phloem cells of Diptera (Suzuki et al. 2015; Bragança et al. 2020a) and Lepidoptera (Ferreira et al. 2015) galls have been associated with the control of cell wall extensibility, acidification and loosening of cellulose microfibrils that influence the direction of cell elongation (Suzuki et al. 2015). Consequently, their presence is expected for the development of the new structural and functional design of the galls.

Alterations in the chemical composition of phloem cell in galls walls have started to be investigated in the last decade. The three only articles report epitopes of xyloglucans labeled by LM15 in Cecidomyiidae induced galls on *I. ingoides* (Rich.) Willd. (Fabaceae) (Bragança et al. 2020b), and epitopes of galactans, and homogalacturonans labeled by LM5, and LM19, respectively in *Clinodiplosis* sp. (Cecidomyiidae) induced galls on *Croton floribundus* Spreng. (Euphorbiaceae) (Teixeira et al. 2017), and in *Nothotrioza catteianii* (Triozidae) induced galls on *Psidium cattleianum* Sabine (Myrtaceae) (Carneiro et al. 2015). However, the functional roles of these cell wall components have not been explored yet. Furthermore, in relation to the biological areas, there are no articles related to genetics and cytology of phloem in insect galls, evidencing topics need to be analyzed in greater detail.

3.2 Phloem Responses to Galling Organisms Other than Insects

Regarding the anatomical responses, nematode galls induced by *Meloidogyne* spp. result from cell proliferation and hypertrophy (Bartlem et al. 2014; Engler et al. 2016). Typically, the galls induced by *M. incognita* in different plant species have disorganized phloem, and atypical phloem formation near the peculiar giant cells (Azam et al. 2011; Robab et al. 2010; Singh and Saxena 2012; Danish et al. 2021). The cells surrounding the giant cells are stimulated to divide, and differentiate into new sieve elements (Jones and Northcote 1972), which are actively involved in the translocation of photosynthates toward the giant cells, as higher metabolic rates require an increased energetic intake (Robab et al. 2010; Azam et al. 2011). Therefore, the formation of atypical phloem cells in nematode-induced galls favors the development of the giant cells. In the case of *Agrobacterium tumefaciens*-induced galls, the main anatomical features regarding the phloem is the formation of anastomoses involving only phloem, as reported on *Ricinus communis* (Euphorbiaceae) (Aloni et al. 1995; Aloni and Ullrich 2008). It is proposed that the dense network of phloem anastomoses, formed by vascular parenchyma cells that redifferentiate, enable a rapid translocation of metabolites (Behnke and Sukkri 1971).

Alterations in the level of cell differentiation are potentiated by the stimuli of gall-inducing organisms but mediated by hormones, mainly auxins, produced both by the host plants or some gall-inducing organisms (Mapes and Davies 2001). Peculiarly, auxin accumulation has been reported for the vascular parenchyma of galls induced by *Cronartizrm ribicola* Fischer, a fungus (Boyer 1967), and the accumulation of cytokinins is typically observed in vascular tissues of galls induced by various organisms, such as nematodes, and bacteria (Knypl et al. 1975; Ferreira et al. 2019b; Pradel et al. 1999). In *M. incognita* galls, the giant cells accumulate auxins, while the phloem around the syncytia in *Heterodera schachtii* galls accumulates both auxins and cytokinins (Absmanner et al. 2013). While auxins allow phloem cell hypertrophy, cytokinins are involved in the hyperplasic process, as observed in galls induced by *A. tumefaciens* (Pradel et al. 1999) and *Pantoea agglomerans* (Best et al. 2004). Therefore, both sieve tube elements and companion cells respond to auxins, whereas only sieve tube elements respond to cytokinins (Absmanner et al. 2013; Baldacci-Cresp et al. 2020). These results suggest that the ratio of auxins to cytokinins is associated with phloem differentiation, as observed in the root-knot nematode feeding sites (Absmanner et al. 2013; Baldacci-Cresp et al. 2020). Furthermore, Knypl and Janas (1976) state that chlorogenic acids inhibit IAA-oxidase and cause an increase in auxin in gall developmental sites.

Mineral accumulation in phloem cells may also occur, as discussed for aluminum in the galls induced by *Ditylenchus gallaeformans* on leaves of *Miconia lacunosa* (Melastomataceae) (Arriola et al. 2020). The histochemical detection of aluminum both in nutritive cells and in the nematode bodies evidenced its translocation by phloem cells, and the bioaccumulating potential of the nematodes (Arriola et al. 2020).

The detection of aluminum in the phloem cell walls is concomitant with pectins, such as the lateral chains of (1 → 4) β-D-galactans linked to RG-I and xyloglucans, detected by LM5 and LM15, respectively, in *Ditylenchus gallaeformans*-induced galls on *M. albicans* and *M. ibaguensis* (Ferreira et al. 2020). The dynamics of galactans and xyloglucans is important in maintaining cell wall structure and occur predominantly in primary cell walls (Marcus et al. 2008). Nevertheless, the functional implications of this dynamics have not been explored for phloem cells in nematode galls.

Changes in the patterns of gene expression regulating phloem development in nematode galls (Baldacci-Cresp et al. 2020), in which, *LBD1* is involved in the formation of secondary phloem ray cells, and *KANADI1* and *KANADI2,* encode positive regulators of phloem formation by regulation of auxin flow (Bartlem et al. 2014; Baldacci-Cresp et al. 2020). To analyze symplastic connectivity between the root phloem and syncytia in nematode-infected *Arabidopsis thaliana* plants, different markers were observed for each cell type. The promoter of the AtSUC2 gene is active exclusively in companion cells (Hoth et al. 2005), and RS6 sieve element marker is detected in the protophloem (Absmanner et al. 2013).

Structural analyses of tumors induced by virus (RBSDV, genus *Fijivirus*) infection in maize plants indicate that they predominantly induce hyperplasia of phloem. Amorphous, electron-dense cytoplasmic bodies with abundant RBSDV virions, specific plastids and the absence of vacuoles are observed in sieve elements as diagnostic features of the virus infection (Shen et al. 2016).

4 Conclusion

As the plant tissue responsible for the translocation of molecules linked to gall development, the phloem exhibits cell proliferation, hypertrophy, and atypical development in response to the gall inducer stimuli facilitating the nutritive support both for the gall structure and the inducers, which has been best explored for hemipteran and nematode galls. The neoformation of vascular bundles and well-developed vascular parenchyma are the most common responses to sap-feeding galling hemiptera. In the case of nematode galls, the striking anatomical feature is the formation of an extensive network of vascular tissues that encases the cluster of the typical giant cells. All these processes involve the association of cytokinins and auxins, which are determinant modulators of phloem development in galls. Accordingly, much is yet to be explored regarding the anatomical changes and implications for the phloem development in galls induced by bacteria, virus and fungi, and approaches on immunocytochemistry, genetics, and cytology for galls of all inducers.

References

Absmanner B, Stadler R, Hammes UZ (2013) Phloem development in nematode-induced feeding sites: the implications of auxin and cytokinin. Front Plant Sci 8:241. https://doi.org/10.3389/fpls.2013.00241

Albert S, Padhiar A, Gandhi D (2011) Morphological, anatomical and biochemical studies on the foliar galls of *Alstonia scholaris* (Apocynaceae). Braz J Bot 34:343–358. https://doi.org/10.1590/S0100-84042011000300009

Aloni R (2021) vascular differentiation and plant hormones. ISBN:978-3-030-53201-7. https://doi.org/10.1007/978-3-030-53202-4

Aloni R, Ullrich CI (2008) Biology of crown gall tumors. In: Tzfira T, Citovsky V (eds) Agrobacterium: from biology to biotechnology. Springer, New York. https://doi.org/10.1007/978-0-387-72290-0_15

Aloni R, Pradel KS, Ullrich CI (1995) The three-dimensional structure of vascular tissues in *Agrobacterium tumefaciens*-induced crown galls and in the host stems of *Ricinus communis* L. Planta 196:597–605. https://doi.org/10.1007/BF00203661

Álvarez R, Encina A, Hidalgo NP (2009) Histological aspects of three *Pistacia terebinthus* galls induced by three different aphids: *Paracletus cimiciformis*, *Forda marginata* and *Forda formicaria*. Plant Sci 176:303–314. https://doi.org/10.1016/j.plantsci.2008.11.006

Álvarez R, Molist P, González-Sierra S, Martinez JJ, Nafría JM (2014) The histo structure of galls induced by aphids as a useful taxonomic character: the case of *Rectinasus* (Hemiptera, Aphididae, Eriosomatinae). Zootaxa:487–492. https://doi.org/10.11646/zootaxa.3861.5.6

Álvarez R, Martinez JJ, Muñoz-Viveros AL, Molist P, Abad-González J, Nieto Nafría JM (2016) Contribution of gall microscopic structure to taxonomy of gallicolous aphids on *Pistacia*. Plant Biol (Stuttg) 18:868–875. https://doi.org/10.1111/plb.12475

Álvarez R, Moreno-González V, Martinez JJI, Ferreira BG, Hidalgo NP (2020) Microscopic study of nine galls induced in *Populus nigra* by aphids of the Iberian Peninsula. Arthropod Plant Interact 14:799–809. https://doi.org/10.1007/s11829-020-09778-1

Amorim DO, Ferreira BG, Fleury G (2017) Plant potentialities determine anatomical and histochemical diversity in *Mikania glomerata* Spreng. galls. Braz J Bot 40:517–527. https://doi.org/10.1007/s40415-016-0357-9

Arriola IA, Figueiredo MA, Boanares D, Costa França MG, Isaias RMS (2020) Apoplast-symplast compartmentalization and functional traits of iron and aluminum in promeristematic tissues of nematode induced galls on *Miconia* spp. Plant Physiol Bioch 154:360–368. https://doi.org/10.1016/j.plaphy.2020.06.031

Azam T, Hisamuddin Singh S, Robab MI (2011) Effect of different inoculum levels of *Meloidogyne incognita* on growth and yield of *Lycopersicon esculentum*, and internal structure of infected root. Arch Phytopathol Pflanzenschutz 44:1829–1839. https://doi.org/10.1080/03235400802678113

Baldacci-Cresp F, Behr M, Kohler A, Badalato N, Morreel K, Goeminne GMA, Engler J, Boerjan W, El Jaziri M, Baucher M (2020) Molecular changes concomitant with vascular system development in mature galls induced by root-knot nematodes in the model tree host *Populus tremula* × *P. alba*. Int J Mol Sci 21:406. https://doi.org/10.3390/ijms21020406

Bartlem DG, Jones MG, Hammes UZ (2014) Vascularization and nutrient delivery at root-knot nematode feeding sites in host roots. J Exp Bot 65:1789–1798. https://doi.org/10.1093/jxb/ert415

Bedetti CS, Modolo LV, Isaias RMS (2014) The role of phenolics in the control of auxin in galls of *Piptadenia gonoacantha* (Mart.) MacBr. (Fabaceae: Mimisoidae). Biochem Syst Ecol 55:53–59. https://doi.org/10.1016/j.bse.2014.02.016

Bedetti CS, Bragança GP, Isaias RMS (2017) Influence of auxin and phenolic accumulation on the patterns of cell differentiation in distinct gall morphotypes on *Piptadenia gonoacantha* (Fabaceae). Aust J Bot 65:411. https://doi.org/10.1071/bt16257

Behnke HD, Sukkri B (1971) Anastomoses in the internode of Dioscorea: their frequency, structure, and function. Z Pflanzenphysiol 66:82–92. https://doi.org/10.1016/s0044-328x(71)80011-9

Best VM, Vasanthakumar A, McManus PS (2004) Anatomy of cranberry stem gall and localization of bacteria in galls. Phytopathology 94:1172–1177. https://doi.org/10.1094/PHYTO.2004.94.11.1172

Boyer MG (1967) The relation of growth regulators to the development of symptoms and the expression of stem resistance in white pine infected with blister rust. Can J Bot 45:501–513. https://doi.org/10.1139/b67-051

Bragança GPP, Freitas MC, Isaias RM (2020a) The influence of gall position over xylem features in leaflets of *Inga ingoides* (Rich.) Willd. (Fabaceae: Caesalpinioideae). Trees 35:199–209. https://doi.org/10.1007/s00468-020-02027-1

Bragança GPP, Alencar CF, Freitas MSC, Isaias RMS (2020b) Hemicelluloses and associated compounds determine gall functional traits. Plant Biol (Stuttg) 22:981–991. https://doi.org/10.1111/plb.13151. PMID: 32597563

Bronner R (1992) The role of nutritive cells in the nutrition of cynipids and cecidomyiids. In: Shorthouse JD, Rohfritsch O (eds) Biology of insect-induced galls. Oxford University Press, New York, pp 118–140

Brown MW, Glenn DM, Wisniewski ME (1991) Functional and anatomical disruption of apple roots by the woolly apple aphid (Homoptera: Aphididae). J Econ Entomol 84:1823–1826. https://doi.org/10.1093/jee/84.6.1823

Carneiro RG, Pacheco P, Isaias RM (2015) Could the extended phenotype extend to the cellular and subcellular levels in insect-induced galls? PLoS One 10:e0129331. https://doi.org/10.1371/journal.pone.0129331

Chauhan SVS, Nisha S, Shyam VSC (2020) Morphological studies of insect-induced galls in flower and fruit of *Alstonia scholaris* (L.) R. Br Biol Sci 90:705–712. https://doi.org/10.1007/s40011-019-01139-0

Chen J, Qiao GX (2012) Galling Aphids (Hemiptera: Aphidoidea) in China: diversity and host specificity. Psyche J Entomol 2012:1–11. https://doi.org/10.1155/2012/621934

Costa EC, Carneiro RGS, Silva JS, Isaias RMS (2018) Biology and development of galls induced by *Lopesia* sp. (Diptera: Cecidomyiidae) on leaves of *Mimosa gemmulata* (Leguminosae: Caesalpinioideae). Aust J Bot 66:161. https://doi.org/10.1071/bt17099

Danish M, Altaf M, Robab MI, Shahid M, Manoharadas S, Hussain SA, Shaikh H (2021) Green synthesized silver nanoparticles mitigate biotic stress induced by *Meloidogyne incognita* in *Trachyspermum ammi* (L.) by improving growth. Biochemical, and Antioxidant Enzyme Activities. ACS Omega 6:11389–11403. https://doi.org/10.1021/acsomega.1c00375

Dorchin N, Freidberg A, Aloni R (2002) Morphogenesis of stem gall tissues induced by larvae of two cecidomyiid species (Diptera: Cecidomyiidae) on *Suaeda monoica* (Chenopodiaceae). Can J Bot 80:1141–1150. https://doi.org/10.1139/b02-104F

Engler JA, Siqueira KMS, Cabral D, Costa T, Engler G (2016) A cellular outlook of galls induced by root-knot nematodes in the model host *Arabidopsis thaliana*. Lit Rev Nematoda 3:e 062016. https://doi.org/10.4322/nematoda.00616

Ferreira BG, Carneiro RG, Isaias RM (2015) Multivesicular bodies differentiate exclusively in nutritive fast-dividing cells in *Marcetia taxifolia* galls. Protoplasma 252:1275–1283. https://doi.org/10.1007/s00709-015-0759-8

Ferreira BG, Alvarez R, Avritzer SC, Isaias RMS (2017) Revisiting the histological patterns of storage tissues: beyond the limits of gall-inducing taxa. Botany 95:173–184. https://doi.org/10.1139/cjb-2016-0189

Ferreira BG, Álvarez R, Bragança GPP, Alvarenga DR, Pérez-Hidalgo N, Isaias RMS (2019a) Feeding and other gall facets: Patterns and determinants in gall structure. Bot Rev The 85:78–106. https://doi.org/10.1007/s12229-019-09207-w

Ferreira BG, Freitas MSC, Bragança GP, Moreira ASFP, Carneiro RGS, Isaias RMS (2019b) Enzyme-mediated metabolism in nutritive tissues of galls induced by *Ditylenchus gallaeformans* (Nematoda: Anguinidae). Plant Biol (Stuttg) 21:1052–1062. https://doi.org/10.1111/plb.13009

Ferreira BG, Bragança GP, Isaias RMS (2020) Cytological attributes of storage tissues in nematode and eriophyid galls: pectin and hemicellulose functional insights. Protoplasma 257:229–244. https://doi.org/10.1007/s00709-019-01431-w

Global Advanced Research Journal of Microbiology (2014) 3(3): 059–063. ISSN: 2315-5116

Guedes LM, Aguilera N, Ferreira BG, Becerra J, Saez K, Péres C, Hernandez VJ, Isaias RMS (2018a) Factors influencing the morphogenesis of galls induced by *Calophya mammifex* (Calophyidae) on *Schinus polygama* (Anacardiaceae) leaves. Botany 96:9–599. https://doi.org/10.1139/cjb-2018-0078

Guedes LM, Aguilera N, Ferreira BG, Becerra J, Hernández V, Isaias RMS (2018b) Anatomical and phenological implications of the relationship between *Schinus polygama* (Cav.) (Cabrera) and the galling insect *Calophya rubra* (Blanchard). Plant Biol (Stuttg) 20:507–515. https://doi.org/10.1111/plb.12696

Guedes LM, Costa EC, Isaias RMS, Sáez-Carillo K, Aguilera N (2023) Structural patterns of Lepidoptera galls and the case of *Andescecidium parrai* (Cecidosidae) galls on *Schinus polygama* (Anacardiaceae). J Plant Res 136:715–728. https://doi.org/10.1007/s10265-023-01472-6

Gullan PJ, Martin JH (2003) Sternorrhyncha (jumping plant-lice, whiteflies, aphids, and scale insects). In: Resh VH, Cardé RT (eds) Encyclopedia of insects. Academic, Amsterdan, pp 1078–1089

Hao Z, Lingjun F-HW, Hongxu Z (2020) Comparative analysis of stylet penetration behaviors of *Eriosoma lanigerum* (Hemiptera: Aphididae) on main apple cultivars in China. J Econ Entomol 113:1761–1767. https://doi.org/10.1093/jee/toaa085

Hatta T, Francki RIB (1976) Anatomy of virus-induced galls on leaves of sugarcane infected with Fiji disease virus and the cellular distribution of virus particles. Physiol Plant Pathal 9:321–330. https://doi.org/10.1016/0048-4059(76)90065-5

Hori K (1992) Insect secretions and their effect on plant growth, with special reference to hemipterans. In: Shorthouse JD, Rohfritsch O (eds) Biology of insect-induced galls. Oxford University Press, New York, pp 157–170

Hoth S, Schneidereit A, Lauterbach C, Scholz-Starke J, Sauer N (2005) Nematode infection triggers the de novo formation of unloading phloem that allows macromolecular trafficking of green fluorescent protein into syncytia. Plant Physiol 138:383–392. https://doi.org/10.1104/pp.104.058800

Hussein AE, Siddig MAE, Yagi SMA, Abdalla AWHA (2014) Nutrients accumulation and vascularization of *Agrobacterium* Tumours induced on *Launaea capitata* Spreng. Plant Glob Adv Res J Microbiol 3:059–063

Isaias RMS, Oliveira DC, Moreira ASFP, Soares GLG, Carneiro RGS (2015) The imbalance of redox homeostasis in arthropod-induced plant galls: mechanisms of stress generation and dissipation. Biochim Biophys Acta 1850:1509–1517. https://doi.org/10.1016/j.bbagen.2015.03.007

Jones MG, Northcote DH (1972) Multinucleate transfer cells induced in coleus roots by the root-knot nematode *Meloidogyne arenaria*. Protoplasma 75:381–395

Jorge NC, Freitas MSC, Caffaro RM, Vale FHA, Lemos-Filho JP, Isaias RMS (2022) Vascular traits of stem galls: cell increment versus morphogenetic constraints in wood anatomy. Plant Biol. ISSN:1435-8603. https://doi.org/10.1111/plb.13392

Knypl JS, Janas KM (1976) Is decreased activity of nitrate reductase in roots of carrot infested with northern Root-Knot nematode related to the increased level of chlorogenic acid? Biochem Physiol Pflanz 169:607–615. https://doi.org/10.1016/s0015-3796(17)30945-9

Knypl JS, Chylinska KM, Brzeski MW (1975) Increased level of chlorogenic acid and inhibitors of indolyl-3-acetic acid oxidase in roots of carrot infested with the northern root-knot nematode. Physiol Plant Pathol 6:51–64. https://doi.org/10.1016/0048-4059(75)90104-6

Kumar PL (2016) Ultrastructural and biochemical studies on insect induced leaf galls in *Alstonia scholaris*. Int J Plant Biol 8:64–81. https://doi.org/10.14787/ijprb.2016

Lev-Yadun S (2003) Stem cells in plants are differentiated too. CurrentOpinion in Plant Biology 4:93–100. https://doi.org/10.1016/j.cell.2005.08.006

Mani MS (1964) Ecology of plant galls. Dr. W. Junk Publishers, The Hague

Mapes CC, Davies PJ (2001) Indole-3-acetic acid and ball gall development on *Solidago altissima*. New Phytol 151:195–202. https://doi.org/10.1046/j.1469-8137.2001.00161.x

Marcus SE, Verhertbruggen Y, Hervé C, Ordaz-Ortiz JJ, Farkas V, Pedersen HL, Knox JP (2008) Pectic homogalacturonan masks abundant sets of xyloglucan epitopes in plant cell walls. Plant Biol 8:60. https://doi.org/10.1186/1471-2229-8-60

Martini VC, Moreira ASFP, Fuzaro L, Oliveira LAG, Oliveira DC (2020) *Pseudophacopteron longicaudatum* (Hemiptera) induces intralaminar leaf galls on *Aspidosperma tomentosum* (Apocynaceae): a qualitative and quantitative structural overview. An Acad Bras Ciênc 92:e20181002

Meyer J (1987) Plant galls and gall inducers. Gerbrüder Borntraeger, Berlin

Meyer J, Maresquelle HJ (1983) Anatomie des galles. Encyclopedia of plant anatomy, Band XIII, xvi, vol 662

Muñoz-Viveros AL, Martinez JJI, Molist P, González-Sierra S, Julián PG, Álvarez R (2014) Microscopic study of galls induced by three species of *Geopemphigus* (Hemiptera: Aphididae: Eriosomatinae) on *Pistacia mexicana*. Arthropod Plant Interact 8:531–538. https://doi.org/10.1007/s11829-014-9333-0

Oliveira DC, Isaias RMS (2010) Cytological and histochemical gradients induced by a sucking insect in galls of *Aspidosperma australe Arg*. Muell (Apocynaceae). Plant Sci 178:350–358. https://doi.org/10.1016/j.plantsci.2010.02.002

Oliveira DC, Christiano JDCS, Soares GLG, Isaias RMS (2006) Reações de defesas químicas e estruturais de *Lonchocarpus muehlbergianus* Hassl. (Fabaceae) à ação do galhador *Euphalerus ostreoides* Crawf. (Hemiptera: Psyllidae). Rev Bras Bot 29:657–667. https://doi.org/10.1590/S0100-84042006000400015

Oliveira DC, Isaias RMS, Fernandes GW, Ferreira BG, Carneiro RGS, Fuzaro L (2016) Manipulation of host plant cells and tissues by gall-inducing insects and adaptive strategies used by different feeding guilds. J Insect Physiol 84:103–113. https://doi.org/10.1016/j.jinsphys.2015.11

Oliveira DC, Burckhardt D, Calácio TF, Kuster VC, Queiroz DL (2019) *Ceropsylla pouteriae* Burckhardt sp. nov. (Hemiptera: Psylloidea: Triozidae), a new species of jumping plant-louse inducing galls on the leaves of *Pouteria ramiflora* (Mart.) Radlk. (Sapotaceae): taxonomy, gall structure and histochemistry. J Nat Hist 53:1923–1950. https://doi.org/10.1080/00222933.2019.1676931

Pradel KS, Ullrich CI, Cruz SS, Oparka KJ (1999) Symplastic continuity in *Agrobacterium tumefaciens*-induced tumours. J Exp Bot 50:183–192. https://doi.org/10.1093/jxb/50.331.183

Prado E, Tjallingii WF (1994) Aphid activities during sieve element punctures. Entomologia Experimentalis et Applicata 72(2):157–165

Richardson RA, Body M, Warmund MR, Schultz JC, Appel HM (2016) Morphometric analysis of young petiole galls on the narrow-leaf cottonwood, *Populus angustifolia*, by the sugarbeet root aphid, *Pemphigus betae*. Protoplasma 254:203–216. https://doi.org/10.1007/s00709-015-0937-8

Robab MI, Hisamuddin AT, Azam T (2010) Histopathology of roots of *Glycine max* (L.) Merrill induced by root-knot nematode (*Meloidogyne incognita*). Arch Phytopathol Plant Protect 43:1758–1767. https://doi.org/10.1080/03235400802678360

Rohfritsch O, Anthony M (1992) Strategies on gall induction by two groups of homopterans. In: Shorthouse JD, Rohfristsch O (eds) Biology of insect-induced galls. Oxford University Press, New York

Sandanayaka WRM, Hale CN (2003) Electronically monitored stylet penetration pathway of woolly apple aphid, *Eriosoma lanigerum* (Homoptera: Aphididae), on apple (*Malus domestica*). N Z J Crop Hortic Sci 31(2):107–113

Sandanayaka WRM, Bus VGM, Connolly P, Newcomb R (2003) Characteristics associated with woolly apple aphid *Eriosoma lanigerum*, resistance of three apple rootstocks. Entomol Exp Appl 109:63–72, 472

Shen J, Chen X, Chen J, Sun L (2016) A phloem-limited fijivirus induces the formation of neoplastic phloem tissues that house virus multiplication in the host plant. Sci Rep 19:29848. https://doi.org/10.1038/srep29848

Silva AFM, Kuster VC, Rezende UC, Oliveira DC (2019) The early developmental stages of gall-inducing insects define final gall structural and histochemical profiles: the case of *Bystracoccus mataybae* galls on *Matayba guianensis*. Botany 97:427–438. https://doi.org/10.1139/cjb-2019-0017

Singh R, Saxena R (2012) Histopathological tudies of *Meloidogyne Incognita* infected roots of *Luffa Cylindrica* Singh & Saxena. Int J Biotech Biosci 1:153–160

Stone GN, Schönrogge K (2003) The adaptive significance of insect gall morphology. Trends Ecol Evol 18:512–522

Suzuki DK, Yukiharu F, Shin-ichi A (2009) Do aphid galls provide good nutrients for the aphids?: Comparisons of amino acid concentrations in galls among *Tetraneura* species (Aphididae: Eriosomatinae). Arthropod Plant Interact 3:241–247. https://doi.org/10.1007/s11829-009-9064-9

Suzuki AYM, Bedetti CS, Isaias RMS (2015) Detection and distribution of cell growth regulators and cellulose microfibrils during the development of *Lopesia* sp. galls on *Lonchocarpus cultratus* (Fabaceae). Botany 93:435–444. https://doi.org/10.1139/cjb-2015-0012

Teixeira CT, Oliveira DC, Kuster VC, Isaias RMS (2017) Immunocytochemical demonstration of cell wall components related to tissue compartments in the globoid galls induced by *Clinodiplosis* sp. (Cecidomyiidae) on *Croton floribundus* Spreng. (Euphorbiaceae). Botany 96:9–18. https://doi.org/10.1139/cjb-2017-0123

Tordable MDC, Andrade AJ, Doucet ME, Lax P (2018) Histopathology of Andean Potato (*Solanum tuberosum* Andigenum group) varieties parasitized by the false root-knot nematode, Nacobbus aberrans. Braz J Biol 78:679–685. https://doi.org/10.1590/1519-6984.172401

Ullrich CI, Aloni R (2000) Vascularization is a general requirement for growth of plant and animal tumours. J Exp Bot 51:1951–1960. https://doi.org/10.1093/jexbot/51.353.1951

Walerowski P, Gündel A, Yahaya N, Truman W, Sobczak M, Olszak M, Rolfe S, Borisjuk L, Malinowski R (2018) Clubroot disease stimulates early steps of phloem differentiation and recruits SWEET sucrose transporters within developing galls. Plant Cell 30:3058–3073. https://doi.org/10.1105/tpc.18.00283

Starch—Histolocalization and Metabolism in Plant Galls

Marian J. Giertych

1 Introduction

Starch is a polysaccharide composed of glucose molecules chemically linked by an alpha glycosidic bond (Bawn et al. 1940). It is the primary storage substance in vascular plants and is synthesized as a result of the photosynthetic activity of chloroplasts and originally constitutes the cell's energy reserve for the time when the photosynthesis process does not take place, i.e. for the period of darkness (Loescher et al. 1990). Some carbohydrates produced in chloroplasts are used for the energy needs of the cell, while the excess is transported to other plant cells, tissues, and organs, where they are used or stored in the form of granules inside specialized organelles, the so-called amyloplasts (Martin and Smith 1995). Starch is an insoluble carbohydrate. Therefore, it does not disturb the osmotic balance of cells, but due to its insolubility, it cannot be directly transported in the plant (Robyt 2008). Therefore, wherever it is produced or used, there must be a complex, sometimes insufficiently recognized, enzymatic system catalyzing the synthesis and decomposition of starch (Goren et al. 2018). The formation of starch granules is multi-stage and the following processes can be distinguished: precursor formation, granule initiation, amylopectin elongation, amylopectin branching, amylose synthesis, debranching, phosphorylation, and dephosphorylation; each of these processes involving several enzymes (Goren et al. 2018). Starch accumulates primarily in chloroplasts and then in many types of storage tissues located in various organs: roots, tubers, rhizomes, and seeds (Zeeman et al. 2010).

M. J. Giertych (✉)
Institute of Dendrology PAS, Kórnik, Poland

Institute of Biological Sciences, University of Zielona Góra, Zielona Góra, Poland
e-mail: giertych@man.poznan.pl

D. C. de Oliveira, R. M. dos Santos Isaias (eds.), *Plant Galls*,
https://doi.org/10.1007/978-3-031-80064-1_6

The metabolic pathway for starch formation is poorly understood in many respects, especially in plants, as well as in gall tissues. The synthesis of starch requires the presence of its soluble precursor ADP-glucose (Goren et al. 2018) and the formation of a starch chain takes place through the addition of subsequent glucose units, which is supported by enzymes generally called starch synthases (Smith 2001). Studies of starch synthesis in galls are scarce, but it is known that the genes responsible for its synthesis in root galls of phylloxera (*Daktulosphaira vitifoliae*) are up-regulated (Griesser et al. 2015). The formation of starch granules requires enzymes responsible for its branching and debranching (Han et al. 2007), yet the presence of these enzymes in gall tissues has not been demonstrated so far, even though starch granules have been described many times.

The breakdown of starch in plant tissues especially leaves and seeds has been quite well-known (Zeeman et al. 2010) due to the utilization of model plants, which has enabled the establishment of a tentative pathway for starch breakdown in chloroplasts, involving primarily hydrolysis to maltose and subsequent transport of maltose to the cytosol (Zeeman et al. 2004). Many enzymes are active in this pathway: glucan, water dikinase, alpha-amylase, and beta-amylase, occurring in various isoforms (Zeeman et al. 2010), some of them were found in gall tissues (Dsouza and Ravishankar 2014; Rohfritsch and Shorthouse 1982). Herein, we revisited and discussed the presence of starch granules in various gall systems, as well as their significance in gall metabolism.

2 Starch Seems to Be Commonly Detected in a Huge Diversity of Galls

Starch has been found in the parenchyma cells of many galls. The history of its examination in galls dates back to the mid-nineteenth century, when, during the analysis of industrially used tannins, starch granules were found in shoot galls induced by species of the Cynipidae family (Fagan 1918). Starch was also found in the galls of many species of insects belonging to several orders: Coleoptera (Murakami et al. 2021), Diptera (Pilichowski and Giertych 2020), Hemiptera (Dardeau et al. 2014; Raman 1991), Hymenoptera (Jankiewicz et al. 2021), Lepidoptera (Ferreira and Isaias 2013), as well as in the galls induced by mites Trombidiformes (Ferreira et al. 2017) or Nematodes (Ferreira et al. 2019). Starch was found in galls developed on various plant organs: the buds (Aguilera et al. 2022), roots (Griesser et al. 2015), leaf petioles (Ugolini et al. 2014), and leaves (Guzicka et al. 2017). The common occurrence of starch in galls might lead one to believe that it occurs in all types of galls, but this is not true. Starch, for example, was not detected in galls induced by *Liodiplosis cylindrica*; Cecidomyiidae (Amorim et al. 2017), *Neuroterus batatus;* Cecidomyiidae (Kostoff and Kendall 1929), *Amphibolips michoacaensis*; Cynipidae (Hernández-Soto et al. 2015) or *Lipara lucens*; Chloropidae (Vandevyvere and De Bruyn 1998). However, there is too little data on the lack of starch in galls to make generalizations about the orders of insects whose galls do not contain this carbohydrate.

The presence of starch in galls was usually confirmed by two methods: histochemically by staining with Lugol's solution (Colin et al. 1814; Stromeyer 1815) or colorimetrically (Gur et al. 1969) by determining its content in dry or fresh matter. The starch content in galls estimated colorimetrically is usually low and amounts to a few percentages. The highest content is approximately 12% dm. found in *Adelges cooley,* Adelgidae galls induced on buds of *Picea pungens* (Shannon and Brewer 1980). The starch content in the galls was repeatedly compared to its level in the tissues near the galls and in the non-galled organs of infected plants. Starch usually accumulates in the galls where there is a few percent more of it than in the adjacent tissues (Huang et al. 2015; Murakami et al. 2021; Oliveira et al. 2017; Ugolini et al. 2014). However, this is not a rule, as there are cases in which there is less starch in the galls (Castro et al. 2012; Dsouza and Ravishankar 2014) or the starch content is similar to other tissues (Pilichowski and Giertych 2020). This lack of clear trends may be due to the large natural seasonal variability of starch content in leaf tissues (Meletiouchristou et al. 1994) or in wood (Yoshimura et al. 2016) and its high variability during the day (Smith et al. 2004). The starch content also changes during gall development. Shannon and Brewer (1980) showed a significant increase in starch content in the multi-chambered galls induced by *Adelges cooley* on *Picea pungens*, whereas Rohfritsch (1999) described it in galls induced by *Physemococecis hartigi* on *Tillia intermedia.* However, there are also opposite situations when the amount of starch in the gall decreases as it grows (Staszak et al. 2023). There are very few studies on changes in starch content in galls over time, which unfortunately, makes it difficult to draw more general conclusions. Certainly, apart from the insect species and the host plant, changes in the starch content of the galls will depend on their origin.

3 The Sink of Carbohydrates and Photosynthesis Determine Starch Accumulation in Galls

In plant tissues, the primary site of starch formation is the chloroplasts (Fig. 1a), which are also found in the tissues of many galls developed on leaves, petioles, and non-woody shoots (Carneiro and Isaias 2015b; Ferreira and Isaias 2013; Huang et al. 2019). It could, therefore, be assumed that carbohydrates and starch are synthesized locally in gall tissues (Fig. 1b), and while they may not fulfill the entire energy requirements of galls, they restrict the uptake of carbohydrates from the host plants (Haiden et al. 2012). The chlorophyll content in galls is usually lower compared with non-galled tissues (Carneiro et al. 2014; Oliveira et al. 2011b). Despite that, the galls can photosynthesize (Martini et al. 2020; Oliveira et al. 2011b), the presence of galls may also reduce the chlorophyll content and photosynthetic activity of adjacent tissues (Kot et al. 2018). It is obvious that in addition to the presence

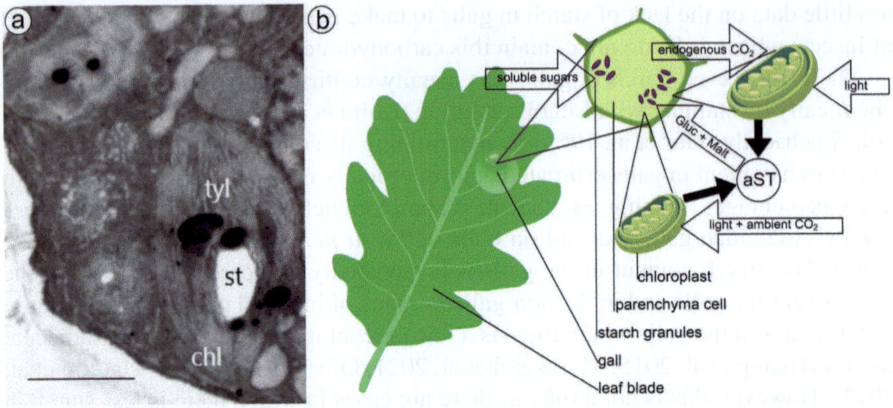

Fig. 1 (**a**) Chloroplast with visible thylakoids and starch granule in the parenchyma tissue of galls induced by *Neruoterus albipes* (Cynipideae) on oak (*Quercus*) leaves. chl chloroplast, tyl thylakoid, st starch granule, TEM, Bar = 2 μm. (Photo M. Guzicka). (**b**) Scheme of the origin of starch in galls. An example of a leaf blade, with a marked gall, an enlarged parenchyma cell and chloroplasts. Dark purple dots – starch granules in amyloplasts; aST - starch briefly accumulated in chloroplasts shortly after CO_2 assimilation; Gluc + Malt - glucose and maltose, i.e. simple sugars that transport assimilated carbon from chloroplasts to the cytoplasm; light + ambient CO_2 – sunlight and carbon dioxide supplied to the gall from the outside; endogenous CO_2 – carbon dioxide from the respiratory processes of larvae and gall tissues; soluble sugars - carbohydrates delivered by conductive bundles from the leaf to the gall

of chloroplasts in plant mesophyll tissues, carbon assimilation requires gas exchange, which brings carbon dioxide into the cells and removes oxygen. In the galls, the situation is the opposite, where, there may be excess CO_2 (hypercarbia) and oxygen deficiency (hypoxia) resulting from respiratory processes and the presence of chloroplasts and photosynthesis within the gall tissues may improve this unfavorable ratio of concentrations of these gases for the larva (Pincebourde and Casas 2016). However, Haiden et al. (2012) have shown that larvae enclosed in galls are resistant to hypercarbia/hypoxia, and even long-term darkening of the galls does not increase their mortality, but only the growth of the galls themselves. The gas exchange process takes place in the leaves, but also in other assimilating organs, using numerous stomata, which may be differentiated in the epidermis (Torii 2015).

The stomata can also be found in the epidermis of galls (Álvarez et al. 2013; Fernandes et al. 2021; Jankiewicz et al. 2021; Liu et al. 2014). This is not a rule, because not all galls have stomata on their surface (Huang et al. 2014) or if present, they can be distorted and have limited functionality (Mani 1964). The limited number, deformations, or complete absence of stomata limit the possibility of gas exchange between ambient air and gall tissues, which is confirmed by lower photosynthetic activity (Castro et al. 2012; Haiden et al. 2012; Inbar et al. 1995; Oliveira et al. 2017). However, there are exceptions to this rule, studies of *Trichilogaster signiventris* galls showed that photosynthesis within galls contributed substantially to the carbon budgets of the galls on *Acacia pycnantha* (Dorchin et al. 2006). The assimilation abilities of galls may be related to their developmental stage. In the early stages of development, *Amphibolips michoacaensis* galls on the leaves on *Quercus castanea*, for example, maybe photosynthetically active (Fig. 1b), and some of the carbon present in the galls is assimilated in situ (Hernández-Soto et al. 2015). In most cases, however, simple carbohydrates directed to starch synthesis must be supplied from other plant parts. The galls are a very strong sink of water and all nutrients, including carbohydrates, which are the main substance that transports carbon in the plant (Haiden et al. 2012).

The presence of vascular bundles penetrating the gall structures has been repeatedly demonstrated (Ferreira et al. 2022; Guzicka et al. 2017; Muñoz-Viveros et al. 2014), and may be involved in carbohydrate transport. In the root galls of *Daktulosphaira vitifoliae,* the transport occurs symplastically (Griesser et al. 2015), but whether this is also the case in leaf galls and whether apoplastic transport occurs in any galls is unknown. A strong sink and transport of carbon-labeled $^{14}CO_2$ assimilates into the galls was demonstrated by Jankiewicz et al. (1970), in which the gall induced by *Cynips quercusfolii* was able to absorb assimilates from all fragments of the leaf blade. Most of the labeled carbon was detected in the larva itself and in the tissues surrounding the larval chamber. Larson and Whitham (1991) showed that *Pemphigus betae* galls generated such a strong sink that the individuals of *P. betae* were able to use assimilates also from neighboring leaves (Fig. 1b). A very strong sink of assimilates is proven by experiments with mechanical damage to the veins of *Quercus petraea* leaves, on which *Cynips quercusfolii* galls occur (Giertych et al. 2023). They showed that cutting the veins near the galls initially reduced their growth, but after a few days the galls began to enlarge again, suggesting that they began to draw resources from the lower rows of veins, which allowed the larvae to complete development.

The origin of the starch in galls can be very remote. For example, research on galls of *Smicronyx madaranus* on the golden dodder *Cuscuta campestris*, an obligate parasitic plant (Murakami et al. 2021) suggests that it may come from the host plant on which the dodder gall is a parasite. It has been shown that the *C. campestris* haustorium penetrates the structures of both the xylem and phloem of the mother plant, from where it can absorb carbohydrates in addition to water and minerals (Shimizu and Aoki 2019). When the golden dodder is in the autotrophic phase, it has a low number of chloroplasts capable of photosynthesize, and the amount of carbohydrates produced is not sufficient to meet the plant's energy needs (Dinelli et al. 1993). In the tissues of galls induced by *Smicronyx madaranus* on the dodder, there are more chloroplasts than in plant tissues, which partially meets the energy needs of the developing larva (Murakami et al. 2021).

4 Starch Metabolism and Histolocalization in Galls

Galls are neoformed structures whose main functions are to feed and protect developing gall-inducers, and the synthesized starch within the galls is related to these purposes. The accumulation of starch in the galls indicates the existence of a set of enzymes in the starch synthesis pathway in the gall cells (Bragança et al. 2017), as well as enzymes that can decompose it (Oliveira et al. 2010). For example, higher activities both of amylase and invertase were found in the tissues of *Pauropsylla depressa* galls than in the non-galled tissues and galled leaves of *Ficus glomerata* (Dsouza and Ravishankar 2014). Comparative studies on oak leaf tissues and galls of several Cynipidae species have shown reduced activity of UDP-glucose pyrophosphorylase in gall cells than in leaf cells, which may be manifested by increased starch content in galls (Pawłowski et al. 2017). The most detailed studies of starch metabolism in galls were carried out for the grapevine pest *Daktulosphaira vitifoliae*, where increased activity of several starch synthesis and degradation enzymes was found in root galls (Griesser et al. 2015). It has also been shown that the genes required for starch biosynthesis e.g. ADP-glucose pyrophosphorylases, and degradation are up-regulated (Forneck et al. 2016). Even though the complete pathway of starch synthesis in galls is not fully known, it has been shown that its synthesis in galls of an unidentified species of Diptera: Cecidomyiidae on *Piper arboreum* takes place, among others, with the participation of sucrose synthase (Bragança et al. 2017). Similarly, galls of *Pseudophacopteron* sp. on *Aspidosperma australe* showed sucrose synthase activity similar in non-galled tissues and in young and mature galls, while in senescent galls, it was not detected (Oliveira and Isaias 2010). The lack of starch in senescent galls may indicate that its resources are being used by the developing nymph.

The analysis of starch distribution in the galls was most often performed using a simple iodine staining reaction, Lugol's solution (Fig. 2a). Most available data on starch distribution concern galls formed by Cecidomyiidae and Cynipidae, but there are also studies covering other groups of insects: Lepidoptera (Rezende et al. 2019) and Hemiptera (Oliveira et al. 2017), mites (Guedes et al. 2023) and nematodes (Ferreira et al. 2019). The distribution of starch in gall tissues (Fig. 3) is not equal and depends on the host plant species, insect species, and type of gall-inducing insect mouthparts. The starch distribution also depends on whether the gall-inducer is alive, because after its death a whole series of rapid metabolic processes take place, disturbing the existing structure of the galls (Bronner 1992). The diversity of gall structure and starch distribution, even within the same insect family and on the same host plant, can be very large (Chao and Liao 2013; Isaias et al. 2013; Pascual-Alvarado et al. 2017). Starch granules are most often found in gall parenchyma forming a centripetal gradient (Fig. 2b), i.e. the closer the gall chamber is, the more parenchyma cells containing starch granules occur (Guzicka et al. 2017; Kostoff and Kendall 1929; Murakami et al. 2021; Oliveira et al. 2011a). Some galls (Fig. 3) contain no starch at all (Aguilera et al. 2022; Amorim et al. 2017; Vandevyvere and De Bruyn 1998), it can be distributed only in the outer parts of the galls (Carneiro

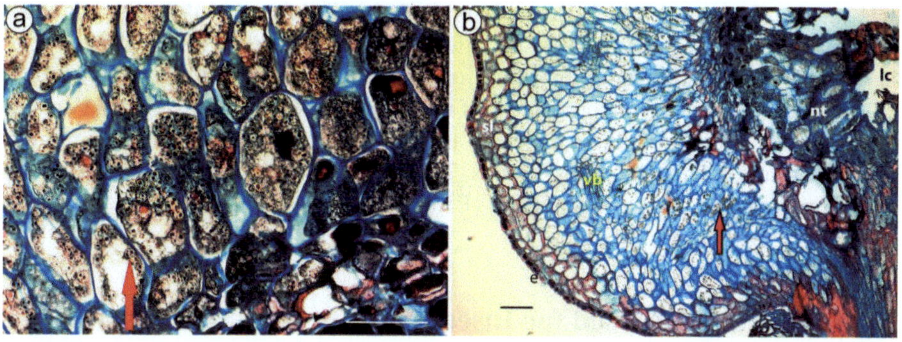

Fig. 2 (**a**) Common storage nutritive tissue of *Cynips quercusfolii* (Cynipideae) gall. The red arrow - plastids with starch grains stained with Lugol's solution, blue colour - cellulose cell walls, LM, Bar = 100 μm. (Photo M. Guzicka). (**b**) Longitudinal section through the *Cynips quercusfolii* (Cynipideae) gall. Single layer, sclerenchymatic epidermis (e), with large vacuoles. An irregular outer ring of sclerenchyma cells (sl) with lignified cell walls (red). Wide layer of parenchyma located between sub-epidermal and inner sclerenchyma layer (cellulose cell walls shown in blue), there are large parenchymatic cells with many amyloplasts with a centripetal gradient of starch. Starch granules are seen in cells in the form of dark spots (red arrow). In parenchyma layer vascular bundles (vb) are visable. In the central part of the gall there is a larval chamber (lc) surrounded by common storage nutritional tissue (nt) also containing starch. LM, Bar = 50 μm. (Photo M. Guzicka)

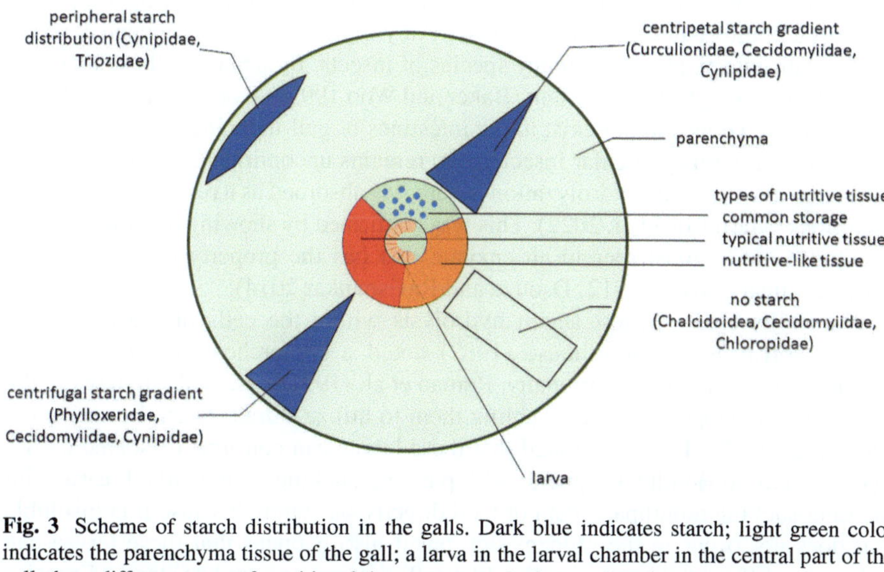

Fig. 3 Scheme of starch distribution in the galls. Dark blue indicates starch; light green color indicates the parenchyma tissue of the gall; a larva in the larval chamber in the central part of the gall, three different types of nutritional tissue were marked around the larval chamber

and Isaias 2015a; Jankiewicz et al. 2021; Shorthouse 1993) or there may form a centrifugal gradient (Bronner 1992; Koncz et al. 2011; Raman 2011). Some authors report that starch is present in the greatest amount in the nutritional layer of galls

(Hernández-Soto et al. 2015; Rezende et al. 2018; Rohfritsch 1999; Staszak et al. 2023). However, it is important to take into account that the term "nutritive tissue" is often too general because the definition itself is not clear and can be defined in various ways, especially in the context of the very diverse structure of galls. Ferreira et al. (2017) distinguish three types of tissues: typical nutritive tissues, common storage tissues and nutritive-like tissues. The starch, if it occurs at all, is most often found only in common storage tissues (Bronner 1992).

5 Starch Can Be Used for Insect's Diet or/and Maintenance of Gall Cell Machinery

The starch storage in galls can be used in four different ways. First, it is for the energy needs of the cells forming the gall, which may be evidenced by a decrease in its amount in the gall growth phase (Dardeau et al. 2014). If starch is in the nutritive tissue, it can serve as food for the growing gall inducer (Costa et al. 2021; Ferreira et al. 2019). This would require mouthparts capable of consuming whole cells with amyloplasts, so it is possible for the chewing (Lepidoptera and Hymenoptera) and scraping–sucking (Diptera: Cecidomyiidae) gall inducers (Ferreira et al. 2017). The use of starch, in addition to the possibility of its absorption, requires the presence of hydrolytic enzymes in the digestive tract of the galling organism allowing the breaking of starch molecules into simple sugars. Amylolytic enzymes have been detected in the gastrointestinal tract of many species of insects, for example, those that feed on seeds with a high starch content (Baker and Woo 1992; Silva et al. 1999). Hence, comparable enzymes might exist in the intestines of gall-inducing insects, yet their presence within this particular insect group remains unconfirmed. The third way to use starch may be by its hydrolyzation before it is absorbed as a result of the insect's secretions (Martinson et al. 2022). This was confirmed by showing that the saliva of Cynipidae larvae must secrete an enzyme that has the property of transforming starch to sugar (Cosens 1912; Dsouza and Ravishankar 2014).

Numerous clues suggest starch hydrolysis within the cells of the nutritional layer. Rohfritsch and Shorthouse (1982) noted a comprehensive array of lytic enzymes in this region. Additionally, Raman et al. (1991) observed amylases in the saliva of Thysanoptera larvae, enabling them to utilize stored starch. Furthermore, Sharma et al. (2014) demonstrated that these larvae can consume dissolved carbohydrates akin to Hemiptera species with piercing-sucking mouthparts. Larvae with piercing-sucking mouthparts cannot feed directly on starch, because it is insoluble and does not occur in the phloem sap, but it also happens that these larvae can induce the differentiation of nutritive-like cells that can accumulate starch (Carneiro and Isaias 2015a). The fact that in many species neither lipids nor starch usually occur in nutritive tissues indicates that the larvae consume simple substances rather than starch (Amorim et al. 2017). A special case of starch use occurs in ambrosia galls, where starch is necessary for the development of specific fungi, the mycelium of which lines the larval chamber and the larva feeds by puncturing the protein-rich mycelium cells (Chao and Liao 2013; Rohfritsch 2008), then the larvae indirectly absorb carbohydrates metabolized by the fungi.

Starch stored in the gall tissues can be directly or indirectly used by the inquilines that inhabit and feed on the gall tissues (Brooks and Shorthouse 1998; Caltagirone 1964). Inquiline larvae of *Periclistis* sp. living in galls induced by *Diplolepis* sp. create their own larval chambers and feed similarly to gall-inducer larvae, taking liquid sap (Shorthouse 1998). Therefore, they can only indirectly use starch. The chewing inquiline larvae of *Pammene amygdalana*, living in the knopper galls of *Andricus quercuscalicis*, have mouthparts adapted to eating solid food and feed on the parenchyma tissue of the galls (Schönrogge et al. 1995), and because the tissues of these galls contain starch (Harper et al. 2004), they can also absorb it directly. However, in most cases, the biology of these species is very poorly understood, and it is not known how they use the substances accumulated in the gall tissues.

6 Conclusions

The presence of starch in galls is quite common having been detected by many researchers in a huge diversity of gall morphotypes induced by different groups of gall-inducing insects. There are many indications that the substrates for its production come from outside the gall tissues by the sink identity of the neoformed structure. Despite that, it is possible that some galls with chloroplasts in their parenchyma can locally synthesize starch granules. The starch content in the galls is not equal and its distribution is uneven, which may indicate the importance of this carbohydrate for the development of the gall-inducing larvae and also for gall growth and metabolism. How the larvae use starch is poorly understood, as so is the use of these resources by inquilines, which are abundant in some galls.

References

Aguilera N, Isaias RMS, Jorge NC, Conejeros MJ, Becerra J, Nieves-Aldrey JL, Guedes LM (2022) Distinctive anatomical and histochemical responses of *Nothofagus obliqua* (Mirb.) Oerst (Nothofagaceae) to two galling Pteromalidae (Hymenoptera: Chalcidoidea) in Chile. Flora 290:152056. https://doi.org/10.1016/j.flora.2022.152056

Álvarez R, González-Sierra S, Candelas A, Martinez J-JI (2013) Histological study of galls induced by aphids on leaves of *Ulmus minor*: *Tetraneura ulmi* induces globose galls and *Eriosoma ulmi* induces pseudogalls. Arthropod Plant Interact 7:643–650. https://doi.org/10.1007/s11829-013-9278-8

Amorim DO, Ferreira BG, Fleury G (2017) Plant potentialities determine anatomical and histochemical diversity in *Mikania glomerata* Spreng. galls. Rev Bras Bot 40:517–527. https://doi.org/10.1007/s40415-016-0357-9

Baker JE, Woo SM (1992) Digestion of starch granules by alpha-amylases from the rice weevil, *Sitophilus-oryzae* - effect of starch type, fat extraction, granule size, mechanical damage, and detergent treatment. Insect Biochem Mol Biol 22:529–537. https://doi.org/10.1016/0965-1748(92)90029-e
Bawn CEH, Hirst EL, Young GT (1940) The nature of the bonds in starch. Trans Faraday Soc 36:880–885. https://doi.org/10.1039/TF9403600880
Bragança GP, Oliveira D, Isaias R (2017) Compartmentalization of metabolites and enzymatic mediation in nutritive cells of Cecidomyiidae galls on *Piper arboreum* Aubl. (Piperaceae). J Plant Stud 6:11
Bronner N (1992) The role of nutritive cells in the nutrition of Cynipids and Cecidomyiids. In: Biology of insect-induced galls, pp 118–140
Brooks SE, Shorthouse JD (1998) Developmental morphology of stem galls of *Diplolepis nodulosa* (Hymenoptera: Cynipidae) and those modified by the inquiline *Periclistus pirata* (Hymenoptera: Cynipidae) on *Rosa blanda* (Rosaceae). Can J Botany-Revue Canadienne De Botanique 76:365–381. https://doi.org/10.1139/b98-001
Caltagirone L (1964) Notes on the biology, parasites, and inquilines of *Pontania pacifica* (Hymenoptera: Tenthredinidae), a leaf-gall incitant on *Salix lasiolepis*. Ann Entomol Soc Am 57:279–291
Carneiro RGD, Isaias RMS (2015a) Gradients of metabolite accumulation and redifferentiation of nutritive cells associated with vascular tissues in galls induced by sucking insects. Aob Plants 7. https://doi.org/10.1093/aobpla/plv086
Carneiro RGS, Isaias RMS (2015b) Cytological cycles and fates in *Psidium myrtoides* are altered towards new cell metabolism and functionalities by the galling activity of *Nothotrioza myrtoidis*. Protoplasma 252:637–646. https://doi.org/10.1007/s00709-014-0709-x
Carneiro RGS, Castro AC, Isaias RMS (2014) Unique histochemical gradients in a photosynthesis-deficient plant gall. S Afr J Bot 92:97–104. https://doi.org/10.1016/j.sajb.2014.02.011
Castro AC, Oliveira DC, Moreira ASFP, Lemos JP, Isaias RMS (2012) Source-sink relationship and photosynthesis in the horn-shaped gall and its host plant *Copaifera langsdorffii* Desf. (Fabaceae). S Afr J Bot 83:121–126. https://doi.org/10.1016/j.sajb.2012.08.007
Chao JF, Liao GI (2013) Histocytological aspects of four types of ambrosia galls on *Machilus zuihoensis* Hayata (Lauraceae). Flora 208:157–164. https://doi.org/10.1016/j.flora.2013.01.001
Colin J-J, Gaultier de Claubry H-Fo, Perronneau VdH-L (1814) Mémoire sur les combinaisons de l'iode avec les substances végétales et animales. Par MM. Colin et H. Gaultier de Claubry. Lu à la Première Classe de l'Institut le 21 mars 1814
Cosens A (1912) A contribution to the morphology and biology of insect galls. University Press, Toronto
Costa EC, Oliveira DC, Ferreira DKL, Isaias RMS (2021) Structural and nutritional peculiarities related to lifespan differences on four *Lopesia* induced bivalve-shaped galls on the single super-host *Mimosa gemmulata*. Front Plant Sci 12. https://doi.org/10.3389/fpls.2021.660557
Dardeau F, Deprost E, Laurans F, Lainé V, Lieutier F, Sallé A (2014) Resistant poplar genotypes inhibit pseudogall formation by the wooly poplar aphid, *Phloeomyzus passerinii* Sign. Trees-Struct Funct 28:1007–1019. https://doi.org/10.1007/s00468-014-1014-1
Dinelli G, Bonetti A, Tibiletti E (1993) Photosynthetic and accessory pigments in *Cuscuta campestris* Yuncker and some host species. Weed Res 33:253–260. https://doi.org/10.1111/j.1365-3180.1993.tb01939.x
Dorchin N, Cramer MD, Hoffmann JH (2006) Photosynthesis and sink activity of wasp-induced galls in *Acacia pycnantha*. Ecology 87:1781–1791. https://doi.org/10.1890/0012-9658(2006)87[1781:Pasaow]2.0.Co;2
Dsouza MR, Ravishankar BE (2014) Nutritional sink formation in galls of *Ficus glomerata* Roxb. (Moraceae) by the insect *Pauropsylla depressa* (Psyllidae, Hemiptera). Trop Ecol 55:129–136
Fagan MM (1918) The uses of insect galls. Am Nat 52:155–176. https://doi.org/10.1086/279661
Fernandes GW, Aguirre-Jaimes A, Contreras-Varela X, Cocoletzi E, de Sousa WO, Araujo L, Nunes B, Angeles G, Quesada M (2021) Morphological and histological characterization

of extrafloral nectaries induced by *Coelocephalapion* galls on *Croton antisyphiliticus* in the Brazilian Cerrado. Flora 274:151758. https://doi.org/10.1016/j.flora.2020.151758

Ferreira BG, Isaias RMS (2013) Developmental stem anatomy and tissue redifferentiation induced by a galling Lepidoptera on *Marcetia taxifolia* (Melastomataceae). Botany 91:752–760. https://doi.org/10.1139/cjb-2013-0125

Ferreira BG, Alvarez R, Avritzer SC, Isaias RMS (2017) Revisiting the histological patterns of storage tissues: beyond the limits of gall-inducing taxa. Botany 95:173–184. https://doi.org/10.1139/cjb-2016-0189

Ferreira BG, Freitas MSC, Bragança GP, Moreira ASFP, Carneiro RGS, Isaias RMS (2019) Enzyme-mediated metabolism in nutritive tissues of galls induced by *Ditylenchus gallaeformans* (Nematoda: Anguinidae). Plant Biol 21:1052–1062. https://doi.org/10.1111/plb.13009

Ferreira BG, Moreira GRP, Carneiro RGS, Isaias RMS (2022) Complex meristematic activity induced by *Eucecidoses minutanus* on *Schinus engleri* turns shoots into galls. Am J Bot 109:209–225. https://doi.org/10.1002/ajb2.1798

Forneck A, Lawo NC, Schoedl-Hummel K, Liebner F, Zweckmayr T, Griesser M (2016) Not just sweet: phylloxerated roots indicate complex plant response mechanisms. In: 1st international symposium on grapevine roots, Rauscedo, Italy, Oct 16 2014. Acta Horticulturae, pp 239–243. https://doi.org/10.17660/ActaHortic.2016.1136.33

Giertych MJ, Łukowski A, Karolewski P (2023) Cynipid galls on oak leaves are resilient to leaf vein disruption. J Plant Res 136:527–534. https://doi.org/10.1007/s10265-023-01462-8

Goren A, Ashlock D, Tetlow IJ (2018) Starch formation inside plastids of higher plants. Protoplasma 255:1855–1876. https://doi.org/10.1007/s00709-018-1259-4

Griesser M, Lawo NC, Crespo-Martinez S, Schoedl-Hummel K, Wieczorek K, Gorecka M, Liebner F, Zweckmair T, Pavese NS, Kreil D, Forneck A (2015) Phylloxera (*Daktulosphaira vitifoliae* Fitch) alters the carbohydrate metabolism in root galls to allowing the compatible interaction with grapevine (*Vitis* ssp.) roots. Plant Sci 234:38–49. https://doi.org/10.1016/j.plantsci.2015.02.002

Guedes LM, Sanhueza C, Torres S, Figueroa C, Gavilán E, Pérez CI, Aguilera N (2023) Gall-inducing *Eriophyes tiliae* stimulates the metabolism of *Tilia platyphyllos* leaves towards oxidative protection. Plant Physiol Biochem 195:25–36. https://doi.org/10.1016/j.plaphy.2022.12.014

Gur A, Cohen A, Bravdo B-A (1969) Colorimetric method for starch determination. J Agric Food Chem 17:347–351. https://doi.org/10.1021/jf60162a009

Guzicka M, Karolewski P, Giertych MJ (2017) Structural modification of *Quercus petraea* leaf caused by *Cynips quercusfolii* - histological study of galls. J Plant Interact 12:7–13. https://doi.org/10.1080/17429145.2016.1269214

Haiden SA, Hoffmann JH, Cramer MD (2012) Benefits of photosynthesis for insects in galls. Oecologia 170:987–997. https://doi.org/10.1007/s00442-012-2365-1

Han Y, Sun F-J, Rosales-Mendoza S, Korban SS (2007) Three orthologs in rice, *Arabidopsis*, and *Populus* encoding starch branching enzymes (SBEs) are different from other SBE gene families in plants. Gene 401:123–130. https://doi.org/10.1016/j.gene.2007.06.026

Harper LJ, Schönrogge K, Lim KY, Francis P, Lichtenstein CP (2004) Cynipid galls: insect-induced modifications of plant development create novel plant organs. Plant Cell Environ 27:327–335. https://doi.org/10.1046/j.1365-3040.2004.01145.x

Hernández-Soto P, Lara-Flores M, Agredano-Moreno L, Jiménez-Garcia LF, Cuevas-Reyes P, Oyama K (2015) Developmental morphology of bud galls induced on the vegetative meristems of *Quercus castanea* by *Amphibolips michoacaensis* (Hymenoptera: Cynipidae). Bot Sci 93:685–693

Huang MY, Huang WD, Chou HM, Lin KH, Chen CC, Chen PJ, Chang YT, Yang CM (2014) Leaf-derived cecidomyiid galls are sinks in *Machilus thunbergii* (Lauraceae) leaves. Physiol Plant 152:475–485. https://doi.org/10.1111/ppl.12186

Huang MY, Huang WD, Chou HM, Chen CC, Chen PJ, Chang YT, Yang CM (2015) Structural, biochemical, and physiological characterization of photosynthesis in leaf-derived cup-shaped galls on *Litsea acuminata*. BMC Plant Biol 15:61. https://doi.org/10.1186/s12870-015-0446-0

Huang MY, Lin KH, Shih TH, Chang YT, Yang CM (2019) Spectral indices to rapidly monitor chloroplast distribution patterns of three cecidomyiid galls. Taiwania 64:438–441. https://doi.org/10.6165/tai.2019.64.438

Inbar M, Eshel A, Wool D (1995) Interspecific competition among phloem-feeding insects mediated by induced host-plant sinks. Ecology 76:1506–1515. https://doi.org/10.2307/1938152

Isaias RMS, Carneiro RGS, Oliveira DC, Santos JC (2013) Illustrated and annotated checklist of brazilian gall morphotypes. Neotrop Entomol 42:230–239. https://doi.org/10.1007/s13744-013-0115-7

Jankiewicz LS, Plich H, Antoszewski R (1970) Preliminary studies on the translocation of ^{14}C-labelled assimilates and $^{32}PO^{3-}$towards the gall evoked by *Cynips quercus-folii* L. on oak leaves. Marcellia (Strasburg) 36:163–172

Jankiewicz LS, Guzicka M, Marasek-Ciolakowska A (2021) Anatomy and ultrastructure of galls induced by *Neuroterus quercusbaccarum* (Hymenoptera: Cynipidae) on oak leaves (*Quercus robur*). Insects 12:850. https://doi.org/10.3390/insects12100850

Koncz NK, Szabó LJ, Máthé C, Jámbrik K, M-Hamvas M (2011) Histological study of quercus galls of *Neuroterus quercusbaccarum* (Linnaeus, 1758) (Hymenoptera: Cynipidae). Acta Biologica Szegediensis 55:247–253

Kostoff D, Kendall J (1929) Studies on the structure and development of certain cynipid galls. Biol Bull 56:402–458. https://doi.org/10.2307/1537132

Kot I, Rubinowska K, Michałek W (2018) Changes in chlorophyll a fluorescence and pigments composition in oak leaves with galls of two cynipid species (Hymenoptera, Cynipidae). Acta Scientiarum Polonorum Hortorum Cultus 17:147–157

Larson KC, Whitham TG (1991) Manipulation of food resources by a gall-forming aphid - the physiology of sink-source interactions. Oecologia 88:15–21

Liu P, Yang ZX, Chen XM, Foottit RG (2014) The effect of the gall-forming aphid *Schlechtendalia chinensis* (Hemiptera: Aphididae) on leaf wing ontogenesis in *Rhus chinensis* (Sapindales: Anacardiaceae). Ann Entomol Soc Am 107:242–250. https://doi.org/10.1603/an13118

Loescher WH, McCamant T, Keller JD (1990) Carbohydrate reserves, translocation, and storage in woody plant-roots. HortScience 25:274–281

Mani MS (1964) Ecology of plant galls. Monographiae Biologicae. W. Junk, Haag. https://doi.org/10.1007/978-94-017-6230-4

Martin C, Smith AM (1995) Starch biosynthesis. Plant Cell 7:971–985. https://doi.org/10.1105/tpc.7.7.971

Martini V, Moreira ASFP, Kuster VC, Oliveira DC (2020) Photochemical performance and source-sink relationships in galls induced by *Pseudophacopteron longicaudatum* (Hemiptera) on leaves of *Aspidosperma tomentosum* (Apocynaceae). Photosynthetica 58:827–835. https://doi.org/10.32615/ps.2020.033

Martinson EO, Werren JH, Egan SP (2022) Tissue-specific gene expression shows a cynipid wasp repurposes oak host gene networks to create a complex and novel parasite-specific organ. Mol Ecol 31:3228–3240. https://doi.org/10.1111/mec.16159

Meletiouchristou MS, Rhizopoulou S, Diamantoglou S (1994) Seasonal-changes of carbohydrates, lipids and nitrogen-content in sun and shade leaves from four Mediterranean evergreen sclerophylls. Environ Exp Bot 34:129–140. https://doi.org/10.1016/0098-8472(94)90032-9

Muñoz-Viveros AL, Martinez JJI, Molist P, González-Sierra S, Julián PG, Alvarez R (2014) Microscopic study of galls induced by three species of *Geopemphigus* (Hemiptera: Aphididae: Eriosomatinae) on *Pistacia mexicana*. Arthropod Plant Interact 8:531–538. https://doi.org/10.1007/s11829-014-9333-0

Murakami R, Ushima R, Sugimoto R, Tamaoki D, Karahara I, Hanba Y, Wakasugi T, Tsuchida T (2021) A new galling insect model enhances photosynthetic activity in an obligate holoparasitic plant. Sci Rep 11:13013. https://doi.org/10.1038/s41598-021-92417-3

Oliveira DC, Isaias RMD (2010) Cytological and histochemical gradients induced by a sucking insect in galls of *Aspidosperma australe* Arg. Muell (Apocynaceae). Plant Sci 178:350–358. https://doi.org/10.1016/j.plantsci.2010.02.002

Oliveira DC, Magalhães TA, Carneiro RGS, Alvim MN, Isaias RMS (2010) Do Cecidomyiidae galls of *Aspidosperma spruceanum* (Apocynaceae) fit the pre-established cytological and histochemical patterns? Protoplasma 242:81–93. https://doi.org/10.1007/s00709-010-0128-6
Oliveira DC, Carneiro RGD, Magalhães TA, Isaias RMD (2011a) Cytological and histochemical gradients on two *Copaifera langsdorffii* Desf. (Fabaceae)-Cecidomyiidae gall systems. Protoplasma 248:829–837. https://doi.org/10.1007/s00709-010-0258-x
Oliveira DC, Isaias RMD, Moreira A, Magalhaes TA, de Lemos JP (2011b) Is the oxidative stress caused by *Aspidosperma* spp. galls capable of altering leaf photosynthesis? Plant Sci 180:489–495. https://doi.org/10.1016/j.plantsci.2010.11.005
Oliveira DC, Moreira ASFP, Isaias RMS, Martini V, Rezende UC (2017) Sink status and photosynthetic rate of the leaflet galls induced by *Bystracoccus mataybae* (Eriococcidae) on *Matayba guianensis* (Sapindaceae). Front Plant Sci 8. https://doi.org/10.3389/fpls.2017.01249
Pascual-Alvarado E, Nieves-Aldrey JL, Castillejos-Lemus DE, Cuevas-Reyes P, Oyama K (2017) Diversity of galls induced by wasps (Hymenoptera: Cynipidae, Cynipini) associated with oaks (Fagaceae: *Quercus*) in Mexico. Bot Sci 95:461–472. https://doi.org/10.17129/botsci.1215
Pawłowski TA, Staszak AM, Karolewski P, Giertych MJ (2017) Plant development reprogramming by cynipid gall wasp: proteomic analysis. Acta Physiol Plant 39:114. https://doi.org/10.1007/s11738-017-2414-9
Pilichowski S, Giertych MJ (2020) Two galling insects (*Hartigiola annulipes* and *Mikiola fagi*), one host plant (*Fagus sylvatica*) - differences between leaf and gall chemical composition. Baltic For 26:209–219. https://doi.org/10.46490/bf474
Pincebourde S, Casas J (2016) Hypoxia and hypercarbia in endophagous insects: larval position in the plant gas exchange network is key. J Insect Physiol 84:137–153. https://doi.org/10.1016/j.jinsphys.2015.07.006
Raman A (1991) Cecidogenesis of leaf galls on *Syzygium cumini* (L) skeels (Myrtaceae) induced by *Trioza jambolanae* crawford (Homoptera, Psylloidea). J Nat Hist 25:653–663. https://doi.org/10.1080/00222939100770421
Raman A (2011) Morphogenesis of insect-induced plant galls: facts and questions. Flora 206:517–533. https://doi.org/10.1016/j.flora.2010.08.004
Raman A, Rajadurai S, Mani T, Balakrishna P (1991) On the salivary enzymes and soluble proteins of the Mimusops gall thrips, Arrhenothrips ramakrishnae Hood (Tubulifera: Thysanoptera: Insecta). Z Angew Zool 78:131–136
Rezende UC, Moreira A, Kuster VC, de Oliveira DC (2018) Structural, histochemical and photosynthetic profiles of galls induced by *Eugeniamyia dispar* (Diptera: Cecidomyiidae) on the leaves of *Eugenia uniflora* (Myrtaceae). Rev Biol Trop 66:1469–1480. https://doi.org/10.15517/rbt.v66i4.32531
Rezende UC, Cardoso JCF, Kuster VC, Gonçalves LA, Oliveira DC (2019) How the activity of natural enemies changes the structure and metabolism of the nutritive tissue in galls? Evidence from the *Palaeomystella oligophaga* (Lepidoptera) - *Macairea radula* (Metastomataceae) system. Protoplasma 256:669–677. https://doi.org/10.1007/s00709-018-1321-2
Robyt JF (2008) Starch: structure, properties, chemistry, and enzymology. In: Fraser-Reid BO, Tatsuta K, Thiem J (eds) Glycoscience: chemistry and chemical biology. Springer, Berlin/Heidelberg, pp 1437–1472. https://doi.org/10.1007/978-3-540-30429-6_35
Rohfritsch O (1999) A so called "rudimentary gall" induced by the gall midge *Physemocecis hartigi* on leaves of *Tilia intermedia*. Can J Botany-Revue Canadienne De Botanique 77:460–470. https://doi.org/10.1139/b98-219
Rohfritsch O (2008) Plants, gall midges, and fungi: a three-component system. Entomol Exp Appl 128:208–216. https://doi.org/10.1111/j.1570-7458.2008.00726.x
Rohfritsch O, Shorthouse JD (1982) 5 - Insect galls. In: Kahl G, Schell JS (eds) Molecular biology of plant tumors. Academic, pp 131–152. https://doi.org/10.1016/B978-0-12-394380-4.50011-6
Schönrogge K, Stone GN, Crawley MJ (1995) Spatial and temporal variation in guild structure - parasitoids and inquilines of *Andricus quercuscalicis* (Hymenoptera, Cynipidae) in its native and alien ranges. Oikos 72:51–60. https://doi.org/10.2307/3546037

Shannon RE, Brewer JW (1980) Starch and sugar levels in three coniferous insect galls. Z Angew Entomol 89:526–533. https://doi.org/10.1111/j.1439-0418.1980.tb03494.x

Sharma A, Khan A, Subrahmanyam S, Raman A, Taylor G, Fletcher M (2014) Salivary proteins of plant-feeding hemipteroids–implication in phytophagy. Bull Entomol Res 104:117–136

Shimizu K, Aoki K (2019) Development of parasitic organs of a stem holoparasitic plant in genus *Cuscuta*. Front Plant Sci 10:1435. https://doi.org/10.3389/fpls.2019.01435

Shorthouse JD (1993) Adaptations of gall wasps of the genus *Diplolepis* (Hymenoptera: Cynipidae) and the role of gall anatomy in cynipid systematics. Memoirs Entomol Soc Canada 125:139–163

Shorthouse JD (1998) Role of *Periclistis* (Hymenoptera: Cynipidae) inquilines in leaf galls of *Diplolepis* (Hymenoptera: Cynipidae) on wild roses in Canada. In: Csoka G, Mattson WJ, Stone GN, Price PW (eds) The biology of gall-inducing arthropods U.S. Dept. of Agriculture, Forest Service, North Central Forest Experiment Station, general technical report NC-199. St. Paul, MN, pp 61–81

Silva CP, Terra WR, Xavier-Filho J, de Sá MFG, Lopes AR, Pontes EG (1999) Digestion in larvae of <i>Callosobruchus maculatus</i> and <i>Zabrotes subfasciatus</i> (Coleoptera: Bruchidae) with emphasis on α-amylases and oligosaccharidases. Insect Biochem Mol Biol 29:355–366. https://doi.org/10.1016/s0965-1748(99)00010-7

Smith AM (2001) The biosynthesis of starch granules. Biomacromolecules 2:335–341. https://doi.org/10.1021/bm000133c

Smith SM, Fulton DC, Chia T, Thorneycroft D, Chapple A, Dunstan H, Hylton C, Zeeman SC, Smith AM (2004) Diurnal changes in the transcriptome encoding enzymes of starch metabolism provide evidence for both transcriptional and posttranscriptional regulation of starch metabolism in *Arabidopsis* leaves. Planrt Physiol 136:2687–2699. https://doi.org/10.1104/pp.104.044347

Staszak AM, Ratajczak E, Lesniewska J, Piotrowska-Niczyporuk A, Kostro-Ambroziak A (2023) A broad spectrum of host plant responses to the actions of the gall midge: case study of *Robinia pseudoacacia* L. and *Obolodiplosis robiniae* (Haldeman). BMC Plant Biol 23:19. https://doi.org/10.1186/s12870-022-03914-0

Stromeyer (1815) Ein sehr empsindliches Reagens für Jodine, aufgefunden in der Stärke (Amidon). Ann Phys 49:146–153. https://doi.org/10.1002/andp.18150490108

Torii KU (2015) Stomatal differentiation: the beginning and the end. Curr Opin Plant Biol 28:16–22. https://doi.org/10.1016/j.pbi.2015.08.005

Ugolini F, Massetti L, Pedrazzoli F, Tognetti R, Vecchione A, Zulini L, Maresi G (2014) Ecophysiological responses and vulnerability to other pathologies in European chestnut coppices, heavily infested by the Asian chestnut gall wasp. For Ecol Manag 314:38–49. https://doi.org/10.1016/j.foreco.2013.11.031

Vandevyvere I, De Bruyn L (1998) Morphological and histochemical analysis of galls of *Lipara lucens* (Diptera, Chloropidae) on *Phragmites australis* (Poaceae). Can J Botany-Revue Canadienne De Botanique 76:1374–1384. https://doi.org/10.1139/b98-143

Yoshimura K, Saiki S-T, Yazaki K, Ogasa MY, Shirai M, Nakano T, Yoshimura J, Ishida A (2016) The dynamics of carbon stored in xylem sapwood to drought-induced hydraulic stress in mature trees. Sci Rep 6:24513. https://doi.org/10.1038/srep24513

Zeeman SC, Smith SM, Smith AM (2004) The breakdown of starch in leaves. New Phytol 163:247–261. https://doi.org/10.1111/j.1469-8137.2004.01101.x

Zeeman SC, Kossmann J, Smith AM (2010) Starch: its metabolism, evolution, and biotechnological modification in plants. Annu Rev Plant Biol 61:209–234

Histochemical Overview of Primary Metabolites in Galls: Tissue Compartmentalization and Taxon-Related Traits

Vinícius Coelho Kuster, Raphaela Lorrayne de Jesus Costa, Vitor Campana Martini, and Maísa Barbosa Santos

1 Introduction

Galls develop from structural and physiological modifications of host organs, which through processes such as hyperplasia and hypertrophy produce newly formed tissues with particular functions (Mani 1964; Isaias et al. 2014; Oliveira et al. 2016). The morphophysiological changes give rise to various gall morphotypes, classified primarily by color, size, and notably shape (Isaias et al. 2013). Clavate, conical, cylindrical, fusiform, globoid, lenticular, and rosette are some morphotypes grouped and standardized in the Brazilian checklist (Isaias et al. 2013). In addition, the gall-inducing organism can manipulate the chemical composition of the host plant during gall formation (Larson and Whitham 1991; Nyman and Julkunen-Tiitto 2000; Kuster et al. 2019), storing primary metabolites. Primary metabolites, such as proteins, lipids, sugars, and nucleic acids, are products of metabolism that are directly associated with the development of plants and are found in all plant species (Salam et al. 2023). The gall-inducer can manipulate gall metabolism through gene regulation, suppressing or increasing the expression of host plant genes according to the galler's needs (Bailey et al. 2015; Shih et al. 2018). These regulations can be explored through histochemical approaches, which indicate the synthesis and remobilization of primary metabolites in gall tissues (Oliveira et al. 2010; Ferreira and Isaias 2013, 2014; Ferreira et al. 2017a, 2019b; Isaias et al. 2018), since they allow the identification of chemical compounds accumulated in plant tissues (Jensen 1962).

V. C. Kuster (✉) · R. L. de Jesus Costa
Instituto de Biociência, Universidade Federal de Jataí (UFJ), Jataí, Brazil
e-mail: viniciuskuster@ufj.edu.br

V. C. Martini
Institute of Biology, Universidade Federal de Uberlândia (UFU), Uberlândia, Brazil

M. B. Santos
Instituto Federal Goiano, Campus Rio Verde, Rio Verde, Brazil

D. C. de Oliveira, R. M. dos Santos Isaias (eds.), *Plant Galls*,
https://doi.org/10.1007/978-3-031-80064-1_7

Sugars are present in all phases of the plant life cycle and act to generate metabolic energy, regulating the growth and development processes on plants and galls (Wind et al. 2010; Stokes et al. 2013; Martini et al. 2020a). Starch is an insoluble and non-structural carbohydrate composed of α-glucose polymers (Pfister and Zeeman 2016). It is composed of two polymers of glucose, amylopectin, and amylose, which together form insoluble semi-crystalline starch granules (Pérez and Bertoft 2010). Plant lipids are essential components of cells since they make up the entire cell membrane system as phospholipids and glycolipids (Hölzl and Dörmann 2007), with a predominance of triacylglycerols packed in lipid droplets as a major form of carbon storage (Masclaux-Daubresse et al. 2020). Animal and plant cells have more than 10.000 types of proteins (Tan et al. 2019), which may be present in intracellular compartments, cell membranes, and organelles (Vitale and Galili 2001), acting as structural components, protective role, enzymatic activity, among others (Hemsley and Grierson 2008). All these compounds from the plant's primary metabolism are restructured in gall formation, creating a new profile associated with the intrinsic demands of the gall-inducing organism and the gall.

The gall structure provides protection, shelter, and nutrition for its inducing organism (Stone and Schönrogge 2003; Oliveira et al. 2016; Isaias et al. 2018), which is related to the distribution of primary and secondary metabolites in the functional tissue compartments. Regarding secondary metabolites, Kuster et al. (2019) demonstrated their association with tissue compartmentalization in galls. Phenolics emerged as the most investigated compound, predominantly detected in the outer cortex (Kuster et al. 2019), which reinforces a potential protective function of phenolics within the gall structure. Herein, we examined the distribution of primary metabolites in galls and associated them with the different tissue compartments. The tissue compartmentalization in galls seems to be less evident in simple galls, where parenchyma homogenization occurs, such as in rolling leaf galls (Ferreira et al. 2019a). In contrast, complex galls exhibit multifunctionally compartmentalized cortex, indicating a more intricate organization of tissues (Bragança et al. 2016; Isaias et al. 2018; Ferreira et al. 2019a; Teixeira et al. 2022). The organization of the gall tissues may reflect the distribution of primary metabolites.

In complex galls, the outer tissues are generally related to the storage of complex sugars, such as starch grains, as occurs in gall induced by *Palaeomystella oligophaga* (Lepidoptera) on *Macairea radula* (Metastomataceae) (Rezende et al. 2019), and inner tissues with the presence of metabolites related to the galler's diet, such as reducing sugars, proteins, and lipids, as in the *Manihot esculenta* (Euphorbiaceae) galls induced by *Iatrophobia brasiliensis* (Diptera, Cecidomyiidae) (Souza et al. 2023). Extra functional layers are also reported, such as the sclerenchymatic layer with mechanical role as in galls induced by *Ceropsylla pouteriae* Burckhardt sp. nov. (Hemiptera: Psylloidea: Triozidae) on leaves of *Pouteria ramiflora* (Oliveira et al. 2019) and by Diptera: Cecidomyiidae on leaves of *Aspidosperma spruceanum* (Formiga et al. 2011). However, the primary metabolite compartmentalization has more complex arrangements than reported here, with the degree of heterogeneity associated with morphogenical constraints.

2 Histochemical Approach of Primary Metabolites in Galls

Carbohydrates, lipids, and proteins are primary metabolites that are routinely reported in galls induced by different gall-inducer taxa (Santos et al. 2014; Rezende et al. 2018; Silva et al. 2019; Guimarães et al. 2021). They are generally related to the maintenance of cellular metabolic activity and/or the supply of the diet of gall inducer (Oliveira et al. 2006; Isaias et al. 2018; Ferreira et al. 2019b; Ferreira et al. 2019a). Herein, we confirm that starch, lipids, reducing sugars, and proteins are the most tested primary metabolites in galls, in which they were reported, overall, 68 times for globoid morphotype, 42 lenticular, 12 fusiform, 8 bivalve-shaped, 6 leaf rolling and cylindrical, 5 horn-shaped and 3 conical galls, induced on leaves, stems, roots, flowers or buds (Table 1). Also, they are highly tested for the mature stage, followed respectively by the young and senescent stages (Fig. 1). Starch grains is the most tested metabolite, followed by lipids, proteins, and reducing sugars, respectively (Fig. 1). The highlight here is the low frequency of negative results for all compounds, which indicates that these metabolites are commonly stored in galls as well as in host plant organs (Fig. 1).

Our findings indicate that primary metabolites are present across various gall morphotypes, exhibiting a universal occurrence in galls. However, significant differences are observed in the histolocalization within gall tissue compartments (Bragança et al. 2016; Rezende et al. 2018; Silva et al. 2019; Martini et al. 2020b), as well as in the quantity produced (Naves et al. 2021). The primary metabolites act directly in plant development (Mahmood et al. 2021), so they are naturally found in plant cells. Also, the qualitative nature of histochemistry (Lyon 1991) may explain the high positive labeling of primary metabolites in galls, since small accumulations are reported as positive.

Different histochemical tests predominate for identifying groups within carbohydrates, proteins, and lipids for galls. Lugol's solution (Johansen 1940) has been used for the detection of starch, as shown for bud galls of *Macairea radula* (Metastomataceae) (Rezende et al. 2019), while for lipids normally Sudan red (Jensen 1962) is used (Rezende et al. 2019; Silva et al. 2019; Bomfim et al. 2020). Black to purple color indicates a positive result for the Lugol test (Fig. 2a, b) and red to Sudan red test (Fig. 2c, d). Bromophenol blue (Baker 1958) is generally used for the detection of proteins, such as in the galls present in the "fruit-like gall" of *Miconia chamissois* (Melastomataceae) (Bomfim et al. 2020), with blue color as a positive result (Fig. 2e, f). For reducing sugars, the Fehling test (Sass 1951) is the most utilized, as in the leaf and stem galls of *Matayba guianensis* (Silva et al. 2019). Orange to reddish indicates a positive result for the Fehling test.

Due to their high energy demand, primary metabolites are generally produced more in young galls than mature ones (Rohfritsch and Anthony 1992; Weis et al. 1988). However, they are quickly used in their development. So, the galls are built with new and fresh metabolite profiles, which are normally distinct among gall development stages. In this sense, the change in the patterns of synthesis and mobilization of reserves in galls compared to the non-galled organ raises the following

Table 1 Main characteristics extracted from scientific articles on primary metabolites in galls based on histochemistry

Primary metabolite	Subtype	Gall (tested?)	Gall presence	Gall region	Non-galled (tested?)	Non-galled presence	Host plant family	Host plant species	Host plant organ	Gall morphotype	Gall stage	Gall inducer species	Order/family gall inducer	References
Lipids		Yes	+	Outer	Yes	+	Fabaceae	*L. muehlbergianus*	Leaf	Bivalve-shaped	M	*E. ostreoides*	Psyllidae	Oliveira et al. (2006)
Carbohydrate	Starch	Yes	+	Both	Yes	+	Fabaceae	*L. muehlbergianus*	Leaf	Bivalve-shaped	M	*E. ostreoides*	Psyllidae	Oliveira et al. (2006)
Carbohydrate	Starch	Yes	+	Both	NR[a]	–	Apocynaceae	*A. sprunceanum*	Leaf	Lenticular	M	NR	Cecidomyiidae	Formiga et al. (2011)
Lipids		Yes	+	Outer	NR	–	Apocynaceae	*A. sprunceanum*	Leaf	Lenticular	M	NR	Cecidomyiidae	Formiga et al. (2011)
Lipids		Yes	+	Outer	Yes	–	Melastomataceae	*M. taxifolia*	Stem	Fusiforms	Y	NR	Lepidoptera	Ferreira and Isaias (2013)
Carbohydrate	Starch	Yes	+	Outer	Yes	–	Melastomataceae	*M. taxifolia*	Stem	Fusiforms	Y	NR	Lepidoptera	Ferreira and Isaias (2013)
Carbohydrate	Reducing sugars	Yes	+	Outer	Yes	–	Melastomataceae	*M. taxifolia*	Stem	Fusiforms	Y	NR	Lepidoptera	Ferreira and Isaias (2013)
Lipids		Yes	+	Inner	Yes	–	Melastomataceae	*M. taxifolia*	Stem	Fusiforms	M	NR	Lepidoptera	Ferreira and Isaias (2013)
Carbohydrate	Starch	Yes	+	Outer	Yes	–	Melastomataceae	*M. taxifolia*	Stem	Fusiforms	M	NR	Lepidoptera	Ferreira and Isaias (2013)
Carbohydrate	Reducing sugars	Yes	+	Outer	Yes	–	Melastomataceae	*M. taxifolia*	Stem	Fusiforms	M	NR	Lepidoptera	Ferreira and Isaias (2013)
Carbohydrate	Starch	Yes	–	NR	Yes	–	Melastomataceae	*M. taxifolia*	Stem	Fusiforms	S	NR	Lepidoptera	Ferreira and Isaias (2013)

Lipids		Yes	–	NR	Yes	–	Melastomataceae	*M. taxifolia*	Stem	Fusiforms	S	NR	Lepidoptera	Ferreira and Isaias (2013)
Lipids		Yes	+	Both	Yes	+	Fabaceae	*B. ungulata*	Leaf	Leaf rolling gall	NR	NR	Lepidoptera	Bedetti et al. (2013)
Carbohydrate	Starch	Yes	+	Inner	Yes	+	Fabaceae	*B. ungulata*	Leaf	Leaf rolling gall	NR	NR	Lepidoptera	Bedetti et al. (2013)
Carbohydrate	Reducing sugars	Yes	+	Inner	Yes	+	Fabaceae	*B. ungulata*	Leaf	Leaf rolling gall	NR	NR	Lepidoptera	Bedetti et al. (2013)
Protein		Yes	–	NR	Yes	+	Fabaceae	*B. ungulata*	Leaf	Leaf rolling gall	NR	NR	Lepidoptera	Bedetti et al. (2013)
Lipids		Yes	+	Both	No	–	Melastomataceae	*M. taxifolia*	Flower bud	Bud-gall	M	NR	Cecidomyiidae	Ferreira and Isaias (2014)
Protein		Yes	+	Inner	No	–	Melastomataceae	*M. taxifolia*	Flower bud	Bud-gall	M	NR	Cecidomyiidae	Ferreira and Isaias (2014)
Carbohydrate	Starch	Yes	+	Outer	No	–	Melastomataceae	*M. taxifolia*	Flower bud axila	Bud-gall	M	NR	Cecidomyiidae	Ferreira and Isaias (2014)
Carbohydrate	Reducing sugars	Yes	+	Outer	No	–	Melastomataceae	*M. taxifolia*	Flower bud	Bud-gall	M	NR	Cecidomyiidae	Ferreira and Isaias (2014)
Carbohydrate	Reducing sugars	Yes	+	Outer	Yes	+	Myrtaceae	*P. cattleianum*	Leaf	Globoid	Y	*N. cattleianum*	Triozidae	Carneiro and Isaias (2015)
Carbohydrate	Starch	Yes	+	Inner	Yes	+	Myrtaceae	*P. cattleianum*	Leaf	Globoid	Y	*N. cattleianum*	Triozidae	Carneiro and Isaias (2015)
Protein		Yes	+	Both	Yes	+	Myrtaceae	*P. cattleianum*	Leaf	Globoid	Y	*N. cattleianum*	Triozidae	Carneiro and Isaias (2015)

(continued)

Table 1 (continued)

Primary metabolite	Subtype	Gall (tested?)	Gall presence	Gall region	Non-galled (tested?)	Non-galled presence	Host plant family	Host plant species	Host plant organ	Gall morphotype	Gall stage	Gall inducer species	Order/family gall inducer	References
Lipids		Yes	+	Outer	Yes	+	Myrtaceae	*P. cattleianum*	Leaf	Globoid	Y	*N. cattleianum*	Triozidae	Carneiro and Isaias (2015)
Lipids		Yes	+	Inner	Yes	+	Piperaceae	*P. arboreum*	Leaf	Intralaminar and lenticular	M	NR	Cecidomyiidae	Bragança et al. (2016)
Carbohydrate	Reducing sugars	Yes	+	Inner	Yes	+	Piperaceae	*P. arboreum*	Leaf	Intralaminar and lenticular	M	NR	Cecidomyiidae	Bragança et al. (2016)
Carbohydrate	Starch	Yes	+	Inner	Yes	+	Piperaceae	*P. arboreum*	Leaf	Intralaminar and lenticular	M	NR	Cecidomyiidae	Bragança et al. (2016)
Protein		Yes	+	Inner	Yes	+	Piperaceae	*P. arboreum*	Leaf	Intralaminar and lenticular	M	NR	Cecidomyiidae	Bragança et al. (2016)
Lipids		Yes	+	Inner	No	–	Myrtaceae	*E. camaldulensis*	Leaf - Petiole	Globoid	M	*L. invasa*	Eulophidae	Isaias et al. (2018)
Protein		Yes	+	Inner	No	–	Myrtaceae	*E. camaldulensis*	Leaf - Petiole	Globoid	M	*L. invasa*	Eulophidae	Isaias et al. (2018)
Carbohydrate	Starch	Yes	+	Outer	No	–	Myrtaceae	*E. camaldulensis*	Leaf - Petiole	Globoid	M	*L. invasa*	Eulophidae	Isaias et al. (2018)
Carbohydrate	Reducing sugars	Yes	+	Outer	Yes	+	Smilacaceae	*S. campestris*	Leaf	Globoid	M	NR	NR	Arriola et al. (2018)
Carbohydrate	Starch	Yes	+	Outer	Yes	+	Smilacaceae	*S. campestris*	Leaf	Globoid	M	NR	NR	Arriola et al. (2018)
Lipids		Yes	–	NR	Yes	+	Smilacaceae	*S. campestris*	Leaf	Globoid	M	NR	NR	Arriola et al. (2018)
Lipids		Yes	+	Outer	Yes	+	Smilacaceae	*S. campestris*	Leaf	Globoid	M	NR	NR	Arriola et al. (2018)
Protein		Yes	–	NR	Yes	+	Smilacaceae	*S. campestris*	Leaf	Globoid	M	NR	NR	Arriola et al. (2018)

Lipids		Yes	–	NR	Yes	+	Anacardiaceae	*S. polygama*	Stem	Globoid	M	*C. rubra*	Calophyidae	Guedes et al. (2018a, b)
Carbohydrate	Starch	Yes	+	Both	Yes	–	Asteraceae	*M. glomerata*	Leaf	Globoid	M	*L. spherica*	Cecidomyiidae	Amorim et al. (2017)
Lipids		Yes	+	Inner	Yes	+	Asteraceae	*M. glomerata*	Leaf	Globoid	M	*L. spherica*	Cecidomyiidae	Amorim et al. (2017)
Carbohydrate	Starch	Yes	–	Both	Yes	–	Asteraceae	*M. glomerata*	Leaf	Globoid	M	*L. cylindrica*	Cecidomyiidae	Amorim et al. (2017)
Lipids		Yes	+	Inner	Yes	+	Asteraceae	*M. glomerata*	Leaf	Globoid	M	*L. cylindrica*	Cecidomyiidae	Amorim et al. (2017)
Carbohydrate	Starch	No	–	NR	Yes	–	Asteraceae	*M. glomerata*	Petiole	Fusiform	M	*Clinodiplosis* sp.	Cecidomyiidae	Amorim et al. (2017)
Lipids		Yes	+	Inner	Yes	+	Asteraceae	*M. glomerata*	Petiole	Fusiform	M	*Clinodiplosis* sp.	Cecidomyiidae	Amorim et al. (2017)
Lipids		Yes	+	Inner	No	–	Clusiaceae	*C. lanceolada*	Leaf	Intralaminar lenticular	M	*C. nitida*	Cecidomyiidae	Guimarães et al. (2013)
Carbohydrate	Starch	Yes	+	Inner	No	–	Clusiaceae	*C. lanceolada*	Leaf	Intralaminar lenticular	M	*C. nitida*	Cecidomyiidae	Guimarães et al. (2013)
Carbohydrate	Starch	Yes	+	NR	Yes	–	Magnoliaceae	*M. champaca*	Stem	NR	M	*Podothrips* sp.	Thysanoptera	Nalini et al. (2015)
Protein		Yes	+	NR	Yes	–	Magnoliaceae	*M. champaca*	Stem	NR	M	*Podothrips* sp.	Thysanoptera	Nalini et al. (2015)
Lipids		Yes	+	NR	Yes	–	Magnoliaceae	*M. champaca*	Stem	NR	M	*Podothrips* sp.	Thysanoptera	Nalini et al. (2015)
Carbohydrate	Starch	Yes	+	Mesophyll	No	–	Anacardiaceae	*P. terebinthus*	Leaf	Leaf rolling gall	M	*P. cimiciformis*	Thysanoptera	Álvarez et al. (2009)
Lipids		Yes	+	Mesophyll	No	–	Anacardiaceae	*P. terebinthus*	Leaf	Leaf rolling gall	M	*P. cimiciformis*	Thysanoptera	Álvarez et al. (2009)
Protein		Yes	+	Mesophyll	No	–	Anacardiaceae	*P. terebinthus*	Leaf	Leaf rolling gall	M	*P. cimiciformis*	Thysanoptera	Álvarez et al. (2009)

(continued)

Table 1 (continued)

Carbohydrate		Yes	+	Mesophyll	No	–	Anacardiaceae	*P. terebinthus*	Leaf	Leaf rolling gall	M	*F. formicaria*	Thysanoptera	Álvarez et al. (2009)
Lipids		Yes	+	Both	No	–	Malpighiaceae	*B. sericea*	Inflorescence	Globoid	M	*B. byrsomine*	Cecidomyiidae	Guimarães et al. (2014)
Carbohydrate		Yes	+	Outer	No	–	Malpighiaceae	*B. sericea*	Inflorescence	Globoid	M	*B. byrsomine*	Cecidomyiidae	Guimarães et al. (2014)
Lipids		Yes	+	Mesophyll	Yes	+	Verbenaceae	*L. camara*	Leaf	Leaf rolling gall	M	*A. lantanae*	Eriophyidae	Moura et al. (2008)
Protein		Yes	+	Outer	Yes	+	Apocynaceae	*A. australe*	Leaf	NR	Y	*Pseudophacopteron* sp.	Phacopteronidae	Oliveira and Isaias (2010b)
Protein		Yes	+	Outer	Yes	+	Apocynaceae	*A. australe*	Leaf	NR	M	*Pseudophacopteron* sp.	Phacopteronidae	Oliveira and Isaias (2010b)
Protein		Yes	+	Both	Yes	+	Apocynaceae	*A. australe*	Leaf	NR	S	*Pseudophacopteron* sp.	Phacopteronidae	Oliveira and Isaias (2010b)
Carbohydrate	Starch	Yes	+	Outer	Yes	+	Apocynaceae	*A. australe*	Leaf	NR	Y	*Pseudophacopteron* sp.	Phacopteronidae	Oliveira and Isaias (2010b)
Carbohydrate	Starch	Yes	+	Outer	Yes	+	Apocynaceae	*A. australe*	Leaf	NR	M	*Pseudophacopteron* sp.	Phacopteronidae	Oliveira and Isaias (2010b)
Carbohydrate	Starch	Yes	+	Both	Yes	+	Apocynaceae	*A. australe*	Leaf	NR	S	*Pseudophacopteron* sp.	Phacopteronidae	Oliveira and Isaias (2010b)
Carbohydrate	Reducing sugars	Yes	+	Outer	Yes	+	Apocynaceae	*A. australe*	Leaf	NR	M	*Pseudophacopteron* sp.	Phacopteronidae	Oliveira and Isaias (2010b)
Carbohydrate	Reducing sugars	No	–		Yes	+	Apocynaceae	*A. australe*	Leaf	NR	Y	*Pseudophacopteron* sp.	Phacopteronidae	Oliveira and Isaias (2010b)
Carbohydrate	Reducing sugars	Yes	+	Outer	Yes	+	Apocynaceae	*A. australe*	Leaf	NR	S	*Pseudophacopteron* sp.	Phacopteronidae	Oliveira and Isaias (2010b)
Lipids		Yes	+	Outer	Yes	+	Apocynaceae	*A. australe*	Leaf	NR	Y	*Pseudophacopteron* sp.	Phacopteronidae	Oliveira and Isaias (2010b)
Lipids		Yes	+	Outer	Yes	+	Apocynaceae	*A. australe*	Leaf	NR	M	*Pseudophacopteron* sp.	Phacopteronidae	Oliveira and Isaias (2010b)

Lipids		No	–		Yes	+	Apocynaceae	*A. australe*	Leaf	NR	S	*Pseudophacopteron* sp.	Phacopteronidae	Oliveira and Isaias (2010b)
Carbohydrate	Starch	Yes	+	Outer	Yes	+	Fabaceae	*C. langsdorffi*	Leaf	Pocket-gall	M	NR	Cecidomyiidae	Oliveira et al. (2011)
Carbohydrate	Starch	Yes	+	Outer	Yes	+	Fabaceae	*C. langsdorffi*	Leaf	Horn-gall	M	NR	Cecidomyiidae	Oliveira et al. (2011)
Protein		Yes	+	Inner	Yes	–	Fabaceae	*C. langsdorffi*	Leaf	Horn-gall	M	NR	Cecidomyiidae	Oliveira et al. (2011)
Protein		Yes	+	Inner	Yes	–	Fabaceae	*C. langsdorffi*	Leaf	Pocket-gall	M	NR	Cecidomyiidae	Oliveira et al. (2011)
Carbohydrate	Reducing sugars	Yes	+	Inner	Yes	–	Fabaceae	*C. langsdorffi*	Leaf	Horn-gall	M	NR	Cecidomyiidae	Oliveira et al. (2011)
Carbohydrate	Reducing sugars	Yes	+	Both	Yes	–	Fabaceae	*C. langsdorffi*	Leaf	Pocket-gall	M	NR	Cecidomyiidae	Oliveira et al. (2011)
Lipids		Yes	+	NR	Yes	+	Fabaceae	*C. langsdorffi*	Leaf	Pocket-gall	M	NR	Cecidomyiidae	Oliveira et al. (2011)
Lipids		Yes	+	NR	Yes	+	Fabaceae	*C. langsdorffi*	Leaf	Horn-gall	M	NR	Cecidomyiidae	Oliveira et al. (2011)
Lipids		Yes	+	Inner	Yes	–	Fagaceae	*Q. robur*	Leaf	Lenticular	M	*N. quercusbaccarum*	Cynipidae	Koncz (2011)
Carbohydrate		Yes	+	Outer	Yes	–	Fagaceae	*Q. robur*	Leaf	Lenticular	M	*N. quercusbaccarum*	Cynipidae	Koncz (2011)
Protein		Yes	+	Inner	Yes	–	Fagaceae	*Q. robur*	Leaf	Lenticular	M	*N. quercusbaccarum*	Cynipidae	Koncz (2011)
Carbohydrate	Starch	Yes	–		Yes	+	Myrtaceae	*P. myrtoides*	Leaf	Globoid	M	*N. myrtoides*	Psylloidea	Carneiro et al. (2014)
Carbohydrate	Reducing sugars		+	Outer	Yes	+	Myrtaceae	*P. myrtoides*	Leaf	Globoid	M	*N. myrtoides*	Psylloidea	Carneiro et al. (2014)
Protein		Yes	+	Inner	Yes	+	Myrtaceae	*P. myrtoides*	Leaf	Globoid	M	*N. myrtoides*	Psylloidea	Carneiro et al. (2014)
Lipids		Yes	+	Both	Yes	+	Myrtaceae	*P. myrtoides*	Leaf	Globoid	M	*N. myrtoides*	Psylloidea	Carneiro et al. (2014)
Protein		Yes	+	Both	Yes	+	Ochnaceae	*O. spectabilis*	Leaf	Conic	M	NR	Cecidomyiidae	Silva and Scareli-Santos (2018)
Carbohydrate	Starch	Yes	+	Inner	Yes	+	Ochnaceae	*O. spectabilis*	Leaf	Conic	M	NR	Cecidomyiidae	Silva and Scareli-Santos (2018)
Lipids		Yes	+	Outer	Yes	+	Ochnaceae	*O. spectabilis*	Leaf	Conic	M	NR	Cecidomyiidae	Silva and Scareli-Santos (2018)
Carbohydrate	Starch	Yes	+	Both	No	–	Fagaceae	*Q. petraea*	Leaf	NR	M	*C. quercusfolii*	NR	Guzicka et al. (2017)

(continued)

Table 1 (continued)

Carbohydrate	Starch	Yes	+	Outer	No	–	NR	NR	Leaf	NR	NR	*B. pallida*	Cynipidae	Bronner (1975)
Lipids		Yes	+	Inner	No	–	NR	NR	Leaf	NR	NR	*B. pallida*	Cynipidae	Bronner (1975)
Lipids		Yes	+	Inner	No	–	NR	NR	Leaf	NR	NR	*B. pallida*	Cynipidae	Bronner (1975)
Carbohydrate	Starch	Yes	+	Inner	No	–	NR	NR	Leaf	NR	NR	*B. pallida*	Cynipidae	Bronner (1975)
Carbohydrate	Starch	Yes	+	Inner	Yes	+	Clusiaceae	*C. fluminensis*	Leaf	Lenticular	M	NR	Cecidomyiidae	Guimarães et al. (2021)
Lipids		Yes	+	Inner	Yes	+	Clusiaceae	*C. fluminensis*	Leaf	Lenticular	M	NR	Cecidomyiidae	Guimarães et al. (2021)
Carbohydrate	Reducing sugars	Yes	+	Inner	Yes	+	Clusiaceae	*C. fluminensis*	Leaf	Lenticular	M	NR	Cecidomyiidae	Guimarães et al. (2021)
Carbohydrate	Starch	Yes	+	Outer	Yes	–	Rosaceae	*R. virginiana*	Leaf	Lenticular	Y	*D. rosaefolii*	Hymenoptera	LeBlanc and Lacroix (2001)
Lipids		Yes	+	Inner	Yes	–	Rosaceae	*R. virginiana*	Leaf	Lenticular	Y	*D. rosaefolii*	Hymenoptera	LeBlanc and Lacroix (2001)
Carbohydrate	Starch	Yes	+	Inner	Yes	–	Rosaceae	*R. virginiana*	Leaf	Lenticular	Y	*P. brandtii*	Hymenoptera	LeBlanc and Lacroix (2001)
Lipids		Yes	+	Inner	Yes	–	Rosaceae	*R. virginiana*	Leaf	Lenticular	Y	*P. brandtii*	Hymenoptera	LeBlanc and Lacroix (2001)
Carbohydrate	Starch	Yes	+	Inner	No	–	Lauraceae	*M. zuihoensis*	Leaf	Urn-shaped gall	M	NR	Cecidomyiidae	Chao and Liao (2013)
Carbohydrate	Starch	Yes	+	Inner	Yes	+	Euphorbiaceae	*C. quercifolius*	Leaf	Globoid	M	NR	Diptera	Santos et al. (2014)
Lipids		Yes	+	Inner	Yes	+	Euphorbiaceae	*C. quercifolius*	Leaf	Globoid	M	NR	Diptera	Santos et al. (2014)
Carbohydrate	Starch	Yes	+	Outer	Yes	–	Euphorbiaceae	*C. blanchetianus*	Leaf	Globoid	M	NR	Diptera	Santos et al. (2014)
Lipids		Yes	+	Inner	Yes	+	Euphorbiaceae	*C. blanchetianus*	Leaf	Globoid	M	NR	Diptera	Santos et al. (2014)
Carbohydrate	Starch	Yes	–		Yes	+	Euphorbiaceae	*M. epruinosa*	Leaf	Clavate	M	NR	Diptera	Santos et al. (2014)
Lipids		Yes	+	Inner	Yes	+	Euphorbiaceae	*M. epruinosa*	Leaf	Clavate	M	NR	Diptera	Santos et al. (2014)
Carbohydrate	Starch	Yes	–		Yes	+	Anacardiaceae	*S. polygama*	Leaf	Globoid	M	*C. mammifex*	Calophyidae	Guedes et al. (2018b)
Carbohydrate	Reducing sugars	Yes	+	Both	Yes	+	Anacardiaceae	*S. polygama*	Leaf	Globoid	M	*C. mammifex*	Calophyidae	Guedes et al. (2018b)

Lipids		Yes	+	Outer	Yes	+	Anacardiaceae	*S. polygama*	Leaf	Globoid	M	*C. mammifex*	Calophyidae	Guedes et al. (2018b)
Protein		Yes	+	Both	Yes	+	Anacardiaceae	*S. polygama*	Leaf	Globoid	M	*C. mammifex*	Calophyidae	Guedes et al. (2018b)
Carbohydrate	Starch	Yes	+	Outer	Yes	+	Sapotaceae	*P. ramiflora*	Leaf	Intralaminar	M	*C. pouteriae*	Triozidae	Oliveira et al. (2019)
Carbohydrate	Reducing sugars	Yes	+	Both	Yes	+	Sapotaceae	*P. ramiflora*	Leaf	Intralaminar	M	*C. pouteriae*	Triozidae	Oliveira et al. (2019)
Protein		Yes	+	Both	Yes	+	Sapotaceae	*P. ramiflora*	Leaf	Intralaminar	M	*C. pouteriae*	Triozidae	Oliveira et al. (2019)
Lipids		Yes	+	Outer	Yes	+	Sapotaceae	*P. ramiflora*	Leaf	Intralaminar	M	*C. pouteriae*	Triozidae	Oliveira et al. (2019)
Carbohydrate	Starch	Yes	+	Outer	Yes	+	Apocynaceae	*A. tomentosum*	Leaf	Intralaminar	Y	*P. longicaudatum*	Hemiptera	Martini et al. (2020b)
Carbohydrate	Starch	Yes	+	Both	Yes	+	Apocynaceae	*A. tomentosum*	Leaf	Intralaminar	M	*P. longicaudatum*	Hemiptera	Martini et al. (2020b)
Carbohydrate	Reducing sugars	Yes	+	Outer	Yes	+	Apocynaceae	*A. tomentosum*	Leaf	Intralaminar	Y	*P. longicaudatum*	Hemiptera	Martini et al. (2020b)
Carbohydrate	Reducing sugars	Yes	+	Outer	Yes	+	Apocynaceae	*A. tomentosum*	Leaf	Intralaminar	M	*P. longicaudatum*	Hemiptera	Martini et al. (2020b)
Protein		Yes	+	Outer	Yes	+	Apocynaceae	*A. tomentosum*	Leaf	Intralaminar	Y	*P. longicaudatum*	Hemiptera	Martini et al. (2020b)
Protein		Yes	+	Outer	Yes	+	Apocynaceae	*A. tomentosum*	Leaf	Intralaminar	M	*P. longicaudatum*	Hemiptera	Martini et al. (2020b)
Protein		Yes	+	Outer	Yes	+	Apocynaceae	*A. tomentosum*	Leaf	Intralaminar	S	*P. longicaudatum*	Hemiptera	Martini et al. (2020b)
Lipids		Yes	+	Inner	Yes	+	Apocynaceae	*A. tomentosum*	Leaf	Intralaminar	M	*P. longicaudatum*	Hemiptera	Martini et al. (2020b)
Lipids		Yes	+	Outer	Yes	+	Apocynaceae	*A. tomentosum*	Leaf	Intralaminar	S	*P. longicaudatum*	Hemiptera	Martini et al. (2020b)
Carbohydrate	Starch	Yes	+	Inner	Yes	–	Myrtaceae	*E. uniflora*	Leaf	Intralaminar	M	*E. dispar*	Cecidomyiidae	Rezende et al. (2018)
Carbohydrate	Reducing sugars	Yes	+	Inner	Yes	–	Myrtaceae	*E. uniflora*	Leaf	Intralaminar	M	*E. dispar*	Cecidomyiidae	Rezende et al. (2018)
Protein		Yes	+	Inner	Yes	–	Myrtaceae	*E. uniflora*	Leaf	Intralaminar	M	*E. dispar*	Cecidomyiidae	Rezende et al. (2018)
Lipids		Yes	+	Outer	Yes	–	Myrtaceae	*E. uniflora*	Leaf	Intralaminar	M	*E. dispar*	Cecidomyiidae	Rezende et al. (2018)
Carbohydrate	Starch	Yes	+	Inner	No	–	Moraceae	*E. racemosa*	Leaf	Globoid	M	NR	NR	Trivedi and Tiagi (1980)
Protein		Yes	+	Inner	Yes	+	Sapotaceae	*M. elengi*	Leaf	Intralaminar	Y	*A. ramakrishnae*	Phlaeothripidae	Gopinathan (1987)
Protein		Yes	+	Inner	Yes	–	Sapotaceae	*M. elengi*	Leaf	Intralaminar	M	*A. ramakrishnae*	Phlaeothripidae	Gopinathan (1987)

(continued)

Table 1 (continued)

Protein		Yes	+	Inner	Yes	–	Sapotaceae	*M. elengi*	Leaf	Intralaminar	S	*A. ramakrishnae*	Phlaeothripidae	Gopinathan (1987)
Lipids		Yes	+	Inner	Yes	–	Sapotaceae	*M. elengi*	Leaf	Intralaminar	M	*A. ramakrishnae*	Phlaeothripidae	Gopinathan (1987)
Carbohydrate	Starch	Yes	+	Inner	Yes	+	Sapotaceae	*M. elengi*	Leaf	Intralaminar	Y	*A. ramakrishnae*	Phlaeothripidae	Gopinathan (1987)
Carbohydrate	Starch	Yes	+	Inner	Yes	+	Sapotaceae	*M. elengi*	Leaf	Intralaminar	M	*A. ramakrishnae*	Phlaeothripidae	Gopinathan (1987)
Carbohydrate	Starch	Yes	+	Inner	Yes	+	Sapotaceae	*M. elengi*	Leaf	Intralaminar	S	*A. ramakrishnae*	Phlaeothripidae	Gopinathan (1987)
Protein		Yes	+	Inner	Yes	+	Melastomataceae	*O. lowsoni*	Stem	Fusiform	M	*Miastor* spp.	Cecidomyiidae	Karun et al. (2010)
Carbohydrate	Starch	Yes	+	Inner	Yes	–	Melastomataceae	*O. lowsoni*	Stem	Fusiform	M	*Miastor* spp.	Cecidomyiidae	Karun et al. (2010)
Carbohydrate	Starch	Yes	+	Inner	No	–	Anacardiaceae	*P. terebinthus*	Leaf	Globoid	M	*G. articularia*	Aphididae	Álvarez (2012)
Carbohydrate	Starch	Yes	+	Inner	No	–	Anacardiaceae	*P. terebinthus*	Leaf	Horn-shaped	M	*B. pistaciae*	Aphididae	Álvarez (2012)
Lipids		Yes	+	Inner	No	–	Melastomataceae	*T. pulchra*	Leaf	Globoid	Y	NR	Lepidoptera	Vecchi et al. (2013)
Carbohydrate	Starch	Yes	+	Inner	Yes	+	Combretaceae	*T. arjilna*	Stem	Globoid	M	*Triozallectcheri* sp.	Homoptera	Kumar and Mathur (2009)
Carbohydrate	Total insoluble polysaccharides	Yes	+	Both	Yes	+	Combretaceae	*T. arjilna*	Stem	Globoid	M	*Triozallectcheri* sp.	Homoptera	Kumar and Mathur (2009)
Lipids		Yes	+	Inner	Yes	+	Combretaceae	*T. arjilna*	Stem	Globoid	M	*Triozallectcheri* sp.	Homoptera	Kumar and Mathur (2009)
Protein		Yes	+	Both	Yes	+	Combretaceae	*T. arjilna*	Stem	Globoid	M	*Triozallectcheri* sp.	Homoptera	Kumar and Mathur (2009)
Protein		Yes	+	Both	Yes	+	Fabaceae	*P. cineraria*	Stem	NR	M	*L. prosopidis*	Cecidomyiidae	Arora and Patni (2001)
Carbohydrate	Starch	Yes	+	Both	Yes	+	Fabaceae	*P. cineraria*	Stem	NR	M	*L. prosopidis*	Cecidomyiidae	Arora and Patni (2001)
Lipids		Yes	+	Both	Yes	+	Fabaceae	*P. cineraria*	Stem	NR	M	*L. prosopidis*	Cecidomyiidae	Arora and Patni (2001)
Carbohydrate	Starch	Yes	+	Both	No	–	Podostemaceae	*N. coulteriana*	Leaf	Fusiform	M	NR	Chironomidae	Jäger-Zürn (2013)
Protein		Yes	+	Inner	No	–	Podostemaceae	*N. coulteriana*	Leaf	Fusiform	M	NR	Chironomidae	Jäger-Zürn (2013)
Carbohydrate	Starch	Yes	+	NR	No	–	Podostemaceae	*N. coulteriana*	Leaf	Fusiform	M	NR	Chironomidae	Jäger-Zürn (2013)
Carbohydrate	Starch	Yes	–	–	Yes	–	Sapindaceae	*M. guianensis*	Stem	NR	Y	*B. mataybe*	Eriococcidae	Silva et al. (2019)

Carbohydrate	Starch	Yes	–	–	Yes	–	Sapindaceae	*M. guianensis*	Stem	NR	M	*B. mataybe*	Eriococcidae	Silva et al. (2019)
Carbohydrate	Starch	Yes	+	Inner	Yes	–	Sapindaceae	*M. guianensis*	Leaf	Globoid	M	*B. mataybe*	Eriococcidae	Silva et al. (2019)
Carbohydrate	Starch	Yes	+	Inner	Yes	–	Sapindaceae	*M. guianensis*	Leaf	Globoid	S	*B. mataybe*	Eriococcidae	Silva et al. (2019)
Carbohydrate	Reducing sugars	Yes	+	Outer	Yes	+	Sapindaceae	*M. guianensis*	Stem	NR	Y	*B. mataybe*	Eriococcidae	Silva et al. (2019)
Carbohydrate	Reducing sugars	Yes	+	Outer	Yes	+	Sapindaceae	*M. guianensis*	Stem	NR	M	*B. mataybe*	Eriococcidae	Silva et al. (2019)
Carbohydrate	Reducing sugars	Yes	+	Outer	Yes	+	Sapindaceae	*M. guianensis*	Leaf	Globoid	M	*B. mataybe*	Eriococcidae	Silva et al. (2019)
Carbohydrate	Reducing sugars	Yes	+	Outer	Yes	+	Sapindaceae	*M. guianensis*	Leaf	Globoid	S	*B. mataybe*	Eriococcidae	Silva et al. (2019)
Protein		Yes	+	Both	Yes	+	Sapindaceae	*M. guianensis*	Stem	NR	Y	*B. mataybe*	Eriococcidae	Silva et al. (2019)
Protein		Yes	+	Both	Yes	+	Sapindaceae	*M. guianensis*	Stem	NR	M	*B. mataybe*	Eriococcidae	Silva et al. (2019)
Protein		Yes	+	Both	Yes	+	Sapindaceae	*M. guianensis*	Leaf	Globoid	M	*B. mataybe*	Eriococcidae	Silva et al. (2019)
Protein		Yes	+	Both	Yes	+	Sapindaceae	*M. guianensis*	Leaf	Globoid	S	*B. mataybe*	Eriococcidae	Silva et al. (2019)
Lipids		Yes	+	Both	Yes	+	Sapindaceae	*M. guianensis*	Stem	NR	Y	*B. mataybe*	Eriococcidae	Silva et al. (2019)
Lipids		Yes	+	Both	Yes	+	Sapindaceae	*M. guianensis*	Stem	NR	M	*B. mataybe*	Eriococcidae	Silva et al. (2019)
Lipids		Yes	+	Both	Yes	–	Sapindaceae	*M. guianensis*	Leaf	Globoid	M	*B. mataybe*	Eriococcidae	Silva et al. (2019)
Lipids		Yes	+	Both	Yes	–	Sapindaceae	*M. guianensis*	Leaf	Globoid	S	*B. mataybe*	Eriococcidae	Silva et al. (2019)
Carbohydrate	Starch	Yes	+	NR	No	–	Asteraceae	*B. concinna*	Leaf	NR	M	NR	Cecidomyiidae	Arduin and Kraus (2001)
Carbohydrate	Starch	Yes	+	Mesophyll	No	–	Asteraceae	*B. dracunculifolia*	Leaf	NR	M	NR	Cecidomyiidae	Arduin and Kraus (2001)
Lipids		Yes	+	Vascular tissues	Yes	+	Asteraceae	*V. polyanthes*	Leaf	Stems and fusiforms	Y	*T. rudolphi*	Tephritidae	Andrade et al. (1995)
Carbohydrate	Starch	Yes	+	Vascular tissues	Yes	+	Asteraceae	*V. polyanthes*	Leaf	Stems and fusiforms	Y	*T. rudolphi*	Tephritidae	Andrade et al. (1995)
Carbohydrate	Starch	Yes	+	Outer	No	–	Malpighiaceae	*B. sericea*	Inflorescence	Piriformis	M	NR	Lepidoptera	Vieira and Kraus (2007)
Lipids		Yes	+	Inner	No	–	Malpighiaceae	*B. sericea*	Inflorescence	Piriformis	M	NR	Lepidoptera	Vieira and Kraus (2007)
Carbohydrate	Starch	Yes	+	Outer	Yes	+	Apocynaceae	*A. scholaris*	Leaf	Globoid	M	*P. tuberculata*	Psyllidae	Raghav et al. (2007)

(continued)

Table 1 (continued)

Carbohydrate		Yes	+	Both	Yes	+	Apocynaceae	*A. scholaris*	Leaf	Globoid	M	*P. tuberculata*	Psyllidae	Raghav et al. (2007)
Protein		Yes	+	Both	Yes	+	Apocynaceae	*A. scholaris*	Leaf	Globoid	M	*P. tuberculata*	Psyllidae	Raghav et al. (2007)
Lipids		Yes	+	Both	Yes	+	Apocynaceae	*A. scholaris*	Leaf	Globoid	M	*P. tuberculata*	Psyllidae	Raghav et al. (2007)
Carbohydrate	Starch	Yes	+	Mesophyll	No	–	Melastomataceae	*M. taxifolia*	Glandular trichomes	NR	M	*NR*	Lepidoptera	Ferreira et al. (2014)
Protein		Yes	+	Outer	No	–	Melastomataceae	*M. taxifolia*	Stem	NR	M	*NR*	Lepidoptera	Ferreira et al. (2014)
Carbohydrate	Reducing sugars	Yes	+	Inner	No	–	Melastomataceae	*M. taxifolia*	Stem	NR	M	*NR*	Lepidoptera	Ferreira et al. (2014)
Lipids		Yes	+	Inner	No	–	Melastomataceae	*M. taxifolia*	Leaf	NR	M	*NR*	Lepidoptera	Ferreira et al. (2014)
Carbohydrate	Starch	Yes	+	NR	No	–	Euphorbiaceae	*C. floribundus*	Leaf	NR	M	*Clinodiplosis* sp.	Cecidomyiidae	Ferreira et al. (2014)
Carbohydrate	Starch	Yes	+	NR	No	–	Euphorbiaceae	*C. floribundus*	Stem	NR	M	*Clinodiplosis* sp.	Cecidomyiidae	Ferreira et al. (2014)
Carbohydrate	Reducing sugars	Yes	+	NR	No	–	Euphorbiaceae	*C. floribundus*	Leaf	NR	M	*Clinodiplosis* sp.	Cecidomyiidae	Ferreira et al. (2014)
Carbohydrate	Starch	Yes	+	Inner	Yes	+	Meliaceae	*G. macrophylla*	Leaf	NR	Y	*NR*	Cecidomyiidae	Kraus et al. (1996)
Carbohydrate	Starch	Yes	+	Outer	Yes	+	Fabaceae	*M. gemmulata*	Leaf	Lenticular bivalve	M	*Lopesia* spp.	Cecidomyiidae	Costa et al. (2021)
Carbohydrate	Reducing sugars	Yes	+	Both	Yes	+	Fabaceae	*M. gemmulata*	Leaf	Lenticular bivalve	M	*Lopesia* spp.	Cecidomyiidae	Costa et al. (2021)
Protein		Yes	+	Both	Yes	+	Fabaceae	*M. gemmulata*	Leaf	Lenticular bivalve	M	*Lopesia* spp.	Cecidomyiidae	Costa et al. (2021)
Lipids		Yes	+	Both	Yes	+	Fabaceae	*M. gemmulata*	Leaf	Lenticular bivalve	M	*Lopesia* spp.	Cecidomyiidae	Costa et al. (2021)
Carbohydrate	Starch	Yes	+	Outer	Yes	+	Fabaceae	*M. gemmulata*	Leaf	Bivalve lanceolate	M	*Lopesia* spp.	Cecidomyiidae	Costa et al. (2021)
Carbohydrate	Reducing sugars	Yes	+	Both	Yes	+	Fabaceae	*M. gemmulata*	Leaf	Bivalve lanceolate	M	*Lopesia* spp.	Cecidomyiidae	Costa et al. (2021)
Protein		Yes	+	Both	Yes	+	Fabaceae	*M. gemmulata*	Leaf	Bivalve lanceolate	M	*Lopesia* spp.	Cecidomyiidae	Costa et al. (2021)

Lipids		Yes	+	Both	Yes	+	Fabaceae	*M. gemmulata*	Leaf	Bivalve lanceolate	M	*Lopesia* spp.	Cecidomyiidae	Costa et al. (2021)
Carbohydrate	Starch	Yes	+	Outer	Yes	+	Fabaceae	*M. gemmulata*	Leaf	Bivalve lanceolate	S	*Lopesia* spp.	Cecidomyiidae	Costa et al. (2021)
Carbohydrate	Reducing sugars	Yes	+	Both	Yes	+	Fabaceae	*M. gemmulata*	Leaf	Bivalve lanceolate	S	*Lopesia* spp.	Cecidomyiidae	Costa et al. (2021)
Protein		Yes	+	Both	Yes	+	Fabaceae	*M. gemmulata*	Leaf	Bivalve lanceolate	S	*Lopesia* spp.	Cecidomyiidae	Costa et al. (2021)
Lipids		Yes	+	Both	Yes	+	Fabaceae	*M. gemmulata*	Leaf	Bivalve lanceolate	S	*Lopesia* spp.	Cecidomyiidae	Costa et al. (2021)
Carbohydrate	Starch	Yes	+	Both	Yes	+	Fabaceae	*M. gemmulata*	Leaf	Bivalve globoid	M	*Lopesia* spp.	Cecidomyiidae	Costa et al. (2021)
Carbohydrate	Reducing sugars	Yes	+	Both	Yes	+	Fabaceae	*M. gemmulata*	Leaf	Bivalve globoid	M	*Lopesia* spp.	Cecidomyiidae	Costa et al. (2021)
Protein		Yes	+	Both	Yes	+	Fabaceae	*M. gemmulata*	Leaf	Bivalve globoid	M	*Lopesia* spp.	Cecidomyiidae	Costa et al. (2021)
Lipids		Yes	+	Both	Yes	+	Fabaceae	*M. gemmulata*	Leaf	Bivalve globoid	M	*Lopesia* spp.	Cecidomyiidae	Costa et al. (2021)
Protein				Inner	Yes	+	Fabaceae	*P. glabra*	Stem	NR	M	NR	Diptera	Gopinathan and Suresh (1985)
Carbohydrate	Starch	Yes	+	Pith	Yes	+	Fabaceae	*P. glabra*	Stem	NR	M	NR	Diptera	Gopinathan and Suresh (1985)
Carbohydrate	Starch	Yes	+	NR	Yes	+	Moraceae	*F. glomerata*	Leaf	Globoid	Y	*P. depressa*	Psyllidae	Dsouza and Ravishankar (2014)
Carbohydrate	Reducing sugars	Yes	+	NR	Yes	+	Moraceae	*F. glomerata*	Leaf	Globoid	Y	*P. depressa*	Psyllidae	Dsouza and Ravishankar (2014)
Protein		Yes	+	NR	Yes	+	Moraceae	*F. glomerata*	Leaf	Globoid	Y	*P. depressa*	Psyllidae	Dsouza and Ravishankar (2014)
Protein		Yes	+	Outer	Yes	+	Apocynaceae	*A. spruceanum*	NR	NR	Y	*NR*	Cecidomyiidae	Oliveira et al. (2010)

(continued)

Table 1 (continued)

Protein		Yes	+	Outer	Yes	+	Apocynaceae	*A. spruceanum*	NR	NR	M	*NR*	Cecidomyiidae	Oliveira et al. (2010)
Protein		Yes	+	Outer	Yes	+	Apocynaceae	*A. spruceanum*	NR	NR	S	*NR*	Cecidomyiidae	Oliveira et al. (2010)
Carbohydrate	Starch	Yes	+	Inner	Yes	+	Apocynaceae	*A. spruceanum*	NR	NR	Y	*NR*	Cecidomyiidae	Oliveira et al. (2010)
Carbohydrate	Starch	Yes	+	Inner	Yes	+	Apocynaceae	*A. spruceanum*	NR	NR	M	*NR*	Cecidomyiidae	Oliveira et al. (2010)
Carbohydrate	Reducing sugars	Yes	+	Inner	Yes	+	Apocynaceae	*A. spruceanum*	NR	NR	M	*NR*	Cecidomyiidae	Oliveira et al. (2010)
Lipids		Yes	+	Both	Yes	+	Fabaceae	*M. tenuiflora*	Leaf	NR	M	*Lopesia* sp.	Cecidomyiidae	Nogueira et al. (2018)
Carbohydrate	Reducing sugars	Yes	+	Outer	Yes	+	Fabaceae	*M. tenuiflora*	Leaf	NR	M	*Lopesia* sp.	Cecidomyiidae	Nogueira et al. (2018)
Lipids		Yes	+	Both	Yes	+	Fabaceae	*M. tenuiflora*	Leaf	NR	M	*Lopesia* sp.	Cecidomyiidae	Nogueira et al. (2018)
Lipids		Yes	+	Both	Yes	+	Fabaceae	*M. tenuiflora*	Leaf	NR	S	*Lopesia* sp.	Cecidomyiidae	Nogueira et al. (2018)
Carbohydrate	Starch	Yes	+	Outer	Yes	+	Solanaceae	*C. annuum*	Root	NR	M	*M. incognita*	Heteroderidae	Trivedi and Tiagi (1980)
Carbohydrate	Starch	Yes	+	Inner	Yes	+	Fabaceae	*G. max*	Root	NR	M	*M. javanica*	Heteroderidae	Vilela et al. (2019)
Protein		Yes	+	Inner	Yes	+	Fabaceae	*G. max*	Root	NR	M	*M. javanica*	Heteroderidae	Vilela et al. (2019)
Carbohydrate	Reducing sugars	Yes	+	Vascular bundles	Yes	+	Ulmaceae	*U. minor*	Leaf	Globoid	M	*E. ulmi*	Aphididae	Ferreira et al. (2017a)
Protein		Yes	+	Vascular bundles	Yes	+	Ulmaceae	*U. minor*	Leaf	Globoid	M	*E. ulmi*	Aphididae	Ferreira et al. (2017a)
Carbohydrate	Starch	Yes	+	Outer	Yes	+	Melastomataceae	*M. albicans*	Leaf	Globoid	M	*D. gallaeformans*	Aphididae	Ferreira et al. (2017a)
Protein		Yes	+	Inner	Yes	+	Melastomataceae	*M. albicans*	Leaf	Globoid	M	*D. gallaeformans*	Aphididae	Ferreira et al. (2017a)

Carbohydrate	Reducing sugars	Yes	+	Inner	Yes	+	Melastomataceae	*M. albicans*	Leaf	Globoid	M	*D. gallaeformans*	Aphididae	Ferreira et al. (2017a)
Carbohydrate	Starch	Yes	+	Outer	Yes	+	Melastomataceae	*M. ibaguensis*	Leaf	Globoid	M	*D. gallaeformans*	Aphididae	Ferreira et al. (2017a)
Carbohydrate	Reducing sugars	Yes	+	Inner	Yes	+	Melastomataceae	*M. ibaguensis*	Leaf	Globoid	M	*D. gallaeformans*	Aphididae	Ferreira et al. (2017a)
Protein		Yes	+	Inner	Yes	+	Melastomataceae	*M. ibaguensis*	Leaf	Globoid	M	*D. gallaeformans*	Aphididae	Ferreira et al. (2017a)
Carbohydrate	Starch	Yes	+	Both	Yes	+	Melastomataceae	*M. ibaguensis*	Leaf	NR	M	*NR*	Eriophyidae	Ferreira et al. (2017a)
Carbohydrate	Reducing sugars	Yes	+	Inner	Yes	+	Melastomataceae	*M. ibaguensis*	Leaf	NR	M	*NR*	Eriophyidae	Ferreira et al. (2017a)
Lipids		Yes	+	Both	Yes	+	Melastomataceae	*M. ibaguensis*	Leaf	NR	M	*NR*	Eriophyidae	Ferreira et al. (2017a)
Protein		Yes	+	Both	Yes	+	Melastomataceae	*M. ibaguensis*	Leaf	NR	M	*NR*	Eriophyidae	Ferreira et al. (2017a)
Carbohydrate	Starch	Yes	+	Outer	Yes	+	Juglandaceae	*J. regia*	Leaf	Lenticular	M	*A. tristiata*	Eriophyidae	Ferreira et al. (2017a)
Carbohydrate	Reducing sugars	Yes	+	Outer	Yes	+	Juglandaceae	*J. regia*	Leaf	Lenticular	M	*A. tristiata*	Eriophyidae	Ferreira et al. (2017a)
Protein		Yes	+	Inner	Yes	+	Juglandaceae	*J. regia*	Leaf	Lenticular	M	*A. tristiata*	Eriophyidae	Ferreira et al. (2017a)
Protein		Yes	+	Inner	Yes	+	Anacardiaceae	*P. terebinthus*	Leaf	Globoid	Y	*G. utricularia*	Aphididae	Ferreira et al. (2017a)
Carbohydrate	Starch	Yes	+	Inner	Yes	+	Anacardiaceae	*P. terebinthus*	Leaf	Globoid	Y	*G. utricularia*	Aphididae	Ferreira et al. (2017a)
Carbohydrate	Reducing sugars	Yes	+	Inner	Yes	+	Anacardiaceae	*P. terebinthus*	Leaf	Globoid	Y	*G. utricularia*	Aphididae	Ferreira et al. (2017a)
Carbohydrate	Reducing sugars	Yes	+	Inner	Yes	–	Apocynaceae	*A. spruceanum*	Leaf	Globoid	Y	NR	Cecidomyiidae	Oliveira et al. (2010)
Carbohydrate	Starch	Yes	+	Both	Yes	+	Apocynaceae	*A. spruceanum*	Leaf	Globoid	Y	NR	Cecidomyiidae	Oliveira et al. (2010)

(continued)

Table 1 (continued)

Protein		Yes	+	Both	Yes	+	Apocynaceae	*A. spruceanum*	Leaf	Globoid	Y	NR	Cecidomyiidae	Oliveira et al. (2010)
Lipids		Yes	+	Outer	Yes	+	Apocynaceae	*A. spruceanum*	Leaf	Globoid	Y	NR	Cecidomyiidae	Oliveira et al. (2010)
Carbohydrate	Reducing sugars	Yes	NR	Inner	Yes	+	Apocynaceae	*A. spruceanum*	Leaf	Globoid	M	NR	Cecidomyiidae	Oliveira et al. (2010)
Carbohydrate	Starch	Yes	NR	Both	Yes	+	Apocynaceae	*A. spruceanum*	Leaf	Globoid	M	NR	Cecidomyiidae	Oliveira et al. (2010)
Protein		Yes	NR	Inner	Yes	+	Apocynaceae	*A. spruceanum*	Leaf	Globoid	M	NR	Cecidomyiidae	Oliveira et al. (2010)
Lipids		Yes	NR	Outer	Yes	+	Apocynaceae	*A. spruceanum*	Leaf	Globoid	M	NR	Cecidomyiidae	Oliveira et al. (2010)
Carbohydrate	Reducing sugars	Yes	NR	Inner	Yes	+	Apocynaceae	*A. spruceanum*	Leaf	Globoid	S	NR	Cecidomyiidae	Oliveira et al. (2010)
Carbohydrate	Starch	Yes	NR		Yes	–	Apocynaceae	*A. spruceanum*	Leaf	Globoid	S	NR	Cecidomyiidae	Oliveira et al. (2010)
Protein		Yes	NR	Inner	Yes	+	Apocynaceae	*A. spruceanum*	Leaf	Globoid	S	NR	Cecidomyiidae	Oliveira et al. (2010)
Lipids		Yes	NR	NR	Yes	+	Apocynaceae	*A. spruceanum*	Leaf	Globoid	S	NR	Cecidomyiidae	Oliveira et al. (2010)
Carbohydrate	Starch	Yes	+	Outer	No	–	Metastomataceae	*M. radula*	Axillary Stem buds	Globoid	Pupa	*P. oligophaga*	Lepidoptera	Rezende et al. (2019)
Carbohydrate	Starch	Yes	+	Outer	No	–	Metastomataceae	*M. radula*	Axillary Stem buds	Globoid	Larva	*P. oligophaga*	Lepidoptera	Rezende et al. (2019)
Protein		Yes	+	Inner	No	–	Metastomataceae	*M. radula*	Axillary Stem buds	Globoid	Larva	*P. oligophaga*	Lepidoptera	Rezende et al. (2019)
Lipids		Yes	+	Inner	No	–	Metastomataceae	*M. radula*	Axillary Stem buds	Globoid	Larva	*P. oligophaga*	Lepidoptera	Rezende et al. (2019)
Carbohydrate	Starch	Yes	–		Yes	–	Melastomataceae	*M. chamissois*	Flower buds	Fruit-like gall	NR	*A. ubernandiensis*	Hymenoptera	Bomfim et al. (2020)
Carbohydrate	Reducing sugars	Yes	–		Yes	–	Melastomataceae	*M. chamissois*	Flower buds	Fruit-like gall	NR	*A. ubernandiensis*	Hymenoptera	Bomfim et al. (2020)

Protein		Yes	–		Yes	–	Melastomataceae	*M. chamissois*	Flower buds	Fruit-like gall	NR	*A. ubernandiensis*	Hymenoptera	Bomfim et al. (2020)
Lipids		Yes	–		Yes	+	Melastomataceae	*M. chamissois*	Flower buds	Fruit-like gall	NR	*A. ubernandiensis*	Hymenoptera	Bomfim et al. (2020)
Carbohydrate	Starch	Yes	–		Yes	–	Melastomataceae	*M. chamissois*	Flower buds	Ovule Gall	NR	*A. ubernandiensis*	Hymenoptera	Bomfim et al. (2020)
Carbohydrate	Reducing sugars	Yes	–		Yes	–	Melastomataceae	*M. chamissois*	Flower buds	Ovule Gall	NR	*A. ubernandiensis*	Hymenoptera	Bomfim et al. (2020)
Protein		Yes	+	Both	Yes	+	Melastomataceae	*M. chamissois*	Flower buds	Ovule Gall	NR	*A. ubernandiensis*	Hymenoptera	Bomfim et al. (2020)
Lipids		Yes	+	Both	Yes	+	Melastomataceae	*M. chamissois*	Flower buds	Ovule Gall	NR	*A. ubernandiensis*	Hymenoptera	Bomfim et al. (2020)
Carbohydrate	Starch	Yes	+	Outer	Yes	+	Melastomataceae	*Miconia* spp.	Inflorescence	NR	M	*D. gallaeformans*	Anguinidae	Ferreira et al. (2017b)
Carbohydrate	Reducing sugars	Yes	+	Inner	Yes	+	Melastomataceae	*Miconia* spp.	Inflorescence	NR	M	*D. gallaeformans*	Anguinidae	Ferreira et al. (2017b)
Protein		Yes	+	Inner	Yes	+	Melastomataceae	*Miconia* spp.	Inflorescence	NR	M	*D. gallaeformans*	Anguinidae	Ferreira et al. (2017b)
Lipids		No	–	Inner	Yes	+	Melastomataceae	*T. pulchra*	Leaf	Globoid	M	NR	Lepidoptera	Vecchi et al. (2013)
Carbohydrate	Starch	Yes	+	–	Yes	+	Apocynaceae	*Aspidosperma* spp.	Leaf	NR	M	NR	Cecidomyiidae	Oliveira et al. (2011)
Protein		Yes	+	Both	Yes	+	Apocynaceae	*A. spruceanum*	Leaf	NR	Y	NR	Cecidomyiidae	Oliveira et al. (2010)
Protein		Yes	+	Both	Yes	+	Apocynaceae	*A. spruceanum*	Leaf	NR	M	NR	Cecidomyiidae	Oliveira et al. (2010)
Protein		Yes	+	Both	Yes	+	Apocynaceae	*A. spruceanum*	Leaf	NR	S	NR	Cecidomyiidae	Oliveira et al. (2010)
Lipids		Yes	+	Outer	Yes	+	Apocynaceae	*A. spruceanum*	Leaf	NR	Y	NR	Cecidomyiidae	Oliveira et al. (2010)
Lipids		Yes	+	Outer	Yes	+	Apocynaceae	*A. spruceanum*	Leaf	NR	M	NR	Cecidomyiidae	Oliveira et al. (2010)
Carbohydrate	Starch	Yes	+	Inner	Yes	+	Apocynaceae	*A. spruceanum*	Leaf	NR	Y	NR	Cecidomyiidae	Oliveira et al. (2010)
Carbohydrate	Starch	Yes	+	Inner	Yes	+	Apocynaceae	*A. spruceanum*	Leaf	NR	M	NR	Cecidomyiidae	Oliveira et al. (2010)
Carbohydrate	Reducing sugars	Yes	+	Inner	Yes	+	Apocynaceae	*A. spruceanum*	Leaf	NR	M	NR	Cecidomyiidae	Oliveira et al. (2010)
Lipids		Yes	+	Inner	No	–	Fabaceae	*I. ingoides*	Leaf	Globoid	M	NR	Cecidomyiidae	Bragança et al. (2020)
Carbohydrate	Reducing sugars	Yes	+	Inner	No	–	Fabaceae	*I. ingoides*	Leaf	Globoid	M	NR	Cecidomyiidae	Bragança et al. (2020)

(continued)

Table 1 (continued)

Protein		Yes	+	Inner	No	–	Fabaceae	*I. ingoides*	Leaf	Globoid	M	NR	Cecidomyiidae	Bragança et al. (2020)
Lipids		Yes	+	Inner	No	–	Fabaceae	*I. ingoides*	Leaf	Fusiform	M	NR	Cecidomyiidae	Bragança et al. (2020)
Carbohydrate	Reducing sugars	Yes	+	Inner	No	–	Fabaceae	*I. ingoides*	Leaf	Fusiform	M	NR	Cecidomyiidae	Bragança et al. (2020)
Protein		Yes	+	Inner	No	–	Fabaceae	*I. ingoides*	Leaf	Fusiform	M	NR	Cecidomyiidae	Bragança et al. (2020)
Lipids		Yes	+	Inner	No	–	Fabaceae	*I. ingoides*	Leaf	Lenticular	M	NR	Cecidomyiidae	Bragança et al. (2020)
Carbohydrate	Reducing sugars	Yes	+	Inner	No	–	Fabaceae	*I. ingoides*	Leaf	Lenticular	M	NR	Cecidomyiidae	Bragança et al. (2020)
Protein		Yes	+	Inner	No	–	Fabaceae	*I. ingoides*	Leaf	Lenticular	M	*NR*	Cecidomyiidae	Bragança et al. (2020)
Carbohydrate	Starch	Yes	+	Outer	Yes	+	Anacardiaceae	*S. engler*	Stem	Globoid	Y	*E. minutanus*	Cecidosidae	Ferreira et al. (2022)
Carbohydrate	Reducing sugars	Yes	+	Outer	Yes	+	Anacardiaceae	*S. engler*	Stem	Globoid	Y	*E. minutanus*	Cecidosidae	Ferreira et al. (2022)
Protein		Yes	+	Both	Yes	+	Anacardiaceae	*S. engler*	Stem	Globoid	Y	*E. minutanus*	Cecidosidae	Ferreira et al. (2022)
Lipids		Yes	+	Both	Yes	+	Anacardiaceae	*S. engler*	Stem	Globoid	Y	*E. minutanus*	Cecidosidae	Ferreira et al. (2022)
Carbohydrate	Starch	Yes	+	Outer	Yes	+	Anacardiaceae	*S. engler*	Stem	Globoid	M	*E. minutanus*	Cecidosidae	Ferreira et al. (2022)
Carbohydrate	Reducing sugars	Yes	+	Outer	Yes	+	Anacardiaceae	*S. engler*	Stem	Globoid	M	*E. minutanus*	Cecidosidae	Ferreira et al. (2022)
Protein		Yes	+	Both	Yes	+	Anacardiaceae	*S. engler*	Stem	Globoid	M	*E. minutanus*	Cecidosidae	Ferreira et al. (2022)
Lipids		Yes	+	Both	Yes	+	Anacardiaceae	*S. engler*	Stem	Globoid	M	*E. minutanus*	Cecidosidae	Ferreira et al. (2022)
Protein		Yes	+	Inner	Yes	–	Poaceae	*P. australis*	Stem	Cylindrical	M	*L. lucens*	Chloropidae	Vandevyvere and Bruyn (1998)
Lipids		Yes	+	Outer	Yes	–	Poaceae	*P. australis*	Stem	Cylindrical	M	*L. lucens*	Chloropidae	Vandevyvere and Bruyn (1998)
Carbohydrate	Starch	Yes	–		Yes	–	Poaceae	*P. australis*	Stem	Cylindrical	M	*L. lucens*	Chloropidae	Vandevyvere and Bruyn (1998)
Lipids		Yes	+	NR	NR	–	Betulaceae	*A. glutinosa*	Leaf	Globoid	NR	*E. laevis*	Eriophyidae	Kane et al. (1997)
Carbohydrate	Polysaccharides	Yes	+	NR	NR	–	Betulaceae	*A. glutinosa*	Leaf	Globoid	NR	*E. laevis*	Eriophyidae	Kane et al. (1997)
Lipids		Yes	+	NR	NR	–	Betulaceae	*A. incana*	Leaf	Globoid	NR	*E. laevis*	Eriophyidae	Kane et al. (1997)
Carbohydrate	Polysaccharides	Yes	+	NR	NR	–	Betulaceae	*A. incana*	Leaf	Globoid	NR	*E. laevis*	Eriophyidae	Kane et al. (1997)

Data collected: (i) scientific name of the host plant; (ii) host plant family; (iii) organ where gall induction occurred; (iv) type and subtype of the primary metabolite stored in the gall; (v) positive or negative test result, both for the gall and the non-galled organ; (vi) scientific name of the gall induce species; (vii) gall inducer taxonomic group; (viii) gall morphotype; (ix) gall development stage; and (x) region of the gall where the metabolite was found

Symbols indicate positive (+) and negative (–) results for the presence of metabolites on non-galled tissues and galls. *M* mature, *S* senescent and *Y* Young

[a]NR means there is no information

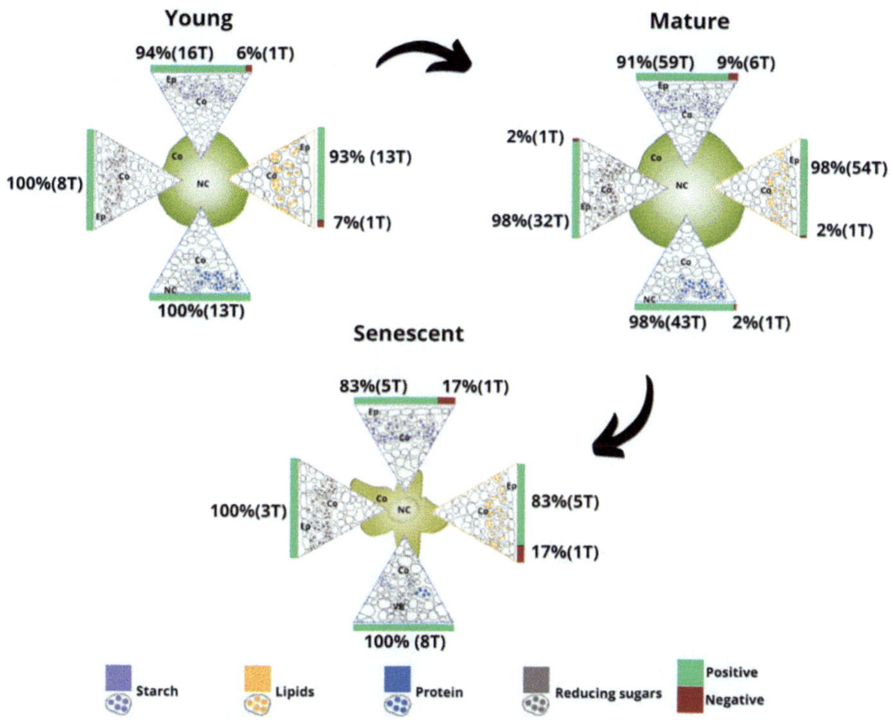

Fig. 1 Scheme of the absolute number (indicated by T = times of occurrence) associated with the percentage of positive (green bars) and negative (red bars) results for histochemical tests for starch grains, lipids, reducing sugars, and proteins in three stages of gall development (young, mature, and senescent) from the data collected. We used the developmental stages of *Sapium glandulosum* galls induced by *Neolithus fasciathus* as a representation. Abbreviations: Co Cortex, Ep Epidermis, NC Nymphal chamber, VB Vascular bundle

question: which stage of gall development is most distinct from the non-galled one? So, we did chi-square tests (see Supplementary material: statistical data) to assess the probability of primary metabolites occurring between host organs and young and mature galls. We found similarity among young galls and non-galled organs (Fig. 3), which demonstrates that despite the structural change at this stage, there is a high probability of occurrence of primary metabolites with the host organ. The opposite pattern was found for mature galls (chi-q = 19.17; $p < 0.001$) (Fig. 4), i.e., with a low probability of starch, reducing sugars, lipids, and proteins occurring between host organs and mature galls. This relationship may be due to the mature galls, in general, present deep structural changes from host organs, which are associated with new metabolite storage and phloem drain patterns (Baggato and Shorthouse 1994; Schoonhoven et al. 2005).

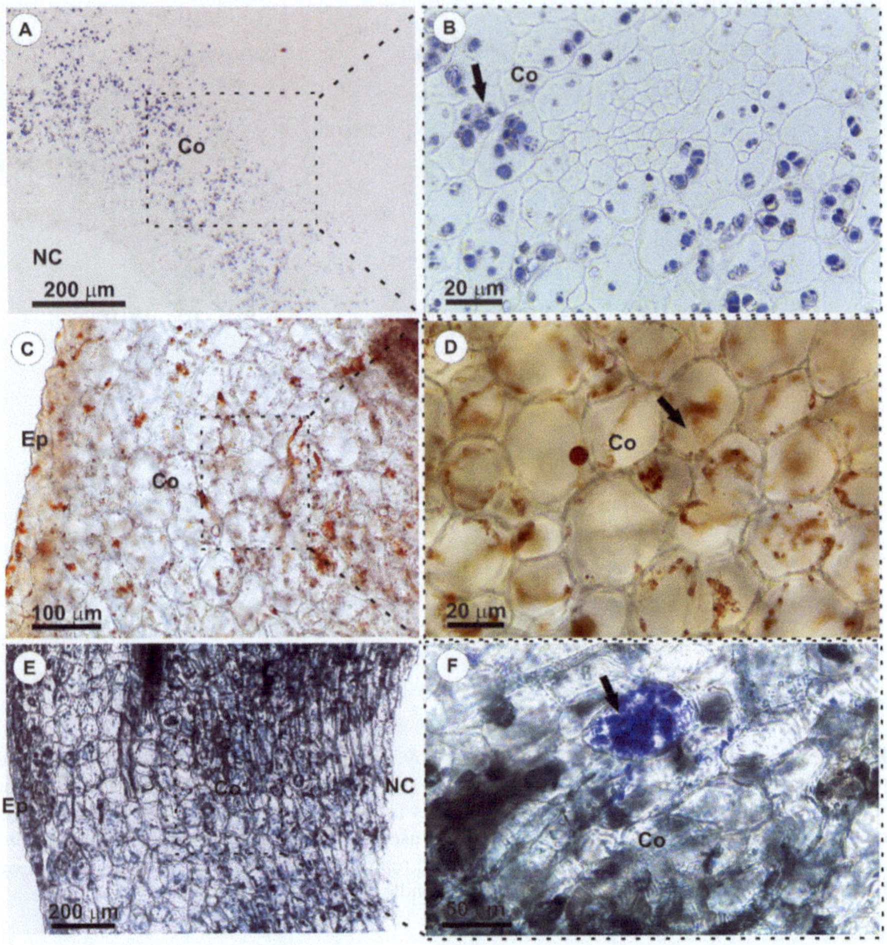

Fig. 2 Results of histochemistry tests with Lugol (**a, b**), Sudan Red (**c, d**), and Bromophenol blue (**e, f**) for mature galls of *Sapium glandulosum* (Euphorbiaceae) induced by *Neolithus fasciatus* (Hemiptera). (**a**) Overview of the starch grains along the cortex. (**b**) Detail of the starch grains in the median region of the cortex (arrow). (**c**) Overview of lipid droplets in the outer and middle regions of the cortex. (**d**) Details of lipid droplets in the medial cortex region (arrow). (**e**) Overview of proteins throughout the cortex. (**f**) Proteins in the medial region of the cortex (arrow). Abbreviations: Co cortex, Ep epidermis, NC nymphal chamber

3 Compartmentalization of Primary Metabolites x Gall Inducer Taxa

Different studies have correlated the histochemical profile of the primary metabolites in galls with the the presence of an outer and an inner cortex. These investigations have revealed significant variability in the labeling of different types of metabolites (Isaias et al. 2015, 2018; Bragança et al. 2016; Ferreira et al. 2017c;

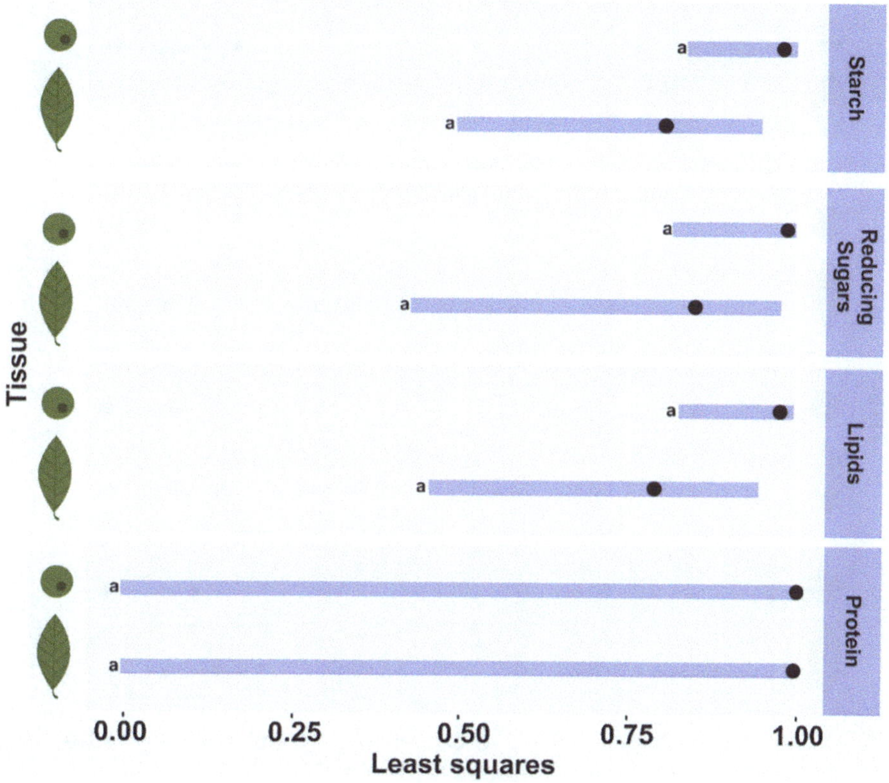

Fig. 3 Chi-square test comparing tissues of young galls (represented by a globoid structure) and mature non-galled organs (represented by a leaf) for the compounds starch, reducing sugars, lipids, and proteins. Black dots indicate forecast averages, while blue bars indicate least squares. *Means followed by similar letters do not differ statistically within the same parameter ($p < 0.05$)

Guedes et al. 2018a; Souza et al. 2023; Rosa et al. 2024). Herein, the primary metabolites were identified in galls induced by 36 species of gall-inducers, which were distributed in 9 orders and 16 families, with some species presenting more than one observation (Table 1). Considering the percentage of occurrence of metabolites in the different tissue regions of mature galls, the proteins were reported 54 times. They were most frequently in the inner tissues of the gall (50%), followed by the widespread presence in the tissues (i.e., in both, with 35,7%) (Fig. 4a). Only 9.5% of protein tests were positive in the outer tissues and 4.7% in the vascular system (Fig. 5a). Lipids was observed 51 times, with 41.4% occurrence in inner tissues, 39.4% in both inner and outer tissues, and 19.2% only in outer tissues (Fig. 5a). Lipids were not reported for the vascular system (Fig. 5a). Reducing sugars were tested a total of 31 times in observations of gall systems. Their occurrence values were similar between inner and outer tissues, with 38.7% and 35.5%, respectively (Fig. 5a). The occurrence of reducing sugars was 19.4% for both regions of the

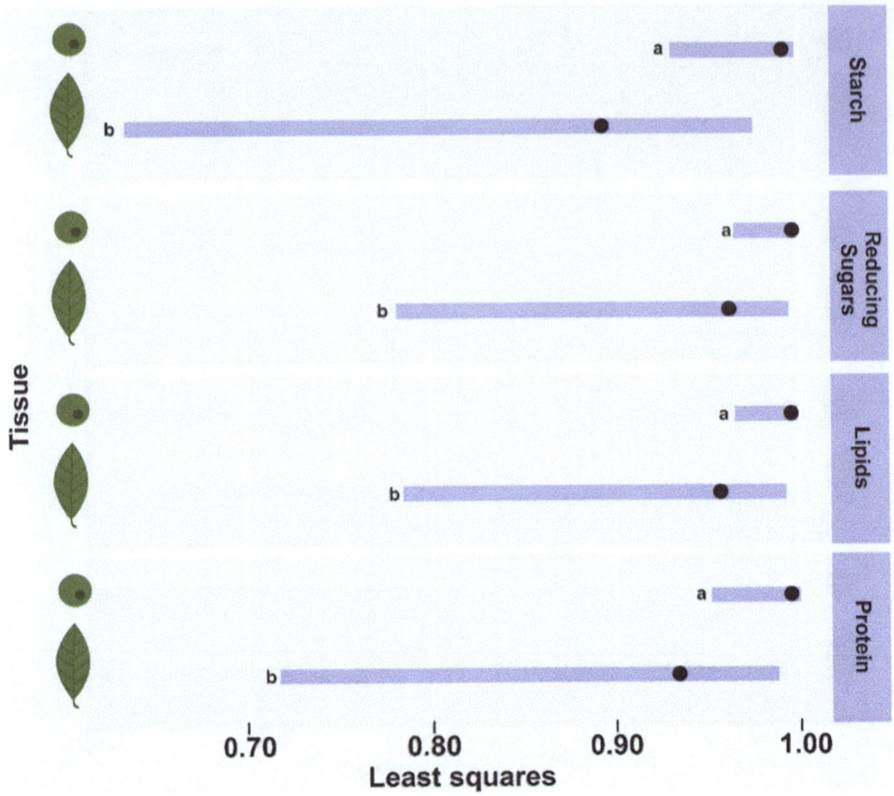

Fig. 4 Chi-square test comparing tissues of mature galls (represented by a globoid structure) and mature non-galled organs (represented by a leaf) for the compounds starch, reducing sugars, lipids, and proteins. Black dots indicate forecast averages, while blue bars indicate least squares. *Means followed by similar letters do not differ statistically within the same parameter ($p < 0.05$)

tissue and 6.5% for the vascular system (Fig. 5a). For starch, the outer tissues had the highest number of observations, with 38.9%, followed by the inner tissues, with 35.2%, and in both, with 25.9% of observations (Fig. 5a).

Galls induced by Diptera, Lepidoptera, Tysanoptera, Hemiptera, Acarina, and Nematoda were those that presented the greatest number of histochemical studies regarding primary metabolites, with a greater concentration of studies on galls induced by Diptera and Hemiptera. These data reveal that more studies are needed on galls induced by organisms present in other taxonomic groups. In Hemiptera, metabolites occurred uniformly between inner and outer tissues, except for reducing sugars that were not reported exclusively for inner tissues (Fig. 5b). Vascular bundles also accumulated metabolites for this group (Fig. 5b). Proteins were most frequent in both tissues (50%), followed by 20% for inner and outer tissues and 10% for vascular bundles (Fig. 5b). Reducing sugars were found mainly in outer tissues, with 62.5% of observations, followed by both tissues (25%) and vascular bundles

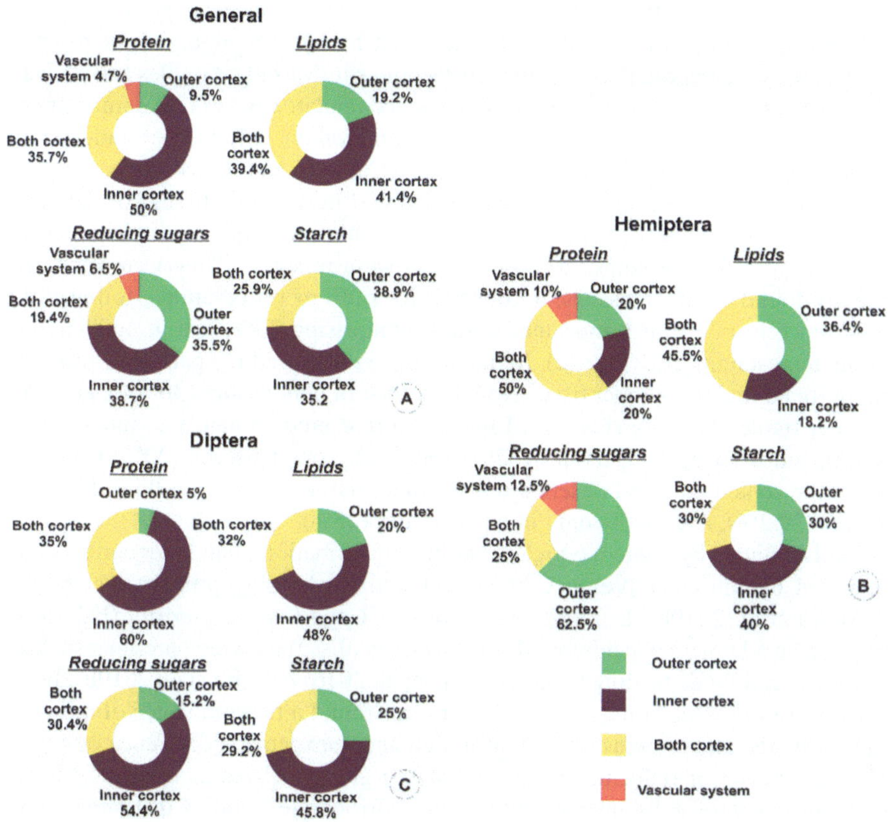

Fig. 5 Proportion of occurrence of starch, reducing sugars, lipids, and proteins in outer and inner tissues, both and in vascular systems for galls induced by different inducer taxa. (**a**) General graphs, composed of compiled results for galls induced all found galling organisms; (**b**) Graphs for Hemiptera order; (**c**) Graphs for Diptera order

(12.5%) (Fig. 5b). Lipids occurred mainly in both tissues, with 45.5% of observations, then with 36.4% presence in outer tissues and 18.2% in inner tissues (Fig. 5b). Starch was the only compound that was found most frequently in inner tissues (40%), followed by 30% only in outer tissues and in both (Fig. 5b).

Galls induced by Hemiptera generally have a predominance of simple parenchymatic cortex and no nutritive tissue, as demonstrated for leaf galls of *Aspidosperma australe* induced by Hemiptera: Psylloidea (Oliveira and Isaias 2010b). These galling insects have a sucking feed habit and, thus, feed directly from the phloem (Oliveira et al. 2016). Galling insects of the order Hemiptera can regularly feed on gall phloem for several days or even during its different developmental stages (Tjallingii 1995). Herein, we demonstrate a tendency of more homogeneous labeling of primary metabolites in the gall tissues induced by Hemiptera, with a lower occurrence in the inner tissue, reinforcing that it has no nutritional function for the galling insect and, rather, acts in the metabolic maintenance of the tissue. Galls induced by Hemiptera were the only ones that had an accumulation of reserves close to the vascular bundles, i.e., in their feeding site. Vascular bundles and perivascular tissues are active in these galls and can accumulate reserves, as demonstrated in our analysis and for galls of *Psidium cattleianum* (Myrtaceae) induced by *Nothotrioza cattleiani* (Hemiptera: Triozidae) (Carneiro and Isaias 2015).

For Diptera, the metabolites were concentrated in the inner tissues of the galls, being non-detected in the vascular bundles and with low frequency in the outer tissues (Fig. 4c). For proteins, 20 observations were reported. They were most frequently found in inner tissues, accounting for 60% of observations, followed by occurrence in both inner and outer tissues, comprising 35%, and in outer tissues alone, representing 5% (Fig. 4c). Reducing sugars followed the pattern of proteins, being found mainly in inner tissues (54.4%), then in both tissues (30.4%), and only in outer tissues (15.2%) (Fig. 4c). Lipids occurred predominantly in inner tissues (48%), followed by both tissues (32%) and finally outer tissues (20%) (Fig. 4c). Starch presented the most observations for inner tissues (45.8%), followed by both tissues (29.2%), and finally only in outer tissues (25%).

Galls induced by Diptera typically exhibit the formation of an inner cortex, characterized by the clear presence of typical nutritive storing primary metabolites (Ferreira et al. 2019b), as reported for galls of *Copaifera langsdorffii* (Fabaceae) induced by a Diptera: Cecidomyiidae (Oliveira et al. 2011), which presents sucking disrupted-cell fluids feeding habit (Oliveira et al. 2016; Ferreira et al. 2019b). In the nutritive tissue of galls induced by Diptera, we found a predominance of reducing sugars, lipids, and proteins, as well as a high and unexpected occurrence of starch. The occurrence of reducing sugars and starch grains in greater frequency in the nutritive tissue of galls induced by Diptera corroborates studies that suggest the large accumulation of carbohydrates in certain families of this order, such as in galls induced by Cecidomyiidae (Bronner 1992; Formiga et al. 2011; Ferreira and Isaias 2014). Carbohydrates and proteins have also been reported for the nutritive tissue of galls induced by Diptera: Cecidomyiidae on leaves of *Aspidosperma spruceanum* (Oliveira and Isaias 2010b). Starch grains are routinely reported for the outer tissues of the cortex of galls, and for this reason, some authors call it the storage cortex (Rezende et al. 2019) or common storage tissue (Ferreira et al. 2019b). Starch can be degraded by enzymatic action, forming simple carbohydrates (e.g. sucrose, fructose, and glucose) that can be sent to the nutritional tissue for consumption by the galling inducer (Oliveira and Isaias 2010a).

Among the other taxonomic groups of gall-inducing organisms highlighted in Table 1, Acarina, Thysanoptera, Nematoda, and Lepidoptera had few observations. In Acarina, 8 histochemical tests were carried out, two tests for each compound. The proteins occurred in the inner or both tissue compartments (50% each). In Thysanoptera, 10 histochemical tests were carried out, 5 for lipids, 3 for starch, and 2 for proteins, with no testing for reducing sugars. The proteins were histolocalized in inner tissues (75%) or both (25%). Lipids were found in both tissue compartments in 60% of observations, and 40% only in inner tissues. Starch was found in inner tissues (66.6%) and both tissues (33.3%). In Nematoda, 9 histochemical tests were carried out, 4 for starch, 2 for reducing sugars, and 3 for proteins, without lipid tests. Proteins and reducing sugars occurred 100% in inner tissues. Starch was found in outer (75%) or inner (25%) tissues. In Lepidoptera, 14 histochemical tests were applied, 5 for starch, 3 for reducing sugars, 4 for lipids, and 2 for proteins. Proteins were histolocalized in outer or inner tissues (50%). Reducing sugars were found in outer (66.6%) or inner (25%) tissues. Lipids were found in inner tissues (75%) or both tissues in 25% of observations. Starch was reported in outer tissues (80%) or both tissues (20%). The collected data of Acarina, Lepidoptera, Nematoda, and Thysanoptera were considered unrepresentative and may just indicate trends.

4 Final Considerations

In the current chapter, the potential of the histolocalization of primary metabolites in galls was revealed, which despite being part of the plant's intrinsic metabolism may be essential to discuss the galler's behavior and feeding habits. Also, the histolocalization of primary metabolites proved to be of great importance for understanding the processes and steps involved in gall formation and development. Our results reinforce that tissue compartmentalization has been associated with the gall inducer taxon, which is evident most for galls induced by Hemiptera and Diptera. Furthermore, we show that mature galls generally have greater primary metabolite complexity than non-galled tissues.

5 Supplementary Material: Statistical Data

A systematic review of scientific articles in Portuguese and English on primary metabolites in galls was made based on histochemistry as a technique for in situ identification of metabolites, with articles using quantitative techniques being excluded from the search. The selection of histochemistry is related to the goals previously described, which are associated with the compartmentalization of gall tissues, something generally identifiable through histological techniques. The articles were extracted from the Google Scholar, Scielo, and CAPES Periodicals Portal database until January 2022, using the following keywords: "galls", "gall starch", "gall protein", "gall carbohydrate", "gall reducing sugar", "gall primary metabolites" and "gall histochemistry". These keywords were also searched in Portuguese.

The following data were extracted from the selected articles: (i) scientific name of the host plant; (ii) host plant family; (iii) organ where gall induction occurred; (iv) type and subtype of the primary metabolite stored in the gall; (v) positive or negative test result, both for the gall and the non-galled organ; (vi) scientific name of the gall inducer species; (vii) gall inducer taxonomic group; (viii) gall morphotype; (ix) gall developmental stage (i.e. young, mature, or senescent); and (x) region of the gall where the metabolite was found (i.e. inner, outer tissues or vascular bundle). Books, book chapters, monographs, dissertations, theses, and prefaces, among other forms of publication than articles, were not used in the present chapter.

The place of the primary metabolites in the tissues was based on the interpretation of the authors of the respective studies or by the authors of this work since some studies did not adequately inform the position of the metabolite in the plant tissue. The epidermis and outer portions of the cortex were considered outer tissues, while inner tissues were those present in the vascular cylinder for roots and in the inner part of the cortex in the galls induced in the other organs. The term "both" was used when the distribution of the primary metabolite was widespread in the gall tissues. For vascular systems, metabolites present in the xylem, phloem, and surrounding parenchyma cells were considered.

To test whether there is a difference in the composition of groups of energetic compounds between gall tissues and non-galled tissues, a secondary database was created following a sequence of filters: (1) Only authors who tested mature or young tissues; (2) Authors who have tested galls and non-galled tissues in the same assay for the same primary metabolite; (3) Authors who have identified at least one of the taxonomic groups of the galling insect (order, family or species). In total, 39 authors were used with 48 host plant-galler systems. With this database, a binomial mixed generalized linear regression analysis was performed using the presence of the compound as a response, the tissue and the compound group as predictors, and the identity of the host plant-galler system as a random variable. A regression was made for the mature stage and one for the young stage using the "lme4" package (Bates et al. 2015). To compare pairwise for each compound, a comparison test was carried out on the average probabilities of occurrence using the "lsmeans" package (Lenth 2016). To assess whether there is a pattern in the distribution of starch, reducing sugars, lipids, or proteins between gall developmental stages and taxonomic groups, graphs were created with the percentage of positive occurrence about the group's total.

References

Álvarez R (2012) Microscopic study of the walls of galls induced by *Geoica utricularia* and *Baizongia pistaciae* in *Pistacia terebinthus*: a contribution to the phylogeny of Fordini. Arthropod-Plant Int 6:137–145. https://doi.org/10.1007/s11829-011-9154-3

Álvarez R, Encina A, Pérez Hidalgo N (2009) Histological aspects of three *Pistacia terebinthus* galls induced by three different aphids: *Paracletus cimiciformis, Forda marginata* and *Forda formicaria*. Plant Sci 176:303–314. https://doi.org/10.1016/j.plantsci.2008.11.006

Amorim D, Ferreira B, Fleury G (2017) Plant potentialities determine anatomical and histochemical diversity in *Mikania glomerata* Spreng. galls. Rev Bras Bot 40:517–527. https://doi.org/10.1007/s40415-016-0357-9

Andrade G, Silva I, Fernandes G et al (1995) Aspectos biológicos das galhas de *Tomoplagia rudolphi* (Diptera: Tephritidae) em *Vernonia polyanthes* (Asteraceae). Rev Bras Biol 55:819–829

Arduin M, Kraus J (2001) Anatomia de galhas de ambrosia em folhas de *Baccharis concinna* e *Baccharis dracunculifolia* (Asteraceae). Revta Bras Bot 24:63–72

Arora D, Patni V (2001) Localisation of metabolites and enzymes in insect induced rachis gall and normal tissues of *Prosopis cineraria* (Linn.) druce. J Phytol Res 14:179–181

Arriola Í, Melo-Júnior J, Ferreira B et al (2018) Galls on *Smilax campestris* Griseb. (Smilacaceae) protect the insects against restinga constraints, but do not provide enriched nutrition. Rev Bras Bot 41:145–153. https://doi.org/10.1007/s40415-017-0423-y

Bagatto G, Shorthouse J (1994) Seasonal acquisition of mineral nutrients by a chalcid gall on lowbush blueberry. Entomol Exp Appl 73:61–66

Bailey S, Percy D, Hefer C et al (2015) The transcriptional landscape of insect galls: Psyllid (Hemiptera) gall formation in Hawaiian *Metrosideros polymorpha* (Myrtaceae). BMC Genomics 16:1–9. https://doi.org/10.1186/s12864-015-2109-9

Baker J (1958) Principles of biological microtechnique. A study of fixation and dyeing

Bates D, Maechler M, Bolker B et al (2015) Package 'lme4'. Convergence 12(1):2

Bedetti C, Ferreira B, Castro N et al (2013) The influence of parasitoidism on the anatomical and histochemical profiles of the host leaves in a galling Lepidoptera - *Bauhinia ungulata* system. Rev Bras Biociênc 11:242–249

Bomfim P, Kuster V, Caetano A et al (2020) Ovule-gall stimulating a large fake fruit on *Miconia chamissois* Naudin (Melastomataceae): a structural overview. Botany 99:115–126

Bragança G, Oliveira D, Isaias R (2016) Compartmentalization of metabolites and enzymatic mediation in nutritive cells of Cecidomyiidae Galls on *Piper Arboreum* Aubl. (Piperaceae). J Plant Stud 6:11. https://doi.org/10.5539/jps.v6n1p11

Bragança G, Alencar C, Freitas M et al (2020) Hemicelluloses and associated compounds determine gall functional traits. Plant Biol 22:981–991. https://doi.org/10.1111/plb.13151

Bronner R (1975) Simultaneous demonstration of lipids and starch in plant tissues. Stain Technol 50:1–4

Bronner R (1992) The role of nutritive cells in nutrition of cynipds and cecidomyiids. In: Shorthouse J, Rohfristsch O (eds) Biology of insect-induced galls. Oxford University Press, New York

Carneiro R, Isaias R (2015) Gradients of metabolite accumulation and redifferentiation of nutritive cells associated with vascular tissues in galls induced by sucking insects. AoB Plants 7. https://doi.org/10.1093/aobpla/plv086

Carneiro R, Castro A, Isaias R (2014) Unique histochemical gradients in a photosynthesis-deficient plant gall. S Afr J Bot 92:97–104. https://doi.org/10.1016/j.sajb.2014.02.011

Chao J, Liao G (2013) Histocytological aspects of four types of ambrosia galls on *Machilus zuihoensis* Hayata (Lauraceae). Flora Morphol Distrib Funct Ecol Plants 208:157–164. https://doi.org/10.1016/j.flora.2013.01.001

Costa E, Oliveira D, Ferreira D et al (2021) Structural and nutritional peculiarities related to lifespan differences on four Lopesia induced bivalve-shaped galls on the single super-host *Mimosa gemmulata*. Front Plant Sci 12. https://doi.org/10.3389/fpls.2021.660557

Dsouza M, Ravishankar B (2014) Nutritional sink formation in galls of *Ficus glomerata* Roxb. (Moraceae) by the insect *Pauropsylla depressa* (Psyllidae, Hemiptera). Trop Ecol 55:129–136

Ferreira B, Isaias R (2013) Developmental stem anatomy and tissue redifferentiation induced by a galling Lepidoptera on *Marcetia taxifolia* (Melastomataceae). Botany 91:752–760

Ferreira B, Isaias R (2014) Floral-like destiny induced by a galling Cecidomyiidae on the axillary buds of *Marcetia taxifolia* (Melastomataceae). Flora Morphol Distrib Funct Ecol Plants 209:391–400. https://doi.org/10.1016/j.flora.2014.06.004

Ferreira B, Teixeira C, Isaias R (2014) Efficiency of the Polyethylene-Glycol (PEG) embedding medium for plant histochemistry. J Histochem Cytochem 62:577–583. https://doi.org/10.1369/0022155414538265

Ferreira B, Álvarez R, Avritzer S et al (2017a) Revisiting the histological patterns of storage tissues: beyond the limits of gall-inducing taxa. Botany 95:173–184. https://doi.org/10.1139/cjb-2016-0189

Ferreira B, Avritzer S, Isaias R (2017b) Totipotent nutritive cells and indeterminate growth in galls of *Ditylenchus gallaeformans* (Nematoda) on reproductive apices of Miconia. Flora Morphol Distrib Funct Ecol Plants 227:36–45. https://doi.org/10.1016/j.flora.2016.12.008

Ferreira B, Falcioni R, Guedes L et al (2017c) Preventing false negatives for histochemical detection of Phenolics and Lignins in PEG-embedded plant tissues. J Histochem Cytochem 65:105–116. https://doi.org/10.1369/0022155416677035

Ferreira B, Álvarez R, Bragança G et al (2019a) Feeding and other gall facets: patterns and determinants in gall structure. Bot Rev 85:78–106. https://doi.org/10.1007/s12229-019-09207-w

Ferreira B, Freitas M, Bragança G et al (2019b) Enzyme-mediated metabolism in nutritive tissues of galls induced by *Ditylenchus gallaeformans* (Nematoda: Anguinidae). Plant Biol 21:1052–1062. https://doi.org/10.1111/plb.13009

Ferreira B, Moreira G, Carneiro R et al (2022) Complex meristematic activity induced by *Eucecidoses minutanus* on *Schinus engleri* turns shoots into galls. Am J Bot 109:209–225. https://doi.org/10.1002/ajb2.1798

Formiga A, Soares G, Isaias R (2011) Responses of the host plant tissues to gall induction in *Aspidosperma spruceanum* Müell. Arg (Apocynaceae). Am J Plant Sci 2:823–834. https://doi.org/10.4236/ajps.2011.26097

Gopinathan K (1987) Morphological patterns and histochemical *Mimusops- Arrhenothrips* gall system. Indian Acad Sci (Plant Sci) 97:203–214

Gopinathan K, Suresh G (1985) On the developmental morphology and histochemistry of the galls induced by an agromyzid on the stems of *Pongamia glabra* Vent. (Fabaceae). Proc Indian Acad Sci (Plant Sci) 95:95–101

Guedes L, Aguilera N, Ferreira B et al (2018a) Anatomical and phenological implications of the relationship between *Schinus polygama* (Cav.) (Cabrera) and the galling insect *Calophya rubra* (Blanchard). Plant Biol 20:507–515. https://doi.org/10.1111/plb.12696

Guedes L, Aguilera N, Ferreira B et al (2018b) Factors influencing the morphogenesis of galls induced by *Calophya mammifex* (Calophyidae) on *Schinus polygama* (Anacardiaceae) leaves. Botany 96:589–599. https://doi.org/10.1139/cjb-2018-0078

Guimarães A, Bizarri C, Barbosa L et al (2013) Characterisation of the effects of leaf galls of *Clusiamyia nitida* (Cecidomyiidae) on *Clusia lanceolata* Cambess. (Clusiaceae): anatomical aspects and chemical analysis of essential oil. Flora Morphol Distrib Funct Ecol Plants 208:165–173. https://doi.org/10.1016/j.flora.2013.03.001

Guimarães A, Cruz S, Vieira A (2014) Structure of floral galls of *Byrsonima sericea* (Malpighiaceae) induced by *Bruggmanniella byrsonimae* (Cecidomyiidae, Diptera) and their effects on host plants. Plant Biol 16:467–475. https://doi.org/10.1111/plb.12060

Guimarães A, Vieira R, Vieira A (2021) Structure of leaf galls in *Clusia fluminensis* planch and triana (Clusiaceae): sex-biased development in a dioecious host plant. Plan Theory 10:1–19. https://doi.org/10.3390/plants10010020

Guzicka M, Karolewski P, Giertych M (2017) Structural modification of *Quercus petraea* leaf caused by *Cynips quercusfolii* - histological study of galls. J Plant Interact 12:7–13. https://doi.org/10.1080/17429145.2016.1269214

Hemsley P, Grierson C (2008) Multiple roles for protein palmitoylation in plants. Trends Plant Sci 13:295–302. https://doi.org/10.1016/j.tplants.2008.04.006

Hölzl G, Dörmann P (2007) Structure and function of glycoglycerolipids in plants and bacteria. Prog Lipid Res 46:225–243. https://doi.org/10.1016/j.plipres.2007.05.001

Isaias R, Carneiro R, Oliveira D et al (2013) Illustrated and annotated checklist of Brazilian Gall Morphotypes. Neotrop Entomol 42:230–239. https://doi.org/10.1007/s13744-013-0115-7

Isaias R, Oliveira D, Carneiro R et al (2014) Developmental anatomy of galls in the Neotropics: arthropods stimuli versus host plant constraints. In: Fernandes G, Santos J (eds) Neotropical insect galls. Springer Science, Netherlands

Isaias R, Oliveira D, Moreira A et al (2015) The imbalance of redox homeostasis in arthropod-induced plant galls: mechanisms of stress generation and dissipation. Biochim Biophys Acta - Gen Subj 1850:1509–1517. https://doi.org/10.1016/j.bbagen.2015.03.007

Isaias R, Ferreira B, Alvarenga D et al (2018) Functional compartmentalisation of nutrients and phenolics in the tissues of galls induced by *Leptocybe invasa* (Hymenoptera: Eulophidae) on

Eucalyptus camaldulensis (Myrtaceae). Aust Entomol 57:238–246. https://doi.org/10.1111/aen.12336

Jäger-Zürn I, Spies M, Philbrick C et al (2013) Plant galls (Cecidia) in the neotropical water plant family Podostemaceae induced by larvae of Chironomidae (Diptera). Spixiana 36:97–112

Jensen W (1962) Botanical histochemistry. WH Freeman, San Francisco

Johansen D (1940) Plant microtechnique. McGraw- Hill, New York

Kane N, Jones C, Vuorisalot T (1997) Development of galls on leaves of *Alnus glutinosa* and *Alnus incana* (Betulaceae) caused by the eriophyid mite *Eriophyes laevis* (Nalepa). Int J Plant Sci 158:13–23

Karun A, Joseph B, Kocheril J et al (2010) Studies on the stem galls of *Osbeckia lowsoni* Linn. Acad Rev 17:8–16

Koncz N (2011) Histological study of quercus galls of *Neuroterus quercusbaccarum* (Linnaeus, 1758) (Hymenoptera: Cynipidae). Acta Biol Szeged 55:247–253

Kraus J, Sugiura H, Cutrupi S (1996) Morfologia e ontogenia em galhas entomógenas de *Guarea macrophylla* subsp. *tuberculata* (Meliaceae). Fitopatol Bras 21:349–356

Kumar S, Mathur A (2009) Localization of metabolites and enzymes in stem galls of *Terminalia arjuna*. Asian J Exp Sci 23:207–213

Kuster V, Rezende U, Cardoso J et al (2019) How galling organisms manipulate the secondary metabolites in the host plant tissues?: a histochemical overview in Neotropical Gall Systems. In: Mérillon J, Ramawat K (eds) Co-evolution of secondary metabolites. Springer, Cham

Larson K, Whitham T (1991) Manipulation of food resources by a gall-forming aphid: the physiology of sink-source interactions. Oecologia 88:15–21. https://doi.org/10.1007/BF00328398

LeBlanc D, Lacroix C (2001) Developmental potential of galls induced by *Diplolepis rosaefolii* (Hymenoptera: Cynipidae) on the leaves of *Rosa virginiana* and the influence of Periclistus species on the *Diplolepis rosaefolii* galls. Int J Plant Sci 162:29–46. https://doi.org/10.1086/317900

Lenth R (2016) Least-squares means: the R package lsmeans. J Stat Softw 69:1–33

Lyon K (1991) Theory and strategy in histochemistry: a guide to the selection and understanding of techniques. Springer, Belin

Mani MS (1964) Ecology of plant galls. Dr. W. Junk Publish, The Hague

Martini V, Moreira A, Kuster V et al (2020a) Photochemical performance and source-sink relationships in galls induced by *Pseudophacopteron longicaudatum* (Hemiptera) on leaves of *Aspidosperma tomentosum* (Apocynaceae). Photosynthetica 58:827–835. https://doi.org/10.32615/ps.2020.033

Martini V, Moreira A, Fuzaro L et al (2020b) *Pseudophacopteron longicaudatum* (Hemiptera) induces intralaminar leaf galls on *Aspidosperma tomentosum* (Apocynaceae): a qualitative and quantitative structural overview. An Acad Bras Cienc 92:1–16. https://doi.org/10.1590/0001-3765202020181002

Masclaux-Daubresse C, D'andrea S, Bouchez I et al (2020) Reserve lipids and plant autophagy. J Exp Bot 71:2854–2861. https://doi.org/10.1093/jxb/eraa082

Moura M, Soares G, Isaias R (2008) Species-specific changes in tissue morphogenesis induced by two arthropod leaf gallers in *Lantana camara* L. (Verbenaceae). Aust J Bot 56:153–160. https://doi.org/10.1071/BT07131

Nalini M, Shilpa K, Basavarajappa S (2015) Stem gall of *Michelia champaca* L. (Magnoliaceae) induced by *Podothrips* sp.: identification, histochemical and phytochemical studies. Trop Palnt Res 2:90–100

Naves G, Kuster V, Machado M et al (2021) Metabolite investments and stress levels among tissue compartments of *Palaeomystella olygophaga* (Lepidoptera) galls on *Macairea radula* (Melastomataceae). Aust J Bot 69:1–17. https://doi.org/10.1071/BT20128

Nogueira R, Costa E, Silva J et al (2018) Structural and histochemical profile of *Lopesia* sp. Rübsaamen 1908 pinnula galls on *Mimosa tenuiflora* (Willd.) Poir. in a Caatinga environment. Hoehnea 45:314–322. https://doi.org/10.1590/2236-8906-80/2017

Nyman T, Julkunen-Tiitto R (2000) Manipulation of the phenolic chemistry of willows by gall-inducing sawflies. Proc Natl Acad Sci USA 97:13184–13187. https://doi.org/10.1073/pnas.230294097

Oliveira D, Isaias R (2010a) Redifferentiation of leaflet tissues during midrib gall development in *Copaifera langsdorffii* (Fabaceae). S Afr J Bot 76:239–248. https://doi.org/10.1016/j.sajb.2009.10.011

Oliveira D, Isaias R (2010b) Cytological and histochemical gradients induced by a sucking insect in galls of *Aspidosperma australe* Arg. Muell (Apocynaceae). Plant Sci 178:350–358. https://doi.org/10.1016/j.plantsci.2010.02.002

Oliveira D, Christiano J, Soares G et al (2006) Reações de defesas químicas e estruturais de *Lonchocarpus muehlbergianus* Hassl. (Fabaceae) à ação do galhador *Euphalerus ostreoides* Crawf. (Hemiptera: Psyllidae). Rev Bras Bot 29:657–667. https://doi.org/10.1590/S0100-84042006000400015

Oliveira D, Magalhães T, Carneiro R et al (2010) Do Cecidomyiidae galls of *Aspidosperma spruceanum* (Apocynaceae) fit the pre-established cytological and histochemical patterns? Protoplasma 242:81–93. https://doi.org/10.1007/s00709-010-0128-6

Oliveira D, Carneiro R, Magalhães T et al (2011) Cytological and histochemical gradients on two *Copaifera langsdorffii* Desf. (Fabaceae)-Cecidomyiidae gall systems. Protoplasma 248:829–837. https://doi.org/10.1007/s00709-010-0258-x

Oliveira D, Isaias R, Fernandes G et al (2016) Manipulation of host plant cells and tissues by gall-inducing insects and adaptive strategies used by different feeding guilds. J Insect Physiol 84:103–113. https://doi.org/10.1016/j.jinsphys.2015.11.012

Oliveira D, Burckhardt D, Calácio T et al (2019) *Ceropsylla pouteriae* Burckhardt sp. nov. (Hemiptera: Psylloidea: Triozidae), a new species of jumping plant-louse inducing galls on the leaves of *Pouteria ramiflora* (Mart.) Radlk. (Sapotaceae): taxonomy, gall structure and histochemistry. J Nat Hist 53:1923–1950. https://doi.org/10.1080/00222933.2019.1676931

Pérez S, Bertoft E (2010) The molecular structures of starch components and their contribution to the architecture of starch granules: a comprehensive review. Starch/Staerke 62:389–420. https://doi.org/10.1002/star.201000013

Pfister B, Zeeman S (2016) Formation of starch in plant cells. Cell Mol Life Sci 73:2781–2807. https://doi.org/10.1007/s00018-016-2250-x

Raghav PS, Lodha P, Arora DK (2007) Histochemical studies on leaf galls and normal leaf of *Alstonia scholaris*(L.) R. BR J Phytol Res 20:193–198

Rezende U, Moreira A, Kuster V et al (2018) Structural, histochemical and photosynthetic profiles of galls induced by *Eugeniamyia dispar* (Diptera: Cecidomyiidae) on the leaves of *Eugenia uniflora* (Myrtaceae). Rev Biol Trop 66:1469–1480. https://doi.org/10.15517/rbt.v66i4.32531

Rezende U, Cardoso J, Kuster V et al (2019) How the activity of natural enemies changes the structure and metabolism of the nutritive tissue in galls? Evidence from the *Palaeomystella oligophaga* (Lepidoptera) -Macairea radula (Metastomataceae) system. Protoplasma 256:669–677. https://doi.org/10.1007/s00709-018-1321-2

Rohfritsch O, Anthony M (1992) Strategies in gall induction by two groups of homopterans. In: Biology of insect-induced galls. Oxford University, New York, pp 102–117

Rosa L, Silva M, Carneiro R et al (2024) Hemiptera-induced galls of *Sapium glandulosum* have histological and cytological compartmentalization created with a large amount of carbohydrate. Protoplasma 1–14. https://doi.org/10.1007/s00709-023-01921-y

Salam U, Ullah S, Tang Z.-H, Elateeq AA, Khan Y, Khan J, Khan A, Ali S (2023) Plant Metabolomics: An Overview of the Role of Primary and Secondary Metabolites against Different Environmental Stress Factors. Life 13:706. https://doi.org/10.3390/life13030706

Santos L, Tavares M, Arruda E et al (2014) Morphological and histochemical study of leaf galls in three species of Euphorbiaceae from north east Brazilian seasonal tropical dry forest. Plant Prot Q 29:2–6

Sass J (1951) Comparative leaf number in the embryos of some types of maize. Lowa State Coll, J. Sci 25:509

Schoonhoven L, Van Loon J, Dicke M (2005) Insect-plant biology. Oxford University Press

Shih T, Lin S, Huang M et al (2018) Transcriptome profile of cup-shaped galls in *Litsea acuminata* leaves. PLoS One 13:1–14. https://doi.org/10.1371/journal.pone.0205265

Silva M, Scareli-Santos C (2018) Descrição morfoanatômica das galhas e sua distribuição nas folhas de *Ouratea spectabilis* (Mart. Ex Engl.) ENGL. (Ochnaceae). Desafios - Rev Interdiscip da Univ Fed do Tocantins 5:178–188. https://doi.org/10.20873/uft.2359-3652.2018v5nespecialp178

Silva A, Kuster V, Rezende U (2019) The early developmental stages of gall-inducing insects define final gall structural and histochemical profiles: the case of *Bystracoccus mataybae* galls on *Matayba guianensis*. Botany 97:427–438. https://doi.org/10.1139/cjb-2019-0017

Souza A, Oliveira D, Dalvi V et al (2023) Nutritive tissue rich in reserves in the cell wall and protoplast: the case of *Manihot esculenta* (Euphorbiaceae) galls induced by *Iatrophobia brasiliensis* (Diptera, Cecidomyiidae). Protoplasma. https://doi.org/10.1007/s00709-023-01912-z

Stokes M, Chattopadhyay A, Wilkins O et al (2013) Interplay between sucrose and folate modulates auxin signaling in *Arabidopsis*. Plant Physiol 162:1552–1565. https://doi.org/10.1104/pp.113.215095

Stone G, Schönrogge K (2003) The adaptive significance of insect gall morphology. Trends Ecol Evol 18:512–522. https://doi.org/10.1016/S0169-5347(03)00247-7

Tan X, Li K, Wang Z et al (2019) A review of plant vacuoles: formation, located proteins, and functions. Plan Theory 8:327

Teixeira C, Kuster V, Carneiro R et al (2022) Anatomical profiles validate gall morphospecies under similar morphotypes. J Plant Res 135:593–608. https://doi.org/10.1007/s10265-022-01397-6

Tjallingii W (1995) Regulation of phloem sap feeding by aphids. In: Chapman R, Boer G (eds) Regulatory mechanisms in insect feeding. Springer, Boston

Trivedi P, Tiagi B (1980) Histochemical study o *Capsicum annuum* L. root galls incited by *Meloidogyne incognita* Chitwood. Proc Indian Acad Sci (Plant sci) 89:109–115

Vandevyvere I, Bruyn L (1998) Morphological and histochemical analysis of galls of Lipara lucens (Diptera, Chloropidae) on Phragmites australis (Poaceae). Can J Bot 76:1374–1384. https://doi.org/10.1139/b98-143

Vecchi C, Menezes N, Oliveira D et al (2013) The redifferentiation of nutritive cells in galls induced by Lepidoptera on *Tibouchina pulchra* (Cham.) Cogn. reveals predefined patterns of plant development. Protoplasma 250:1363–1368. https://doi.org/10.1007/s00709-013-0519-6

Vieira A, Kraus J (2007) Biologia e estrutura da galha do pedicelo de *Byrsonima sericea* DC. (Malpighiaceae) induzida por Lepidoptera. Rev Bras Biociências 5:402–404

Vilela R, Martini V, Gonçalves L et al (2019) Structure and development of root gall induced by *Meloidogyne javanica* in *Glycine max* L. Semin Agrar 40:1033–1048. https://doi.org/10.5433/1679-0359.2019v40n3p1033

Vitale A, Galili G (2001) The endomembrane system and the problem of protein sorting. Plant Physiol 125:115–118. https://doi.org/10.1104/pp.125.1.115

Weis A, Walton R, Crego C (1988) Reactive plant tissue sites and the population biology of gall makers. Annu Rev Entomol 33:467–486. https://doi.org/10.1146/annurev.en.33.010188.002343

Wind J, Smeekens S, Hanson J (2010) Sucrose: metabolite and signaling molecule. Phytochemistry 71:1610–1614. https://doi.org/10.1016/j.phytochem.2010.07.007

Green Galls Can Photosynthesize: Gall Metabolism and Host Organ Impact

Ana Silvia Franco Pinheiro Moreira, Vinicius Coelho Kuster, and Jessica Ferreira de Lima

1 Introduction

Photosynthesis is a crucial process in plants, using light to convert water and carbon dioxide into glucose and oxygen. In the photochemical phase, the sunlight energy is captured by chlorophylls (Chl) *a* and *b* in the chloroplasts and directed by resonance to the reaction center of the photosystems. At the reaction center, a pair of special Chl *a* releases an electron, initiating the electron transport chain that will result in $NADP^+$ to NADPH reduction. In this process, O_2 is produced by water photolysis, and an electrochemical gradient is formed in the thylakoid membrane, allowing the reduction of ADP to ATP (Eberhard et al. 2008; Evans 2013). The NADPH and ATP will be used in the carboxylation phase (Calvin cycle) for CO_2 transformation into carbohydrates (Evans 2013). The energy not used in the photochemical phase can be dissipated through fluorescence (the excess light energy absorbed by Chls is reemitted as light) or by accessory pigments (referred to as non-photochemical quenching—NPQ), the carotenoids (Demmig-Adams and Adams 2000; Banks 2017). Increasingly modern equipment captures Chl *a* fluorescence in photosystem II (PSII) and allows us to evaluate the photochemical efficiency of photosynthesis (Banks 2017). Furthermore, variations in Chl fluorescence resulting from environmental changes, whether abiotic or through the action of other organisms, such as gall-inducing insects, have been used as an indicator of stress in plant tissues by a wide range of researchers (Lichtenthaler and Miehé 1997; Kalaji et al. 2016; Rezende et al. 2018; Kuster et al. 2022).

A. S. F. P. Moreira (✉) · J. F. de Lima
Institute of Biology, Universidade Federal de Uberlândia (UFU), Uberlândia, Brazil
e-mail: anasilviamoreira@ufu.br

V. C. Kuster
Instituto de Biociência, Universidade Federal de Jataí (UFJ), Jataí, Brazil

D. C. de Oliveira, R. M. dos Santos Isaias (eds.), *Plant Galls*,
https://doi.org/10.1007/978-3-031-80064-1_8

Galls represent a biotic stress capable of altering the structure and metabolism of the organ where they develop (Mani 1964; Oliveira et al. 2016), most commonly the primary photosynthesizing plant organ, the leaves (Mani 1964). In a checklist of Neotropical galls, slightly over 50% of the 806 gall systems studied develop on leaves, with many stem-gall systems displaying green or red colors (Isaias et al. 2013). The green color maintenance in galls indicates that photosynthesis is conserved. However, the neoformed condition of the gall tissues and the relationship between the galling inducer and the gall totally alters the metabolism, suggesting singular features in gall photosynthesis. So, the analyses of quantitative and qualitative changes in the photosynthetic process provide insight into the physiological profile of the new plant organ, offering a potential condition to understand the galling fitness and the impact of the galls on plant productivity. In this chapter, we propose evaluating photosynthesis's role in gall development and maintenance. Additionally, the oxidative stress generated by the action of the gall-inducing insect will be discussed, considering its relationship with the photosynthetic process.

2 Photosynthesis in Galls

2.1 Changes in Photosynthetic Pigment Content from Non-galled Organs to Galls

The extent of changes in the photosynthetic pigment content and distribution in the galls varies depending on the plant-gall inducer system. In leaves, the gall development is often linked to reduced photosynthetic pigment content (Oliveira et al. 2011; Castro et al. 2012; Huang et al. 2015), which is related to cellular reprogramming that usually involves a shift in leaf cell function from photosynthesis to support, protection, and storage, orientated by cellular hypertrophy and tissue hyperplasia (Mani 1964; Ferreira et al. 2019). These processes tend to decrease the number of Chl parenchyma layers, from the largest tissue occupancy in the leaf to a few subepidermal layers in galls, as reported for galls on *Aspidosperma tomentosum* leaves (Apocynaceae) induced by *Pseudophacopteron longicaudatum* (Hemiptera) (Martini et al. 2020). So, the disproportional increase in non-Chl tissues in galls, as reported for galls on *Matayba guianensis* (Oliveira et al. 2017) and *Aspidosperma* spp. (Kuster et al. 2022), lead to the general reduction of photosynthetic pigments. In fact, green galls have chloroplasts usually more abundant in the outer gall tissues, while these organelles and their associated pigments are lower in the innermost gall tissues (Oliveira et al. 2011; Castro et al. 2012; Huang et al. 2015). Nabity et al. (2013) and Huang et al. (2015) suggested not only compartmentalization of the pigment contents but also a functional perspective in which the outer gall tissues are more autotrophic (i.e., with higher photosynthetic capacity) than the tissues closer to the gall chamber, which are more heterotrophic (i.e., with the lowest photosynthetic capacity).

The effect of the gall-inducing organism may be confined to the gall structure without altering the photosynthetic pigments of adjacent tissues or even of the non-galled leaves on the same branch, as observed in the *Machilus thunbergii* galls (Lauraceae) induced by Cecidomyiidae (Huang et al. 2014). However, the impact of the gall-inducing organism can extend to the non-galled tissues adjacent to the gall, reducing photosynthetic pigment levels and affecting its local photosynthesis rate. For instance, non-galled tissues of leaves infected by mite galls may contain 22% less chlorophyll than non-galled leaves (Pestov and Ogorodnikova 2020). According to these authors, the impact on the chlorophyll content also affects the balance between chlorophyll *a* and *b* differently, with Chl *b* levels decreasing more significantly. In contrast, the alterations in photosynthetic pigment production have been addressed differently in other systems. In *Copaifera langsdorffii* galls (Fabaceae) induced by a Cecydomiidae, a decrease in the chlorophyll *a* and *b* ratio was observed (Castro et al. 2012), with a higher proportion of Chl *b* content, which is related to absorbing a broader range of the visible light spectrum (Kume et al. 2018). This strategy, common in shaded plants (Valladares and Niinemets 2008), enhances light absorption and improves photosynthetic efficiency.

Other pigments associated with photosynthetic metabolism are found in the galls. The reddish color of many galls, for example, is often correlated with higher levels of carotenoids and anthocyanins (Bomfim et al. 2019; Cardoso et al. 2022). Carotenoids are pigments, mainly C40 lipophilic isoprenoids, synthesized in all photosynthetic organisms and responsible for the yellow, orange, and red colors, especially in fruits and flowers (Nisar et al. 2015). They are important for plant metabolic processes, photoprotection, and growth and development regulation (Havaux 2014). Xanthophylls, a type of carotenoid, play a vital role in dissipating excess light through the xanthophyll cycle (Demmig-Adams and Adams 2000; Demmig-Adams et al. 2019). In a study involving mite galls on rowan and elm trees, the leaves exhibited a decrease in carotenoid content (by 1–16%), while other tree species studied did not show changes in the carotenoid levels (Pestov and Ogorodnikova 2020). In such cases, alternative systems for reactive oxygens species (ROS) scavenging in chloroplasts and dissipating light exist, including the presence of plastoglobules, tocopherol and antioxidants, high carbohydrates accumulation, and the presence of phenolic compounds (Isaias et al. 2015).

Due to the low chlorophyll and carotenoid contents and subsequent decrease in photosynthetic light-use efficiency, anthocyanins play a prominent role and absorb visible light not utilized in the electron transport chain, reducing photodamage and the production of ROS (Huang et al. 2011; Santos et al. 2021). Anthocyanins are the class of flavonoids responsible for the red, pink, purple, and blue colors in plants (Grotewold 2006). They are a hydrosoluble pigment that accumulates in vacuoles of different plant cell types (Hatier and Gould 2009), acting as sunlight attenuators and ROS scavengers (Landi et al. 2015). In *Qualea parviflora* galls (Vochysiaceae), the higher concentrations of anthocyanins in adaxial galls were associated with light stress, with anthocyanin accumulation changing according to gall exposition to the sun (Bomfim et al. 2019). The *Qualea parviflora* gall results indicate a photoprotective role of anthocyanins in some gall morphotypes.

2.2 Photochemical Performance of Green Galls

The decrease in photosynthetic pigments is usually associated with lower photosynthetic rates in galls than in non-galled tissues (Castro et al. 2012; Huang et al. 2014; Oliveira et al. 2017; Rezende et al. 2018), as demonstrated in galls induced by *Eugeniamyia dispar* (Cecidomyiidae) on the leaves of *Eugenia uniflora* (Myrtaceae) (Rezende et al. 2018). Chl *a* fluorescence of the PSII has been used as a non-invasive method for assessing the photochemical performance in galls, aiding in the detection of biotic stress (see Rezende et al. 2018; Oliveira et al. 2017; Martini et al. 2020; Kuster et al. 2022). The evaluation of Chl *a* fluorescence is based on parameters derived from the Kautsky effect (named in honor of the word of Kautsky and Hirsch 1931). The induced kinetics of Chl fluorescence in dark-adapted samples is induced by applying a saturating pulse of light (Banks 2017), with the parameters most frequently used in gall studies being F_0 (minimum fluorescence in dark-adapted state), F_m (maximum fluorescence in dark-adapted state), F_v/F_m (maximum PSII quantum yield in dark-adapted state), $\Delta F/F_m'$ (PSII quantum yield in light), and Rfd (fluorescence decline ratio) (see Rezende et al. 2018; Oliveira et al. 2017; Martini et al. 2020; Kuster et al. 2022).

The F_0 has been observed to be lower in galls compared to non-galled leaves, potentially indicating damage to the photosynthetic apparatus (Bolhar-Nordenkampf et al. 1989), likely due to oxidative stress induced by the gall-inducing insect, as seen in galls induced by *Eugeniamyia dispar* (Cecidomyiidae) on leaves of *Eugenia uniflora* (Myrtaceae) (Rezende et al. 2018). However, the relationship between F_0 and stress in galls appears somewhat contradictory, as conflicting patterns have been reported in the literature, such as in *Aspidosperma spruceanum*, *A. macrocarpon*, and *A. australe*, where F_0 was higher in non–galled leaves than in galls (Kuster et al. 2022). The F_m, F_v/F_m, $\Delta F/F_m'$, and Rfd are typically reduced in galls, indicating the impact of biotic stress caused by the feeding activity of the gall-inducing insect (Oliveira et al. 2011, 2017; Carneiro et al. 2014; Martini et al. 2020), and supporting the assumptions of Lichtenthaler and Miehé (1997) for plants under abiotic stress conditions. Also, the reduction of quantum efficiency in galls is related to decreases in the number of PSII units responsible for PAR absorption (Yang et al. 2007). However, some studies have reported a similar quantum yield between galls and non-galled organs, suggesting distinct pathways of stress level in galls that create new photosynthetic profiles. In some *Aspidosperma* gall systems, the nutritive tissue formed during gall development remains separated from the chlorophyllous tissue by a layer of lignified cells, preserving the PSII operating efficiency at the outermost layers of the galls, even with a reduction of the maximum quantum yield (F_v/F_m) (Oliveira et al. 2011; Kuster et al. 2022). For instance, the quantum yield was similar between non-galled leaves and galls induced by Cecidomyiidae in *Clusia arrudae* (Clusiaceae) (Fernandes et al. 2010), possibly indicating a compensatory response that mitigates the negative effect of galls on plant development (Fay et al. 1993).

The significant decrease in the electron transport rate (ETR) (Castro et al. 2012) and the F_v/F_m of the gall surface and nearby leaf tissues (including the adaxial and abaxial surfaces) indicates that galls have much lower PSII quantum efficiency compared to non-galled leaves (Aldea et al. 2006; Huang et al. 2011, 2014; Carneiro et al. 2014). The decline in PSII efficiency in galls and surrounding tissues is associated with physical damage to the PSII reaction centers (Aldea et al. 2006; Huang et al. 2011, 2014), with the lower content of the chlorophyll-protein complex in the PSI and PSII (Yang et al. 2007; Huang et al. 2015) or other photosynthesis-related proteins such as RuBisCo (Huang et al. 2014). The insufficiency in pigment content and protein complexes related to photosynthesis leads to reduced light-harvesting by the photosystems in the thylakoid membrane of the gall chloroplasts, thereby diminishing their photosynthetic capacity (Huang et al. 2015). The horn-shaped gall of *Copaifera langsdorffii* has a large size and exhibits a decrease in F_v/F_m and PSII operating efficiency values from the outer periphery toward the inner cortex. These lower values are attributed to the limited light intensity reaching the innermost layers and to the underdeveloped photosynthetic apparatus of the meristem-like cell groups near the larval chamber (Carneiro et al. 2017). Some studies showed similar values of ETR (Oliveira et al. 2011) and F_v/F_m between galls and non-galled tissues (Huang et al. 2015), and in some cases, an increase in the PSII efficiency of the galls compared to non-galled leaves occur (Retuerto et al. 2004). In the laminar gall systems of *Aspidosperma* examined by Oliveira et al. (2011) and Kuster et al. (2022), despite a decrease in gall chlorophyll content, the chlorophyllous tissue remains preserved at the gall surface in a *continuum* with the palisade parenchyma, which seems to result in minimal damage to the photosynthetic apparatus (see Fig. 1). In that case, even with a decrease in the overall chlorophyll content in the galls, the maintenance of photochemical activity is attributed to the peripheral positioning of the pigments near the leaf surface, where the chlorophyll fluorescence meters can reach.

2.3 Carboxylation Performance in Galls and Carbohydrate Dynamic

The development of galls not only affects the photochemical yield of the plant tissue but also disrupts the O_2 and CO_2 balance, including carboxylation and respiration activity. Hypoxia often occurs when O_2 is consumed without replenishment, and simultaneous large amounts of CO_2 are produced (Pincebourde and Casas 2015). In galls, gas diffusion is reduced by different structural modifications. For example, the neoformed organ may alter the epidermis development, often resulting in a lack of stomata on both gall surfaces (Huang et al. 2014, 2015), with the presence of anomalous and non-functional stomata (Isaias et al. 2015) or the formation of a periderm (Castro et al. 2012). These changes lead to reduced evapotranspiration, a decline in gas exchange capacity (CO_2 and O_2), and decreased CO_2 assimilation

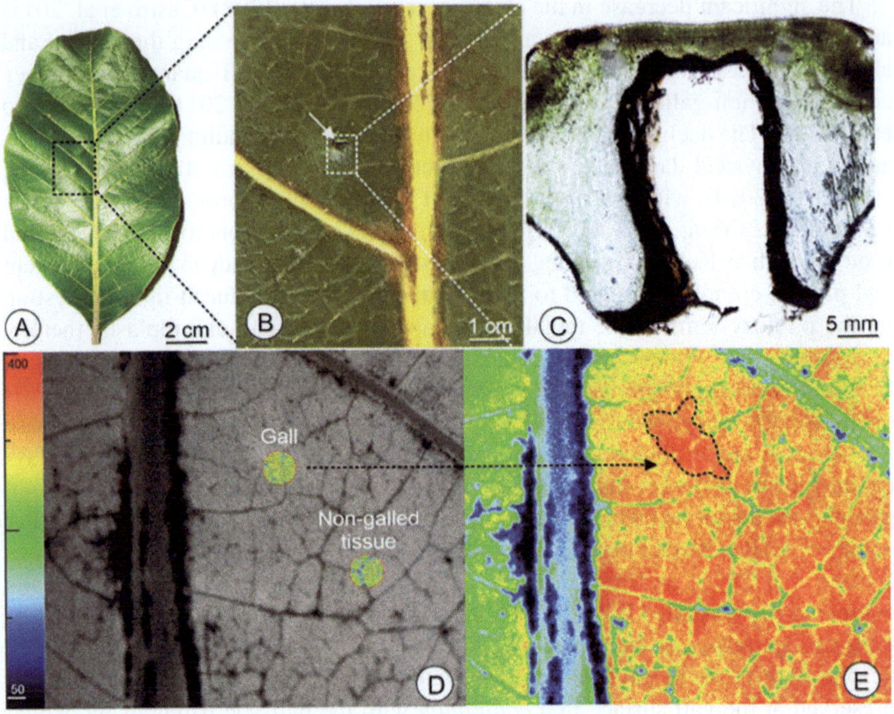

Fig. 1 *Aspidosperma macrocapon* gall induced by *Pseudophacopteron longicaudatum* (Hemiptera). (**a**) Mature leaf with galls. (**b**) Fragment of a leaf with galls (white arrow) used to perform the Kautscky effect, in which many different parameters are used to evaluate the impact of the gall on the photochemical phase of photosynthesis. (**c**) The transverse section of a gall shows preserved palisade parenchyma on the adaxial side of the leaves. This tissue may be conserved during gall development but can often show reduced cell layers. (**d**, **e**) Chlorophyll fluorescence measured using the Handy FluorCam (Photon Systems Instruments, Czech Republic) device. In (**d**), the measures were made on a delimitated area under the gall and non-galled tissues, and the maintenance of the PSII quantum yield ($\Delta F/F_m'$) showed minimal damage to the photosynthetic apparatus. However, the measures showed high values of maximum fluorescence (F_m) (**e**) and a reduction in the maximum quantum yield of the galls (F_v/F_m), and the site of action of the galling insect may be higher than that of the own gall

(Kar et al. 2013; Huang et al. 2014, 2015). Additionally, tissue hyperplasia and cell hypertrophy decrease intercellular spaces within galls, limiting oxygen and CO_2 diffusion and contributing to anaerobic conditions (Castro et al. 2012). Besides the limiting $O_2$and CO_2 diffusion, low levels of O_2 and high levels of CO_2 are favored by gall tissue and insect respiration (Pincebourde and Casas 2015). In this low-oxygen environment, the photosynthesis in galls can contribute to alleviating hypoxia and hypercarbia, ensuring the maintenance of aerobic metabolism and gall structure stability (Castro et al. 2012; Isaias et al. 2015).

Despite all we have discussed thus far, the main question concerning the significance of photosynthesis for galls remains: what role does photosynthesis play in gall development and maintenance? The answer to this question remains unanswered. However, we know that the low photosynthetic activity in galls often fails to meet the gall and, sometimes, the gall-inducing demand for sugars (Carneiro et al. 2014; Huang et al. 2014). Therefore, the gall development, the metabolic upkeep, and the heightened metabolism of the gall-inducing herbivore make this structure a sink for the host photoassimilates, mainly carbohydrates (Castro et al. 2012; Carneiro et al. 2014; Huang et al. 2014). Also, the variability in gall photosynthesis responses is influenced by insect species diversity, their feeding habits, and microenvironments shaped by gall structure (Kuster et al. 2022). For example, the gall-inducing insect's phloem-sucking habit might enhance the sink of photoassimilates, thereby increasing photosynthetic rates. Retuerto et al. (2004) extend this increase as compensatory photosynthesis when comparing scales feeding habits with phloem-sucking habits of some gall-inducing insects, or even increase compared to the non-galled host organ. For instance, the gall inducer *Antisttrophus silphii* (Cynipidae) on *Silphium integrifolium* (Asteraceae) shoots increased leaf photosynthesis, stomatal conductance, and water potential on leaves over ungalled shoots (Fay et al. 1993). This gall-induced increase in leaf photosynthesis may be interpreted as a compensatory response against the negative effect of galls on plant performance. However, the authors discuss the real benefit of this increase in leaf photosynthesis, which may be carefully interpreted and ponder that it depends on the entire gall impact on the plant. For instance, the *Silphium integrifolium* shoot gall (Asteraceae) induced by cynipid wasps can enhance CO_2 assimilation by up to 36% in nearby leaves, but they also lead to a 37% decrease in shoot leaf numbers and result in a 1/3 reduction in biomass production (Fay and Hartnett 1991). Conversely, scraping feeding habit increases ROS formation in the surrounding nutritive tissue (Isaias et al. 2015; Kuster et al. 2022), as shown here for galls of *Manihot esculenta* (Euphorbiaceae) galls induced by *Iatrophobia brasiliensis* (Diptera, Cecidomyiidae), which presents nutritive tissue rich in reserves in the cell wall and protoplast (Fig. 2) (Souza et al. 2023).

2.4 *The Complex Relation Between Oxidative Stress and Photosynthesis in Galls*

Oxidative stress is inherent to plant cellular metabolism and is primarily generated by the activities of mitochondria, chloroplasts, and peroxisomes, wherein the electron transport chain mainly contributes to ROS formation (Harir and Mittler 2009; Apel and Hirt 2004; Anjum et al. 2020). ROS are overproduced due to environmental stresses like high salinity, water deficit, extreme temperatures, and biotic stresses (Raza et al. 2019), such as those induced by gall development (Isaias et al. 2015). These highly reactive products, resulting from the stepwise reduction of molecular oxygen (O2) through high-energy exposure or electron-transfer chemical reactions, encompass both free radicals and non-radical molecules (Anjum et al. 2020). Major

Fig. 2 Oxidative stress labeled by 3,3′ diaminobenzidine (DAB), which labels H_2O_2 (Rosseti and Bonnatti 2001), on *Manihot esculenta* (Euphorbiaceae) galls induced by *Iatrophobia brasiliensis* (Diptera, Cecidomyiidae). (**a**) Overview of leaf galls in the field, and (**b**) Cross-section of the leaf galls subjected to the DAB, with oxidative stress marked by the brown color. There is greater labeling in the nutritive tissue (inner cortex) and vascular bundles. Abbreviation: Ep Epidermis, IC Inner cortex, OC Outer cortex, VB Vascular bundle

ROS in plants include superoxide anion ($O_2^{\bullet-}$), singlet oxygen ($^1O_2$), hydroxyl radical (•OH), and hydrogen peroxide (H_2O_2) (Hasanuzzaman et al. 2020; Anjum et al. 2020). While ROS serve as signaling molecules in plant development (Apel and Hirt 2004; Mishra et al. 2023), their unregulated levels can become harmful agents, inducing oxidative damage to plant tissues and cells, particularly in lipid membranes and genetic material (Jaleel et al. 2009; Mishra et al. 2023). Imbalanced ROS production can oxidize polyunsaturated fatty acids, i.e., lipid components, forming lipid hydroperoxides (LOOH) and generating various short-chain oxidation products known as reactive carbonyl species (RCS), such as malondialdehyde (MDA), 4-hydroxy-(E)-2-nonenal (HNE), and acrolein (Alché 2019).

Gall development imposes a biotic stress on host leaves, and the galling insect's action involves cell restructuring mediated by a new redox balance caused by the accumulation of ROS (Isaias et al. 2015). This accumulation is responsible for new cell development, integrated into new functional tissue compartments. Labeling ROS in Neotropical galls usually employs 3.3′ diaminobenzidine (DAB) histochemical method, as observed in galls of *Aspidosperma* spp. (Kuster et al. 2022), *Eugenia uniflora* (Rezende et al. 2018) and other systems. The homeostasis of redox balance is maintained by an efficient antioxidant system, which involves the efficient dissipation of energy, antioxidant enzymes, and the production of other metabolites, such as phenolics (Isaias et al. 2015). This cascade of biochemical changes ensures the proper development and functioning of the gall structure, avoiding hypersensitivity response by high levels of oxidative stress (Isaias et al. 2015). Furthermore, the topographical analyses of histochemical reactions have revealed the association of ROS with metabolites in a more complex relationship. For instance, in *Piptadenia gonoacantha* galls, the ROS occurred in the same cell layers of phenols, flavonoids, indole-3-acetic acid, and indole-3-acetaldehyde, suggesting the potential for forming auxin-(poly)phenol adducts, and contributing to the determination of gall shapes (Bedetti et al. 2014).

The photosynthetic performance of gall tissues depends on the level of oxidative stress generated by the insect's activity and may vary depending on the galler's feeding habits (Kuster et al. 2022). Consequently, it was reasonable to anticipate investment in stress dissipation pathways like non-photochemical quenching (NPQ) and phenolics. NPQ is a secondary pathway for stress dissipation, expending energy as heat (Demmig-Adams and Adams 2000; Maxwell and Johnson 2000) and maintaining tissue homeostasis within galls (Isaias et al. 2015; Oliveira et al. 2017). However, in some systems, NPQ levels are lower in galls than in non-galled organs (as evidenced in *Aspidosperma tomentosum*, *A. macrocarpon*, and *A. australe* (see Kuster et al. 2022). Phenolics play a pivotal role in shielding plant tissues from oxidative stress by scavenging free radicals like ROS, thereby reducing lipid peroxidation (Gottlieb and Kaplan 1993). These phenolics may accumulate in the cell wall, enhancing its structural support capacity, or within the protoplast, often stored in vacuoles (Lewis 1993). Regardless of the location, their function remains to capture free radicals and mitigate oxidative stress. In galls induced by *Espinosa nothofagi* on *Nothofagus obliqua* (Mirb.) Oerst (Fagaceae) buds (Guedes et al. 2021), there was a notable increase in the production of phenolics compared to the non-galled organ, suggesting effective scavenging of ROS. However, this pattern is not universal across all gall systems evaluated, as observed in *Aspidosperma* spp. galls (Kuster et al. 2022). In these galls, the site of occurrence of phenolics seems to be pretty important, with a preference for tissues traditionally exposed to higher levels of oxidative stress, such as nutritive tissues and vascular bundles.

The augmentation in both quantity and dimensions of plastoglobules at chloroplasts serves as cytological evidence of the heightened oxidative stress in gall tissues. Plastoglobules are lipoproteic corpuscules associated with the thylakoids and are involved in responses to stressful conditions, such as excess of light (Kessler and Vidi 2007) and gall development (Rosa et al. 2024), as demonstrated here (Fig. 3). The production of plastoglobules in *Aspidosperma australe* (Apocynaceae) and *A. spruceanum* galls was related to an upsurge in plastoglobulin synthesis and tocopherol cyclase activity (Oliveira et al. 2011). These molecules help safeguard thylakoid membranes against photooxidation and prevent photoinactivation of photosystem II (Oliveira et al. 2011).

3 Final Considerations

Leaves are the primary photosynthetic organ and the preferred choice for gall-inducing insects. The gall development on leaves is directly influenced by tissue redifferentiation, which changes the photosynthesis profile. In general, hyperplasia and hypertrophy decrease photosynthetic pigment content and the quantum yield of the PSII. However, in some cases, these changes may be related to photosynthetic pigment compartmentalization. The low content of pigments in intralaminar galls

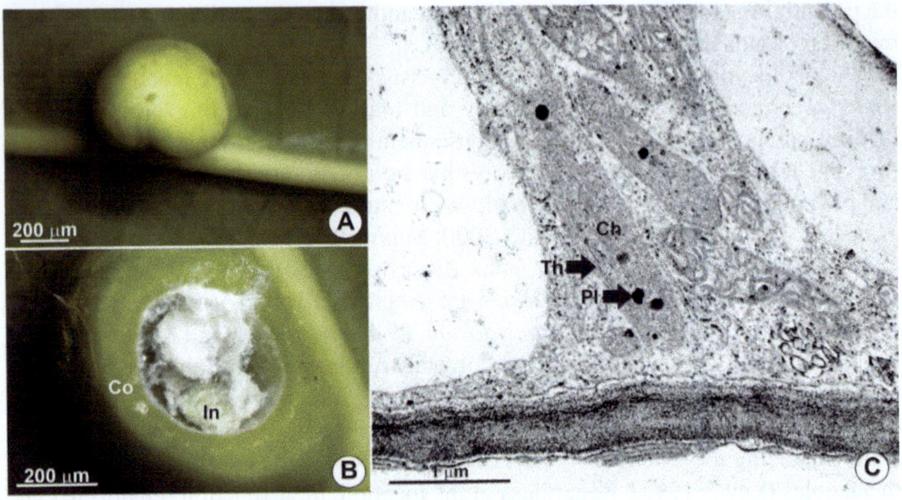

Fig. 3 Mature leaf galls of *Sapium glandulosum* (Euphorbiaceae) induced by *Neolithus fasciatus* (Hemiptera). (**a**) Front view of galls. (**b**) Gall cut transversely showing the nymphal chamber and the inducer covered with wax. (**c**) Transmission electron microscopy showing chloroplasts with plastoglobules. Abbreviations: Co Cortex, Ch Chloroplast, In Inducer, Pl Plastoglobule, Th Thylacoids

would not result from a reduction in their synthesis or degradation but from the position of the chlorophyllian tissue that is preserved only in the outer gall tissues, with lower pigment content in the gall innermost tissues. Besides the location and the total chlorophyll content, the balance between the photosynthetic pigments and their function is also observed. Changes in the proportion of chlorophyll *a* and *b*, as well as the imbalance in chlorophyll, carotenoids, and anthocyanins, are shown in many studies. Finally, we show that photosynthesis in gall tissues seems to play a crucial role in maintaining tissue homeostasis, as even minimal levels of photosynthesis may reduce hypoxia and hypercarbia, besides carbohydrate production.

References

Alché JD (2019) A concise appraisal of lipid oxidation and lipoxidation in higher plants. Redox Biol 23:101–136. https://doi.org/10.1016/j.redox.2019.101136

Aldea M, Hamilton JG, Resti JP, Zangerl AR, Berenbaum MR, Frank TD, DeLucia EH (2006) Comparison of photosynthetic damage from arthropod herbivory and pathogen infection in understory hardwood saplings. Oecologia 149:221–232

Anjum NA, Amreen TAY, Khan NA, Ahmad A (2020) Reactive oxygen species detection-approaches in plants: Insights into genetically encoded FRET-based sensors. J Biotechnol 308:108–117

Apel K, Hirt H (2004) Reactive oxygen species: metabolism, oxidative stress, and signal transduction. Annu Rev Plant Biol 55:373–399. https://doi.org/10.1146/annurev.arplant.55.031903.141701

Banks JM (2017) Continuous excitation chlorophyll fluorescence parameters: a review for practitioners. Tree Physiol 37:1128–1136

Bedetti CS, Modolo LV, Isaias RMS (2014) The role of phenolics in the control of auxin in galls of *Piptadenia gonoacantha* (Mart.) MacBr (Fabaceae: Mimosoideae). Biochem Syst Ecol 55:53–59. https://doi.org/10.1016/j.bse.2014.02.016

Bolhar-Nordenkampf HR, Long SP, Baker NR, Oquist G, Schreiber U, Lechner EG (1989) Chlorophyll fluorescence as a probe of the photosynthetic competence of leaves in the field: a review of current instrumentation. Funct Ecol 3:497–514. https://doi.org/10.2307/238962

Bomfim PMS, Cardoso JCF, Rezende UC, Martini VC, Oliveira DC (2019) Red galls: the different stories of two gall types on the same host. Plant Biol 21:284–291

Cardoso JCF, Gonçalves PHP, Oliveira DC, Rezende UC (2022) Host plant intraspecific variation determines gall traits. Plant Biol 25:208–214

Carneiro RGS, Castro AC, Isaias RMS (2014) Unique histochemical gradients in a photosynthesis-deficient plant gall. S Afr J Bot 92:97–104

Carneiro RGS, Isaias RMS, Moreira ASFP, Oliveira DC (2017) Reacquisition of new meristematic sites determines the development of a new organ, the Cecidomyiidae gall on *Copaifera langsdorffii* Desf. (Fabaceae). Front Plant Sci 8:1622

Castro AC, Oliveira DC, Moreira ASFP, Lemos-Filho JP, Isaias RMS (2012) Source–sink relationship and photosynthesis in the horn-shaped gall and its host plant *Copaifera langsdorffii* Desf. (Fabaceae). S Afr J Bot 83:121–126

Demmig-Adams B, Adams WWIII (2000) Harvesting sunlight safely. Nature 403:371–373

Demmig-Adams B, Stewart JJ, Adams WWIII (2019) Less photoprotection can be good in some genetic and environmental contexts. Biochem J 476:2017–2029

Eberhard S, Finazzi G, Wollman F-A (2008) The dynamics of photosynthesis. Annu Rev Genet 42:463–515

Evans JR (2013) Improving photosynthesis. Plant Physiol 162:1780–1793

Fay PA, Hartnett DC (1991) Constraints on growth and allocation patterns of *Silphium integrifolium* (Asteraceae) caused by a cynipid gall wasp. Oecologia 88:243–250

Fay PA, Hartnett DC, Knapp AK (1993) Increased photosynthesis and water potentials in *Silphium integrifolium* galled by cynipid wasps. Oecologia 93:114–120

Fernandes GW, Coelho MS, Lüttge U (2010) Photosynthetic efficiency of *Clusia arrudae* leaf tissue with and without Cecidomyiidae galls. Braz J Biol 70:723–728. https://doi.org/10.1590/S1519-69842010000400004

Ferreira BG, Álvarez R, Bragança GP, Alvarenga DR, Pérez-Hidalgo N, Isaias RM (2019) Feeding and other gall facets: patterns and determinants in gall structure. Bot Rev 85:78–106

Gottlieb OR, Kaplan MAC (1993) Phytochemical evolution: the redox theory. Nat Prod Lett 2:171–176

Grotewold E (2006) The genetics and biochemistry of floral pigments. Annu Rev Plant Biol 57:761–780

Guedes LM, Torres S, Saéz–Carilho K, Becerra J, Pérez CI, Aguilera N (2021) High antioxidant activity of phenolic compounds dampers oxidative stress in *Espinosa nothofagi* galls induced on *Nothofagus obliqua* buds. Plant Sci 3214:111114

Harir Y, Mittler R (2009) The ROS signaling network of cells. In: Del Río LA, Puppo A (eds) Reactive oxygen species in plant signaling. Springer, Berlin, Heidelberg, pp 165–174

Hasanuzzaman M, Bhuyan MHMB, Zulfiqar F, Raza A, Mohsin SM, Mahmud JA, Fujita M, Fotopoulos V (2020) Reactive oxygen species and antioxidant defense in plants under abiotic stress: revisiting the crucial role of a universal defense regulator. Antioxidants 9:681. https://doi.org/10.3390/antiox9080681

Hatier JHB, Gould KS (2009) Anthocyanin function in vegetative organs. In: Gould KS, Davies KM, Winefield C (eds) Anthocyanins: biosynthesis, functions, and applications. Springer, New York, pp 1–20

Havaux M (2014) Carotenoid oxidation products as stress signals in plants. Plant J 79:597–606

Huang MY, Lin KH, Yang MM, Chou HM, Yang CM, Chang YT (2011) Chlorophyll fluorescence, spectral properties, and pigment composition of galls on leaves of *Machilus thunbergii*. Int J Plant Sci 172(3):323–329

Huang MY, Huang WD, Chou HM, Lin KH, Chen CC, Chen PJ, Chang YT, Yang CM (2014) The leaf-derived cecidomyiid galls are sinks in *Machilus thunbergia* (Lauraceae) leaves. Physiol Plant 152:475–485

Huang MY, Huang WD, Chou HM, Chen CC, Chen PJ, Chang YT, Yang CM (2015) Structural, biochemical, and physiological characterization of photosynthesis in leaf-derived cup-shaped galls on *Litsea acuminata*. BMC Plant Biol 15:61

Isaias RMS, Carneiro RGS, Oliveira DC, Santos JC (2013) Illustrated and annotated checklist of Brazilian gall morphotypes. Neotrop Entomol 42:230–239

Isaias RMS, Oliveira DC, Moreira ASFP, Soares GLG, Carneiro RGS (2015) The imbalance of redox homeostasis in arthropod-induced plant galls: mechanisms of stress generation and dissipation. Biochim Biophys Acta 1850:1509–1517

Jaleel CA, Riadh K, Gopi R, Manivannan P, Ines J, Aljuburi HJ, Chang-Xing Z, Hong-Bo S, Pannerselvam R (2009) Antioxidant defense responses: physiological plasticity in higher plants under abiotic constraints. Acta Physiol Plant 31:427–436

Kalaji HM, Jajoo A, Oukarroum A et al (2016) Chlorophyll a fluorescence as a tool to monitor physiological status of plants under abiotic stress conditions. Acta Physiol Plant 38:102

Kar PK, Jena K, Srivastava AK, Sinha MK (2013) Gall-induced stress in the leaves of *Terminalia arjuna*, food plant of tropical tasar silkworm, Antheraeamylitta. Emir J Food Agric 25:205–210

Kautsky H, Hirsch A (1931) Neue Versuche zur Kohlensäureassimilation. Naturwissenschaften 19: 964

Kessler F, Vidi PA (2007) Plastoglobule lipid bodies: their functions in chloroplasts and their potential for applications. In: Fiechter A, Sautter C (eds) Green gene technology, Advances in biochemical engineering/biotechnology, vol 107. Springer, Berlin, Heidelberg, pp 153–172. https://doi.org/10.1007/10_2007_054

Kume A, Akitsu T, Nasahara HN (2018) Why is chlorophyll b only used in light-harvesting systems? J Plant Res 131:961–972

Kuster VC, Cardoso JCF, Moreira ASFP, Isaias RMS, Oliveira DC (2022) Assessing oxidative stress and photosynthetic activity in leaf galls induced by different species of galling insect on *Aspidosperma* spp. Biologia 77:3441–3454

Landi M, Tattini M, Gould Kevin S (2015) Multiple functional roles of anthocyanins in plant-environment interactions. Environ Exp Bot 119:4–17

Lewis NG (1993) Plant phenolics. In: Alscher RG, Hess JL (eds) Antioxidants in higher plants, 1st edn. CRC Press, Boca Raton. https://doi.org/10.1201/9781315149899

Lichtenthaler HK, Miehé JA (1997) Fluorescence imaging as a diagnostic tool for plant stress. Trends Plant Sci 2:316–319. https://doi.org/10.1016/S1360-1385(97)89954-2

Mani MS (1964) Ecology of plant galls. Dr. W. Junk Publis, The Hague

Martini V, Moreira ASFP, Kuster VC, Oliveira DC (2020) Photochemical performance and source–sink relationships in galls induced by *Pseudophacopteron longicaudatum* (Hemiptera) on leaves of *Aspidosperma tomentosum* (Apocynaceae). Photosynthetica 58:827–835

Maxwell K, Johnson GN (2000) Chlorophyll fluorescence—a practical guide. J Exp Bot 51:659–668

Mishra N, Jiang C, Chen L, Paul A, Chatterjee A, Shen G (2023) Achieving abiotic stress tolerance in plants through antioxidative defense mechanisms. Front Plant Sci 14:1110622

Nabity PD, Miranda MJ, Berenbaum MR, Delucia EH (2013) Leaf-galling phylloxera on grapes reprograms host metabolism and morphology. Proc Natl Acad Sci USA 110:16663–16668

Nisar N, Li L, Lu S, Khin NC, Pogson BJ (2015) Carotenoid metabolism in plants. Mol Plant 8:68–82

Oliveira DC, Isaias RMS, Moreira ASFP, Magalhães TA, Lemos-Filho JP (2011) Is the oxidative stress caused by *Aspidosperma* spp. galls capable of altering leaf photosynthesis? Plant Sci 180:489–495

Oliveira DC, Isaias RMS, Fernandes GW, Ferreira BG, Carneiro RGS, Fuzaro L (2016) Manipulation of host plant cells and tissues by gall-inducing insects and adaptive strategies used by different feeding guilds. J Insect Physiol 84:103–113

Oliveira DC, Moreira AS, Isaias RMS, Martini V, Rezende UC (2017) Sink status and photosynthetic rate of the leaflet galls induced by *Bystracoccus mataybae* (Eriococcidae) on *Matayba guianensis* (Sapindaceae). Front Plant Sci 8:1–12

Pestov SV, Ogorodnikova SY (2020) Status of the photosynthetic apparatus of woody plants damaged by gall mites. Biol Bull 47(10):1392–1397

Pincebourde S, Casas J (2015) Hypoxia and hypercarbia in endophagous insects: larval position in the plant gas exchange network is key. J Insect Physiol 84:137–153

Raza A, Razzaq A, Mehmood SS, Zou X, Zhang X, Lv Y, Xu J (2019) Impact of climate change on crops adaptation and strategies to tackle its outcome: a review. Plan Theory 8:34

Retuerto R, Fernandez-Lema B, Rodriguez-Roiloa S, Obeso JR (2004) Increased photosynthetic performance in holly trees infested by scale insects. Funct Ecol 18:664–669

Rezende UC, Moreira ASFP, Kuster VC, Oliveira DC (2018) Structural, histochemical and photosynthetic profiles of galls induced by *Eugeniamyia dispar* (Diptera, Cecidomyiidae) on the leaves of *Eugenia uniflora* (Myrtaceae). Rev Biol Tropical 66:1469–1480

Rosa LMP, Silva MS, Carneiro RGS, Machado M, Kuster VC (2024) Hemiptera-induced galls of *Sapium glandulosum* have histological and cytological compartmentalization created with a large amount of carbohydrate. Protoplasma 260:593–606

Rosseti S, Bonnatti PM (2001) In situ histochemical monitoring of ozone-and TMV- induced reactive oxygen species in tobacco leaves. Plant Physiol Biochem 39:433–442

Santos GF, Kuster VC, Silva AFM, Faria AP, Moreira ASFP (2021) Changes in color during leaf development of *Eugenia uniflora* (Myrtaceae): different strategies for integrity conservation of the photosynthetic apparatus. Aust J Bot 69:1–11

Souza AP, Oliveira DC, Dalvi VC, Kuster VC (2023) Nutritive tissue rich in reserves in the cell wall and protoplast: the case of *Manihot esculenta* (Euphorbiaceae) galls induced by *Iatrophobia brasiliensis* (Diptera, Cecidomyiidae). Protoplasma 261:513–525

Valladares F, Niinemets U (2008) Shade tolerance, a key plant feature of complex nature and consequences. Annu Rev Ecol Evol Syst 39:237–257

Yang CM, Yang MM, Huang MY, Hsu JM, Jane WN (2007) Life-time deficiency of photosynthetic pigment-protein complexes CP1, A1, AB1, and AB2 in two cecidomyiid galls derived from *Machilus thunbergii* leaves. Photosynthetica 45:589–593

Physiological and Biochemical Properties of Aphid Galls

Katarzyna Kmieć

1 Introduction

In some parts of the world, especially in the temperate climate, the aphids (Hemiptera: Sternorrhyncha: Aphidoidea and Phylloxeroidea) appear to represent one of the most important group of gall-inducing insects. Approximately 10% of known aphid species are true gall formers (Blackman and Eastop 1994; Wool 2004; Chen and Qiao 2012). Gall-inducing aphids manipulate the host plant tissues, stimulating the new organ formation, namely the gall (Stone and Schönrogge 2003; Oliveira et al. 2014; Giron et al. 2016; Kmieć et al. 2018a; Takeda et al. 2019, 2021). The aphids are sap-sucking hemipterans with complex life cycles (Wool 2004). They possess specialized mouthparts known as stylets, enabling them to feed on plant sap. These stylets consist of long, slender and flexible needle-like structures designed for piercing plant tissues and feeding exclusively on phloem cells (Tjallingii 2006; Giordanengo et al. 2010). During the feeding activity, they secrete saliva into plant tissues, which contains biomolecules that facilitate sap extraction and manipulate host plant physiology (Tjallingii 2006; Morkunas et al. 2011; Will et al. 2012; Rodriguez and Bos 2013; Wang et al. 2023). It is known, that the aphid's saliva also acts as a mitigating agent against plant defenses, ensuring uninterrupted sap flow (Walling 2008; Giordanengo et al. 2010; Eitle et al. 2019). Most gall-inducing aphid species alternate sexual and parthenogenetic generations, as well as hosts during their life cycles (Blackman and Eastop 1994). With few exceptions (Wool 2004; Álvarez et al. 2014; Kurzfeld-Zexer et al. 2015), only the fundatrix stage, hatching from an overwintering egg, is capable of inducing galls (Wool 2004; Kmieć et al. 2018a, 2022). Young fundatrix larva, an extremely specialized stage,

K. Kmieć (✉)
Faculty of Horticulture and Landscape Architecture, Department of Plant Protection, University of Life Sciences in Lublin, Lublin, Poland
e-mail: katarzyna.kmiec@up.lublin.pl

D. C. de Oliveira, R. M. dos Santos Isaias (eds.), *Plant Galls*,
https://doi.org/10.1007/978-3-031-80064-1_9

intensively punctures plant tissue in the initial period of gall formation. After several days of sucking, the larva settles and feeds in one place, and the gall that encloses the fundatrix is rapidly formed and continuously developed (Kmieć et al. 2018a). When the gall is fully grown parthenogenetic reproduction of daughters and granddaughters starts. Two or three generations of offspring are produced in the gall, sometimes resulting in clone sizes of hundreds or thousands (Suzuki et al. 2009; Rehill and Schultz 2012; Kmieć et al. 2018a; Lu et al. 2021). Therefore, the galls serve as a steady nutrient supply for aphids, ensuring their continued development and reproduction. Also, the constant food intake promotes key phases of gall development.

Success in gall initiation and development requires sophisticated interaction between a plant and a gall-inducing aphid, and involves complex networks of molecular and physiological processes within and between the organisms (Giron et al. 2016; Chen et al. 2020). Gall-inducing aphids activate specific developmental pathways of each host plant and tightly control them. The following steps of gall formation are proposed: (1) induction (with the secretion of signalling molecules from insects, and perception of the signals by plants); (2) growth (with plant cell regeneration, qualitative differentiation, and organization of gall tissues); (3) maturation; (4) senescence (Raman 2011, 2012). To induce complex gall structures in plant tissues, gall-inducing aphids must control the cellular machinery of their host plants, including development, metabolism, chemistry, and physiology to hijack host gene expression programs in favor of those of the aphids (Takeda et al. 2021). Aphids also need to suppress the plant's defense responses (Giron et al. 2016).

The manipulation of phytohormone concentrations, particularly auxins and cytokinins, facilitates gall induction and growth (Isaias et al. 2014; Andreas et al. 2020). Moreover, phytohormones present in aphid saliva may also stimulate gall growth (Tooker and Helms 2014). It is suggested that at early stages of gall development the feeding behaviour of fundatrix larva and/or biomolecules in their saliva trigger developmental processes in host plants to generate tissues for aphid use (Wool 2004; Andreas et al. 2020). Histochemical analyses indicate that cells at gall sites are under oxidative stress, which increase along with gall development and maturation due to the impacts of nymphs activity within the gall (Isaias et al. 2015; Kmieć et al. 2018a). The balance between reactive oxygen species (ROS) production and scavenging by antioxidants modulates cell reprogramming in response to the biotic stimuli. However gall development alters cellular carbohydrate metabolism, inducing gall tissues to act as sink for photosynthates (Rehill and Schultz 2003; Castro et al. 2012; Andreas et al. 2020). Additionally, gall formation changes the allocation and chemical composition of the primary metabolites and establishes a new profile of secondary metabolites, with phenolic derivatives playing a pivotal role (Isaias et al. 2015; Rosa et al. 2024).

2 Phytohormones in Aphid-Induced Galls

Plant hormones, including cytokinins (CKs), abscisic acid (ABA), gibberellins (GA), and auxins (AUX), are pivotal regulators of plant growth and division, with emerging roles in plant defense responses (Body et al. 2019; Choi 2024). Manipulating plant hormone pathways represents a potential strategy to redirect plant development to accommodate the needs of gall-inducing aphids (Isaias et al. 2014; Giron et al. 2016). While traditionally high CKs and AUX concentrations were thought to provoke plant tissue proliferation, recent researches also underscores the ubiquitous presence of ABA across insect species, challenging their exclusive association with gall-inducing organisms (Takei et al. 2015; Andreas et al. 2020; Seng et al. 2023).

Galling aphids target meristematic plant regions capable of cell division and tissue growth, essential for gall formation (Silvia and Connor 2017; Andreas et al. 2020). Aphid-induced galls develop mostly on young, actively growing plant tissues (Wool 2004; Oliveira et al. 2016; Kmieć et al. 2018a). Early plant responses to aphid feeding trigger protein phosphorylation, membrane depolarization, calcium influx, and ROS release, which leads to the activation of phytohormone-dependent pathways (Giordanengo et al. 2010; Tooker and Helms 2014). Auxin and CKs are the key phytohormones in gall initiation and development, as well as other physiological processes in plants including cell division, cell regeneration, and differentiation (Takei et al. 2015; Argueso and Kieber 2024). They accumulate in galls, especially during initial stages, acting as primary chemical messengers, and play a crucial role in the induction, development, and establishment of aphid galls (Takei et al. 2015; Wang et al. 2016; Body et al. 2019). Studies suggest that AUX and CKs regulate cell proliferation and vascular differentiation during gall development, with several genes responsive for cell division and cytokinesis activated in aphid-induced galls (Body et al. 2019; Takeda et al. 2019, 2021). Genes associated with AUX and CKs metabolic and signal transduction pathways are substantially up-regulated in galls formed on *Rhus chinensis* and *Rhus javanica* (Wang et al. 2017; Hirano et al. 2020). Also, up–regulation of WRKY23 can be involved in vascular patterning in galls through regulation of AUX flux (Takeda et al. 2019). In aphid-induced galls the principal feeding sites are the newly formed vascular bundles around the gall chamber. The presence of nutritive-like cells originating from the vascular and perivascular parenchyma is observed (Álvarez et al. 2016). Vascular hypertrophy and hyperplasia represent pivotal features facilitating the redirection of metabolites toward the gall (Ferreira et al. 2019; Isaias et al. 2024), which is strong physiological sink for nutrients (Larson and Whitham 1991, 1997; Burstein et al. 1994; Rehill and Schultz 2003).

The formation of mobilizing sinks, facilitating the translocation of photosynthates to the feeding site of the galling insect demands the mediation of Cks (Andreas et al. 2020; Hua et al. 2023). Localized production and secretion of exogenous CKs by galling insects likely contribute to the establishment of mobilizing sinks, ensuring a steady supply of sugars and other vital compounds to sustain the insect's nutritional needs (Andreas et al. 2020). The initial CK signal is reinforced by the concentration gradient associated with sucrose accumulation, thereby facilitating the continued transportation of sucrose towards the source of the initial signal (Connor et al. 2012). The metabolism of glucose, the main product of photosynthesis, is the source of sucrose and various phytohormones (Choi 2024). However, galling aphid infestation induced significant alterations in the photochemical activity of host trees (Gailite et al. 2005; Samsone et al. 2012; Kmieć et al. 2018b). Low content of photosynthetic pigments, namely chlorophylls and carotenoids in gall tissues, indicates a reduction of photosynthesis in gall tissues—a widespread phenomenon observed across various studies (Castro et al. 2012; Carneiro et al. 2014; Oliveira et al. 2017; Kuster et al. 2022). Recent research has identified downregulation of photosynthetic genes in *Schlechtendalia chinensis* galls, particularly those encoding components involved in carbon fixation (Lu et al. 2021). Downregulation of photosynthetic genes in galls corroborates a shift towards sink organ characteristics, similar to buds, flowers, and fruits (Takeda et al. 2021). On the other hand, in the same biological system *S. chinensis*—*R. chinensis*, photosynthesis related genes involved in photosystem I and II, electron transport chain, Rubisco components, light harvesting, and chloroplast organization have been up-regulated. Moreover, galls exhibited measurable levels of photosynthesis, and this activity declined along with gall increase in size (Chen et al. 2020). According to Isaias et al. (2015), the reduction in photosynthesis rates is likely due to the diminished chlorophyllous parenchyma area during gall growth and development, as well as to the processes of lignification or suberization occurring from gall maturation to senescence.

Apart from the well-documented influence of insect-derived phytohormones on gall formation, other effector molecules are believed to be needed for constructing intricate gall structures (Giron et al. 2016). Recently, a novel insect secretory protein, named BICYCLE, originating from the gall-inducing aphid *Hormaphis cornu*, which is responsible for the creation of distinctive cone galls on the leaves of *Hamamelis virginiana*, was identified. Through comprehensive analysis, Korgaonkar et al. (2021) discovered that BICYCLE genes exhibit heightened expression levels in the salivary glands of gall-inducing generations. These findings suggest that BICYCLE proteins play a regulating role in the process of gall development in *H. virginiana*, shedding new light on the molecular mechanisms underlying gall formation.

3 Metabolism of Stress Molecules on Aphid Galls: The Role of Reactive Oxygen Species (ROS)

Cell reprogramming, triggered by biotic stimuli during gall initiation and development, relies on the balance between ROS generation and scavenging (Isaias et al. 2015; Choudhury et al. 2017). The ROS can provoke reversible or irreversible protein modifications, thereby altering plant metabolism regulation and activating transcriptional processes (Morkunas et al. 2011; Das et al. 2015; Choudhury et al. 2017; Martin et al. 2022). Recent research has revealed enhanced accumulation of hydrogen peroxide (H_2O_2) in the galls of the aphid *Pemphigus spyrothecae* (Kot and Kmieć 2020) and *Pemphigus phenax* (Kmieć and Kot 2021) induced on *Populus nigra*, and in mature galls of *Tetraneura ulmi* induced on *Ulmus pumila* (Kmieć et al. 2018a). However, the H_2O_2 content in the galls of *Pemphigus bursarius* induced on *P. nigra* and pseudogalls of *Eriosoma ulmi* on *Ulmus glabra* was similar to that of intact leaves, while in galls induced by *Colopha compressa* it was even lower than that of *U. glabra* leaves (Kmieć et al. 2022). Gall cells may be involved in ROS production, which in turn may trigger gall development (Oliveira et al. 2014). Elevated ROS concentrations serve as direct defenses and can lead to programmed cell death (PCD), while lower levels are employed as signalling molecules, mediating stress responses (Morkunas et al. 2011; Das et al. 2015). Thus, specific ROS levels in galled tissues could lead to biochemical alterations that modulate plant cell responses (Oliveira et al. 2016) and the formation of cytological and histochemical gradients in the gall (Oliveira et al. 2014). The variations in ROS levels across the gall cortex likely drive the formation of gall final shapes. These changes result from alterations in cell elongation patterns induced by ROS, leading to increased elasticity of the cell walls, facilitating cell expansion and elongation (Oliveira and Isaias 2010). Additionally, the ROS influence the arrangement of microtubules and cellulose microfibrils. Consequently, cell division orientation and shape are modified in response to the ROS gradient accumulated within the gall cortex (Oliveira et al. 2014). Disturbed balance between ROS production and scavenging results in lipid peroxidation, increased membrane fluidity and permeability. Electrolyte leakage (E_L) analysis from the gall cells of several aphid species has indicated cell membrane damage (Kot and Kmieć 2020; Kmieć and Kot 2021; Kmieć et al. 2022).

Aphids, as exclusive phloem feeders, puncture cells with stylets, causing tissue injury during feeding, while continuously injecting salivary secretions to keep phloem cells alive (Tjallingii 2006; Giordanengo et al. 2010). Mature galls of different aphid species can host dozens, hundreds or even thousands of feeding individuals, which can feed continuously for several weeks, degrading leaf proteins and utilizing increased photosynthesis product translocation (Wool 2004; Chen et al. 2020; Kot and Kmieć 2020). Galls induced by aphid *Pemphigus betae* act as strong sinks by reversing normal transport patterns and inducing mature leaves to import photoassimilates produced in neighbouring leaves (Larson and Whitham 1991). Sugars drained to gall sites may play secondary roles in the dissipation of free radical species accumulated in gall tissues (Isaias et al. 2015). However, malondialdehyde (MDA) content, a widely used indicator of plant membrane damage, shows quite opposite trend to E_L in several aphid galls (Kot and Kmieć 2020; Kmieć and Kot 2021; Kmieć et al. 2022). Only in the galls of *T. ulmi* and *P. bursarius* increased membrane damage estimated by both E_L and MDA accumulation were found. It could suggest that MDA elimination and redox signalling regulation maintains redox homeostasis in aphid galls. The low level of MDA was also indicated in the simple galls of the psyllid *Pseudophacopteron* spp. induced on leaves of *Aspidosperma* spp. (Kuster et al. 2022). On the other hand, across subsequent stages of *T. ulmi* gall development, H_2O_2 level, membrane permeability, and lipid peroxidation gradually increased in gall tissues (Kmieć et al. 2018a). The increase in MDA content (via ROS excess) may indicate plant membrane damage (trigger cell death) or represent an acclimation process, since MDA can activate regulatory genes involved in plant defence and development, depending on its cellular concentration (Morales and Munné-Bosch 2019). Reactive molecule production occurs throughout gall formation, peaking during maturation. Effective gall growth and development requires the control of toxic and signalling effect of ROS within plant cell compartments (Isaias et al. 2015; Kmieć et al. 2018a). Transcriptional analyses performed by Takeda and co-workers (2019) confirmed the participation of ROS during gall development. The up-regulation of AtMYB77 (At3g50060) involved in ROS metabolism by direct transcriptional regulation of the ORBITALLY MANIFESTED GENE 1 (OMG1) has been detected. Moreover, up-regulation of putative cationic cell-wall-bound peroxidase (AtPrx25) and NADPH oxidase (At5g51060) involved in lignin biosynthesis and ROS generation suggests that active ROS production is crucial in cell wall modification and lignification within the gall and/or cellular signalling (Takeda et al. 2019).

3.1 Enzymatic Ways of Stress Dissipation in Aphid-Induced Galls

Under normal plant developmental conditions, a low level of ROS is maintained through the presence of an efficient antioxidant defensive system formed by secondary plant metabolites and scavenging enzymes (Waszczak et al. 2018; Wang et al. 2024). Oxidative enzymes, such as superoxide dismutase (SOD), peroxidases (POD) and catalase (CAT), play pivotal roles in detoxifying ROS generated in response to biotic stresses (Wang et al. 2024). Ascorbate peroxidase (APX) is regarded as one of the most efficient scavengers of H_2O_2 under stress conditions (Sharma et al. 2012). Guaiacol peroxidase (GPOD) is implicated in essential biosynthetic processes, including cell-wall lignification, IAA degradation, ethylene biosynthesis, wound healing, and stress defense mechanisms (Mika et al. 2010). In turn, various isoforms of CAT help in eliminating H_2O_2 generated during photorespiration, β-oxidation of fatty acids, and purine catabolism in peroxisomes (Corpas et al. 2017). Notably, the activity of GPOD was significantly reduced in aphid galls formed on elm and poplar trees as compared to leaves without galls (Gailite et al. 2005; Samsone et al. 2012; Kot and Kmieć 2020; Kmieć and Kot 2021; Kmieć et al. 2022). A similar trend was observed with APX activity in poplar galls, while in galls formed on elm, its activity was increased compared to the non-galled leaves (Kot and Kmieć 2020; Kmieć and Kot 2021; Kmieć et al. 2022). Significantly enhanced peroxidase activity in galls formed on *Pistacia atlantica* was also detected by Inbar and co-authors (2003). Detailed analyses of antioxidant enzyme activities in *T. ulmi* galls during subsequent developmental stages revealed elevated activity of APX and CAT in young galls, followed by a decrease in APX activity and CAT inactivation in mature galls, which was correlated with high levels of free radicals, lipid peroxidation, and membrane permeability (Kmieć et al. 2018a). Additionally, a unigene encoding β-(1,3)-galactosyltransferase, an enzyme involved in the modification of arabinogalactan-proteins (AGPS), has been identified in *S. chinensis* galls. The involvement of this enzyme in various processes, such as growth and development, programmed cell death, and signalling pathways, has been identified (Takeda et al. 2019).

In the tissues of galls, the presence of pathogenesis-related (PR) proteins, namely chitinase and β-1,3-glucanase has been identified. Elevated chitinase activity has been detected in mature final galls of *Smynthurodes betae* and *Forda riccobonii*, young galls of *Slavum wertheimae* formed on *P. atlantica*, and only in mature galls of *T. ulmi* on *U. pumila*. Conversely, significantly higher β-1,3-glucanase activity in galls compared to intact leaves was identified only in the case of T. *ulmi*. It is noteworthy that an increased activity of both enzymes occurred when the galls were filled with numerous feeding aphids (Inbar et al. 2003; Kmieć et al. 2018d). The local accumulation of phenolic compounds in galls, combined with the high activity of PR proteins such as β-1,3-glucanase and chitinase in the stage of mature galls, may contribute to the protection of both, the galls and the aphids. Mature galls release emigrants that are exposed to predation, parasitism, and pathogenic infection. Additionally, the activation of PR proteins associated with a high level of phenolics, lipid peroxidation, and ROS accumulation in the tissue of mature galls appears to resemble the plant hypersensitivity reaction in pathogen infection.

3.2 Phenolics in Aphid-Induced Galls

Aphid-induced galls exhibit remarkable accumulation of phenolic compounds, especially detected in the outer tissue layers (Kuster et al. 2022; Rosa et al. 2024). Tannins, a phenolic derivative, constitute up to 60–70% of the total dry weight in galls induced by the aphid *S. chinensis* on *R. chinensis* and *R. javanica* (Chen et al. 2018). It was evidenced that the level of phenols and their derivatives increased over time in gall tissues. When galls of *H. hamamelidis* reach full size, the fundatrix begin to reproduce parthenogenetically, displaying nearly double the concentration of condensed tannins (Rehill and Schultz 2012). In mature *T. ulmi* galls, soluble phenols accumulate over 6.6-fold more than the control leaves, and over 3-fold more than gall tissues in the initial stage of gall development (Kmieć et al. 2018d). In turn, polyphenol oxidase activity gradually declines in galls induced by *Tetraneura fusiformis* on *Ulmus davidiana* from the second week of fundatrix feeding within the gall (Hori et al. 1997). A low activity of this enzyme is also observed in *T. ulmi*, *E. ulmi*, *C. compressa* (Gailite et al. 2005), and *P. spyrothecae* (Samsone et al. 2012) galls. Such profound alterations in phenolic metabolism occur due to the interaction of the fundatrix with the host plant tissues.

The influx of carbohydrates with phloem sap supports the production of plant phenolics, thus phenolic compounds might accumulate in galls due to enhanced sink strength (Rehill and Schultz 2012). Alternatively, upregulated phenolic synthesis could induce sink enzyme activity (Isaias et al. 2014; Carneiro and Isaias 2015; Oates et al. 2016). Phenolics are not only involved in developmental processes by controlling cell expansion and division, by interacting with auxin oxidases and modulation of IAA level (Oliveira et al. 2016), but also mediate vascular tissue morphogenesis (Aloni 2001). The co-localization of hydrogen peroxide and phenolics in gall tissues suggests the involvement of phenolic derivatives in a conventional pathway for oxidative stress mitigation. The interplay between phenolics, proanthocyanidins, and hydrogen peroxide implies their role in scavenging ROS and maintaining homeostasis within gall tissues (Oliveira et al. 2017). Acting as antioxidants, phenolic compounds likely prevent oxidative stress in gall tissues by distributing along the gradient of ROS concentration (Carneiro and Isaias 2015; Rosa et al. 2024). Moreover, the concurrent lignification of plant cell walls appears to complement this stress-alleviating mechanism (Kuster et al. 2022). Genes related to gallotannin biosynthesis (Chen et al. 2018), gallic acid synthesis (Wang et al. 2017), and lignin biosynthesis (Hirano et al. 2020) have been identified in *S. chinensis* galls, indicating lignin biosynthesis through oxidation of phenolic compounds and/or ROS generation, underscoring the role of ROS during gall development as regulatory and/or signalling molecules (Takeda et al. 2019). Galls induced by *Tetraneura akinire* on elm leaves express genes encoding lignocellulose synthase, suggesting the reinforcement of cell walls (Lu et al. 2019). This positive relationship between phenolic accumulation and the performance of galling aphids demonstrates an adaptive strategy beyond plant defense.

3.3 Primary Metabolites in Aphid-Induced Galls

The phenomenon of up-regulated primary metabolism and nutrient transport to gall-inducing sites is widely observed across various plant-aphid interactions. For instance, *P. betae* galls formed on *Populus angustifolia* exhibit increased accumulations of sugars, carbon- and nitrogen-containing compounds, as reported by Compson et al. (2011). Similarly, galls induced by several *Tetraneura* species on *U. davidiana* var. *japonica* leaves significantly accumulate higher amounts of amino acids compared to intact leaves, with asparagine being the predominant amino acid, followed by glutamic acid, glutamine, and threonine (Suzuki et al. 2009). Nevertheless, the aphids cannot utilize free amino acids contained in gall tissues, and it is hypothesized that these amino acids might be translocated into the phloem circulating within the gall. Notably, asparagine and glutamine are crucial for nitrogen synthesis in actively growing plant parts and are involved in nitrogen remobilization during leaf senescence (Suzuki et al. 2009). The phloem exudates of leaves with pseudogalls induced by the aphid *Sorbaphis chaetosiphon* on *Sorbus commixta* also exhibit high concentrations of glutamine and glutamic acid (Koyama et al. 2004). Although glutamine and asparagine are not essential for aphids, their intracellular bacterial symbiont *Buchnera* can utilize them to produce essential amino acids (Feng et al. 2019). The contents of free and essential amino acids in the gall of *S. chinensis* formed on *R. chinensis* were significantly higher than those in non-galled leaves. The transcriptomic analyses revealed the up-regulation of genes involved in the synthesis of essential amino acids, such as histidine, leucine, lysine, methionine, tryptophan, and valine in the galls (Lu et al. 2021). High protein content was also noted in *T. ulmi* galls. However, the amount of total proteins decreased along with gall development (Kmieć et al. 2018d). In turn, a low content of extractable proteins was found in mature galls of *E. ulmi* and *C. compressa* (Gailite et al. 2005). Interestingly, *T. ulmi* galls exhibit enhanced accumulation of polyamines (PAs) and their biosynthesis enzymes (Kmieć et al. 2018c). Enhanced accumulation of histamine, spermine (SPM), spermidine (SPD), and cadaverine (CAD) in the galls was found in the initial stage and in fully developed galls. However, none of the polyamines were detected in mature galls (Kmieć et al. 2018c). The growing body of evidence suggests that PAs are involved in the regulation of various processes, such as cell division, xylem differentiation, leaf senescence, and plant defence signalling (Asija et al. 2023; Kolesnikov et al. 2024). For instance, exogenously applied SPM and SPD reduced the levels of MDA, facilitating the scavenging of H_2O_2, and protecting plant cell membranes (Velikova et al. 2000). It has been documented that PAs possess dual ability in the regulation of ROS metabolism. They are involved in the production of signalling molecules, such as H_2O_2, NO, ethylene, and GABA, as well as the expression of SOD, CAT, APX and GR enzymes required for ROS detoxification (Asija et al. 2023). Peptide signalling pathways have been suggested in the galls. On the basis of GO analyses, Takeda and co-authors (2019) found that CLE family genes, especially CLE44, have been up-regulated in *S. chinensis* galls. The CLE peptides are small ligands that bind to the CLV1/CLV2 proteins and are involved in cell–cell communication during development, symbiosis, and parasitism, among others. Vascular development in galls may be regulated by CLE and LRR-RLK genes, along with factors like auxin in *S. chinensis* and *D. vitifoliae* galls (Schultz et al. 2019; Takeda et al. 2019).

Low concentrations of glucose, fructose, sucrose, and fats were found in *S. chinensis* galls, although the biosynthesis of monosaccharides was enhanced. In contrast, starch content was significantly higher in galls than in non-galled leaf tissues (Lu et al. 2021). The presence of starch grains was also reported by Alvarez et al. (2009) in galls induced by aphids on *Pistacia terebinthus*, and various studies indicated that the accumulation of this polysaccharide is typical for galls induced by sap-sucking insects. The role of starch is related to gall maintenance, and local response to the production of ROS in galls, rather than the feeding behavior (Oliveira and Isaias 2010; Oliveira et al. 2011, 2014). Unigenes involved in amino acid, sugar, and lipid pathways, as well as in plant resistance showed selective expression in galls. These selective expressed unigenes played a wide range of regulatory roles in plants, such as enhancing the enrichment and transportation of amino acids and lipids, enhancing glycoside hydrolysis and biosynthesis of starch. The up-regulation of these genes may provide the basis for continuous growth of galls through the growing season (Lu et al. 2021). The unbalanced shift of the tricarboxylic acid (TCA) cycle might be responsible for reducing the metabolism of simple sugars and enhanced accumulation of NAD+/NADH. Aphids are able to utilize simple sugar and NAD+/NADH as nutrients (Lu et al. 2021).

It has been proposed that the relationship between the production of ROS and the sugars could initiate and maintain the functionality of the gall (Oliveira et al. 2014). In the galls induced by the psyllid *Pseudophacopteron* sp. on *A. australe*, the activity of invertases and sucrose synthase were related to the sites of high respiratory stress and cell division (Oliveira and Isaias 2010). The significant correlation with the activity of invertase and the growth of galls induced by the aphids *H. hamamelidis* on *H. virginiana* and *S. chinensis* on *R. chinensis* have been detected (Rehill and Schultz 2003; Ruan et al. 2017). Increased sink strength of galls required both cell wall and vacuolar invertases. Findings of Ruan et al. (2017) demonstrated that vacuolar invertase exerted a significant influence during early development, while ionically bound cell wall invertase exhibited notably increased activity during the gall's rapid growth phase, suggesting its involvement in tissue expansion. Apart from catalyzing the breakdown of sucrose, the invertases take part in the biosynthesis of phytohormones as AUX, ABA and CKs (Oliveira and Isaias 2010). The heightened activity of sink enzymes, particularly the increase in cell wall invertases, proves advantageous for the fundatrix, as observed enhancements precede the presence of offspring within the galls (Rehill and Schultz 2003).

Recent studies have unveiled intriguing findings regarding gene expression patterns within leaf galls induced by *D. vitifoliae* on *Vitis riparia*. Schultz et al. (2019) demonstrated the activation of genes associated with flower and fruit development within these leaf galls, suggesting a remarkable reprogramming of developmental pathways. Particularly noteworthy is the activation of class-1 KNOX transcription factors, which are known for their role in orchestrating de novo meristematic structures within leaves. This observation supports the hypothesis proposed by Takeda et al. (2021) that galling aphids induce the transformation of source tissues into meristematic tissues by stimulating the expression of class-1 KNOX genes within leaf cells. Moreover, selective expression of unigenes involved in DNA methylation has been observed within the galls, highlighting the intricate molecular regulation underlying gall formation. The DNA methylation, a pivotal mechanism in epigenetics, exerts regulatory control over diverse biological processes, including plant resistance and flowering phase (Zhang et al. 2018).

4 Concluding Remarks

Aphid galls develop mostly on young plant tissues, and the colonization activity of insects is timed with phenology of host plants. In general, every gall is induced by a single female hatching from a fertilized egg, which propagates in the gall by parthenogenetic reproduction. The mature colonial aphid galls reach different shape and volume, and contain various number of aphid individuals, depending on the aphid and host plant species. The induction of the gall causes a change in the design of plant tissues, thereby generating complex structures in the host plant with a typical and definitive shape and size. Constant food intake promotes key phases of aphid-induced gall development, such as increment in mass and qualitative differentiation, maturation and senescent. The activity of the gall-inducing aphids causes several structural and physiological changes in the host plant tissues. During the successive stages of gall development, dynamic changes in the activity of various biomolecules are observed. Most evidence suggests that manipulation of local phytohormone concentrations, auxins and CKs in particular, facilitates gall induction and growth. The feeding of fundatrix triggers a cascade of events that results in cell hypertrophy and tissue hyperplasia with the gradient accumulation of ROS, primary and secondary metabolites, and activity of enzymes. The presence of chlorophylls and carotenoids in galls enabled the photosynthesis performance, however, their low contents indicated that the galls act as assimilation products acceptors. Molecular studies have revealed that gall formation is similar to the development of acceptor organs like flowers or fruits. Initially, the activity of antioxidant enzymes, polyamines and phenolic compounds in aphid galls increases, and in subsequent stages an increasing number of feeding aphids generates an increase in the level of ROS and final products of unsaturated fatty acid destruction with a decrease in the activity of antioxidant enzymes and accumulation of phenolic compounds and PR proteins in gall tissues.

Most of the existing studies provided findings from analyses conducted only at one stage of gall development, typically focusing on mature galls. However, the cytological, physiological, biochemical, and molecular alterations in developing galls are highly dynamic. Hence, to facilitate comparative analysis and draw substantiated conclusions, it seems necessary to replicate analyses across different stages of gall development, while specifying the precise developmental stage of the gall. Consequently, there is a distinct demand for expanded research efforts, particularly those delving into molecular aspects at various developmental stages of aphid galls.

References

Aloni R (2001) Foliar and axial aspects of vascular differentiation: hypotheses and evidence. J Plant Growth Regul 20:22–34

Álvarez R, Encina A, Hidalgo NP (2009) Histological aspects of three Pistacia terebinthus galls induced by three different aphids: Paracletus cimiciformis, Forda marginata and Forda formicaria. Plant Sci 176(2):303–314

Álvarez R, Molist P, González-Sierra S, Martinez JJI, Nieto-Nafría JM (2014) The histo structure of galls induced by aphids as a useful taxonomic character: the case of Rectinasus (Hemiptera, Aphididae, Eriosomatinae). Zootaxa 3861:487–492

Álvarez R, Martinez J-JI, Muñoz-Viveros AL, Molist P, Abad-González J, Nieto Nafría JM (2016) Contribution of gall microscopic structure to taxonomy of gallicolous aphids on Pistacia. Plant Biol 18:868–875

Andreas P, Kisiala A, Emery RJN, De Clerck-Floate R, Tooker JF, Price PW, Miller DG III, Chen MS, Connor EF (2020) Cytokinins are abundant and widespread among insect species. Plants 9:208

Argueso CT, Kieber JJ (2024) Cytokinin: from autoclaved DNA to two-component signalling. Plant Cell 36(5):1429–1450. https://doi.org/10.1093/plcell/koad327

Asija S, Seth T, Umar S, Gupta R (2023) Polyamines and their crosstalk with phytohormones in the regulation of plant defense responses. J Plant Growth Regul 42:5224–5246

Blackman RL, Eastop VF (1994) Aphids on the world's trees. An identification and information guide. CAB International, London, 986pp

Body MJA, Zinkgraf MS, Whitham TG, Lin C-H, Richardson RA, Appel HM, Schultz JC (2019) Heritable phytohormone profiles of poplar genotypes vary in resistance to a galling aphid. Mol Plant-Microbe Interact 32:654–672. https://doi.org/10.1094/MPMI-11-18-0301-R

Burstein M, Wool D, Eshel A (1994) Sink strength and clone size of sympatric, gall forming aphids. Eur J Entomol 91:57–61

Carneiro RGS, Isaias RMS (2015) Gradients of metabolite accumulation and redifferentation of nutritive cells associated with vascular tissues in galls by sucking insects. AoB Plants 7:plv086

Carneiro RGS, Castro AC, Isaias RMS (2014) Unique histochemical gradients in a photosynthesis-deficient plant gall. S Afr J Bot 92:97–104

Castro AC, Oliveira DC, Moreira ASFP, Lemos-Filho JP, Isaias RMS (2012) Source–sink relationship and photosynthesis in the horn-shaped gall and its host plant *Copaifera langsdorffii* Desf. (Fabaceae). S Afr J Bot 83:121–126

Chen J, Qiao GX (2012) Galling aphids (Hemiptera: Aphidoidea) in China: diversity and host specificity. Psyche-J Entomol 2012(1):621934

Chen H, Liu J, Cui K, Lu Q, Wang C, Wu H, Yang Z, Ding W, Shao S, Wang H (2018) Molecular mechanisms of tannin accumulation in Rhus galls and genes involved in plant-insect interactions. Sci Rep 8:9841

Chen X, Yang Z, Chen H, Qi Q, Liu J, Wang C, Shao S, Lu Q, Li Y, Wu H, King-Jones K, Chen M-S (2020) A complex nutrient exchange between a gall-forming aphid and its plant host. Front Plant Sci 11:811. https://doi.org/10.3389/fpls.2020.00811

Choi HW (2024) From the photosynthesis to hormone biosynthesis in plants. Plant Pathol J 40(2):99–105. https://doi.org/10.5423/PPJ.RW.01.2024.0006

Choudhury FK, Rivero RM, Blumwald E, Mittler R (2017) Reactive oxygen species, abiotic stress and stress combination. Plant J 90(5):856–867. https://doi.org/10.1111/tpj.13299

Compson ZG, Larson KC, Zinkgraf MS, Whitham TG (2011) A genetic basis for the manipulation of sink-source relationships by the galling aphid *Pemphigus betae*. Oecologia 167:711–721

Connor EF, Bartlett L, O'Toole S, Byrd S, Biskar K, Orozco J (2012) The mechanism of gall induction makes galls red. Arthropod Plant Interact 6:489–495. https://doi.org/10.1007/s11829-012-9210-7

Corpas FJ, Barroso JB, Palma JM, Rodriguez-Ruiz M (2017) Plant peroxisomes: a nitro-oxidative cocktail. Redox Biol 11:535–542

Das P, Nutan KK, Singla-Pareek SL, Pareek A (2015) Oxidative environment and redox homeostasis in plants: dissecting out significant contribution of major cellular organelles. Front Environ Sci 2:70. https://doi.org/10.3389/fenvs.2014.00070

Eitle MW, Carolan JC, Griesser M, Forneck A (2019) The salivary gland proteome of root-galling grape phylloxera (*Daktulosphaira vitifoliae* Fitch) feeding on Vitis spp. PLoS One 14(12):e0225881. https://doi.org/10.1371/journal.pone.0225881

Feng H, Edwards N, Anderson CMH, Althaus M, Duncan RP, Hsu Y-C, Luetje CW, Price DRG, Wilson ACC, Thwaites DT (2019) Trading amino acids at the aphid–Buchnera symbiotic interface. Proc Natl Acad Sci U S A 116:16003–16011. www.pnas.org/cgi/doi/10.1073/pnas.1906223116

Ferreira BG, Álvarez R, Bragança GP, Alvarenga DR, Pérez-Hidalgo N, Isaias RM (2019) Feeding and other gall facets: patterns and determinants in gall structure. Bot Rev 85:78–106

Gailite A, Andersone U, Ievinsh G (2005) Arthropod-induced neoplastic formations on trees change photosynthetic pigment levels and oxidative enzyme activities. J Plant Interact 1:61–67

Giordanengo P, Brunissen L, Rusterucci C, Vincent C, van Bel A, Dinant S, Girousse C, Faucher M, Bonnemain JL (2010) Compatible plant-aphid interactions: how aphids manipulate plant response. C R Biol 333:516–523

Giron D, Huguet E, Stone GN, Body M (2016) Insect-induced effects on plants and possible effectors used by galling and leaf-mining insects to manipulate their host-plant. J Insect Physiol 84:70–89

Hirano T, Kimura S, Sakamoto T, Okamoto A, Nakayama T, Matsuura T, Ikeda Y, Takeda S, Suzuki Y, Ohshima I et al (2020) Reprogramming of the developmental program of *Rhus javanica* during initial stage of gall induction by *Schlechtendalia chinensis*. Front Plant Sci 11:471

Hori K, Wada A, Shibuta T (1997) Changes in phenoloxidase activities of the galls on leaves of *Ulmus davidiana* formed by *Tetraneura fusiformis* (Homoptera: Eriosomatinae). Appl Entomol Zool 32:365–371

Hua J, Liu J, Zhou W, Ma C, Luo S (2023) A new perspective on plant defense against foliar gall-forming aphids through activation of the fruit abscission pathway. Plant Physiol Biochem 196:1046–1054. https://doi.org/10.1016/j.plaphy.2023.03.007

Inbar M, Mayer R, Doostdar H (2003) Induced activity of pathogenesis related (PR) proteins in aphid galls. Symbiosis 34:1–10

Isaias RMS, Oliveira DC, Carneiro RGS, Kraus JE (2014) Developmental anatomy of galls in the Neotropics: arthropods stimuli versus host plant constraints. In: Fernandes G, Santos J (eds) Neotropical insect galls. Springer, Dordrecht. https://doi.org/10.1007/978-94-017-8783-3_2

Isaias RMS, Oliveira DC, Moreira ASFP, Soares GLG, Carneiro RGS (2015) The imbalance of redox homeostasis in arthropod-induced plant galls: mechanisms of stress generation and dissipation. Biochim Biophys Acta 1850:1509–1517

Isaias RMDS, Kraus JE, Costa EC, Carneiro RGDS (2024) The anatomy of neotropical galls and the untold lessons about the morphogenetical potentialities of plants. Rodriguésia 75:e01542023

Kmieć K, Kot I (2021) Physiological response of *Populus nigra* "Italica" to galling aphids feeding. Plant Biol 23:675–679. https://doi.org/10.1111/plb.13265

Kmieć K, Rubinowska K, Golan K (2018a) *Tetraneura ulmi* (Hemiptera: Eriosomatinae) induces oxidative stress and alters antioxidant enzyme activities in elm leaves. Environ Entomol 47(4):840–847

Kmieć K, Rubinowska K, Michałek W, Sytykiewicz H (2018b) The effect of galling aphids feeding on photosynthesis photochemistry of elm trees (*Ulmus* sp.). Photosynthetica 56(4):989–997

Kmieć K, Sempruch C, Chrzanowski G, Czerniewicz P (2018c) The effect of *Tetraneura ulmi* L. galling process on the activity of amino acid decarboxylases and the content of biogenic amines in Siberian elm tissues. Bull Entomol Res 108(1):69–76

Kmieć K, Złotek U, Jakubczyk A, Karaś M (2018d) Biochemical alterations in *Ulmus pumila* L. leaves induced by galling aphid *Tetraneura ulmi* L. Acta Sci Pol Hortorum Cultus 17(6):175–183

Kmieć K, Kot I, Rubinowska K, Górska-Drabik E, Golan K, Sytykiewicz H (2022) The variation of selected physiological parameters in elm leaves (*Ulmus glabra* Huds.) infested by gall inducing aphids. Plants 11(3):244. https://doi.org/10.3390/plants11030244

Kolesnikov YS, Kretynin SV, Filepova R, Dobrev PI, Martinec J, Kravets VS (2024) Polyamines metabolism and their biological role in plant cells: what do we really know? Phytochem Rev 23:997–1026. https://doi.org/10.1007/s11101-024-09913-3

Korgaonkar A, Han C, Lemire AL, Siwanowicz I, Bennouna D, Kopec RE, Andolfatto P, Shigenobu S, Stern DL (2021) A novel family of secreted insect proteins linked to plant gall development. Curr Biol 31:1836–1849

Kot I, Kmieć K (2020) Poplar tree response to feeding by the petiole gall aphid *Pemphigus spyrothecae* Pass. Insects 11(5):282. https://doi.org/10.3390/insects11050282

Koyama Y, Yao I, Akimoto SI (2004) Aphid galls accumulate high concentrations of amino acids: a support for the nutrition hypothesis for gall formation. Entomol Exp Appl 113:35–44

Kurzfeld-Zexer L, Lev-Yadun S, Inbar M (2015) One aphid species induces three gall types on a single plant: comparative histology of one genotype and multiple extended phenotypes. Flora 210:19–30

Kuster VC, Cardoso JCF, Moreira ASFP, Isaias RMS, Oliveira DC (2022) Assessing oxidative stress and photosynthetic activity in leaf galls induced by different species of galling insect on *Aspidosperma* spp. Biologia 77:3441–3454. https://doi.org/10.1007/s11756-022-01167-8

Larson KC, Whitham TG (1991) Manipulation of food resources by a gall-forming aphid: the physiology of sink-source interactions. Oecologia 88:15–21

Larson KC, Whitham TG (1997) Competition between gall aphids and natural plant sinks: plant architecture affects resistance to galling. Oecologia 109:575–582

Lu HB, Jin LP, Wei D, Huang ZH (2019) Study on the differential gene expression of elm leaves fed on by *Tetraneura akinire* Sasaki. Genes Genomics 41:1505–1516

Lu Q, Chen X, Yang Z, Bashir NH, Liu J, Cui Y, Shao S, Chen M-S, Chen H (2021) Molecular and histologic adaptation of horned gall induced by the aphid *Schlechtendalia chinensis* (Pemphigidae). Int J Mol Sci 22:5166. https://doi.org/10.3390/ijms22105166

Martin RE, Postiglione AE, Muday GK (2022) Reactive oxygen species function as signaling molecules in controlling plant development and hormonal responses. Curr Opin Plant Biol 69:102293

Mika A, Boenisch MJ, Hopff D, Lüthje S (2010) Membrane-bound guaiacol peroxidases from maize (*Zea mays* L.) roots are regulated by methyl jasmonate, salicylic acid, and pathogen elicitors. J Exp Bot 61:831–841. https://doi.org/10.1093/jxb/erp353

Morales M, Munné-Bosch S (2019) Malondialdehyde: facts and artifacts. Plant Physiol 180:1246–1250

Morkunas I, Mai VC, Gabryś B (2011) Phytohormonal signaling in plant responses to aphid feeding. Acta Physiol Plant 33:2057–2073. https://doi.org/10.1007/s11738-011-0751-7

Oates CN, Denby KJ, Myburg AA, Slippers B, Oates SN (2016) Insect gallers and their plant hosts: from omics data to systems biology. Int J Mol Sci 17:1891. https://doi.org/10.3390/ijms17111891

Oliveira DC, Isaias RMS (2010) Cytological and histochemical gradients induced by a sucking insect in galls of *Aspidosperma australe* Arg. Muell (Apocynaceae). Plant Sci 178:350–358

Oliveira DC, Isaias RMS, Moreira ASFP, Magalhães TA, Lemos-Filho JP (2011) Is the oxidative stress caused by *Aspidosperma* spp. galls capable of altering leaf photosynthesis? Plant Sci 180:489–495

Oliveira DC, Moreira ASFP, Isaias RMS (2014) Functional gradients in insect gall tissues: studies on Neotropical host plants. In: Fernandes G, Santos J (eds) Neotropical insect galls. Springer, Dordrecht. https://doi.org/10.1007/978-94-017-8783-3_3

Oliveira DC, Isaias RMS, Fernandes GW, Ferreira BG, Carneiro RGS, Fuzaro L (2016) Manipulation of host plant cells and tissues by gall-inducing insects and adaptive strategies used by different feeding guilds. J Insect Physiol 84:103–113

Oliveira DC, Moreira ASFP, Isaias RMS, Martini V, Rezende UC (2017) Sink status and photosynthetic rate of the leaflet galls induced by *Bystracoccus mataybae* (Eriococcidae) on *Matayba guianensis* (Sapindaceae). Front Plant Sci 8:1249. https://doi.org/10.3389/fpls.2017.01249

Raman A (2011) Morfogenesis of insect-induced plant galls: facts and questions. Flora 206:517–533

Raman A (2012) Gall induction by hemipteroid insects. J Plant Interact 7:29–44

Rehill BJ, Schultz JC (2003) Enhanced invertase activities in the galls of *Hormaphis hamamelidis*. J Chem Ecol 29:2703–2720

Rehill BJ, Schultz JC (2012) *Hormaphis hamamelidis* fundatrices benefit by manipulating phenolic metabolism of their host. J Chem Ecol 38:496–498. https://doi.org/10.1007/s10886-012-0115-9

Rodriguez PA, Bos JIB (2013) Toward understanding the role of aphid effectors in plant infestation. Mol Plant Microbe Interact 26:25–30

Rosa LMP, Silva MS, da Silva Carneiro RG, Machado M, Kuster VC (2024) Hemiptera-induced galls of *Sapium glandulosum* have histological and cytological compartmentalization created with a large amount of carbohydrate. Protoplasma 261(3):593–606. https://doi.org/10.1007/s00709-023-01921-y

Ruan ZY, Chen XM, Yang P, Wang BY (2017) Roles played by invertase and gene expression in the development of the horn-shaped gall on leaves of Rhus chinensis. Funct Plant Biol 44(12):1160–1170

Samsone I, Andersone U, Ievinsh G (2012) Variable effect of arthropod-induced galls on photochemistry of photosynthesis, oxidative enzyme activity and ethylene production in tree leaf tissues. Environ Exp Biol 10:15–26

Schultz JC, Edger PP, Body MJA, Appel HM (2019) A galling insect activates plant reproductive programs during gall development. Sci Rep 9:1833

Seng S, Ponce GE, Andreas P, Kisiala A, De Clerck-Floate R, Miller DG III, Chen M-S, Price PW, Tooker JF, Emery RJN, Connor EE (2023) Abscisic acid: a potential secreted effector synthesized by phytophagous insects for host-plant manipulation. Insects 14:489. https://doi.org/10.3390/insects14060489

Sharma P, Jha AB, Dubey RS, Pessarakli M (2012) Reactive oxygen species, oxidative damage, and antioxidative defense mechanism in plants under stressful conditions. J Bot 2012:26. https://doi.org/10.1155/2012/217037, https://www.hindawi.com/journals/jb/2012/217037/

Silvia MS, Connor EF (2017) Differences in meristems between monocots and dicots and susceptibility to attack by gall-inducing insects. Arthropod Plant Interact 11:485–494

Stone GN, Schönrogge K (2003) The adaptive significance of insect gall morphology. Trends Ecol Evol 18:512–522. https://doi.org/10.1016/S0169-5347(03)00247-7

Suzuki DK, Fukushi Y, Akimoto SI (2009) Do aphid galls provide good nutrients for the aphids?: comparisons of amino acid concentrations in galls among *Tetraneura* species (Aphididae: Eriosomatinae). Arthropod Plant Interact 3:241–247

Takeda S, Yoza M, Amano T, Ohshima I, Hirano T, Sato MH et al (2019) Comparative transcriptome analysis of galls from four different host plants suggests the molecular mechanism of gall development. PLoS One 14(10):e0223686. https://doi.org/10.1371/journal.pone.0223686

Takeda S, Hirano T, Ohshima I, Sato MH (2021) Recent progress regarding the molecular aspects of insect gall formation. Int J Mol Sci 22:9424. https://doi.org/10.3390/ijms22179424

Takei M, Yoshida S, Kawai T, Hasegawa M, Suzuki Y (2015) Adaptive significance of gall formation for a gall-inducing aphids on Japanese elm trees. J Insect Physiol 72:43–51

Tjallingii WF (2006) Salivary secretions by aphids interacting with proteins of phloem wound responses. J Exp Bot 57:739–745

Tooker JF, Helms AM (2014) Phytohormone dynamics associated with gall insects, and their potential role in the evolution of the gall-inducing habit. J Chem Ecol 40:742–753

Velikova V, Yordanov I, Edreva A (2000) Oxidative stress and some antioxidant systems in acid rain-treated bean plants: protective role of exogenous polyamines. Plant Sci 151:59–66. https://doi.org/10.1016/S0168-9452(99)00197-1

Walling LI (2008) Avoiding effective defences: strategies employed by phloem-feeding insects. Plant Physiol 146:859–866

Wang H, Liu J, Cui K et al (2016) Gibberellic acid is selectively downregulated in response to aphid-induced gall formation. Acta Physiol Plant 38:214. https://doi.org/10.1007/s11738-016-2224-5

Wang H, Cui K, Shao S, Liu J, Chen H, Wang C, Yang Z, Lu Q, King-jones K, Chen X et al (2017) Molecular response of gall induction by aphid *Schlechtendalia chinensis* (Bell) attack on *Rhus chinensis* Mill. J Plant Interact 12:465–479

Wang H, Shi S, Hua W (2023) Advances of herbivore-secreted elicitors and effectors in plant-insect interactions. Front Plant Sci 14:1176048. https://doi.org/10.3389/fpls.2023.1176048

Wang P, Liu W-C, Han C, Wang S, Bai M-Y, Song C-P (2024) Reactive oxygen species: multidimensional regulators of plant adaptation to abiotic stress and development. J Integr Plant Biol 66:330–367

Waszczak C, Carmody M, Kangasjärvi J (2018) Reactive oxygen species in plant signaling. Annu Rev Plant Biol 69:209–236

Will T, Steckbauer K, Hardt M, van Bel AJE (2012) Aphid gel saliva: sheath structure, protein composition and secretory dependence on stylet-tip milieu. PLoS One 7(10):e46903. https://doi.org/10.1371/journal.pone.0046903

Wool D (2004) Galling aphids: specialization, biological complexity, and variation. Annu Rev Entomol 49:175–192

Zhang H, Lang Z, Zhu J-K (2018) Dynamics and function of DNA methylation in plants. Nat Rev Mol Cell Biol 19:489–506

Plant Galls: Implications for the Chemical Profile and Biological Activity of Their Host Plants

Lubia María Guedes, Narciso Aguilera, and Lorena Rodríguez-Cerda

1 Introduction

Secondary metabolites (SMs) emerged early in the evolutionary history of land plants (Zunjarrao et al. 2020). During the colonization of land, plants had to cope with attacks by herbivores animals, and pathogens, such as bacteria, viruses, and fungi (Wink 2020) Under this situation, as sessile organisms, plants produce bioactive compounds as a defense mechanism against herbivores and pathogens. These bioactive compounds, typically lineage-specific, assist plants in interacting with the biotic and abiotic environment (Hartmann 2007). The diversity of SMs not only protect plants against herbivores and pathogens, but also perform non-defensive functions (Wink 2020, 2003). Certain SMs have been associated with ecological and physiological functions, participating in plant-plant communication, attraction of pollinators, seed dispersers, and protection against abiotic stress, as UV-B radiation protectors (Ramawat and Goyal 2020).

There are many SMs, with more than 200,000 described structures, whose precursors are derived from metabolic pathways, such as glycolysis, the Krebs cycle, and the shikimate pathway (Wink 2020). Traditionally, SMs have been classified according to their structure and chemical composition, including phenolic compounds, nitrogen-containing compounds, and terpenes (Erb and Kliebenstein 2020). The synthesis of SMs is resource-intensive for plant growth and reproduction (Paul--Victor et al. 2010; Zust et al. 2011; Huber et al. 2016); thus, their concentration in plants is not constant. Therefore, the profile and concentration of SMs varies because of numerous biotic and abiotic factors (Tollrian and Harvell 1999).

L. M. Guedes (✉) · N. Aguilera · L. Rodríguez-Cerda
Facultad de Ciencias Forestales, Laboratorio de Semioquímica Aplicada, Universidad de Concepción (UdeC), Concepción, Chile

D. C. de Oliveira, R. M. dos Santos Isaias (eds.), *Plant Galls*,
https://doi.org/10.1007/978-3-031-80064-1_10

Biotic and abiotic stresses are crucial factors influencing the chemical composition and biological activity of plants (Naik and Al-Khayri 2016), with herbivory being one of them. During herbivores attack, both primary and secondary metabolism undergo major changes due to a transcriptional reconfiguration, which is specific for each insect species (Schwachtje and Baldwin 2008; Heidel-Fischer et al. 2014). Plants respond by diverting their metabolism towards the synthesis of specialized metabolites to defend against herbivores or activating tolerance mechanisms, including the mobilization of nutrients away from the site of attack or storage in inaccessible form to the herbivore (Schuman and Baldwin 2016). However, gall--inducing insects, the most specialized group within herbivorous insects, can circumvent the defenses of their host plants to establish their galls. Galling organisms induce morphological and physiological alterations in the host plant (Oliveira et al. 2016; Giron et al. 2016; Ferreira et al. 2018; Costa et al. 2021) manipulating the synthesis of secondary metabolites and their metabolic pathways (Hall et al. 2017). Gall induction can stimulate, inhibit, or maintain the synthesis of SMs of the host organ, and these effects could be reflected in the biological activity of host organs and plants. Tannins and other phenol types are the most detected compounds in galls, while volatiles and other classes of secondary metabolites have received less attention (Hall et al. 2017).

2 Different Polyphenol Compounds Detected in Gall-Systems

Phenolic compounds have been associated with defensive functions against natural enemies (Nyman and Julkunen-Tiitto 2000; Guedes et al. 2019) and gall morphogenesis, through interaction with plant hormones regulating cell hypertrophy processes (Bedetti et al. 2014; Li et al. 2017). However, more recently phenolic compounds in galls have been linked to antioxidant functions due to the increased oxidative stress during the gall establishment and development (Isaias et al. 2015; Oliveira et al. 2016; Guedes et al. 2022).

Quantitative studies have revealed that tannins and other phenolic compounds are the most abundant in the galls, and their concentrations generally increase under the influence of gall inducers (Hall et al. 2017) (Table 1). However, the distribution of these compounds and their concentration will be influence by the physiological status of host plant, the gall developmental stages (Yamaguchi et al. 2012; Tanaka et al. 2013; Wang et al. 2016), and galling insect species (Li et al. 2017; Yang et al. 2018; Guedes et al. 2019).

A comparative study of the total soluble phenol content in galled leaves and stems induced by two calophyid species (Hemiptera: Calophyidae) on *Schinus polygama* (Cav.) Cabrera (Anacardiaceae) (Fig. 1a, b), revealed a distinctive polyphenol dynamic through the gall developmental stages. The two calophyid species affect the total phenol contents during the gall development, inducing a significant

Table 1 Main galls of plant recognized for their high biological activity

Host Plant species	Galling insect species	Host organ	Reported biological activity	Responsible compounds	References
Terminalia chebula (Gaertn.) Retz. (Combretaceae)	*Dixothrips onerosus Karny* (Thysanoptera)	Leaves	Antiaging, antifungal, antioxidant, antiinflammatory, and tyrosinase inhibition activities	Phenols, such as gallic acid, chebulagic acid, apigenin, and terpenoids, such as phytol, stigmasterol	Eshwarappa et al. (2016), Debnath et al. (2011), Srivastava et al. (2012), and Kabir et al. (2017)
Quercus infectoria Olivier (Fagaceae)	*Adleria gallae--tinctoriae* L. (Cynipidae)	Buds	Anticancer, antibacterial, anti-MRSA, anti-viral, antifungal, larvicide, anti-inflammatory and antioxidant activities, and analgesic effects	Tannins, such as gallotannins and ellagitannins, phenolic acids, such as gallic acid, flavonoids, suchas amentoflavone and isocryptomerin, triterpenoids, such as roburic acid, nyctanthic acid, methyl-oleanolate, and methyl-betulate, steroids as beta-sitosterol, vitamin A, and vitamin C	Elham et al. (2021), Redwane et al. (2002), Chusri and Voravuthikunchai (2009), Hussein et al. (2000), Basri et al. (2005), Kamarudin et al. (2021), Mohd-Nasir et al. (2021), and Kaur et al. (2008)
Pistacia atlantica Desf. (Anacardiaceae)	*Slavum wertheimae* HRL (Homoptera; Pemphigidae), *Baizongia pistaciae* L. (Hemiptera, Aphididae)	Leaves	Antioxidant, cytotoxyc and antimycobacterial activities and α-amylase inhibitory, anti-Alzheimer effects	Phenolic compounds, such as quinic acid, valoneic acid dilactone, methyl gallate, and gallic acid; essential oils as α-pinene, Δ3-carene; fatty acid, such as palmitic, oleic, and linoleic acids	Ben Ahmed et al. (2022a), Ben Ahmed et al. (2022b), Yoram and Inbar (2011), Sifi et al. (2015), and Hefied et al. (2020)

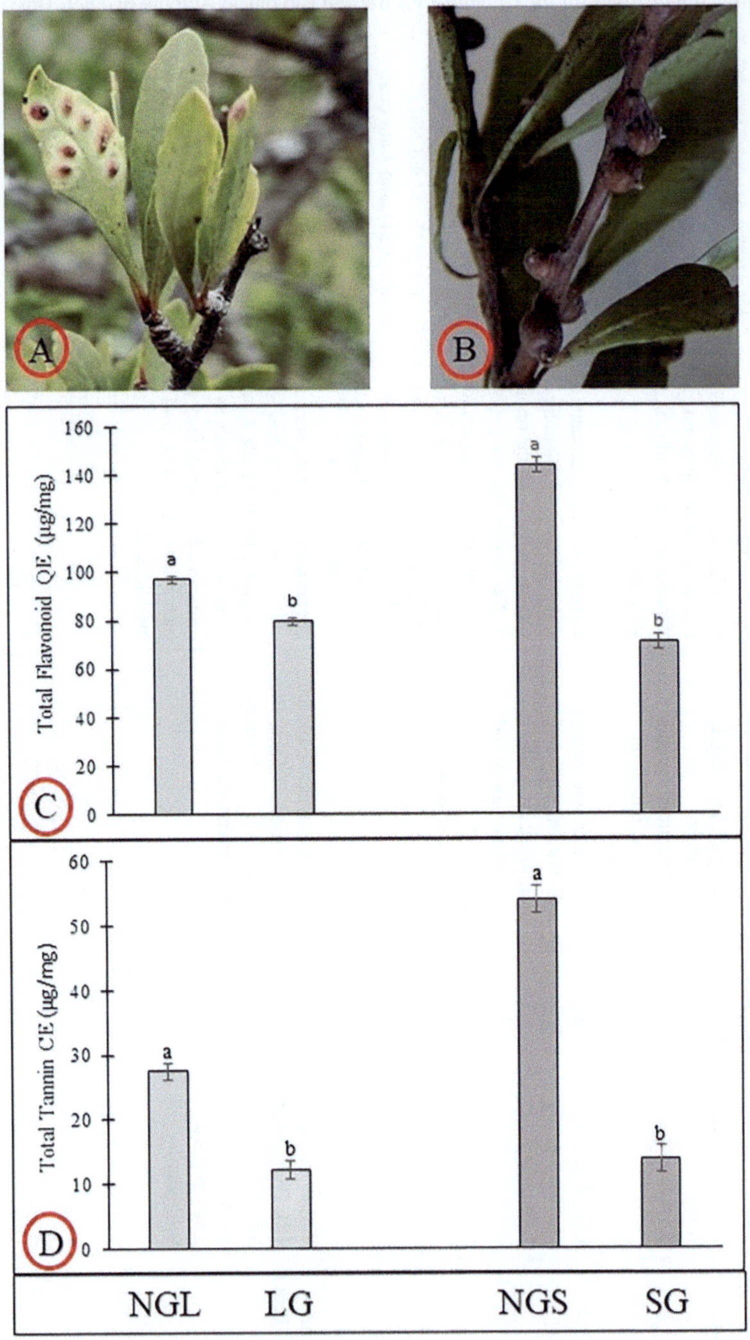

Fig. 1 Galls induced by two species of Calophyidae on *Schinus polygama* (Cav.) Cabrera (Anacardiaceae). (**a**) Leaf galls induced by *Calophya mammifex* Burckhardt & Basset. (**b**) Stem galls induced by *C. rubra* Blanchard. (**c**) Total flavonoid content (expressed as ug of quercetin equivalent per mg of fresh weight). (**d**) Total tannin content (expressed as ug of catechin equivalent per mg of fresh weight). NGL non-galled leaves, LG leaf galls, NGS non-galled stems, SG stem galls. Different letters indicate significant differences after a Student t-test for $p < 0.05$

decrease of these compounds compared with non-galled host organs, but with opposite dynamics (Guedes et al. 2019). Polyphenol content decreased in leaf galls and increased in stem galls from maturation towards senescence. To illustrate what happens to the tannin and flavonoid contents, we determined the concentrations of these two phenol groups in the same study system during the maturate stage, using a spectrophotometric method. The results indicated that both total flavonoid (Fig. 1c) and total tannin contents (Fig. 1d) also decreased during the gall maturation stage compared to the non-galled leaves and stems.

Using chromatographic techniques, such as high-performance liquid chromatography (HPLC) and gas chromatography coupled with mass spectrophotometry (GC--MS), researchers have been able to separate and identify phenolic compounds in galls and their host organs. Gallic acid (GA) and its derivatives are one of the most detected phenols in galls (Guedes et al. 2016; Upadhye and Rajopadhye 2010; Patni et al. 2012; Syukriah et al. 2014; Mahjoub et al. 2018; Manosroi et al. 2010). Gallic acid is formed by an aromatic ring, to which three hydroxyl groups are attached in an ortho position to each other and a carboxyl group (Badhani et al. 2015). Along with gallotannins and ellagic acid, GA is the largest component of Gall Oak (*Quercus infectoria* Olivier; Fagaceae) (Syukriah et al. 2014; Elham et al. 2021) (Table 1). Through thin-layer chromatography, high-performance liquid chromatography, and infrared spectra studies, Patni et al. (2012) confirmed the presence of more gallic acid in cecidomyiid and eriophyid galls on *Q. leucotrichophora* A. Camus leaves than in non-galled leaves.

Terminalia chebulata Retzius (Combretaceae) is another species rich in phenolic compounds, including phenolic acids such as gallic acid, chebulinic acid, chebulagic acid, chebulic acid, ellagic acid, and hydrolyzable tannins like corilagin (Table 1) (Kumar 2006). Manosroi et al. (2010) also identified six phenolic compounds in aqueous extracts of *T. chebulata* galls: gallic acid, punicalagin, isoterchebulin, 1,3,6-tri-O-galloyl- -D-glucopyranose, chebulagic acid, and chebulinic acid. In gall methanol extract, in addition to gallic acid, O-methylgallic acid and seven hydrolyzable tannins (Methyl (S)-flavogallonate, 6-TriO-galloylb-d-glucopyranoside, chebulinic acid, methyl neochebulagate, punicalagin) were isolated (Manosroi et al. 2013).

The aphid galls induced by *Pemphigus* sp. (Aphididae) on genus *Pistacia* (Anacardiaceae) are also rich in flavonoids and tannins; however, other SMs, such as alkaloids and terpenoids have been detected (Grover 2021) (Table 1). Pyrogallol and quercetin have been identified in the *P. integerrima* Shringi gall extracts (Zahoor et al. 2018). In the acetone fraction of the methanol extracts of these galls, quercitin and naringenin were isolated (Kumar et al. 2022). Similarly, four other flavonoids are present in the chloroform fraction of *P. integerrima* (Rauf et al. 2017), while naringenin and 3,5,7,4′-tetrahydroxy-flavanone have been isolated from the methanol gall extract (Rauf et al. 2015).

3 Impact of the Gall-Inducing Organism in a Gall Volatile Compound Emissions

Plant volatiles are normal plant constituents released into the air under different conditions, one of them being herbivory. Volatile compounds (VOC) are a mixture of various classes of compounds, mainly terpenoids, compounds with aromatic rings, fatty acid derivatives, and volatiles derived from amino acids (Baldwin 2010). However, terpenoids play a central role in the synthesis of other plant volatiles. After phenols, terpenoids have been the most studied compounds, both histochemically and quantitatively, in host plant-gall-inducing organism systems (Hall et al. 2017; Kuster et al. 2019).

Galling insects could stimulate the emission of volatile compounds throughout every stage of gall development, extending beyond mere feeding activities impacting various ecological interactions (Barônio and Oliveira 2019). These interactions not only affect plant-galling herbivores but also have repercussions on other organisms, including pollinators and non-galling herbivores, altering their behavior, visitation frequency, and selection of oviposition sites (Hilker and Fatouros 2016; Borges 2018). Furthermore, these volatile compounds serve as crucial cues for activities such as female location for mating (Tooker et al. 2002; Tooker and Hanks 2004; Germinara et al. 2011) and the galling female's selection of suitable oviposition sites (Tooker et al. 2005; Kehl et al. 2010).

Regardless of the specific volatile function in the plant-galling insect relationship, gall inducers can modify the profile and concentration of volatiles in their host plants. For example, it has been postulated that galling organisms can suppress the emission of volatiles (Hall et al. 2017; Tooker et al. 2008), related to an increase in indole-3-acetic acid during gall development, which induces the suppression of the jasmonic acid, a pathway that produces VOCs (Borges 2018). Some evidence demonstrates crosstalk between auxins and cytokinins, hormones essential for gall development, and the production of jasmonic acid and salicilic acid, which ultimately result in the suppression of volatile synthesis (Berens et al. 2017). Depletion in the volatile synthesis in the host plant has been demonstrated in *Solidago altissima* L. (Asteraceae), where two gall inducers, *Eurosta solidaginis* Fitch (Tephritidae) and *Gnorimoschema gallaesolidaginis* Riley (Gelechiidae) induced a significant decrease in the volatile emissions (Tooker et al. 2008). Similarly, in *Myrcia splendens* (Sw) DC (Myrtaceae) leaves, there is a significant decrease of volatile emission under the effect of the gall inducer *Nexothrips* sp. (Thysanoptera), related to a decrease in the area and size of the leaf oil glands (Jorge et al. 2018).

Galling organisms not only manipulate the concentration of the VOC of their host plants, but also their composition towards the synthesis of new compounds, as has been demonstrated in the *Eucalyptus camaldulensis* Dehnh (Myrtaceae)-*Leptocybe invasa* Fisher and La Salle (Eulophidae) system. This gall wasp manipulated the *E. camaldulensis* oil composition towards the synthesis of aliphatic non-oxygenated monoterpenes, such as myrcene and phellandrenes, while oxygenated and aromatic compounds, like p-cymene, decrease (Mohamed 2016).

Regardless of these examples of content decrease, the increase of VOC contents in galls and surrounding tissues also occurs. Although the galling *Neuroterus quercusbaccarum* L. (Hymenoptera: Cynipidae) does not significantly affect the total production and diversity of volatiles from the *Q. robur* L. leaves, it induced a significant increase in the concentration of eucalyptol, β-bourbonene, and farnesene in galled leaves (Klimm et al. 2020). The aphid *Baizongia pistaciae* L. induces a significant increase in volatile monoterpene production on *Pistacia palaestina* Boiss leaves (Rand et al. 2017). This increase is dependent on the gall development stage, as previously observed for phenolic compounds in other gall systems (Li et al. 2017; Yang et al. 2018; Guedes et al. 2019, 2024).

The changes induced by gall-inducing organisms on volatile emissions do not seem to be related to the gall-inducing taxa or volatile types. An example that illustrates this statement is the production of isoprene in galls. Isoprene synthesis is strongly correlated with leaf photosynthesis rates; and as galls can modify gas exchange and photosynthetic rate (Dorchin et al. 2006; Ye et al. 2019; Jiang et al. 2021), a decrease in isoprene emission in the galls would be expected (Tooker et al. 2008; Noe and Niinemets 2020). However, the results of different studies do not show a single trend in this regard. For example, the infection by galling eriophyoids (Acari, Eriophyoidea) on *Alnus glutinosa* (L.) Gaertn (Betulaceae) and *Tilia cordata* Mill. (Tiliaceae) improved the isoprene emission rates compared to the emissions from non-galled leaves (Swanson et al. 2021). Similarly, in *Populus × petrovskiana* Dippel (*P. deltoides × P. laurifolia*) petioles, the aphid *Pemphigus spyrothecae* Passerini (Hemiptera, Aphididae) also enhanced the emission of this organic compound (Ye et al. 2019). In contrast, other gall-forming eriophyoid mite (*Aculus tetanothrix* Napela) did not affect isoprene emission in *Salix myrsinites* L. (Salicaceae) in an Arctic tundra ecosystem (Swanson et al. 2021). Noe and Niinemet (2020) constructed a quantitative model that predicts the reduction of constitutive VOC emissions and an increase of induced VOCs in plants infected by gall-inducing insects. This model suggests a global-scale decrease in isoprene emissions in deciduous tree species under the influence of galling organisms. However, due to the few studies that exist to date on the effect of galling organisms on volatile emissions, or on a specific volatile, we consider that no conclusions can be reached on this topic.

4 Useful Chemical Properties of Galls: Ethnobotanical Use

Due to the rich secondary metabolite composition, the galls are used in numerous cultures as Chinese and Greek, and in Indian folklore and pharmacopoeia (Patel et al. 2018). The folkloric use of some galls due to medicinal properties is based on their chemical properties. Among the most studied plant galls used in traditional medicine, there are those on *T. chebula*, *Rhus* spp., *Pistacia* spp., and *Q. infectoria* (Table 1).

The high antioxidant capacity of some plant galls, particularly those with abundant polyphenol contents confers their main therapeutic properties. For example, *T. chebula* galls, as previously described, are rich in phenols, and have been extensively used for medicinal purposes, especially in Ayurvedic medicines. In this culture, galls are known as Karkatshringi and are used to combat asthma, digestive and liver disorders, lung and heart diseases, fever, and bleeding from the nose and gums (Santha et al. 1991; Sukh 1997; Singh and Kumar 2013). *Terminalia chebula* galls are known for their antioxidant (Srivastava et al. 2012; Eshwarappa et al. 2016), anti-aging (Manosroi et al. 2010; Debnath et al. 2011) and antifungal (Kabir et al. 2017) properties. A study carried out by Eshwarappa et al. (2016) demonstrated the inhibitory capacity of the lipoxygenase enzyme, indicating the promising use of ethanol extracts of these galls as an anti-inflammatory to combat diseases such as asthma, cancer, lymphoma, leukemia, and autoimmune disorders. The anti-cancer properties of the ethanol extract of *T. chebula* galls have also been corroborated in the study of Shankara et al. (2016).

Extensive reviews by Elham et al. (2021) and Alam et al. (2022) have summarized the pharmacological properties of *Q. infectoria* galls (Table 1) for the control of diarrhea, hemorrhage, and skin diseases, as well as their antiviral, antifungal, antibacterial, larvicidal, and anti-inflammatory properties, mainly related to their antioxidant capacity due to the high content of flavonoids and tannins. In Uyghur and Unani medicine, *Q. infectoria* galls are extensively used for the treatment of numerous oral and intestinal conditions (Wang et al. 2007), so it is used in various types of formulations, such as enema, ointments, mouthwashes, and powders (Dar et al. 2010). The anti-inflammatory and healing properties of these galls position them as a promising candidate for the fight against colorectal and cervical cancer (Elham et al. 2021; Zhang et al. 2020; Ismail et al. 2021). References to the use of *Q. infectoria* galls in different Asian cultures and trial studies of their properties, mostly as anti-cancer, are extensive; however, most of their properties have been tested in in vitro assays and in some animal models (Elham et al. 2021; Zhang et al. 2020; Ismail et al. 2021), but more in vivo studies are needed to validate the anti-cancer use.

The galls of the *Pistacia* genus have also been used in traditional Chinese culture and in the Ayurveda and Iranian cultures for their biological properties related to the abundance of terpenic compounds (Rauf et al. 2017). The most studied galls of the genus are those of *P. integerrima* and *P. atlantica*. The essential oil and methanol extracts of the *P. integerrina* galls have been promisingly used for the control of asthma (Rauf et al. 2015; Shirol et al. 2014; Rana et al. 2016) due to their anti--inflammatory properties (Mehla et al. 2011). Other uses of these galls have been reported for the mitigation of hepatitis, control of gum and nose bleeding, treating otitis, and as antipyretic and analgesic (Dev 1997; Padulosi et al. 2002; Ahmad et al. 2006; Bibi et al. 2015; Kumari et al. 2021). Numerous studies have tested the medicinal properties of the galls of both species, *P. integerrina* and *P. atlantica*, including reports in reviews (Rauf et al. 2017; Mahjoub et al. 2018; Ben Ahmed et al. 2022a; Grover 2021). For *P. atlantica* galls, the anti-inflammatory (Ibrahim et al. 2022) and antioxidant (Chelghoum et al. 2020) potential of the essential oil has been proven. Aqueous, ethanol, and methanol extracts have also demonstrated antioxidant effects (Rezaie et al. 2015; Ben Ahmed et al. 2022b; Toul et al. 2017). The test of aqueous gall extract also reports an inhibitory effect on bacteria of the genus *Bacillus* and *Pseudomonas* (Gerchman and Inbar 2011). Although the literature on the folkloric use of galls is abundant, through the examples described above, we wanted to illustrate the wide spectrum of use of galls in different cultures, with promising therapeutic effects.

5 Useful Structural and Chemical Properties of Galls as Food Resources

Plant galls function as nutrient sinks from their host organs, turning them into rich nutrient sources, not only for the inducing organism, but also for other associated organisms, i.e., parasitoids, opportunists, and inquilines (Silva et al. 2019). This greater amount of nutrients exceeds what is normally offered by the host plants and turns the gall into a food source for associated organisms and for human consumption. Although the insect galls are not frequently consumed by humans, mainly due to their hardness degree conferred by the presence of lignified tissues, in some cultures, galling fungi are highly consumed. For example, in Mexican culture, huitlacoche, a dish based on the galling fungus *Ustilago maydis* (DC.) Corda (Ustilaginaceae) is widely consumed (López-Martínez et al. 2022). Huitlacoche, also known as corn smut, develops as gray galls on young maize plants (*Zea mays* L.) and turns black during maturation (Villagrán et al. 2023). This fungus has been consumed in Mexico since pre-Hispanic times due to its exquisite flavor. Recent studies have demonstrated its nutritional quality, supported by the presence of essential fatty acids rich in omega 3 and 6, such as oleic and linoleic acids, essential amino acids, soluble sugars, fibers, vitamin C, and minerals (Lizárraga-Guerra and López 1996; Juárez-Montiel et al. 2011; López-Martínez et al. 2022).

Another species of edible galling fungus within the genus *Ustilago* is *U. esculenta* (Guo et al. 2015), which develops on *Zizania latifolia* (Griseb.) Stapf stems. The wild rice, as *Z. latifolia* is known, is currently cultivated for the consumption of its fungus galls (Li et al. 2021) in China and some southeastern countries, where it is known as "Jiaobai". *Ustilago esculenta* galls are rich in nutrients, such as protein, amino acids, polysaccharides, fat, minerals, and fibers (Jain et al. 2012; Wu et al. 2023).

The consumption of fungus of the genus *Cyttaria* (Cyttariaceae) has been known since pre-Hispanic times by indigenous cultures of South America. These fungi form galls on the branches and trunks of *Nothofagus* trees (Nothofagaceae), from which edible stroma-shaped fructifications emerge (Fig. 2a, b). At least seven species of *Cyttaria* have been described in South America (Gamundí et al. 2004), the most common in Chile are *C. espinosae* and *C. berteroi* (Salazar-Vidal et al. 2020). Regardless of the species, these fungi are known as "digüeñes" "quireñes" o

Fig. 2 Galls of economic importance. (**a, b**) Galls induced by two different species of fungi of the genus *Cyttaria* on *Nothofagus* spp. from the Central-South Region of Chile. Note the edible stroma known as "Digüeñes". (**c**) Galls induced by *Eriophyes tiliae* Pagenstecher on *Tilia platyphyllos*. This species is used for medicinal purposes due to its high antioxidant potential. (Photography in **a** and **b** courtesy of Dra. Camila Fuica)

"lihueñes" and are highly appetizing due to their fruity and mild flavor (Salazar-Vidal et al. 2020). Digüeñes have been consumed raw, cooked, and fermented since ancient times (Elliott and Elliott 2019), although their consumption and sale continue to be a frequent and profitable activity for some rural populations in central--southern Chile. The rich polysaccharide composition of these fungi is known, especially β-glucans, although high protein and lipid contents have also been described (Salazar-Vidal et al. 2020).

6 Gall-Inducing Organisms Affect Biological Activity of Their Host Plant Organs

As we have previously described, gall induction can stimulate, inhibit, or maintain the synthesis of secondary metabolites in their host organ, which may imply a positive or negative effect on the properties of its the host organs. These secondary metabolites can exhibit various biological activities, such as antioxidant, antimicrobial, anti-inflammatory, or allelopathic properties, depending on the specific compounds involved and their concentrations. However, it is important to note that only a few studies have compared the biological activity of the non-galled organs with the galls, and there is no general trend in the effects of the galling organisms on the host plant properties in terms of biological activity.

Balanites aegyptiaca Del. (Zygophyllaceae), the "desert date", is a wild plant species common to the arid areas of Africa and southern Asia (Murthy et al. 2020). Its leaves and galls, rich in phenolic compounds (Meda et al. 2011), have numerous medicinal uses. However, when the *in vivo* anti-inflammatory activities of galls and leaves of *B. aegyptiaca* were compared, it was found that gall extracts were more effective in reducing nitric oxide synthesis (Meda et al. 2020). The authors concluded that aqueous acetone extracts of galls were better than leaf extracts in reducing oxidative stress and were even more potent than meloxicam, a non-steroidal anti-inflammatory drug. The potent antioxidant and anti-inflammatory activity of *B. aegyptiaca* has been correlated with the phenolic content reported for the species (Meda et al. 2011).

The galls induced by *Procontarinia matteiana* Kieffer & Cecconi (Diptera, Cecidomyiidae) on the mango leaves (*Mangifera indica* L., Anacardiaceae) (Augustyn et al. 2013) are another example of a positive effect of gall-inducing organisms on the chemical profile and biological activity of their host organs. The aqueous extracts of the leaf galls showed higher iron chelating activity and ABTS (2,2- Azinobis-3-ethylbenzothiazoline-6- sulfonic acid) scavenging activity than the non-galled leaves, which is strongly correlated with a higher tannin content of galls (Rajasekaran and Mahimaidoss 2016). Similarly, the mite *Eriophyes tiliae* Pagenstecher (Acari, Eriophydae) induced a significant increase in the content of anthocyanin and polyphenols on *T. platyphyllos* Scop (Malvaceae) leaves, where the mite induces conspicuous red galls (Fig. 2c). Although *T. platyphyllos* has recognized antioxidant activity (Jabeur et al. 2017; Pavlovića et al. 2020), the galling mite increased its antioxidant capacity by about 400% (Guedes et al. 2023a).

Salvia is one of the most important genera within the Lamiaceae. Due to their rich metabolic composition, some *Salvia* species are used in the cosmetic, food, and pharmaceutical industries (Carović-Stanko et al. 2016). In the Mediterranean region, *S. fruticosa* Miller is one of the most valued species due to the high content of eucalyptol (1,8-cineole) in the essential oil of its leaves (Figueredo et al. 2012). The composition of the essential oil and hexane extracts of *S. fruticosa* leaves are modified by the action of a gall-inducing species of cynipid wasp (Hymenoptera: Cynipidae) (Topçu et al. 2013; Süzgeç-Selçuk et al. 2021). The methanol extract and essential oil of *S. fruticosa* and the gall hexane extracts exhibit high in vitro inhibition values of the acetylcholinesterase enzyme (AChE) (Topçu et al. 2013). However, the gall hexane extract showed quite strong inhibition of AChE compared to the essential oil from the non-galled leaves. Topçu et al. (2013) indicated that gall extract was rich in unsaturated oils (75.4%), such as oleic acid, palmitoleic acid, and linoleic acid.

Peumus boldus Molina (Mominaceae), a Chilean native tree, is renowned for its digestive and sedative properties, primarily attributed to its rich alkaloid composition. However, recent studies have highlighted the presence of a considerable number of phenolic compounds in *P. boldus*, potentially responsible for its enhanced antioxidant activity (Pastene et al. 2014). On boldo stems, an unidentified species of Diptera, Cecidomyiidae, induces fusiform galls. The structural alterations caused by this gall-inducing insect (Guedes et al. 2023b, c) coincide with changes in the phenolic profile of the stems (Guedes et al. 2024). These changes in the composition and concentration of soluble polyphenol induced by the Cecidomyiidae result in a decline in the antioxidant capacity of gall extracts during the maturation and senescence stages (Guedes et al. 2024). The antioxidant activity of hydroalcoholic extracts from galls at various developmental stages and non-galled stems was assessed by measuring the quenching activity of DPPH and ABTS radicals—a common technique of *in vitro* antioxidant activity studies.

Some studies have demonstrated that the essential oil (Verdeguer et al. 2011; Blázquez and Carbó 2015), and the aqueous (Toledo et al. 2016) and methanol extracts (Jara 2021) of *P. boldus* leaves have phytotoxic properties. To analyze the effect of the galling insect on the phytotoxic activity of *P. boldus* stems, a phytotoxicity study out to analyze the effect of *P. boldus* phenolic extracts from galled (SG) and non-galled stems (NGS) on germination and early growth of *Portulaca oleracea* L. (Portulacaceae), according to the methodology described by Aguilera et al. (2015) was carried (Fig. 3a). Non-galled stem extract decreases the germination of *P. oleracea* (Fig. 2b) but did not affect the length of the radicle and hypocotyl (Fig. 2c–e). On the contrary, the extract of SG did not affect germination (Fig. 2b), but the length of the hypocotyl and radicle did decrease significantly (Fig. 2c–e). These results could be related to changes induced by the gall inducer in the phenol profile of *P. boldus* stems, as previously discussed Guedes et al. (2023b).

The phytotoxic properties of some plant species are due to the presence of SMs, namely allelochemicals, which are released into the environment through the processes of volatilization, root exudation, decomposition and/or leaching, thereby affecting the growth of adjacent plants (Begna and Yali 2021). Allelochemicals are synthesized in all plant organs and affect, negatively or positively, the functioning of other organisms in their vicinity, a phenomenon named as allelopathy (Bajwa et al. 2013). Another example is *Lonchocarpus muehlbergianus* Hassl. (Fabaceae), which exhibits phytotoxicity due to its high flavonoid content (Oliveira et al. 2008). The aqueous extract of the galls induced by *Euphalerus ostreoides* Crawf. (Psyllideae) significantly decrease the radicle and shoot length of *Lactuca sativa* L.; indicating that the galls potentiate the phytotoxic effects of *L. muehlbergianus* (Oliveira et al. 2008).

Not only polyphenols, as flavonoids, are responsible for the phytotoxic effects of plants. In *Eremanthus erythropappus* (DC) McLeisch (Asteraceae), the phytotoxic power of the essential oils (Pinto et al. 2018), terpenes, and alkanes (de Castro Jorge et al. 2022) has been demonstrated. However, *Neolasioptera* sp. (Cecidomyiidae) induced changes in the chemical profile of the methanol extract of *E. erythropappus* stem, where it induces galls. The gall inducer elicited the neosynthesis and accumulation of two triterpenes (γ-taraxasterol, and β-amyrone) and one alkane (tricosane) and inhibited the synthesis of compounds previously reported to the extracts of *E. erythropappus* stems, as the triterpenes A-neooleana-3(5),12-diene, lupenone tritepernes, and 2(1H) naphthalenone ketone (Jorge et al. 2022). Even though, non-galled and galled extracts altered the differentiation of root parenchyma cells of *L. sativa* seedlings, the galling activity decreased the phytotoxic effects of *E. erythropappus* stems.

Fig. 3 (continued) of *Portulaca oleracea* L. (Portulacaceae). (**a**) Assays in Petri dishes (control and extract from non-galled stems). (**b**) Germination percentage. (**c**) Comparative morphology of the seedlings at the end of the assay. (**d**) Length of hypocotyl. (**e**) Length of the radicle. Co: control (plants grown in water), non-galled stems (NGS): methanol extracts of NGS; Galled stems: methanol extract of mature galls. Different letters indicate significant differences after a one-way ANOVA followed by Tukey's test for $p < 0.05$

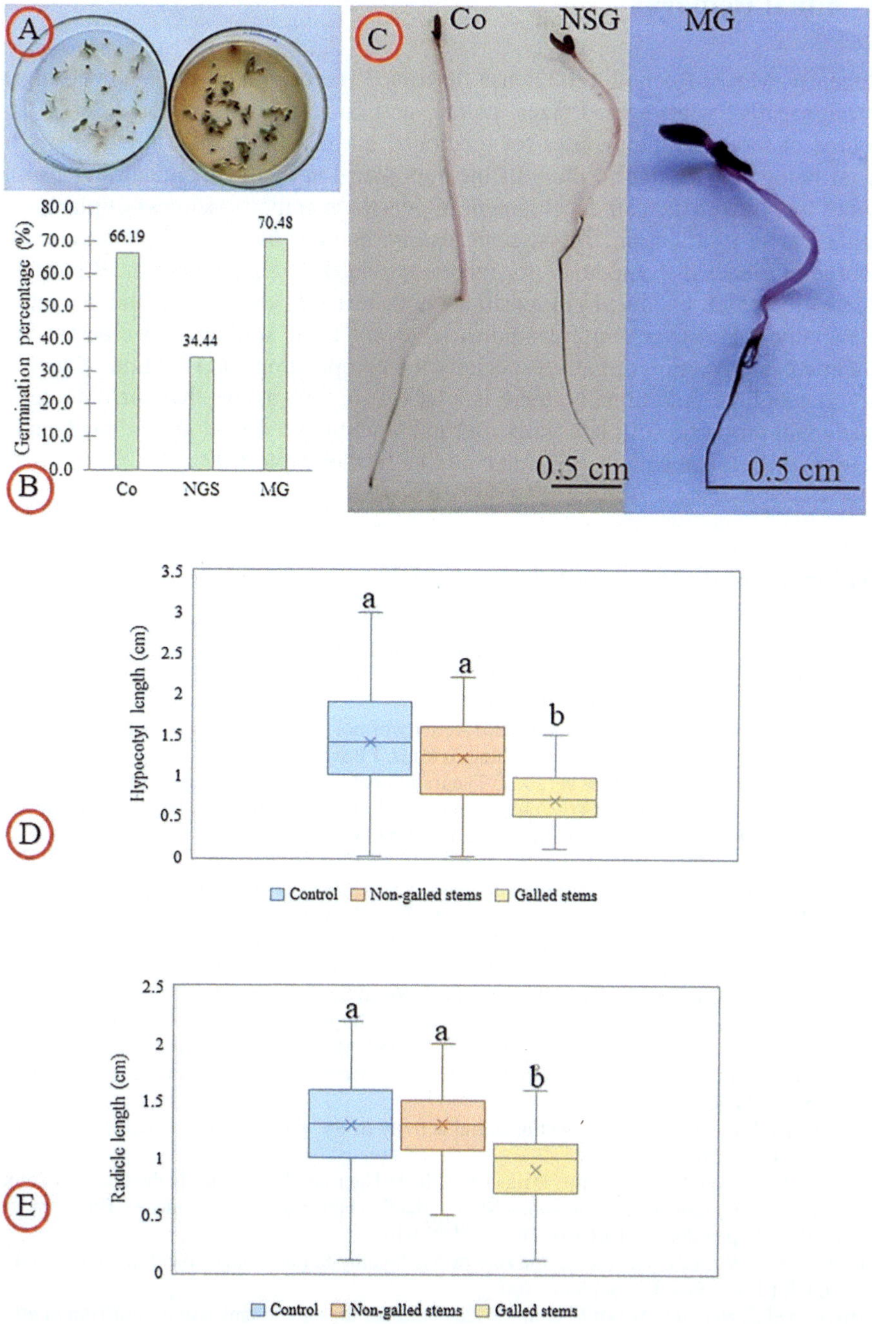

Fig. 3 Effects of methanol extracts of non-galled stems and mature galls induced by Cecidomyiidae sp. (Diptera) on *Peumus boldus* Molina (Mominaceae) stems on the germination and early growth

7 Final Remarks

Recently, Meyer-Rochow (2022) has discussed the use of genetic engineering to "design galls" with desired sizes, colors, and contents, which would serve as a source of food and metabolites for medicinal purposes. Although this issue is far from being settled, due to the still incomprehension of the chemical signals and genes involved in the gall development, it puts the metabolic richness of plant galls in the center of attention. Throughout history, humans hasve used plant galls as a source of bioactive compounds for medicinal, nutritional, and herbicidal purposes due to their rich chemical composition, both in terms of primary and secondary metabolites. However, current studies have not been sufficient to elucidate the chemical composition of galls, especially the composition and isolation of secondary metabolites. Additionally, there is a lack of *in vivo* assays that corroborate the medicinal properties of plant galls. But galls, without a doubt, are an incalculable source of chemical compounds that could be further exploited.

References

Aguilera N, Guedes LM, Becerra J et al (2015) Morphological effects at radicle level by direct contact of invasive *Acacia dealbata* link. Flora: Morphol Distrib Funct Ecol Plants 215:54–59. https://doi.org/10.1016/j.flora.2015.07.007

Ahmad N, Farman M, Najmi M et al (2006) Activity of polyphenolic plant extracts as scavengers of free radicals and inhibitors of xanthine oxidase. J Basic Appl Sci 2:1–6

Alam A, Siddiqui JI, Viquar U et al (2022) The traditional uses, phytochemistry, pharmacological, and botanical description of *Quercus infectoria galls* (Mazuphal): a review. J Drug Deliv Ther 12:214–221. https://doi.org/10.22270/jddt.v12i1.5178

Augustyn WA, Du Plooy WD, Botha BM, Van Wilpe EV (2013) Infestation of mangifera indica by the mango gall fly, *Procontarinia matteiana*, (Kieffer & Cecconi) (Diptera: Cecidomyiidae). Afr Entomol 21:79–88. https://doi.org/10.4001/003.021.0103

Badhani B, Sharma N, Kakkar R (2015) RSC advances. Royal Society of Chemistry Advances

Bajwa AA, Khalid S, Sadia S et al (2013) Influence of combinations of allelopathic water extracts of different plants on wheat and wild oat. Pak J Weed Sci Res 19:157–166

Baldwin IT (2010) Plant volatiles. Curr Biol 20:392–397. https://doi.org/10.1016/j.cub.2010.02.052

Barônio GJ, Oliveira DC (2019) Eavesdropping on gall–plant interactions: the importance of the signaling function of induced volatiles. Plant Signal Behav 14:1665454. https://doi.org/10.1080/15592324.2019.1665454

Basri D, Ha FS, Zin NM (2005) Antibacterial activity of the galls of *Quercus infectoria*. Malays J Sci 24:257–262

Bedetti CS, Modolo LV, dos Santos Isaias RM (2014) The role of phenolics in the control of auxin in galls of *Piptadenia gonoacantha* (Mart.) MacBr (Fabaceae: Mimosoideae). Biochem Syst Ecol 55. https://doi.org/10.1016/j.bse.2014.02.016

Begna T, Yali W (2021) Review on the role of Allelopathy in pest management and crop production. Int J Adv Res Biol Sci 8:88–100

Ben Ahmed Z, Hefied F, Yousfi M et al (2022a) Study of the antioxidant activity of *Pistacia atlantica* Desf. Gall extracts and evaluation of the responsible compounds. Biochem Syst Ecol 100:104358. https://doi.org/10.1016/j.bse.2021.104358

Ben Ahmed Z, Hefied F, Mahammed TH et al (2022b) Identification of potential anti-Alzheimer agents from *Pistacia atlantica* Desf. galls using UPLC fingerprinting, chemometrics, and molecular docking analyses. J Food Process Preserv 46:1–12. https://doi.org/10.1111/jfpp.16916
Berens ML, Berry HM, Mine A et al (2017) Evolution of hormone signaling networks in plant defense. Rev Phytopathol 55:401–425
Bibi Y, Zia M, Qayyum A (2015) An overview of *Pistacia integerrima* a medicinal plant species: ethnobotany, biological activities and phytochemistry. Pak J Pharm Sci 28:1009–1013
Blázquez MA, Carbó E (2015) Control of *Portulaca oleracea* by boldo and lemon essential oils in different soils. Ind Crop Prod 76:515–521. https://doi.org/10.1016/j.indcrop.2015.07.019
Borges RM (2018) The galling truth: limited knowledge of gall-associated volatiles in multitrophic interactions. Front Plant Sci 9:1–8. https://doi.org/10.3389/fpls.2018.01139
Carović-Stanko K, Petek M, Grdiša M et al (2016) Medicinal plants of the family Lamiaceae asfunctional foods – a review. Czech J Food Sci 34:377–390
Chelghoum M, Guenane H, Harrat M, Yousfi M (2020) Total tocopherols, carotenoids, and fatty acids content variation of *Pistacia atlantica* from different organs' crude oils and their antioxidant activity during development stages. Chem Biodivers 17:e2000117
Chusri S, Voravuthikunchai SP (2009) Detailed studies on *Quercus infectoria* Olivier (nutgalls) as an alternative treatment for methicillin-resistant Staphylococcus aureus infections. J Appl Microbiol 106:89–96. https://doi.org/10.1111/j.1365-2672.2008.03979.x
Costa EC, Oliveira DC, Ferreira DKL, Isaias RMS (2021) Structural and nutritional peculiarities related to lifespan differences on four *Lopesia* induced bivalve-shaped galls on the single super-host mimosa gemmulata. Front Plant Sci 12. https://doi.org/10.3389/fpls.2021.660557
Dar MS, Ikram M, Fakouhi T (2010) Pharmacology of Quercus infectoria. J Pharm Sci 65:1791–1794. https://doi.org/10.1002/jps.2600651224
Jorge NC, Guedes LM, Aguilera N, Becerra J, dos Santos Isaias RMS (2022) Allelopathic potential of the extracts of non-galled stems and globoid stem galls of *Eremanthus erythropappus* (DC) McLeish (Asteraceae). Biochem Syst Ecol 100:104379
Debnath J, Prakash T, Karki R et al (2011) An experimental evaluation of anti-stress effects of *Terminalia chebula*. J Physiol Biomed Sci 24:13–19
Dev S (1997) Ethnotherapeutics and modern drug development: the potential of Ayurveda. Curr Sci 73:909–928
Dorchin N, Cramer M, Hoffmann J (2006) Photosynthesis and sink activity of wasp-induced galls in *Acacia pycnantha*. Ecology 87:1781–1791
Elham A, Arken M, Kalimanjan G et al (2021) A review of the phytochemical, pharmacological, pharmacokinetic, and toxicological evaluation of *Quercus infectoria* galls. J Ethnopharmacol 273:113592. https://doi.org/10.1016/j.jep.2020.113592
Elliott TF, Elliott K (2019) Vertebrate consumption and dispersal of the Nothofagaceae associated ascomycete *Cyttaria*. Austral Ecol 44:747–751. https://doi.org/10.1111/aec.12711
Erb M, Kliebenstein DJ (2020) Plant secondary metabolites as defenses, regulators, and primary metabolites: the blurred functional trichotomy. Plant Physiol 184:39–52
Eshwarappa R, Ramachandra Y, Subaramaihha S et al (2016) Anti-lipoxygenase activity of leaf gall extracts of *Terminalia chebula* (Gaertn.) Retz. (Combretaceae). Pharm Res 8:78–82. https://doi.org/10.4103/0974-8490.171103
Ferreira BG, Oliveira DC, Moreira ASFP et al (2018) Antioxidant metabolism in galls due to the extended phenotypes of the associated organisms. PLoS One 13. https://doi.org/10.1371/journal.pone.0205364
Figueredo G, Ünver A, Chalchat JC et al (2012) A research on the composition of essential oil isolated from some aromatic plants by microwave and hydrodistillation. J Food Biochem 36:333–343
Gamundí IJ, Minter DW, Romero AI et al (2004) Catálogo de los Discomycetes (fungi) de Patagonia, Tierra del Fuego y áreas antárticas adjacentes. Darwiniana, nueva serie 42:63–164

Gerchman Y, Inbar M (2011) Distinct antimicrobial activities in aphid galls on *Pistacia atlantica*. Plant Signal Behav 6:2008–2012. https://doi.org/10.4161/psb.6.12.18031
Germinara GS, De Cristofaro A, Rotundo G (2011) Chemical cues for host location by the chestnut gall wasp, *Dryocosmus kuriphilus*. J Chem Ecol 37:49–56
Giron D, Huguet E, Stone GN, Body M (2016) Insect-induced effects on plants and possible effectors used by galling and leaf-mining insects to manipulate their host-plant. J Insect Physiol 84:70–89. https://doi.org/10.1016/j.jinsphys.2015.12.009
Grover M (2021) Pistacia integerrima (Shringi)- a plant with significant pharmacological activities. J Phytopharmacol 10:323–330. https://doi.org/10.31254/phyto.2021.10508
Guedes LM, Aguilera N, Becerra J et al (2016) Leaf and stem galls of *Schinus polygamus* (Cav.) Cabr (Anacardiaceae): anatomical and chemical implications. Biochem Syst Ecol 69:266–273. https://doi.org/10.1016/j.bse.2016.10.012
Guedes LM, Aguilera N, Ferreira BG et al (2019) Spatiotemporal variation in phenolic levels in galls of calophyids on *Schinus polygama* (Anacardiaceae). J Plant Res 132:509–520. https://doi.org/10.1007/s10265-019-01118-6
Guedes LM, Torres S, Sáez-Carillo K et al (2022) High antioxidant activity of phenolic compounds dampens oxidative stress in *Espinosa nothofagi* galls induced on *Nothofagus obliqua* buds. Plant Sci 314. https://doi.org/10.1016/j.plantsci.2021.111114
Guedes L, Sanhueza C, Torres S et al (2023a) Gall-inducing *Eriophyes tiliae* stimulates the metabolism of *Tilia platyphyllos* leaves towards oxidative protection. Plant Physiol Biochem 195:25–36. https://doi.org/10.1016/j.plaphy.2022.12.014
Guedes L, Gavilán E, Pérez C et al (2023b) Impact of the galling *Dasineura* sp. on the structural and chemical profile of *Peumus boldus* stems. Trees 37:545–553. https://doi.org/10.1007/s00468-022-02368-z
Guedes LM, Aguilera N, Gavilán E et al (2023c) Vascular implications of *Dasineura* sp. galls' establishment on *Peumus boldus* stems. Plant Biol:1–8. https://doi.org/10.1111/plb.13561
Guedes LM, Aguilera N, Torres S et al (2024) Gall inducer *Dasineura* sp. alters the polyphenol profile and antioxidant activity of *Peumus boldus* stems. Tree Physiol. https://doi.org/10.1093/treephys/tpae024/7607185.
Guo L, Qiu J, Han Z et al (2015) A host plant genome *Zizania latifolia* after a century-long endophyte infection. Plant J 83:600–609
Hall CR, Carroll AR, Kitching RL (2017) A meta-analysis of the effects of galling insects on host plant secondary metabolites. Arthropod Plant Interact 11:463–473. https://doi.org/10.1007/s11829-016-9486-0
Hartmann T (2007) From waste products to ecochemicals: fifty years research of plant secondary metabolism. Phytochemistry 68:2831–2846
Hefied F, Ben Ahmed Z, Yousfi M (2020) In vitro antioxidant and α-amylase inhibitory potential of methanolic and lipid fractions from *Pistacia atlantica* Desf. galls. J Food Process Preserv 44:1–16. https://doi.org/10.1111/jfpp.14956
Heidel-Fischer HM, Musser RO, Vogel H (2014) Plant transcriptomic responses to herbivory. Ann Plant Rev: Insect-Plant Interact 47:155–196
Hilker M, Fatouros NE (2016) Resisting the onset of herbivore attack: plants perceive and respond to insect eggs. Curr Opin Plant Biol 32:9–16
Huber M, Bont Z, Fricke J et al (2016) A below-ground herbivore shapes root defensive chemistry in natural plant populations. Proc R Soc B Biol Sci 283:20160285
Hussein G, Miyashiro H, Nakamura N et al (2000) Inhibitory effects of Sudanese medicinal plant extracts on hepatitis C virus (HCV) protease. Phytother Res 14:510–516. https://doi.org/10.1002/1099-1573(200011)14:7<510::AID-PTR646>3.0.CO;2-B
Ibrahim S, Mohamed Y, Khedidja B et al (2022) Essential oil from galls formed on leaves of Pistacia atlantica Desf.: new in-vitro and in-silico studies of anti-inflammatory activities. S Afr J Bot 144:464–470

Isaias RMS, Oliveira DC, Moreira ASFP et al (2015) The imbalance of redox homeostasis in arthropod-induced plant galls: mechanisms of stress generation and dissipation. Biochim Biophys Acta Gen Subj 1850. https://doi.org/10.1016/j.bbagen.2015.03.007
Ismail I, Kasiraja V, Abdullah H (2021) A review on anticancer potential of Quercus infectoria and its bioactive compounds. Biomedicine (India) 41:701–705. https://doi.org/10.51248/.V41I4.298
Jabeur I, Martins N, Barros L et al (2017) Contribution of the phenolic composition to the antioxidant, anti-inflammatory and antitumor potential of *Equisetum giganteum* L. and *Tilia platyphyllos* Scop. Food Funct 8:975–984. https://doi.org/10.1039/c6fo01778a
Jain A, Singh HB, Bhattacharyya PR (2012) The ethnobotany and nutritional values of wild rice [*Zizania latifolia* (Griseb.) Turcz. ex Stapf] (Poaceae) in Manipur. Indian J Tradit Knowl 1:66–69
Jara B (2021) Efectos de aleloquímicos extraídos de *Acacia dealbata* (Link) y *Peumus boldus* (Mol.) sobre la germinación y morfología de la radícula de *Lactuca sativa* (L.). Universidad de Concepción
Jiang Y, Jiayan Y, Veromann-Jürgenson L-L, Niinemets Ü (2021) Gall- and erineum-forming *Eriophyes* mites alter photosynthesis and volatile emissions in an infection severity-dependent manner in broad-leaved trees *Alnus glutinosa* and *Tilia cordata*. Tree Physiol 41:1122–1142. https://doi.org/10.1093/treephys/tpaa173
Jorge NC, Souza-silva ÉA, Alvarenga DR et al (2018) Structural and chemical profiles of *Myrcia splendens* (Myrtaceae) leaves under the influence of the galling *Nexothrips* sp. (Thysanoptera). Front Plant Sci 9:1–14. https://doi.org/10.3389/fpls.2018.01521
Juárez-Montiel M, Ruiloba de León S, Chávez-Camarillo, Hernández-Rodríguez G, Villa-Tanaca CL (2011) Huitlacoche (corn smut), caused by the phytopathogenic fungus *Ustilago maydis*, as a functional food. Rev Iberoam Micol 28:69–73
Kabir ML, Ali MT, Haque ME (2017) Antimicrobial, antifungal & cytotoxic activities screening of stem bark fractions from *Terminalia chebula*. Univ J Plant Sci 5:41–44
Kamarudin NA, Muhamad N, Salleh NNHN, Tan SC (2021) Impact of solvent selection on phytochemical content, recovery of tannin and antioxidant activity of *Quercus infectoria* galls. Pharm J 13:1195–1204. https://doi.org/10.5530/pj.2021.13.153
Kaur G, Athar M, Alam MS (2008) *Quercus infectoria* galls possess antioxidant activity and abrogates oxidative stress-induced functional alterations in murine macrophages. Chem Biol Interact 171:272–282. https://doi.org/10.1016/j.cbi.2007.10.002
Kehl A, Dötterl S, Aas G, Rambold G (2010) Is flower scent influencing host plant selection of leaf-galling sawflies (Hymenoptera, Tenthredinidae) on willows? Chemoecology 20:215–221
Klimm FS, Weinhold A, Volf M (2020) Volatile production differs between oak leaves infested by leaf-miner *Phyllonorycter harrisella* (Lepidoptera: Gracillariidae) and galler *Neuroterus quercusbaccarum* (Hymenoptera: Cynipidae). Eur J Entomol 117:101–109. https://doi.org/10.14411/eje.2020.011
Kumar KJ (2006) Effect of geographical variation on contents of tannic acid, gallic acid, chebulinic acid and ethyl gallate in *Terminalia chebula*. Nat Prod 2:170–175
Kumar S, Baldi A, Sharma DK (2022) Characterization and in vitro investigation of antiscabietic effect of phytosomes assimilating quercetin and naringenin rich fraction of *Pistacia integerrima* galls extract against Sarcoptes scabiei. J Drug Deliv Sci Technol 67:102851
Kumari I, Kaurav H, Chaudhary G (2021) Punica granatum L.(Dadim) Punica granatum L.(Dadim), therapeutic importance of World's most ancient fruit plant. J Drug Deliv Ther 11:113–121
Kuster VC, Rezende UC, Custódio J et al (2019) How galling organisms manipulate the secondary metabolites in the host plant tissues?: a histochemical overview in Neotropical gall systems. In: Co-evolution of secondary metabolites. Springer, pp 1–20
Li XQ, Liu YZ, Guo WF et al (2017) The gall wasp *Leptocybe invasa* (Hymenoptera: Eulophidae) stimulates different chemical and phytohormone responses in two *Eucalyptus* varieties that vary in susceptibility to galling. Tree Physiol 37:1208–1217. https://doi.org/10.1093/treephys/tpx098

Li J, Lu Z, Yang Y et al (2021) Transcriptome analysis reveals the symbiotic mechanism of ustilago esculenta-induced gall formation of *Zizania latifolia*. Mol Plant-Microbe Interact 34:168–185. https://doi.org/10.1094/MPMI-05-20-0126-R

Lizárraga-Guerra R, López MG (1996) Content of free amino acids in Huitlacoche (*Ustilago maydis*). J Agric Food Chem 44:2556–2559

López-Martínez LX, Aguirre-Delgado A, Saenz-Hidalgo HK et al (2022) Bioactive ingredients of huitlacoche (*Ustilago maydis*), a potential food raw material. Food Chem: Mol Sci 4:1–7. https://doi.org/10.1016/j.fochms.2022.100076

Mahjoub F, Rezayat KA, Yousefi M et al (2018) *Pistacia atlantica* Desf. A review of its traditional uses, phytochemicals and pharmacology. J Med Life 11:180

Manosroi A, Jantrawut P, Akazawa H et al (2010) Biological activities of phenolic compounds isolated from galls of *Terminalia chebula* Retz. (Combretaceae). Nat Prod Res 24:1915–1926. https://doi.org/10.1080/14786419.2010.488631

Manosroi A, Jantrawut P, Ogihara E et al (2013) Biological activities of phenolic compounds and triterpenoids from the galls of *Terminalia chebula*. Chem Biodivers 10:1448–1463. https://doi.org/10.1002/cbdv.201300149

Meda RNT, Vlase L, Lamien-Meda A et al (2011) Identification and quantification of phenolic compounds from *Balanites aegyptiaca* (L) Del (Balanitaceae) galls and leaves by HPLC-MS. Nat Prod Res 25:93–99. https://doi.org/10.1080/14786419.2010.482933

Meda RNT, Guenne S, Mariama Combassere-Cherif K et al (2020) Antioxidant and anti--inflammatory effects of *Balanites aegyptiac* (L) Del (Balanitaceae) galls and leaves. J Dis Med Plant 6:82. https://doi.org/10.11648/j.jdmp.20200604.13

Mehla K, Balwani S, Kulshreshtha A et al (2011) Ethyl gallate isolated from *Pistacia integerrima* Linn inhibits cell adhesion molecules by blocking AP-1 transcription factor. J Ethnopharmacol 137:1345–1352

Meyer-Rochow VB (2022) Can molecularly engineered plant galls help to ease the problem of world food shortage (and our dependence on pollinating insects)? Food Secur 11. https://doi.org/10.3390/foods11244014

Mohamed MES (2016) The interaction between the gall wasp leptocybe invasa and eucalyptus camaldulensis leaves: a study of phyto-volatile metabolites. J Pharmacogn Phytother 8:90–98. https://doi.org/10.5897/JPP2015.0384

Mohd-Nasir H, Putra NR, Chuo SC et al (2021) Optimization of the supercritical carbon dioxide extraction of *Quercus infectoria* galls extracts and its bioactivities. J Food Process Preserv 45. https://doi.org/10.1111/jfpp.15156

Murthy HN, Yadav GG, Dewir YH, Ibrahim A (2020) Phytochemicals and biological activity of desert date (*Balanites aegyptiaca* (L.) Delile). Plan Theory 10:32

Naik PM, Al-Khayri JM (2016) Abiotic and biotic elicitors–role in secondary metabolites production through in vitro culture of medicinal plants. In: Abiotic and biotic stress in plants – recent advances and future perspectives. InTech. https://doi.org/10.5772/61442

Noe SM, Niinemets Ü (2020) Impact of gall-forming insects on global BVOC emissions and climate: a perspective. Front For Glob Change 3:1–8. https://doi.org/10.3389/ffgc.2020.00009

Nyman T, Julkunen-Tiitto R (2000) Manipulation of the phenolic chemistry of willows by gall-inducing sawflies. Proc Natl Acad Sci USA 97:13184–13187. https://doi.org/10.1073/pnas.230294097

Oliveira DC, Soares GLG, Isaias RMDS (2008) Phytotoxicity of the extracts of *Lonchocarpus muehlbergianus* Hassl. (Fabaceae) leaflets and galls on seed germination and early development of lettuce. Acta Bot Bras 22:1095–1100. https://doi.org/10.1590/s0102-33062008000400020

Oliveira DC, Isaias RMS, Fernandes GW et al (2016) Manipulation of host plant cells and tissues by gall-inducing insects and adaptive strategies used by different feeding guilds. J Insect Physiol 84:103–113. https://doi.org/10.1016/j.jinsphys.2015.11.012

Padulosi S, Hodgkin T, Williams J, et al (2002) Underutilized crop and opportunities in the 21st century. Managing plant genetic diversity. Managing plant genetic diversity. Proceedings of an international conference, Kuala Lumpur, 12–16 June 2000. CABI Publishing, Wallingford, pp 323–338

Pastene E, Parada V, Avello M et al (2014) Catechin-based Procyanidins from *Peumus boldus* Mol. aqueous extract inhibit helicobacter pylori urease and adherence to adenocarcinoma gastric cells. Phytother Res 28:1637–1645. https://doi.org/10.1002/ptr.5176

Patel S, Rauf A, Khan H (2018) The relevance of folkloric usage of plant galls as medicines: finding the scientific rationale. Biomed Pharmacother 97:240–247

Patni V, Sharma N, Mishra P (2012) Oak *(Quercus leucotrichophora)* galls, as an intense source of natural gallic acid. LS Int J Life Sci 1:186. https://doi.org/10.5958/j.2319-118x.1.3.013

Paul-Victor C, Zust T, Rees M et al (2010) A new method for measuring relative growth rate can uncover the costs of defensive compounds in *Arabidopsis thaliana*. New Phytol 187:1102–1111. https://doi.org/10.1111/j.1469-8137.2010.03325.x

Pavlovića T, Dimkića I, Andrićb S et al (2020) Linden tea from Serbia – an insight into the phenolic profile, radical scavenging and antimicrobial activities. Ind Crops Prod 154:112639

Pinto APR, Seibert JB, Dos Santos ODH et al (2018) Chemical constituents and allelopathic activity of the essential oil from leaves of *Eremanthus erythropappus*. Aust J Bot 66:601–608. https://doi.org/10.1071/BT18138

Rajasekaran A, Mahimaidoss M (2016) In vitro antioxidant activity in leaf galls of *Mangifera indica* L. Asian J Pharm Clin Res 9:198–201

Ramawat KG, Goyal S (2020) Co-evolution of secondary metabolites during biological competition for survival and advantage: an overview. In: Mérillon JM, Ramawat KG (eds) Co-evolution of secondary metabolites, Reference series in phytochemistry, pp 3–17. https://doi.org/10.1007/978-3-319-96397-6_2

Rana S, Shahzad M, Shabbir A (2016) Pistacia integerrima ameliorates airway inflammation by attenuation of TNF-a, IL-4, and IL-5 expression levels, and pulmonary edema by elevation of AQP1 and AQP5 expression levels in mouse model of ovalbumin-induced allergic asthma. Phytomedicine 23:838–845. https://doi.org/10.1016/j.phymed.2016.04.006

Rand K, Bar E, Ben AM et al (2017) Differences in monoterpene biosynthesis and accumulation in Pistacia palaestina leaves and aphid-induced galls. J Chem Ecol 43:143–152. https://doi.org/10.1007/s10886-016-0817-5

Rauf A, Saleem M, Uddin G et al (2015) Phosphodiesterase-1 inhibitory activity of two flavonoids isolated from pistacia integerrima J.L. Stewart Galls. Evid Based Complement Altern Med 2015:506564. https://doi.org/10.1155/2015/506564

Rauf A, Patel S, Uddin G et al (2017) Phytochemical, ethnomedicinal uses and pharmacological profile of genus *Pistacia*. Biomed Pharmacother 86:393–404. https://doi.org/10.1016/j.biopha.2016.12.017

Redwane A, Lazrek HB, Bouallam S et al (2002) Larvicidal activity of extracts from *Quercus lusitania* var. infectoria galls (Oliv.). J Ethnopharmacol 79:261–263. https://doi.org/10.1016/S0378-8741(01)00390-7

Rezaie M, Farhoosh R, Sharif A et al (2015) Chemical composition, antioxidant and antibacterial properties of Bene (*Pistacia atlantica* subsp. mutica) hull essential oil. J Food Sci Technol 52:6784–6790

Salazar-Vidal V, Figueroa F, Soto L et al (2020) Nutritional characteristics and cytotoxic effect of polysaccharides extracted from the digüeñes *Cyttaria berteroi* and *Cyttaria hariotii* present in Chile. Rev Chil Nutr 47:750–756. https://doi.org/10.4067/S0717-75182020000500750

Santha T, Shetty J, Yoganarasimhan S, Sudha R (1991) Farmacognostical studies on the South Indian market sample of karkatasringi (kadukkaipoo) – Terminalia chebul (gaertn. Leaf gall). Anc Sci Life 11:16–22

Schuman MC, Baldwin IT (2016) The layers of plant responses to insect herbivores. Annu Rev Entomol 61:373–394

Schwachtje J, Baldwin IT (2008) Why does herbivore attack reconfigure primary metabolism? Plant Physiol 146:845–851

Shankara BR, Ramachandra YL, Rajan SS, Ganapathy PS et al (2016) Evaluating the anticancer potential of ethanolic gall extract of *Terminalia chebula* (Gaertn.). Pharm Res 8:209

Shirol RL, Shirole NL, Kshatriya AA et al (2014) Investigation into the mechanism of action of essential oil of *Pistacia integerrima* for its antiasthmatic activity. J Ethnopharmacol 153. https://doi.org/10.1016/j.jep.2014.02.009

Sifi I, Dzoyem JP, Ouinten M et al (2015) Antimycobacterial, antioxidant and cytotoxic activities of essential oil of gall of *Pistacia atlantica* desf. from Algeria. Afr J Tradit Complement Altern Med 12:150–155. https://doi.org/10.4314/ajtcam.v12i3.19

Silva AFM, Kuster VC, Rezende UC, Oliveira DC (2019) The early developmental stages of gall--inducing insects define final gall structural and histochemical profiles: the case of *Bystracoccus matayba*e galls on matayba guianensis. Botany 97. https://doi.org/10.1139/cjb-2019-0017

Singh G, Kumar P (2013) Extraction, gas chromatography-mass spectrometry analysis and screening of fruits of Terminalia chebula Retz. for its antimicrobial potential. Pharmacognosy Res 5:162–168

Srivastava P, Raut HN, Wagh RS et al (2012) Purification and characterization of an antioxidant protein (~16 kDa) from *Terminalia chebula* fruit. Food Chem 131:141–148

Sukh D (1997) Ethanotherapeutics and modern drug development: the potential of ayurveda. Curr Sci 73:909–928

Süzgeç-Selçuk S, Özek T, Özek G et al (2021) The leaf and the gall volatiles of *Salvia fruticosa* miller from Turkey: chemical composition and biological activities. Rec Nat Prod 15:10–24. https://doi.org/10.25135/rnp.185.20.03.1579

Swanson L, Li T, Rinnan R (2021) Contrasting responses of major and minor volatile compounds to warming and gall-infestation in the Arctic willow *Salix myrsinites*. Sci Total Environ 793:148516. https://doi.org/10.1016/j.scitotenv.2021.148516.

Syukriah AN, Liza MS, Harisun Y, Fadzillah AAM (2014) Effect of solvent extraction on antioxidant and antibacterial activities from *Quercus infectoria* (Manjakani). Int Food Res J 21:1031

Tanaka Y, Okada K, Asami T, Suzuki Y (2013) Phytohormones in Japanese mugwort gall induction by a gall-inducing gall midge. Biosci Biotechnol Biochem 77:1942–1948

Toledo AMO, Ulguim PSB, dos Santos Kaeser S, Gomes FT (2016) Interferência alelopática do chá de boldo-do-chile (*Peumus boldus* Molina, Monimiaceae) sobre sementes de alface e pepino. Revista de Ciências Agroveterinárias 15:180–187. https://doi.org/10.5965/223811711532016180

Tollrian R, Harvell C (1999) The ecology and evolution of inducible defenses. Princeton University Press

Tooker JF, Hanks LM (2004) Stereochemistry of host plant monoterpenes as mate location cues for the gall wasp *Antistrophus rufus*. J Chem Ecol 30:473–477

Tooker JF, Koenig WA, Hanks LM (2002) Altered host plant volatiles are proxies for sex pheromones in the gall wasp Antistrophus rufus. Proc Natl Acad Sci 99:15486–15491

Tooker J, Crumrin A, Hanks L (2005) Plant volatiles are behavioral cues for adult females of the gall wasp Antistrophus rufus. Chemoecology 15:85–88

Tooker JF, Rohr JR, Abrahamson WG, De Moraes CM (2008) Gall insects can avoid and alter indirect plant defenses. New Phytol 178:657–671. https://doi.org/10.1111/j.1469-8137.2008.02392.x

Topçu G, Öztürk M, Kuşman T et al (2013) Terpenoids, essential oil composition, fatty acid profile, and biological activities of *Anatolian Salvia fruticosa* Mill. Turk J Chem 37:619–632. https://doi.org/10.3906/kim-1303-25

Toul F, Belyagoubi-Benhammou N, Zitouni A, Atik-Bekkara F (2017) Antioxidant activity and phenolic profile of different organs of *Pistacia atlantica* Desf. subsp. atlantica from Algeria. Nat Prod Res 31:718–723

Upadhye AS, Rajopadhye AA (2010) Pharmacognostic and phytochemical evaluation of leaf galls of Kakadshringi used in Indian system of medicine. J Sci Ind Res (India) 69:700–704

Verdeguer M, García-Rellán D, Boira H et al (2011) Herbicidal activity of *Peumus boldus* and *Drimys winterii* essential oils from Chile. Molecules 16:403–411. https://doi.org/10.3390/molecules16010403

Villagrán Z, Martínez-Reyes M, Gómez-Rodríguez H et al (2023) Huitlacoche *(Ustilago maydis)*, an iconic Mexican fungal resource: biocultural importance, nutritional content, bioactive compounds, and potential biotechnological applications. Molecules 28:4415
Wang JY, Zhang QH, Deng XM (2007) Progress on pharmacological activities of Quercus infectoria galls. Shizhen Guoyi Guoyao 18:2570–2572
Wang H, Liu J, Cui K et al (2016) Gibberellic acid is selectively downregulated in response to aphid-induced gall formation. Acta Physiol Plant 38:214
Wink M (2003) Evolution of secondary metabolites from an ecological and molecular phylogenetic perspective. Phytochemistry 64:3–19. https://doi.org/10.1016/S0031-9422(03)00300-5
Wink M (2020) Evolution of the angiosperms and co-evolution of secondary metabolites, especially of alkaloids. In: Mérillon J-M, Ramawa KG (eds) Co-evolution of secondary metabolites, Reference series in phytochemistry. Springer Nature Switzerland AG, pp 1–24. https://doi.org/10.1007/978-3-319-96397-6
Wu HC, His HY, Hsiao G et al (2023) Chemical constituents and bioactive principles from the Mexican truffle and fermented products of the derived fungus *Ustilago maydis* MZ496986. J Agric Food Chem 71:1122–1131
Yamaguchi H, Tanaka H, Hasegawa M et al (2012) Phytohormones and willow gall induction by a gall-inducing sawfly. New Phytol 196:586–595
Yang L, Yin P, Li K et al (2018) Seasonal dynamics of constitutive levels of phenolic components lead to alterations of antioxidant capacities in *Acer truncatum* leaves. Arab J Chem 11:14–25. https://doi.org/10.1016/j.arabjc.2017.01.009.
Ye J, Jiang Y, Veromann-Jürgenson LL, Niinemets Ü (2019) Petiole gall aphid (*Pemphigus spyrothecae*) infestation of *Populus* × *petrovskiana* leaves alters foliage photosynthetic characteristics and leads to enhanced emissions of both constitutive and stressinduced volatiles. Trees-Struct Funct 33:37–51. https://doi.org/10.1007/s00468-018-1756-2
Yoram G, Inbar M (2011) Distinct antimicrobial activities in aphid galls on *Pistacia atlantica*. Plant Signal Behav. https://doi.org/10.4161/psb.6.12.18031
Zahoor M, Zafar R, Rahman NU (2018) Isolation and identification of phenolic antioxidants from *Pistacia integerrima* gall and their anticholine esterase activities. Heliyon 4:e01007
Zhang H, Wang Y, Liu J et al (2020) Traditional Uyghur medicine *Quercus infectoria* galls water extract triggers apoptosis and autophagic cell death in colorectal cancer cells. BMC Complement Med Ther 20:1–12. https://doi.org/10.1186/s12906-020-03167-0.
Zunjarrao SS, Tellis MB, Joshi SN, Joshi RS (2020) Plant-insect interaction: the saga of molecular coevolution. In: Mérillon JM, Ramawat KG (eds) Co-evolution of secondary metabolites, Reference series in phytochemistry. Springer Nature Switzerland AG, pp 19–45. https://doi.org/10.1007/978-3-319-96397-6_42
Zust T, Joseph B, Shimizu KK, Kliebenstein DJ, Turnbull LA et al (2011) Using knockout mutants to reveal the growth costs of defensive traits. Proc R Soc B 278:2598–2603

Structural and Physiological Aspects of Aphid-Induced Galls

Rafael Álvarez, Antonio Encina-García, Víctor Moreno-González, and Bruno Garcia Ferreira

1 Introduction

The main difference of the aphid galls to other insect galls lies in the distinct feeding behavior of the gall-inducing aphids. As the free-living aphids, they feed on the nutrient-rich phloem sap (Spiller et al. 1985; Tjallingii and Hogen-Esch 1993). Therefore, the cells that serve as food for aphids are not destroyed during their feeding behavior, distinct from chewer or scrapper gall inducers. The gall-inducing aphids stimulate a notable hypertrophy of the vascular bundles, particularly of the phloem, caused by the cell hypertrophy and hyperplasia (Fig. 1a, b) (Álvarez et al. 2008, 2009, 2013, 2014, 2016, 2020; Álvarez 2012; Muñoz-Viveros et al. 2014; Mellah et al. 2016; Martinez et al. 2018; Ferreira et al. 2019; Hemmam et al. 2022). A particular case occurs with host plants of Anacardiaceae, which exhibit intraphloematic schizogenous channels, as a synapomorphy. Different species of *Pistacia* and *Rhus* are superhosts of aphid galls, in which the phloem cell hypertrophy is accompanied by a conspicuous increase in intraphloematic schizogenous channels (Fig. 1d, e) (Álvarez et al. 2009, 2014, 2016; Álvarez 2012; Muñoz-Viveros et al. 2014; Mellah et al. 2016; Martinez et al. 2018; Hemmam et al. 2022). As discussed below, these hypertrophied ducts secrete a resin toxic to natural enemies or generalist cecidophages that could attack the galls.

R. Álvarez · A. Encina-García · V. Moreno-González
Universidad de León, León, Spain
e-mail: ralvn@unileon.es

B. G. Ferreira (✉)
Department of Botany, Universidade Federal do Rio de Janeiro (UFRJ), Rio de Janeiro, Brazil
e-mail: bgf@biologia.ufrj.br

D. C. de Oliveira, R. M. dos Santos Isaias (eds.), *Plant Galls*,
https://doi.org/10.1007/978-3-031-80064-1_11

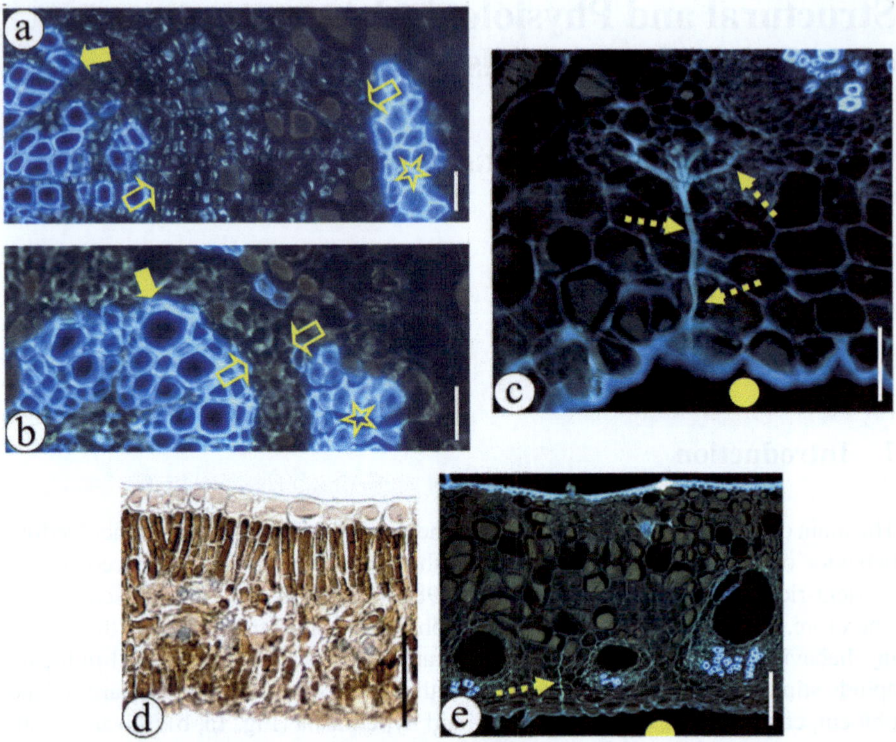

Fig. 1 Altered vascular bundles in aphid galls. (**a**) Hypertrophy and hyperplasia of phloem. Petiole gall induced by *Pemphigus vesicarius*. (**b**) The same region above in the control petiole. *Populus nigra* (control). (**c**) Aphid probing in search of phloem feeding sites. Gall induced by *Geopemphigus blackmani*. (**d**) Control leaf blade, showing two parenchyma: palisade parenchyma and aeriferous parenchyma. *Pistacia terebinthus* (control). (**e**) Wall of a petiole aphid gall in *Pistacia*. Note the increase in cellularity compared to the previous one, as well as vascular bundles with intraphloematic ducts. Gall induced by *Geopemphigus floccosus*. *Symbols*: arrow—xylem, circle and semicircle—gall chamber, dotted arrow—saliva sheath through which the stylet penetrates, hollow arrow—phloem, star—sclerenchyma. *Bars (micrometers)*: 30 – 30 – 50 – 100 – 100

Inside the aphid galls, the inhabitants probe various feeding points using their stylets (Fig. 1c). The access of the stylets towards the phloematic elements—through which the elaborated sap circulates—is accompanied by salivary secretion (Miles 1999; Álvarez et al. 2016). This secretion can be observed in histological sections analyzed under an epifluorescence microscope, after the procedures of fixation and histological processing of the samples for microscopic observation. The intervention of gall-inducing aphids on the vascular bundles impacts similarly the differentiation of xylem and phloem cells. As both vascular tissues are inevitably together, the hypertrophy and hyperplasia of one also impact the other vascular tissues, including the parenchymatic cells, essential components of both tissues.

1.1 Gall-Inducing Aphid Life Cycles

Aphids (Hemiptera: Aphididae) are a family of phytophagous, fluidophagous and colonial insects with a subcosmopolitan distribution, consisting of approximately 4700 species (Blackman and Eastop 2020; Douglas 2003; Nieto Nafría and Mier Durante 1998). In Aphididae, the ability to induce galls is found in species belonging to the subfamilies Hormaphidinae and Eriosomatinae (Blackman and Eastop 2020; Chen and Qiao 2009). However, as lack of monophyly has been documented in Eriosomatinae, as well as in some tribes within both subfamilies (Nováková et al. 2013; Ortiz-Rivas and Martínez-Torres 2010; Sano and Akimoto 2011; Zhang and Qiao 2007; Chen et al. 2014), it can be argued that the ability to induce galls in aphids would have arisen multiple times independently.

Aphid life cycles are among the most complex in the kingdom Animalia, as they are highly intricate (Wool 2005). All generations in the cycle are comprised of viviparous and parthenogenetic females, excluding the amphigonic generation, which includes males and oviparous females (Moran 1992). In the subfamilies Eriosomatinae and Hormaphidinae, the presence of gallicolous morphs is critical for completing the holocycle (Blackman and Eastop 2020; Nieto Nafría and Mier Durante 1998). Focusing on the tribes Fordini and Pemphigini (Eriosomatinae), whose galls will be the focus of this chapter, they present a holocycle with alternating hosts (Pérez Hidalgo and Nieto Nafría 2003). While both tribes' cycles are very similar, Pemphigini species complete it in 1 year, whereas Fordini species complete it in two.

The resistance egg is deposited on the primary host (Anacardiaceae species in Fordini; *Populus* species in most Pemphigini), and from this egg hatches a viviparous, apterous female, known as a the fundatrix, which triggers gall formation. Subsequently, one or two generations of apterous or winged forms, called fundatrigeniae, develop parthenogenetically within the gall. Once the gall opens, the winged fundatrigeniae migrate to the secondary host, where they establish non-gallicolous colonies. Finally, the viviparous females give birth to the sexuparae, which return to the primary host and complete the cycle producing sexual males and oviparous females. These offspring then reproduce, and the female lay the resistance egg, restarting the cycle.

The breadth of gall-inducing aphid diversity becomes apparent through the varied gall shapes that distinguish them across different species, even when they induce multiple galls within the same host plant species (Wool 2004; Álvarez 2012; Muñoz-Viveros et al. 2014; Álvarez et al. 2009, 2013, 2014, 2016, 2020). The examination of morphological and microscopic aspects of aphid galls offers a comprehensive understanding not only of the alterations in the functionality of host plant tissues but also sheds light on the phylogenetic relationships among the galling organisms (Muñoz-Viveros et al. 2014; Álvarez et al. 2014, 2016, 2020), as elucidated in this chapter.

2 Structure of Aphid-Induced Galls

The galls induced by aphids are extended phenotypes of the gall-inducing species. The control of gall shape is credited to the aphid, while the plant contributes to attributes involved in gall protection. The aphid utilizes the plant defense structures for its own protection (Dawkins 1982; Stern 1995; Stone and Schönrogge 2003; Sano and Akimoto 2011; Álvarez et al. 2013, 2014; Martinez et al. 2018). Thus, inside such galls, aphids find shelter from possible climatic adversities, safe places from predators, and access to phloem cells, their feeding site (Price et al. 1986, 1987; Stone and Schönrogge 2003; Koyama et al. 2004; Miller et al. 2009; Martinez 2010; Martinez et al. 2018). Following this premise, the comparison between galls induced by various species of aphids on close species of *Pistacia* was proposed. Samples of galls induced by the same species on both *Pistacia terebinthus* and *Pistacia palaestina* were collected in two regions separated by the Mediterranean Sea: western Spain and eastern Israel, respectively (Álvarez et al. 2009, 2016; Martinez et al. 2018). The two host plant species were synoymized as *Pistacia terebinthus* (Hessler 2017; Martinez et al. 2018). Along the same lines, galls from eastern Israel and western Algeria were studied (Mellah et al. 2016; Martinez et al. 2018; Hemmam et al. 2022). Such microscopic studies showed that all the galls induced by the same species on these two varieties of host plants were indistinguishable from each other, despite originating from two different continents. Thus, the galls retained their morphological and histological characteristics (for example, the presence of microcrystals in the outer epidermis of *Paracletus cimiciformis* galls, see Fig. 4c), corroborating the initial premise that the galls induced by aphids are extended phenotypes of their aphid-inducers. Despite the similarities in anatomical aspects, the quantitative features differed among galls induced on Israel and Spain, which was determined by the variations of host plant developmental responses to different environments (Martinez et al. 2018).

2.1 Morphological Diversity of Galls Induced by Aphids

The aphid galls may be classified as histioid galls due to the production of abnormal tissues, due to tissue hyperplasia and cell hypertrophy. They are also categorized as prosoplasmic galls because they exhibit definite size and shape according to the aphid-inducing species, have evident tissue differentiation, and are distinct from the host organ. They are monothalamic galls because they have a single chamber and are initially inhabited by a single individual (Küster 1911; Mani 1964; Meyer 1987; Arduin and Kraus 1995). Aphid-galls can be a result of a totally modified organ or of a partially modified organ, and shelter dozens, hundreds, or thousands of aphids (Price et al. 1987; Hartley and Lawton 1992; Stone and Schönrogge 2003; Sano and Akimoto 2011; Aoyama et al. 2012; Chen and Qiao 2012). Aphid-induced galls include a diversity of morphological and histological features, both related to each other and evidently related to the host plant structural potentials and constraints imposed on gall development (Isaias et al. 2014).

The first approach to study the galls induced by aphids is morphological, once the aphid galls may be classified in three main types:

- Galls that are observed as rolled-up leaves (rolling galls). They are always open galls in which the gall chamber is not completely isolated from the external environment (Fig. 2a). Some authors consider them as pseudogalls (Chen and Qiao 2012; Álvarez et al. 2013; Hałaj and Osiadacz 2017). In them, non-severe and intermittent or partial histological alterations are observed (Fig. 2b, c): the rolling galls are composed of modified leaf areas which altern with others unaltered leaf areas (Álvarez et al. 2013, 2020). The latter are histologically similar to the aphid-free leaf.

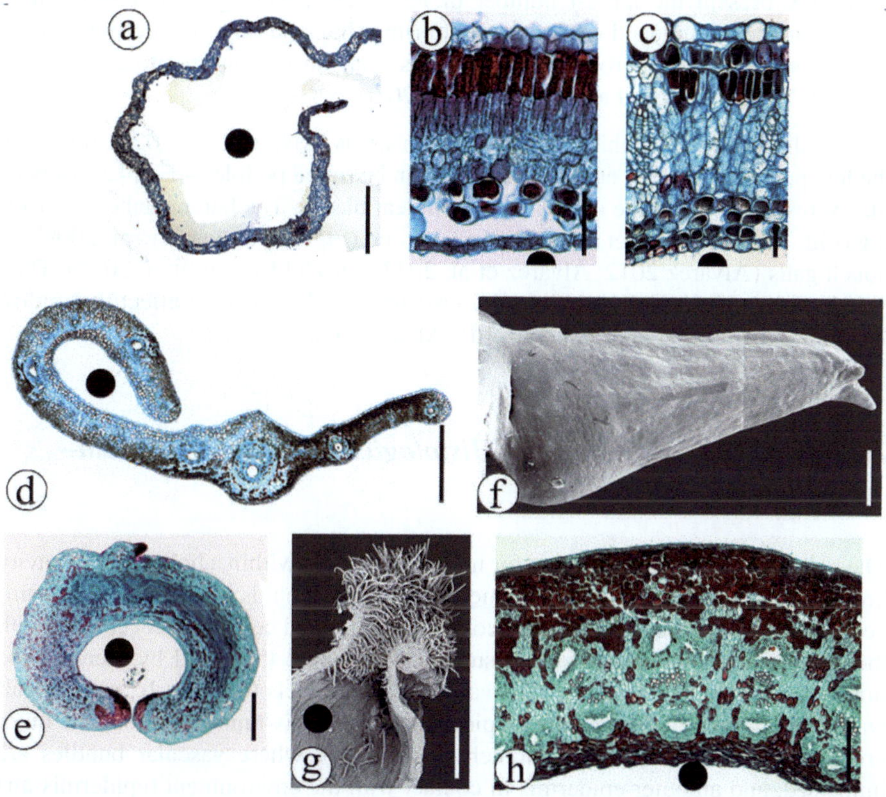

Fig. 2 Diversity of aphid galls. (**a**) Pseudogall (rolled leaf). Gall induced by *Eriosoma ulmi*. (**b**) A region of little modified wall of a pseudogall. Temporal gall induced by *Thecabius affinis*. (**c**) A region with a modified wall of a pseudogall. Temporal gall induced by *Thecabius affinis*. (**d**) Gall formed as a leaf fold. Young gall induced by *Forda marginata*. (**e**) Petiole gall. Gall induced by *Pemphigus spyrothecae*. (**f**) Gall of whole organ. Young gall induced by *Baizongia pistaciae*. (**g**) Closing zone with trichomes. Gall induced by *Geopemphigus morral*. (**h**) Wall with many layers of cells and great vascularity. Gall induced by *Baizongia pistaciae*. *Symbol*: circle and semicircle—gall chamber. *Bars (micrometers)*: 1000 – 50 – 50 – 500 – 1000 – 1000 – 300 – 200

- Open galls, appearing as leaf folds (of the leaf blade or petiole) (Fig. 2d, e, g) or as excrescences on young branches. Such galls remain open at all times of its existence, with a closure area (or door). These galls have a chamber where the aphids live (Álvarez et al. 2009, 2013, 2020; Muñoz-Viveros et al. 2014; Mellah et al. 2016; Ferreira et al. 2019; Hemmam et al. 2022). The closure zone usually has microscopic attributes different from those of the rest of the gall, such as a concentration of trichomes (Fig. 2g), crystal sand, or sclereids.
- Closed galls, which have tissues closing a large gall chamber occupied by the aphids, and usually show highly vascularized and particularly thick gall walls (Fig. 2h). These galls close in the early stages of development (Fig. 2f), and stay closed until the end of gall maturation, when they open by tissue disruption. Once open, they allow the exit of the winged individuals. They are the galls that usually present the largest number of inhabitants. Among these closed galls, there are those induced on buds or leaf primordia, resulting in entirely modified organs, with few microscopic similarities with host plant structures (Álvarez 2011, 2012; Álvarez et al. 2014, 2016, 2020).

Some galls induced by aphids are observed on stems (Álvarez et al. 2020), others on the leaf petioles (Álvarez et al. 2020) (they can be rolled petioles – Fig. 1e – or petiole excrescences), and the majority occur in leaf blades. The latter can be: leaf rolling (Fig. 1a–c) (Álvarez et al. 2013), leaf folding (Fig. 1d) (Álvarez et al. 2009), or pouch galls (Álvarez 2012; Álvarez et al. 2013; Kurzfeld-Zexer et al. 2015). They may be oriented adaxially or abaxially (Álvarez 2012), and may affect the midrib (Fig. 1f) (Álvarez 2011) or not (Fig. 1d) (Álvarez et al. 2009).

2.2 *Parts to Consider in the Histological Study of the Aphid-Induced Gall Wall*

The gall structure provides refuge for numerous aphids within a hollow inner cavity known as the gall chamber or gall lumen, wherein the aphids reside. The gall chamber is lined by the inner gall surface, coated by epidermal cells in direct contact with the aphids. The aphids rest on this surface, from there they feed by piercing the inner gall surface and cross the gall wall with their stylets (see Fig. 1c, e). The gall wall is then formed by an inner epidermis (epidermis-lumen), inner and outer ground tissue layers (usually parenchymatic cells), where vascular bundles are immersed, and an outer epidermis in contact with the environment (epidermis-air) (Fig. 3).

2.2.1 Epidermis-Air

The outer epidermis, or epidermis-air, of the galls induced by aphids does not show a drastic change from the non-galled host organ. It is a typical compact epidermis, which does not present discontinuities or fissures (Álvarez et al. 2013, 2016, 2020)

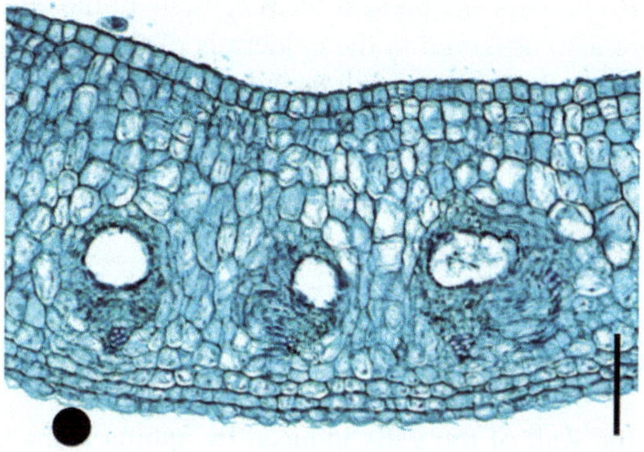

Fig. 3 Parts to consider in the histological study of the aphid–induced gall wall. From top to bottom, epidermis–air (always uniseriate), wall itself (with reserve parenchyma and vascular bundles), and epidermis–lumen (generally multiseriate and sometimes absent). *Symbol*: circle—gall chamber. Gall induced by *Forda riccobonii*. *Bar (micrometers)*: 100

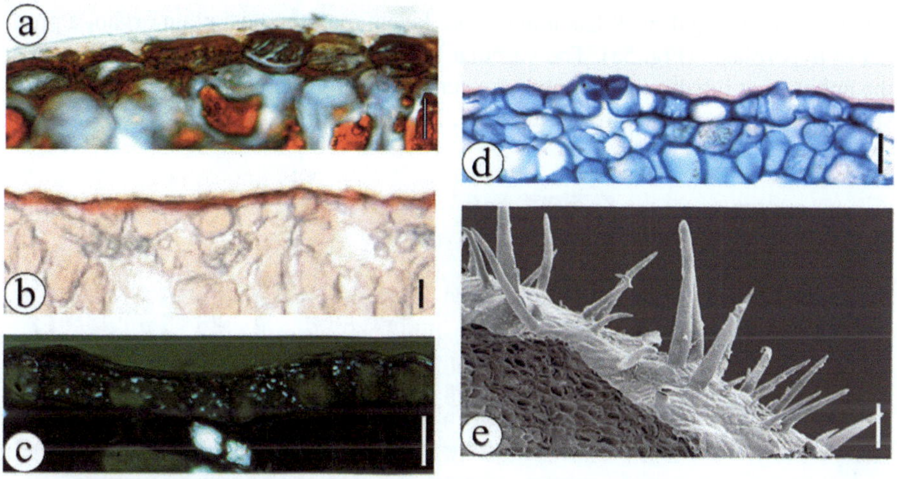

Fig. 4 Epidermis–air. (**a**) Uniseriate epidermis with cuticle, and tannins in the cells. Gall induced by *Forda formicaria*. (**b**) Lipid cuticle. Gall induced by *Forda formicaria*. (**c**) Microcrystals in epidermal cells. Gall induced by *Paracletus cimiciformis*. (**d**) Stoma. Gall induced by *Paracletus cimiciformis*. (**e**) Trichomes. Gall induced by *Pemphigus immunis*. *Bars (micrometers)*: 20 – 20 – 20 – 20 – 100

(Fig. 4a). The outer epidermis is commonly uniseriate, has stomata (Fig. 4d), is covered by a more or less thickened cuticle, and its cells have tannin inclusions (Fig. 4a, b). In some galls it may also have trichomes (Fig. 4e).

The epidermis-air of the leaf margin fold galls induced by *Paracletus cimiciformis* is remarkable. It is a gall induced on leaves of *Pistacia terebinthus*, a superhost of galling aphids (Álvarez et al. 2016). Of all the galls induced on these leaves, only those induced by *P. cimiciformis* present microcrystals in the outer epidermal cells (Fig. 4c), which are also observed in the epidermis of the non-galled leaf (Álvarez et al. 2009; Martinez et al. 2018). As will be indicated later, this presence of crystals can be related to the defense of the gall against possible intruders who intend to access it from the outside, that is, through the epidermis-air. The outer epidermis of galls may also accumulate mucilage and other ergastic compounds, usually related to the host plant capabilities (Álvarez et al. 2013).

2.2.2 Gall Wall

In general terms, the wall of the galls induced by aphids is made up—apart from vascular bundles—of hypertrophied cells that make up a more or less homogeneous storage parenchyma with starch, crystals, and tannin inclusions (Fig. 5a–d) with interspersed sclereids in some galls (Fig. 5e) (Álvarez et al. 2016, 2020). The sclereids originated by the deposition of the secondary cell wall and a subsequent lignification of parenchymatic cells, usually the outermost ones (Fig. 5e). The group of sclereids usually surrounds the whole gall wall, rigidifying the gall (Álvarez et al. 2009, 2016). The gall wall has much more layers than the host plant organ, a product of hyperplasia (Fig. 5a). The vascular bundles that represent the food source of the aphids that inhabit the chamber are surrounded by such layers of hypertrophied storage parenchyma (Álvarez et al. 2013, 2014; Ferreira et al. 2017).

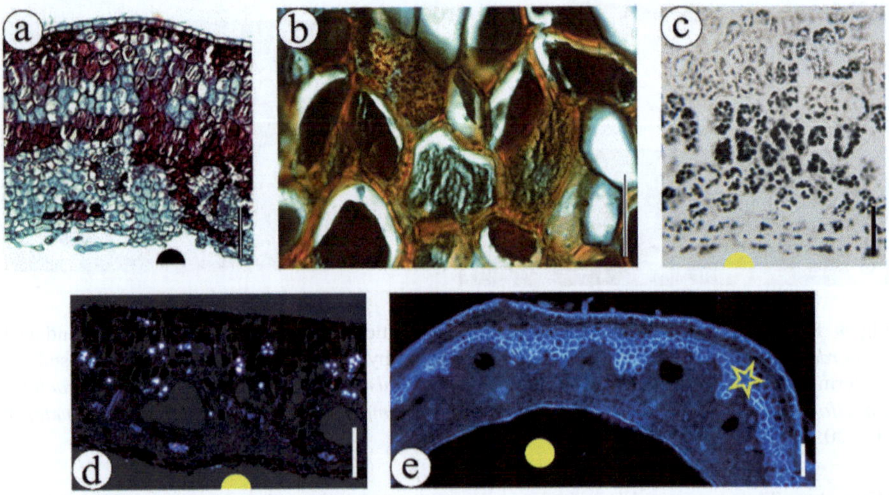

Fig. 5 Gall wall. (**a**) Gall wall showing a large number of storage parenchyma layers. Gall induced by *Tetraneura ulmi*. (**b**) Various conformations of tannins. Gall induced by *Forda marginata*. (**c**) Amyloplasts. Gall induced by *Paracletus cimiciformis*. (**d**) Crystals arranged in the upper part of the wall. Gall induced by *Forda formicaria*. (**e**) Sclereid layer. Gall induced by *Smynthurodes betae*. *Symbols*: circle and semicircle—gall chamber, star—sclerenchyma. *Bars (micrometers)*: 100 – 50 – 50 – 200

In young galls, a subepidermal spongy chlorophyll parenchyma can be observed (Richardson et al. 2017), which is generally not present in mature galls. Also, it is common for the parenchyma to be rich in starch (Fig. 5c), although its greater or lesser presence must be attributed to physiological issues. In general, the gall wall presents calcium oxalate crystals, particularly druses, often located especially in the outermost part of the wall (Fig. 5d), but they can be distributed on the whole wall.

The presence of tannins in distinct conformations is conspicuous in aphid galls (Fig. 5a, b). They may be associated with the parenchyma cell walls, deposited as fine or thick vacuolar granules, either as threads, and can also be observed as homogeneous and dense inclusions that occupy practically the entire vacuoles. These last two conformations are more frequently observed in mature galls (Laakso et al. 2000; Álvarez et al. 2009, 2016, 2020; Álvarez 2012).

2.2.3 Vascular Bundles

Not only changes in vascular bundle sizes, but also in their arrangement, may be observed in aphid galls (Fig. 6b). In some cases, collateral vascular bundles of the host organ are changed to a practically anfiphloematic (amphicribral) vascular bundle in the gall wall (Fig. 6a). Additionally, in galls formed on stems or by the rolling up the petioles, amphicribral vascular bundles may be observed, if they occur in non-galled petioles and stems (Fig. 6e, f) (Álvarez et al. 2020). The bundle sheath, groups of sclerenchyma fibers that surround the vascular bundles in some host plants (Arduin et al. 2005; Álvarez et al. 2009), are usually absent in their associated aphid galls. It is noteworthy that the galls induced by aphids do not usually have specially developed supporting tissues, even despite the large size of some galls. This may be explained by the significant development of the vascular bundles since xylem also assumes the support function in plants (Evert 2006).

In leaf galls, the orientation of vascular bundles are determined by the side where the gall is induced. In the leaves of the host plants, specifically in the leaf blade, the second order veins appear as small collateral vascular bundles, where the xylem is oriented toward the adaxial surface and the phloem towards the abaxial surface (Álvarez et al. 2008; Hemmam et al. 2022). If the gall is formed by an adaxial folding of the blade (see Fig. 9), the xylem is closer than the phloem to the gall lumen. If the fold is on the abaxial surface, the phloem is the vascular tissue closest to the lumen (Fig. 6c, d) (Álvarez et al. 2016, 2020). Additionally, if the gall is formed by an invagination of the leaf blade, forming a bulge or a pouch, the phloem will be oriented toward the gall chamber, when the gall appears on the leaf adaxial surface, since the inner layers of the gall will be formed by the abaxial layers of the leaf (Álvarez et al. 2013, 2020).

The number of vascular bundles in galls is also determined by their organization in non-galled organs. In the galls induced by *Geoica utricularia* on the midrib of *Pistacia terebinthus*, two rows of vascular bundles occur on the wall, specularly arranged: phloem—xylem/xylem—phloem (Fig. 7a) (Álvarez 2012; Álvarez et al. 2014, 2016). Similarly, two specular bundles occur in the non-galled leaf, but with

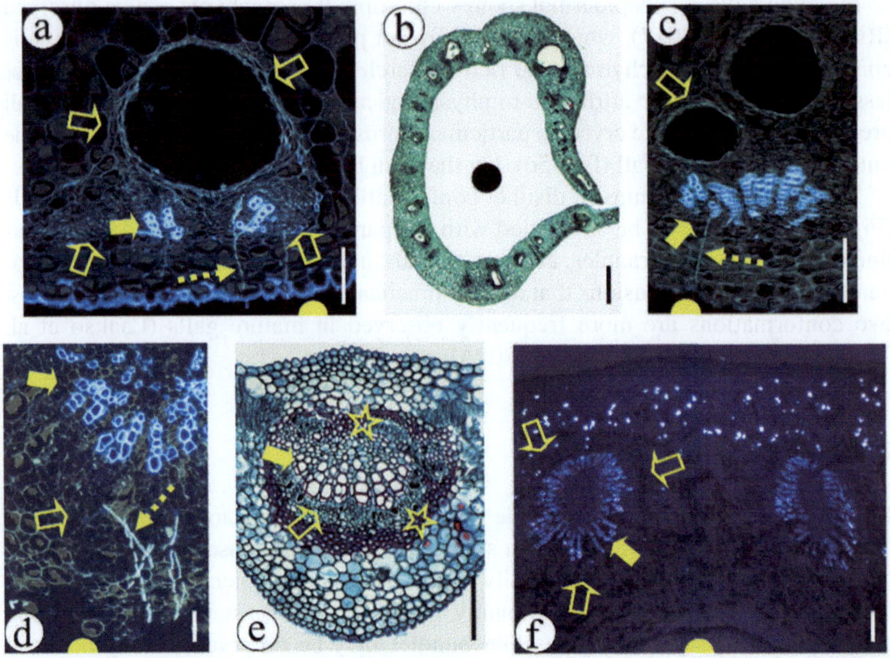

Fig. 6 Vascular bundles in aphid galls. (**a**) Collateral vascular bundle that appears amphicribral due to the great development of the phloem (with an intraphloematic duct in the center). Gall induced by *Geopemphigus torsus*. (**b**) Gall in Anacardiaceae presenting many intraphloematic ducts in the wall. Gall induced by *Forda riccobonii*. (**c**) Collateral vascular bundle close to the chamber. The xylem is shown oriented towards the light. Gall induced by *Geopemphigus floccosus*. (**d**) Vascular collateral bundle. The phloem is shown closer to the chamber. Gall induced by *Pemphigus vesicarius*. (**e**) Midrib with bundle sheath. *Populus nigra* (control). (**f**) Amphicribral vascular bundles without bundle sheath. Gall induced by *Pemphigus spyrothecae*. *Symbols*: arrow—xylem, circle and semicircle—gall chamber, dotted arrow—saliva sheath through which the stylet penetrates, hollow arrow—phloem, star—sclerenchyma. *Bars (micrometers)*: 100 – 500 – 100 – 50 – 100 – 200

clear distinct sizes (Fig. 7b). In the leaf, the supernumerary (adaxial) vascular bundle has few vascular elements, the schizogenous duct is absent, and the phloem oriented toward the adaxial surface, while the main vascular bundle (abaxially positioned) is highly developed, with schizogenous ducts, and phloem oriented abaxially (Álvarez et al. 2008). During gall development, cell hypertrophy and hyperplasia occur in the two vascular bundles. In the galls induced by *Geoica utricularia*, the midrib of *Pistacia terebinthus* suffer an invagination toward the abaxial surface of the leaf. Therefore, the hypertrophied main vascular bundle is situated in the outer portion of the gall wall, in the part furthest from the lumen (Fig. 7c). This main vascular bundle is large and opposite to the usually a smaller supernumerary bundle. The hypertrophied supernumerary vascular bundle (which now presents conspicuous schizogenous ducts) is located in the inner portion of the gall wall, closest to the gall chamber (Álvarez 2011).

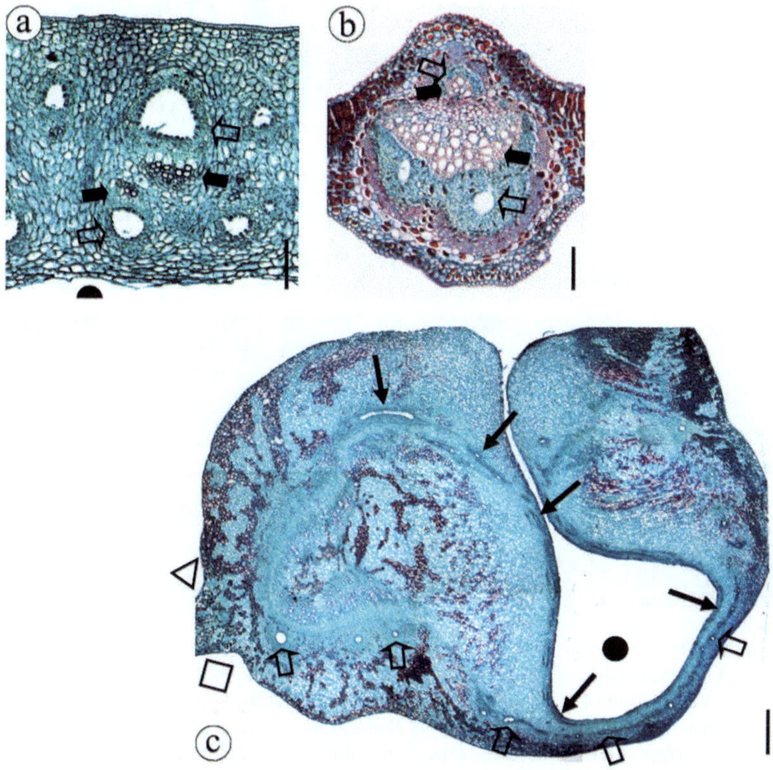

Fig. 7 Vascular bundles of aphid-galls on Anacardiaceae. (**a**) Gall wall with two specularly arranged vascular bundles: phloem—xylem/xylem—phloem. Gall induced by *Baizongia pistaciae*. (**b**) Midrib. Note above the small supernumerary vascular bundle without schizogenous intraphloematic ducts, and the large lower vascular bundle with conspicuous ducts. The arrangement of the vascular tissues is specular: phloem—xylem/xylem—phloem. *Pistacia terebinthus* (control). (**c**) The hypertrophied supernumerary vascular bundle occupies the inner part of the gall wall. Gall in formation induced by *Geoica utricularia*. *Symbols*: circle and semicircle—gall chamber, hollow arrow—phloem, square—abaxial zone, thin arrow—hypertrophy of the supernumerary bundle, triangle—adaxial zone. *Bars (micrometers)*: 200 – 50 – 100

2.2.4 Inner Gall Surface

Beyond the fact that the interior surface of the chamber in which the aphids live is lined by an inner epidermis (or epidermis-lumen)—generally multiseriate with about 3–5 cell layers, rich in tannins—better or worse defined, two situations can be observed: the inner surface has not holes (Fig. 8b) or the inner surface has holes between epidermal cells (Fig. 8g) (Álvarez et al. 2016). Generally, a centripetal gradient of tannins, from homogeneous and dense tannins in epidermal cells and reducing toward the cells in direct contact with the chamber, where the tannins are fine-grained or absent (Fig. 8c). The inner surface is usually formed by cells with protuberances, which provide a padded appearance on the surface (Fig. 8b, c)

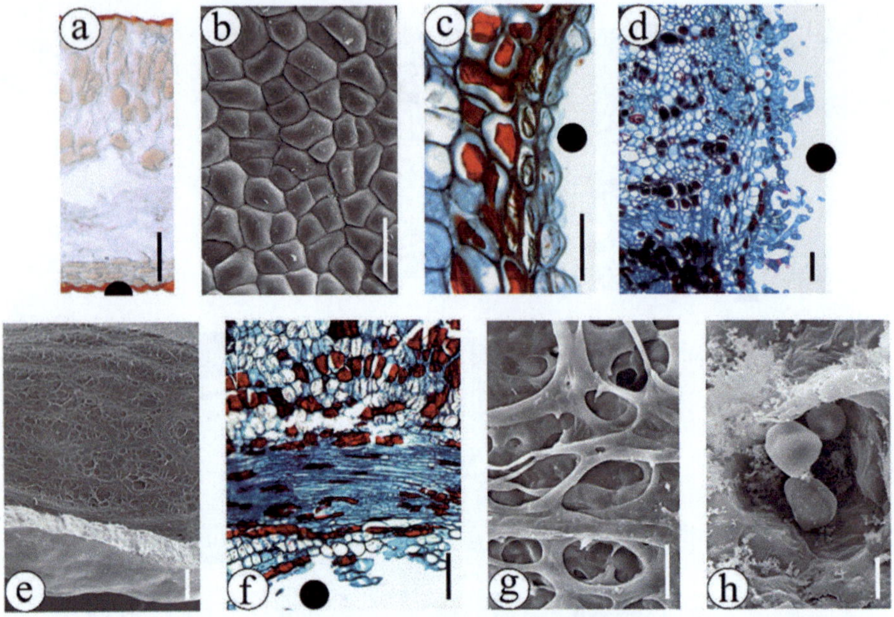

Fig. 8 Epidermis–lumen. (**a**) Presence of lipid cuticle in both the epidermis–air and the epidermis–lumen. Gall induced by *Forda formicaria*. (**b**) Quilted look, no gaps. Gall induced by *Forda marginata*. (**c**) Gradation of tannins in the multiseriate epidermis. Gall induced by *Forda marginata*. (**d**) Presence of trichomes. Gall induced by *Pemphigus populinigrae*. (**e**) Interior of gall, showing unpadded surface. Gall induced by *Tetraneura ulmi*. (**f**) Discontinuity in the epidermis. Gall induced by *Pemphigus populi*. (**g**) Characteristic cavities of the chamber surface. Gall induced by *Tetraneura ulmi*. (**h**) Presence of molasses balls in cavities. Gall induced by *Baizongia pistaciae*. *Symbol*: circle and semicircle—gall chamber. *Bars (micrometers)*: 100 – 50 – 50 – 100 – 500 – 100 – 50 – 10

(Álvarez et al. 2009). The epidermis usually has a cuticle (Fig. 8a), sometimes trichomes (Fig. 8d), but stomata are not observed. When present, the multiseriate and irregular epidermis may help the aphids to avoid the water entrance from the gall wall to the gall chamber, since in this surface, the aphids secrete waxes, creating a hydrophobic surface (Pike et al. 2002; Álvarez et al. 2020). Trichomes may also be observed in the epidermis-lumen, and they are live, unicellular, or multicellular. As the multiseriate epidermis, the trichomes tend to be superficially impregnated with the wax secreted by aphids, potentiating the hydrophobicity of the inner surface (Uematsu et al. 2018; Kutsukake et al. 2019; Álvarez et al. 2020).

A poorly defined or absent epidermis-lumen and the presence of gaps on the inner surface of the gall (Fig. 8e–g) are related to the issue posed by honeydew and the molts of many individuals within a closed space in aphid-induced galls (Fig. 13f–h). Therefore, a closed gall is commonly associated with small cavities (holes) on the inner surface (Álvarez et al. 2016, 2020), from where the honeydew and debris may be evacuated until reaching the vascular bundles (Kutsukake et al. 2012, 2019; Álvarez et al. 2014, 2016). Additionally, gall-inducing aphids have wax glands with which they encase the honeydew they expel, preventing the gall inhabitants drowning in their own honeydew. Thus, in some microscopic images, rounded forms of molasses (molasses balls) are observed in the cavities (Fig. 8h). By surrounding the molasses with wax, they also prevent it from becoming a breeding ground, particularly for fungi (Heie 1987).

In open galls there are several alternatives to the accumulation of the honeydew and debris: the honeydew surrounded by wax can be crushed and thus contribute to the padding of the interior; it can also be pushed to the door; and also, in large galls, it can be pushed toward the center of the chamber, thus remaining, as a more or less large mass inside the gall (Álvarez et al. 2009, 2020). In closed galls, the storage abovementioned in the center of the chamber can also be noticed, although it is common for debris to be evacuated through the cavities (small holes) present on the inner surface.

2.3 *Galls on Leaf-Blade, Petiole and Stem*

2.3.1 Leaf-Blade Galls

Most aphid–induced galls are folded leaf-blades (laminar galls) or are located in the midrib (vascular galls). In laminar galls formed as a folding of the leaf blade, the midrib is usually not involved, and the histological organization of the host leaf blade is not recognized. The characteristic mesophyll of the non-galled leaf is modified in a more or less homogeneous set of hyperplasic storage parenchyma with hypertrophied cells (Fig. 9a, b) (Álvarez et al. 2008, 2009, 2013, 2020). The vascular bundles of the galls will follow the organization in the host leaf blade. Commonly, the xylem is oriented adaxially and the phloem oriented abaxially. Depending on whether the fold occurs toward the upper surface or toward the underside of the leaf, the xylem or the phloem will be oriented to the gall chamber (Álvarez et al. 2008, 2020) (Fig. 9c, d). Similarly, the leaf adaxial or abaxial epidermis will cover the gall chamber, depending on whether the fold is abaxial or adaxial. The pseudogalls, or rolling galls, are also laminar galls, and distinct levels of tissue modification may be observed (Chen and Qiao 2012; Álvarez et al. 2013, 2020) (see Fig. 2a–c). The galls occurring on the midribs can be called vascular galls or midrib galls (to differentiate them from the previous ones: the laminar galls). The inducer intervenes on the main vein, the midrib, and the gall is a result of cell hypertrophy and hyperplasia of the midvein tissues (Fig. 9e, f) (Álvarez 2011, 2012). As in the laminar galls, the wall of the midrib galls is formed by large storage parenchyma cells, and the vascular bundles orientation is determined according to the position of gall formation—adaxial or abaxial.

A bundle sheath formed by sclerenchyma fibers with a supporting function usually surrounds the vascular bundle of the non-galled midribs (Fig. 9g). Such bundle sheath usually does not appear on the galls (Fig. 9h), because if it occurred, it would be a physical impediment for aphids to feed (Álvarez et al. 2020). The sclerenchyma

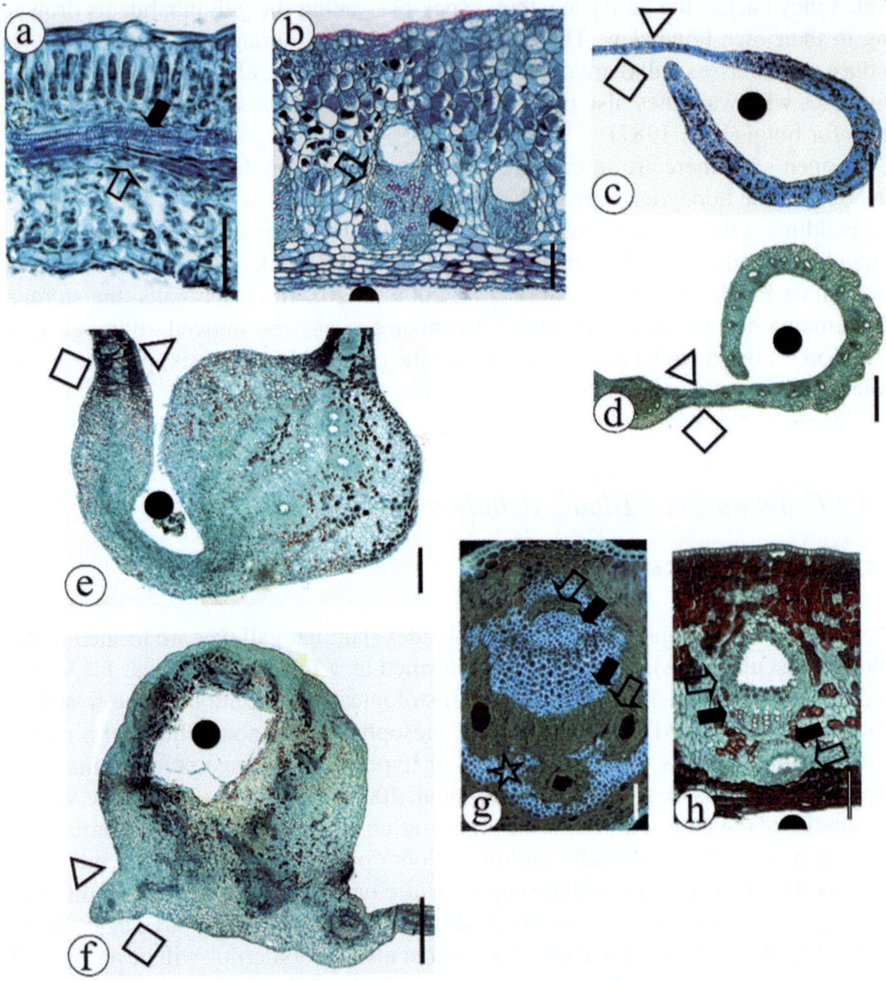

Fig. 9 Leaf-blade galls. (**a**) In the leaf blade, the second order veins present poorly developed vascular tissues. The xylem is arranged towards the upper side and the phloem towards the underside. Note the presence of palisade chlorophyll parenchyma above and aeriferous parenchyma below. *Pistacia atlantica* (control leaf blade). (**b**) The laminar gall wall is occupied by storage parenchyma and hypertrophic and hyperplasic vascular bundles. The parenchyma of the leaf blade is not observed. Gall induced by *Geopemphigus torsus*. (**c**) Laminar gall formed by a fold towards the leaf underside. Temporal gall induced by *Thecabius affinis*. (**d**) Laminar gall formed by a fold towards the leaf bundle. Gall induced by *Smynthurodes betae*. (**e**) Vascular gall induced towards the leaf underside (abaxial surface). Young gall induced by *Geoica utricularia*. (**f**) Vascular gall induced towards the leaf bundle. Gall induced by *Pemphigus populi*. (**g**) Midrib with two vascular bundles surrounded by bundle sheath. Above the small supernumerary vascular bundle. *Pistacia terebinthus* (control midrib). (**h**) Gall wall with two vascular bundles and no bundle sheath. Gall induced by *Baizongia pistaciae*. *Symbols*: arrow—xylem, circle and semicircle—gall chamber, hollow arrow—phloem, star—sclerenchyma, square—abaxial zone, triangle—adaxial zone. *Bars (micrometers)*: 50 – 100 – 500 – 500 – 200 – 1000 – 100 – 100

fibers with lignified secondary walls (practically without pits) involves the vascular bundles in clusters or continuously, forming a hard wall that aphid stylets would have to pass through to access the phloem.

It is noteworthy that there are host plants in which two (or more) vascular bundles are located (phloem-xylem/xylem-phloem) separated by parenchymatic cells, vascularizing the midvein of the leaf (Fig. 9g). In the wall of the galls induced there, two vascular bundles (phloem-xylem/xylem-phloem) will also be observed (Fig. 9h), forming the vascularization of the wall (Álvarez 2011, 2012; Álvarez et al. 2014, 2016).

2.3.2 Petiole and Stem Galls

Aphid galls may be formed in petioles because of its rolling, that is, the petiole suffers several turns around itself, which leaves an interior space in which the aphids live (Fig. 10a, b). On other interactions, the galls appear as lateral petiole or stem neoformations, with a chamber where the aphids live (Álvarez et al. 2020) (Fig. 10c). In all instances, the regions proximal to the gall chamber undergo morphological

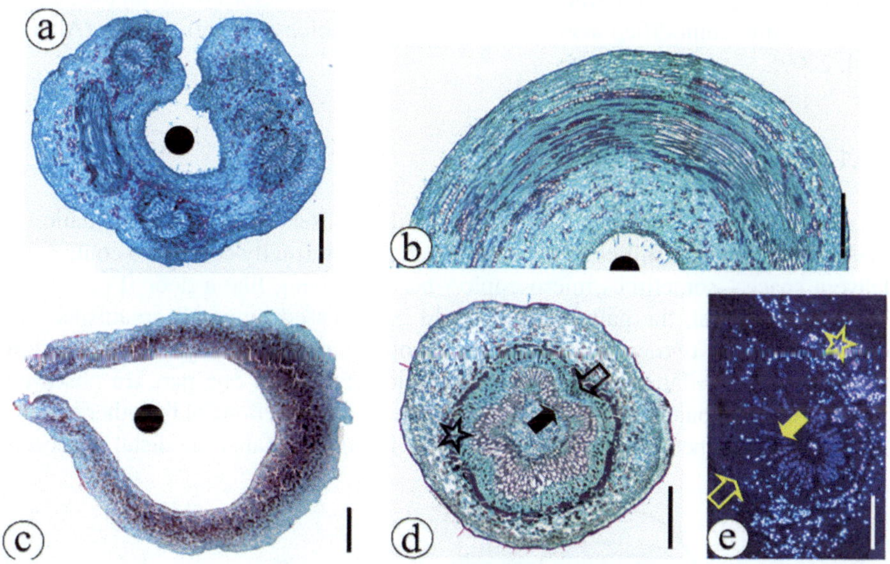

Fig. 10 Galls on petiole and stem. (**a**) Petiole gall. Gall induced by *Pemphigus spyrothecae*. (**b**) The area close to the chamber presents altered morphology compared to the outer part. Gall induced by *Pemphigus protospirae*. (**c**) Stem gall. Gall induced by *Pemphigus immunis*. (**d**) Stem with bundle sheath surrounding the vascular system. *Populus nigra* (control). (**e**) Periphloematic vascular bundle without bundle sheath in the area facing the gall chamber (left). Gall induced by *Pemphigus bursarius*. *Symbols*: arrow—xylem, circle and semicircle—gall chamber, hollow arrow—phloem, star—sclerenchyma. *Bars (micrometers)*: 500 – 500 – 1000 – 500 – 500

alterations, whereas the areas distal from the gall lumen exhibit morphological similarities to the non-galled petiole or stem (Fig. 10d).

Generally, the vascular bundles of the petioles and stems present a bundle sheath (Fig. 10d), which is not observed intact in the galls. In certain galls, instead of a compact cluster of sclerenchyma fibers, bundles of fibers or groups of sclereids separated from each other by parenchyma cells are observed (Fig. 10e) (Álvarez et al. 2020). These groups of lignified cells are not supposed to hinder the access of the aphid stylets since they use the parenchymatic spaces between the hard cells to advance toward the phloem.

2.4 *Parts to Consider in the Histological Study of Open Galls*

In a study of an open gall induced by aphids, one must consider (I) the gall chamber where aphids live, (II) the closure zone, and (III) the reaction zone (Fig. 11). The closure zone—which characterizes this type of galls—is an area that allows the communication of the chamber with the outside, and through which the winged individuals emerge. It is also a possible access route for intruders into the gall (Álvarez et al. 2009). The reaction zone is a part adjacent to the closure zone. In some cases, it presents intermediate histological characteristics between the closure zone and the unmodified zone of the organ in which the gall occurs (Álvarez et al. 2009).

2.4.1 Closure Zone

The closure zone of the galls induced by aphids is observed as an approximation of the walls at two extremities of the gall, which do not usually come into contact and leave a space—sometimes microscopic—between them, like a door (Fig. 12a–f). Through this door, the gall chamber could be accessed by small organisms. The approximation of two wall extremities is accompanied by an orographic adaptation of the two parts: flattening or protrusions and recesses of one part are observed, which are accompanied by flattening or recesses and projections of the other extremity (Fig. 12a). When the closure zone occurs in laminar galls, the distal portion of

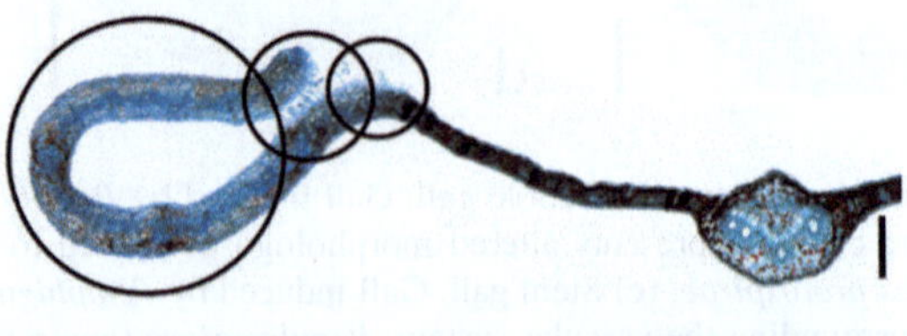

Fig. 11 Parts of open aphid gall. From left to right: gall chamber, closure zone, reaction zone. On the right, unmodified leaf blade and midrib. Young gall induced by *Forda formicaria*. *Bar (micrometers)*: 300

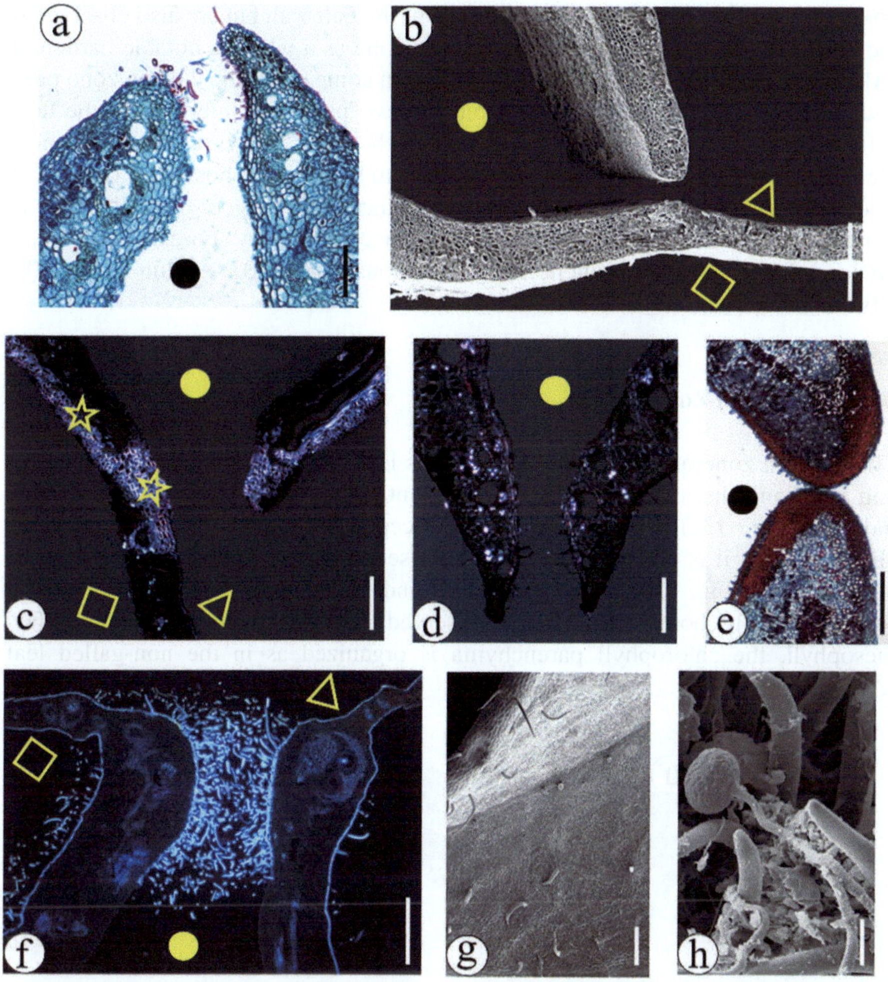

Fig. 12 Closure zone. (**a**) The two parts that intervene in the closure zone adapt orographically to each other. Note the scarce presence of trichomes. Gall induced by *Geopemphigus floccosus*. (**b**) Closure zone without trichomes. Gall induced by *Forda marginata*. (**c**) Sclereids at the top and bottom of the wall. Gall induced by *Forda marginata*. (**d**) Abundant presence of crystals. Gall induced by *Geopemphigus floccosus*. (**e**) Accumulation of tannins. Gall induced by *Pemphigus spyrothecae*. (**f**) Lots of trichomes in the closing zone. Gall induced by *Asiphonella cynodonti*. (**g**) Host plant trichomes. *Pistacia terebinthus* (control). (**h**) Spore retained between trichomes. Gall induced by *Forda formicaria*. *Symbols*: circle—gall chamber, star—sclerenchyma, square—abaxial zone, triangle—adaxial zone. *Bars (micrometers)*: 200 – 500 – 300 – 200 – 500 – 500 – 200 – 20

the host leaf in the gall loses the microscopic features of the leaves (Álvarez et al. 2008, 2009, 2016, 2020). As in the gall wall, in the closure zone it is possible to observe accumulation of crystals and tannins (Fig. 12d, e), related as there, with defense. In galls particularly hardened by the presence of sclereids on the wall, sclereids are also observed in the closure zone (Fig. 12c). These sclereids are not observed restricted to the upper part (distal) of the gall wall but are also observed in the lower part (proximal), giving the closure zone as a whole a notable hardening (Álvarez et al. 2009, 2016). It is frequent that in some galls, the closure zone presents trichomes that usually emerge from the two parts that intervene in the door (Fig. 12a, d–f). They tend to mingle with each other. They are commonly non-glandular, living unicellular trichomes, similar to those of the host plant (Fig. 12g). The participation of these trichomes in the mechanical defense of the gall, by forming a plug and, of course, as an impediment to access (also by spores) (Fig. 12h), is obvious (Meyer and Maresquelle 2003; Simmons et al. 2003; Arduin et al. 2005; Álvarez et al. 2009, 2013, 2016, 2020).

2.4.2 Reaction Zone

The reaction zone of galls induced by aphids is an intermediate zone between the gall itself and the unaffected part of the plant organ on which the gall has been induced (Fig. 13a). The reaction zone occurs particularly in laminar galls. Histologically, it is characterized by the presence of trichomes (if present in the closure zone) (Fig. 13b, c) and a compact and thickened mesophyll. Beyond the reaction zone, trichomes are no longer observed (Fig. 13b, c), and in the unmodified mesophyll, the chlorophyll parenchyma is organized as in the non-galled leaf (Álvarez et al. 2009).

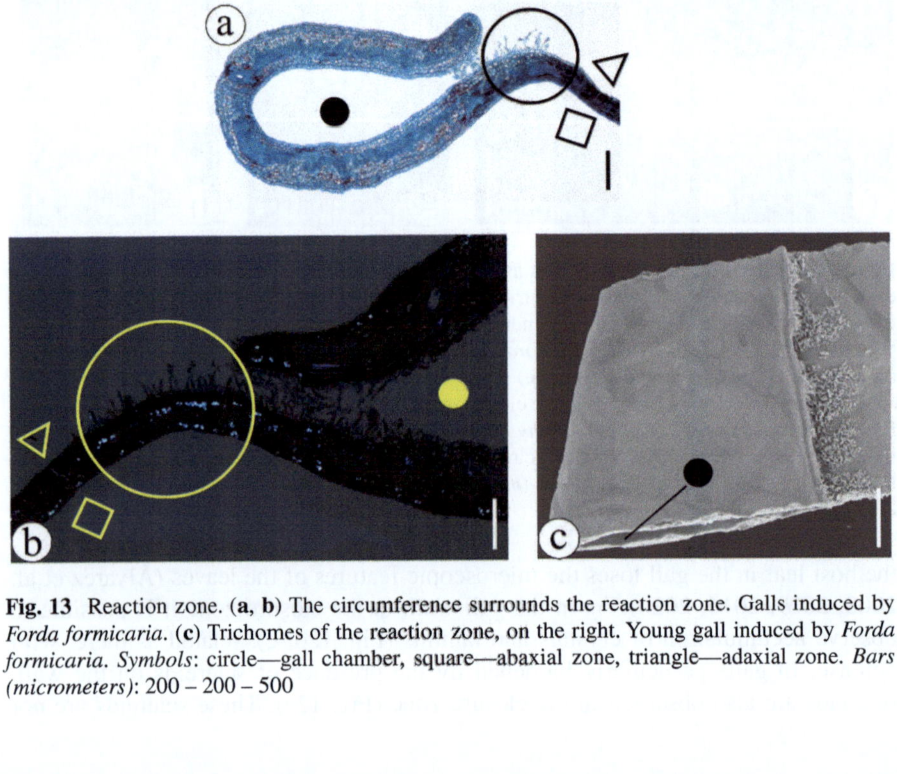

Fig. 13 Reaction zone. (**a, b**) The circumference surrounds the reaction zone. Galls induced by *Forda formicaria*. (**c**) Trichomes of the reaction zone, on the right. Young gall induced by *Forda formicaria*. *Symbols*: circle—gall chamber, square—abaxial zone, triangle—adaxial zone. *Bars (micrometers)*: 200 – 200 – 500

2.5 The Defense Mechanisms of Aphid-Induced Galls

The histological aspects of the aphid galls serve as defense mechanisms against potential natural enemies. On the one hand, they deter possible phytophagous insects, and on the other hand, they thwart potential predators attempting to gain access to the interior of the gall chamber to feed on aphids. To gain access to the space in which aphids live, intruders must break through the gall wall. In open galls, the enemies may access the gall chamber through the closure zone. The trichomes in the closure zone of certain galls are arranged as a physical barrier that prevents or hinders the intruder's access to the interior of the gall. In any case, the gall inducer uses the defense structures of the plant for its own protection against its own enemies (Martinez et al. 2018).

The galls' defense mechanisms against possible attack can be divided into two types: physical defense and chemical defense. The first factor to consider within the defense is the great thickness of the gall wall, especially for intruders who intend to access the interior of the gall by breaking through the wall (Wool and Ben Zvi 1998). Crystals and lignified cells can also help in physical defense. These are non-digestible elements, which must be separated by the phytophagous or the possible intruder. The presence of microcrystals in the epidermis-air, prismatic crystals, and druses (very frequently more abundant in the part of the wall closest to the outside) play this role (Fernandes et al. 1989; Arduin and Kraus 1995; Álvarez et al. 2009, 2020). The lignified cells serve as physical barrier, comprising an indigestible element known as lignin. This defense mechanism involves the presence of hard sclereid layers in certain galls, and in all cases, the hypertrophied and hyperplasic vascular bundles, especially their xylem component (Álvarez et al. 2009, 2016; Álvarez 2012).

Within the chemical defense of the galls induced by aphids, tannins play a decisive role. In general, tannins precipitate proteins (for example those of the saliva of herbivores) and therefore prevent or make it difficult for insects (and fungi) to feed on the walls of the galls, which are very rich in tannins (Nyman and Julkunen-Tiitto 2000; Álvarez et al. 2009; Inbar et al. 2003; Martinez 2010; Ferreira et al. 2019). The galls induced in species belonging to the Anacardiaceae family deserve special mention. As already indicated, these galls have a hypertrophied phloem, with schizogenic intraphloematic ducts, through which a sticky resin, rich in terpenes, circulates, which can be a natural repellent for herbivores (Fahn 1990; Watson and Dallwitz 1992; Álvarez et al. 2009; Álvarez 2012).

2.6 Galls as Taxonomic Tools

Assuming that the galls are extended phenotypes of the inducing species, their macroscopic and microscopic study is one more tool to carry out taxonomic and phylogenetic studies. This is the case of the doubtful taxonomic location of certain genera of aphids that induce galls on *Pistacia* species. In a certain phylogenetic arrangement, the species in question are grouped with others, whose galls present few histological similarities to each other. In another taxonomic arrangement, however, they appear grouped with microscopically similar gall-inducing species.

Given such a doubtful taxonomic position of the genera, the histological characteristics of the galls they induce were considered, to finally create groups (clusters): considering the morphological and histological attributes as variables, the statistical study was carried out, which led to obtaining hierarchical clusters based on to histology. The groups obtained were consistent with one of the taxonomic classifications that are considered when classifying the species studied (Álvarez et al. 2009, 2014, 2016; Álvarez 2011, 2012; Muñoz-Viveros et al. 2014).

Subsequently, in galls induced by aphids in the genus *Populus*, the importance of microscopic studies of galls induced by aphids in their taxonomy was confirmed. The clusters obtained, coinciding with the classification in use, allow the microscopic attributes of the galls they induce to be added to the description of said species (Álvarez et al. 2020).

2.7 *Natural Enemies in Aphid Galls. The Case of* **Alophia combustella**

Aphid–induced galls on plants of the genus *Pistacia* are attacked by few insects belonging to the orders Hymenoptera, Diptera and Lepidoptera (Wool 2005; Martinez et al. 2018). Among these insects, there is the larval phase of the lepidopteran *Alophia combustella* (Garrido Briñón et al. 2009). The data on the biology and ecology of *A. combustella* are scarce and even contradictory regarding the feeding habits of its larva: some authors defend that it is exclusively phytophagous, while others affirm that its diet is basically predatory. *Alophia combustella* larva is found with some frequency within galls induced by various species of aphids. In some cases, they are galls that have a single vascular bundle on the wall, and in some of them, a hard layer of sclereids, and in other cases, these galls have two vascular bundles. In all cases –as indicated in a previous section—the wall is composed mainly by storage parenchyma (Álvarez et al. 2016). After histologically studying the galls in which the larva was found, absent wall areas were observed, presumably eaten by the intruder (Garrido Briñón et al. 2009). Absence of the epidermis and of a portion of the parenchyma is observed (Fig. 14a–c). Such cells, eaten by the larva, are softer living cells with primary walls and particularly rich in amyloplasts. However, the larva does not seem to interfere with the vascular bundles, stopping before the hard layer of sclereids, formed by dead cells with lignified secondary walls (Fig. 14d). It is appropriate to indicate that the larva reaches the chamber where the aphids are, from the outside of the gall, that is, crossing the wall. The fact that the apparent herbivorous action stops before the barrier that the layer of sclereids supposes allows us to venture that the larva penetrates young galls when they do not yet have sclereids. The intervention of the larva in the galls that present

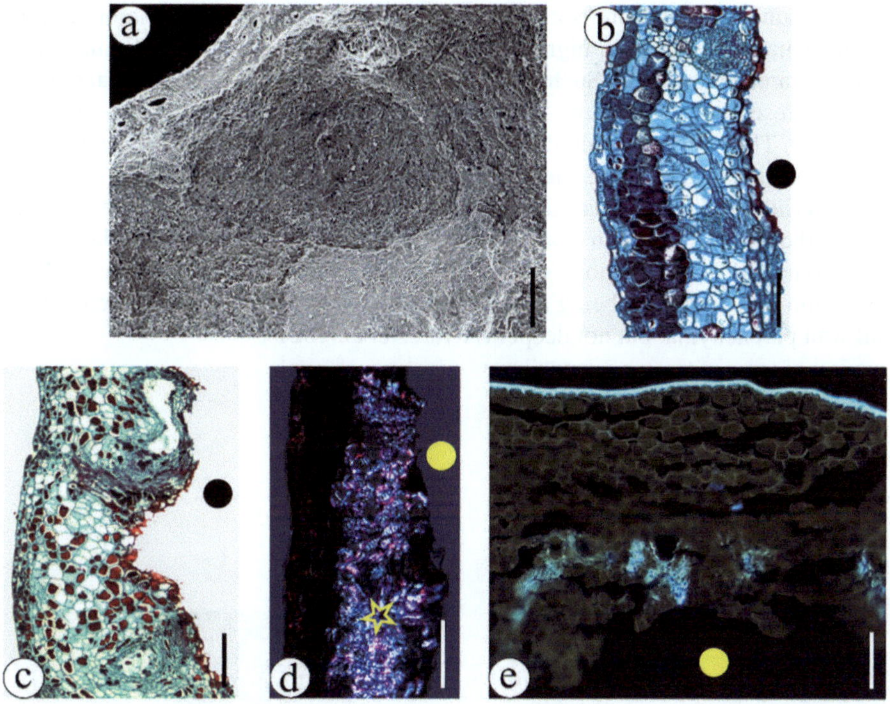

Fig. 14 Gall walls altered by the action of *Alophia combustella* larvae. (**a**) Inside of a gall with devastated areas. Gall induced by *Geoica utricularia*. (**b**) Absence of epidermis. Gall induced by *Paracletus cimiciformis*. (**c**) Absence of storage parenchyma. Gall induced by *Forda riccobonii*. (**d**) The action of the larva appears to stop at the sclereid layer. Gall induced by *Smynthurodes betae*. (**e**) Absence of vascular bundles close to the chamber. Gall induced by *Geoica utricularia*. *Symbols*: circle—gall chamber, star—sclerenchyma. *Bars (micrometers)*: 500 – 100 – 200 – 100 – 100

two vascular bundles is also characterized by the absence of parenchyma, as well as the small vascular bundle closest to the gall chamber (Fig. 14e).

2.8 Similarities Between Aphid and Phylloxerid Galls

Phylloxera is a disease that decisively affected viticulture throughout the world, since the mid–nineteenth century. It produced devastating effects on cultivars (Benheim et al. 2012). Today it is considered a controlled disease. It is a condition caused by a phylloxerid hemipterous insect: *Daktulosphaira vitifoliae*, which feeds on the roots of *Vitis*, sucking the phloem sap. Some generations of the insect induce the formation of galls on the leaves (Kingston et al. 2007).

The leaf galls are macroscopically and microscopically similar to those induced by other aphids: they are open galls that show hypertrophy and hyperplasia of the vascular bundles (which are conspicuous far from the chamber), and homogeneous parenchyma (Fig. 15), with high concentrations of starch and lipids (Raman et al. 2009; Álvarez et al. 2021). Such galls are formed toward the abaxial surface of the leaves, presenting the xylem—and not the phloem—closest to the chamber (Fig. 15e). This fact, which could cause difficulties for aphids as their food source is far from the place where they live, does not affect *Daktulosphaira vitifoliae*: Phylloxeriidae (like adelgids), as it feeds by sucking the contents of the nutritive cells in the gall wall (Raman et al. 2009; Álvarez et al. 2021). It is not trivial that, in the entomological descriptions of *D. vitifoliae*, it is indicated that its stylets have 60–65 μm in length, thus being able to reach up to five or six layers of cells of the gall wall parenchyma, but not deeper (Álvarez et al. 2021).

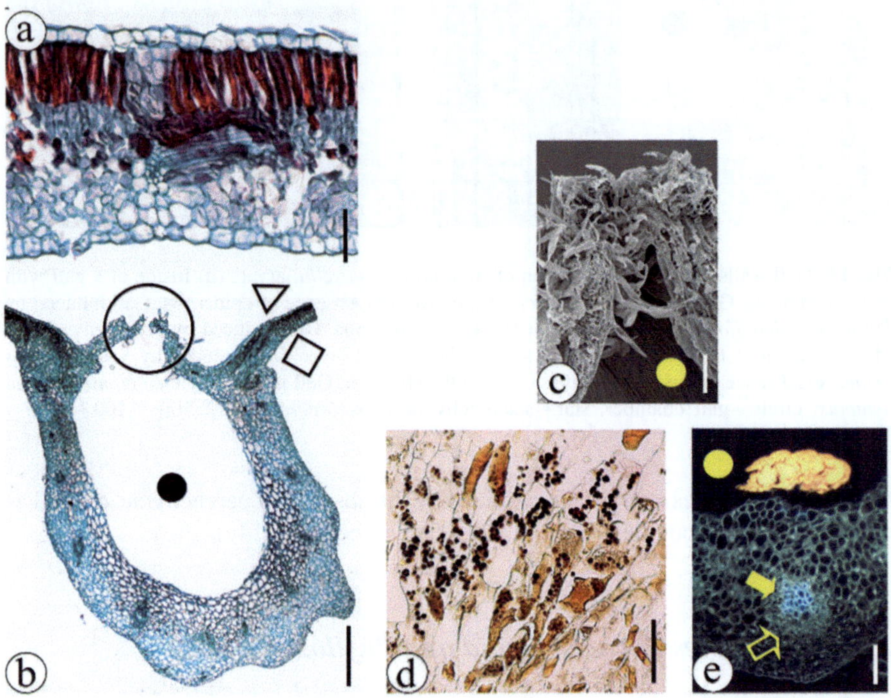

Fig. 15 Phylloxera leaf galls. (**a**) Leaf lamina of the host plant, presenting palisade chlorophyll parenchyma in an adaxial arrangement and aeriferous parenchyma in an abaxial arrangement (Álvarez et al. 2022). *Vitis vinifera* (control). (**b–e**) Gall induced by *Daktulosphaira vitifoliae*. (**b**) Open gall facing the underside of the leaf. The wall is not reminiscent of the *Vitis* leaf blade. The circumference marks the closure zone. (**c**) Detail of the closure zone with trichomes. (**d**) Presence of amyloplasts in the wall. (**e**) The vascular bundle is shown away from the gall chamber, with the xylem being above the phloem. *Symbols*: arrow—xylem, circle—gall chamber, hollow arrow—phloem, square—abaxial zone, triangle—adaxial zone. *Bars (micrometers)*: 50 – 500 – 200 – 100 – 100

2.9 *Methods Used to Study the Microscopy Morphology of Aphid-Induced Galls*

To the morphological and histological studies of aphid galls, the samples obtained in the field are fixed in some fixative, usually FAA (formalin, acetic acid and ethyl alcohol), and processed to be embedded in a paraffin (e.g., Álvarez et al. 2009). Unfixed or physically fixed samples in liquid nitrogen may be cut directly in a cryostat.

The preparation of the blocks is carried out by orienting the samples to obtain cross sections of the midrib of the leaf, petiole or branch where the galls appear. The sections obtained are stained with routine techniques, for example, with Safranin–Fast Green (Fig. 16a), which allows distinguishing between cells with primary walls and cells with secondary walls. Histochemical techniques are used for the detection of certain elements such as starch, lipids, and tannins (Fig. 16b–d). For histological staining, sections are sometimes treated with sodium hypochlorite to remove cell contents (see Fig. 3).

Observation is carried out with a bright field optical microscope. Polarized light microscope is also used because it allows the observation of anisotropic elements, particularly crystals and cells with secondary walls (xylem and sclerenchyma) (Fig. 16e). In addition, an epifluorescence microscope is used, attending to the

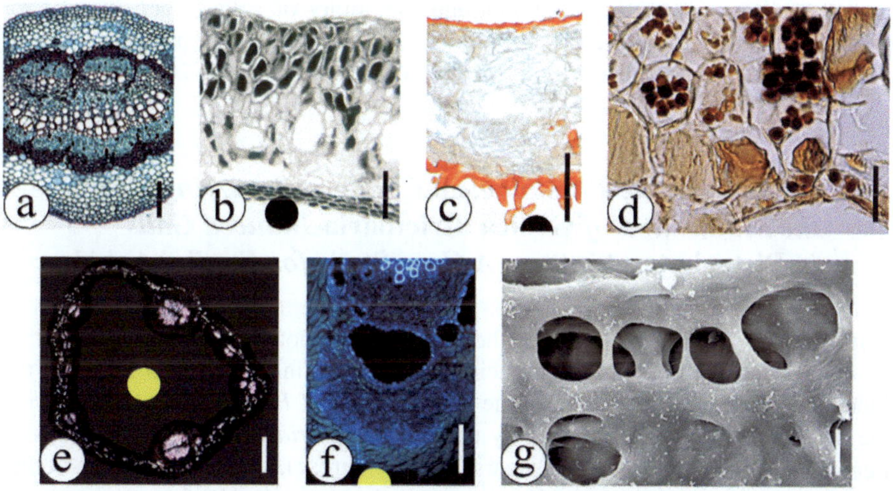

Fig. 16 Methods used to histological study aphid-induced galls. (**a**) Safranin–Fast Green stain. Midrib of *Pistacia terebinthus* (control). (**b**) Histochemical detection of tannins. Gall induced by *Forda formicaria.* (**c**) Histochemical detection of lipids. Gall induced by *Forda formicaria.* (**d**) Histochemical detection of starch. Gall induced by *Forda formicaria.* (**a–d**) Brightfield optical microscope. (**e**) Polarized light microscope. Gall induced by *Slavun wertheimae*. (**f**) Epifluorescence microscope. Gall induced by *Baizongia pistaciae*. (**g**) Scanning electron microscope. Gall induced by *Geoica utricularia. Symbols*: circle and semicircle—gall chamber. *Bars (micrometers)*: 100 – 200 – 100 – 300 – 500 – 100 – 20

autofluorescence of plant cells (Fig. 16f). The scanning electron microscope allows the study of surfaces (Fig. 16g) (Álvarez et al. 2009, 2020).

3 Physiological Aspects of Aphid Galls

Various hypotheses have been proposed to explain the adaptive value of the cecidogenic habit. The most experimentally supported hypothesis is the nutritional one (Price et al. 1987; Stone and Schönrogge 2003; Wool 2005). According to it, the gall-inducing aphid would induce multi-level changes (cytological, morphological, and biochemical) in the host plant during the cecidogenesis, which provide itself a rich source of food for its development inside the gall. In line with the nutritional hypothesis, several studies carried out in recent years indicate that the phytophagous galling insect induces the formation of the galls, which are phloem sinks for photo-assimilates and nutrients (Carneiro et al. 2014; Castro et al. 2012; Compson et al. 2011; Dsouza and Ravishankar 2014; Hartley 1998; Huang et al. 2014, 2015; Kmiec et al. 2018; Larson 1998; Larson and Whitham 1991; Oliveira et al. 2017).

Previous studies on the physiology and development of aphid-induced galls have focused on species of the tribes Pemphigini (Kot and Kmiec 2020; Larson and Whitham 1991) and Eriosomatini (Kot and Kmiec 2020; Larson and Whitham 1991). These studies demonstrated the negative impact of galling action on the photosynthetic capacity of the host plant and the ability of galls to act as sinks for photo-assimilates. Nevertheless, only limited studies of this type have been carried out in Fordini (Burstein et al. 1994; Inbar et al. 2004; Chen et al. 2020).

3.1 *Monitoring the Physiological Changes Occurring During the Maturation of* Geoica utricularia*-Induced Galls in* Pistacia terebinthus*: A Case Study for Fordini Aphids*

With the aim to demonstrate that the nutritional hypothesis fits the relationship among galling Fordini aphids and their hosts, a preliminary characterization of the biochemical changes that occur in the leaflet tissues of *Pistacia terebinthus* during the development of galls induced by *Geoica utricularia* was carried out. For this purpose, levels of water-soluble sugars (mainly sucrose), starch, nitrogen and other mineral nutrients were compared between galled and ungalled leaf tissues (controls) throughout gall development. In this study, two types of ungalled controls were considered: the gall-free part of the leaflet in a leaflet containing a gall (control 1-C1-; Fig. 17) and distant leaf tissues: the gall-free leaflets in leaves where at least one leaflet was affected by a gall (control 2 -C2-; Fig. 17).

At the morphological and histological level, the differences between gall and non-gall-bearing leaflet tissues are very evident (Álvarez et al. 2008), as discussed

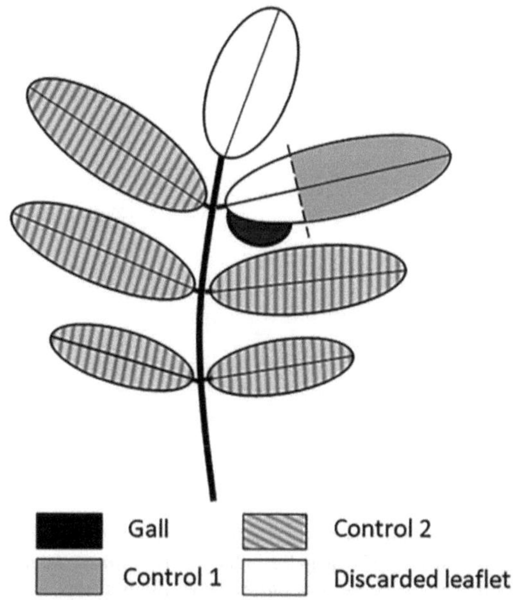

Fig. 17 Experimental scheme of galled leaf. *Pistacia terebinthus* L. leaf with a gall induced by *Geoica utricularia* distinguishing the gall tissue (Gall), the leaflet affected by the gall (Control 1, C1), the healthy leaflets (Control 2, C2), and the leaflets discarded for physiological analysis

before. Comparison of the galls with the selected controls allow us to find out changes at the biochemical level and to what extent these changes are localized. In addition, the sampling procedure was defined so that the young, semi-mature, and mature galls were collected to coincide with different milestones of insect colony life cycle (Fig. 18).

Water soluble sugars significantly increased throughout gall development (Fig. 19a; Table 1), as already observed in other insect galls (Giron et al. 2016; Oliveira et al. 2017). Additionally, soluble sugars sampled in gall tissues were significantly higher than those from ungalled tissues (both C1 and C2), independently of the stage of gall development. At late stages of gall development, soluble sugars assayed in mature galls were 2.3–1.5-fold higher than that assayed in ungalled tissues, which reflect the sink effect of the galls (Inbar et al. 2004). Galls of *G. wertheimae* import photosynthates from leaflets below their position and from nearby leaves (Burstein et al. 1994), although the source of resources varies among Fordini species, as their sink strength is different (Burstein et al. 1994; Larson 1998).

In contrast to soluble sugars, starch did not vary much during gall development in *G. utricularia* induced-galls (Fig. 19b; Table 1). Throughout all sampling periods, even at very early stages of gall development, starch accumulated to a greater extent in galling tissues than in ungalled leaflet tissues (Both C1 and C2). It appears that galls accumulate starch as a reserve to translocate sugars to the phloem to serve as food for galling insects as already observed by Oliveira et al. (2017). By using lugol-staining, Álvarez (2012) observed that the relative amount of amyloplasts was higher in galls induced by *G. utricularia* than in the ungalled tissues, although they did not follow a clear trend during their development, which agrees with the results

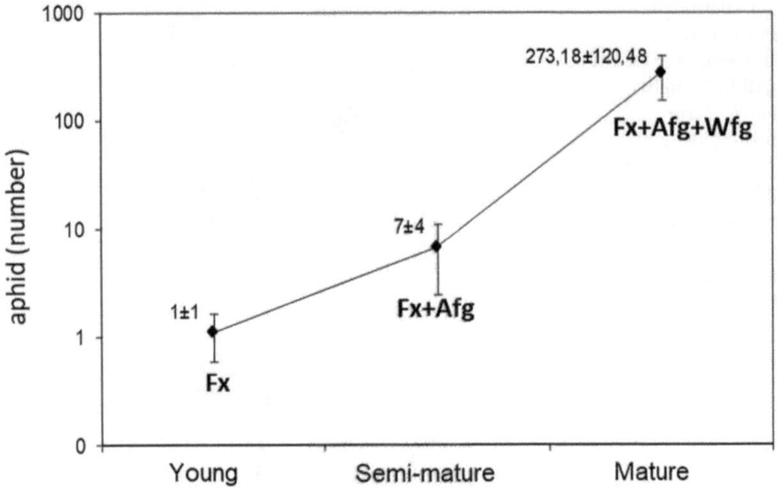

Fig. 18 Population of aphids in young, semi-mature and mature galls. Population dynamics of *Geoica utricularia* in galls induced on ***Pistacia terebinthus*** (Young and semi-mature, n = 61; Mature, n = 52). *G. utricularia* generations found in galls sampled: Fx fundatrix, Afg apterous fundatrigeniae, Wfg winged fundatrigeniae

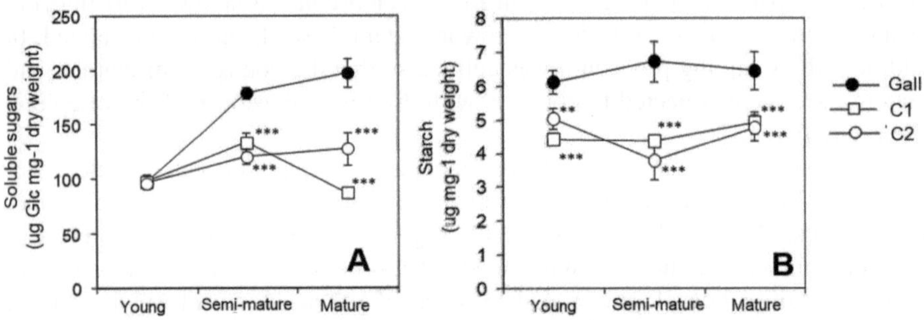

Fig. 19 Carbohydrate content in galls and controls. Variation of (**a**) soluble sugar and (**b**) starch content on *G. utricularia* induced galls, control 1 (C1; see Fig. 17) and control 2 (C2; see Fig. 17) at different stages of development (young, semi-mature and mature). Mean ± standard deviation is represented (n = 3). Asterisk indicate significant differences when compared with gall tissue at the same developmental stage (*** p < 0.001; ** p < 0.01; * p < 0.05)

Table 1 Statistics of the analysis of variance to test the effect of the tissue analyzed (gall, C1, C2) and developmental stage (young, semimature and mature) and on the starch content, soluble sugars. Significant differences are indicated with asterisks

	Starch		Ln (Soluble sugars)	
	F	*p*	*F*	*p*
Tissue	85.17	<0.001*	7.34	<0.001*
Stage	3.05	0.064	79.39	<0.001*
Interaction	5.85	0.002*	32.33	<0.001*

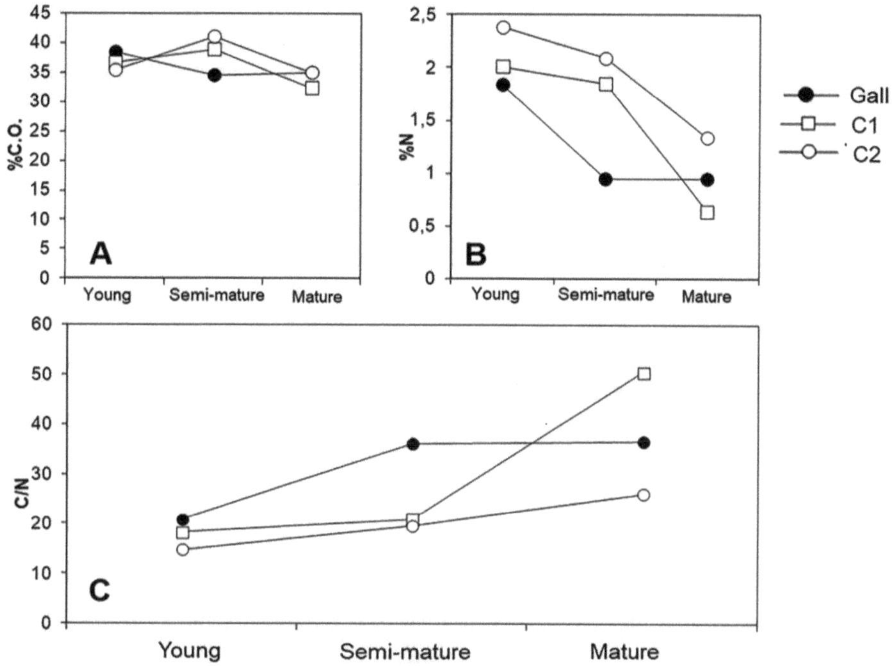

Fig. 20 Carbon and nitrogen content in galls and controls. Variation in (**a**) C, (**b**) N, and (**c**) C/N ratio on *G. utricularia* induced galls, control 1 (C1; see Fig. 17) and control 2 (C2; see Fig. 17) at different stages of development (young, semi-mature and mature). Abbreviation: % C.O. = organic carbon, percentage of dry matter

presented here. Other Fordini species that induce semi-closed galls on the leaf margins, such as *P. cimiciformis, F. formicaria and F. marginata*, show a constant relative amount of amyloplasts, but starch seems to be more abundant in young galls than in mature ones (Álvarez et al. 2009), highlighting different physiological behavior depending on the phylogenetic group (Dsouza and Ravishankar 2014; Oliveira and Isaias 2010).

Taken together, the increased content of water-soluble sugars and starch in gall tissues compared to nearby (C1) and distant (C2) ungalled tissues reinforce the notion that *G. utricularia* galls function as an active sink (Burstein et al. 1994; Inbar et al. 2004). The significant decrease in water soluble sugars in leaflet tissues close to the gall (C1) may reflect that *G. utricularia*-induced galls use the unaffected area of the leaflet on which they settle as a preferential nutrient source.

The time-course of C and N variation in galling tissues, C1 and C2, was also quantified (Fig. 20). Total C content did not vary between galled and ungalled tissues or along gall development (Fig. 20a). However, a marked decrease of N was assayed in galled and ungalled tissues along gall development (Fig. 20b). As a result, the C/N ratio increases in all tissues analyzed (Fig. 20c). The C/N ratio was higher in the gall compared to C2 controls (Fig. 20c) throughout the whole gall development (Fig. 20c). This result is a consequence of the marked decrease in N content observed in the galls, agreeing with previous findings based on a survey of several galls from different galling insects (Hartley 1998). Whether the decrease in N is reflecting the extractive activity of the aphid colony or is a consequence of the colony regulating the gall N content to avoid a possible negative effect of excess N on its survival (Martinez et al. 2021) is something that would deserve to be considered. N content decreases markedly in the C1 controls (Fig. 20b) in the phase coinciding with mature galls (which is reflected in an increased C/N ratio in same tissues). This result could be explained by an active N translocation effect into the gall from neighboring leaflet tissues, coinciding with the final stages of gall development.

In the absence of statistical treatment, the variation in macronutrients (Fig. 21) shows that the composition and temporal variation of the mineral elements analysed is similar in both controls and different from that found in gall tissues. In other

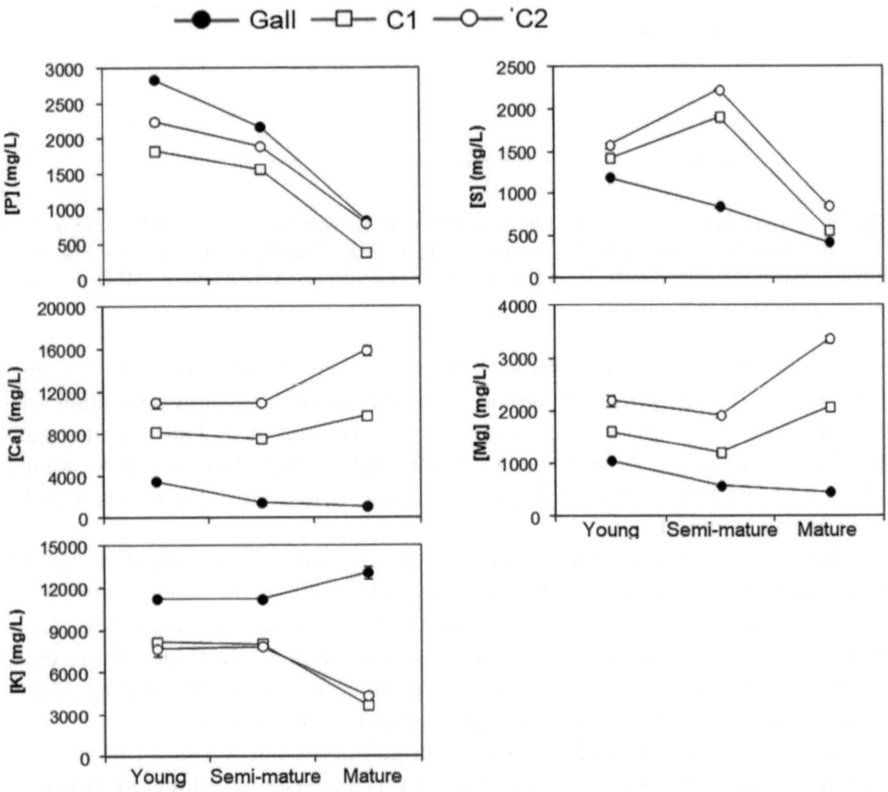

Fig. 21 Mineral content in galls and controls. Variation in mineral nutrients on *G. utricularia* induced galls, control 1 (C1; see Fig. 17) and control 2 (C2; see Fig. 17) at different stages of development (young, semi-mature and mature)

words, it is likely that the aphid colony activity modulates the mineral composition of a very localised area of the leaflet.

According to the results obtained, K and P preferentially accumulate in the galls, whereas galls are depleted in Ca, S and Mg compared to C1 and C2 controls (Fig. 21). The accumulation in galls of mineral nutrients, mobile and abundant in phloem such as P, and specially K, would be consistent with *G. utricularia* gall acting as a phloem sink. The reduction in Mg levels in gall tissues (Fig. 21) could be related to a lower chlorophyll content. Ca content is markedly reduced in gall tissues compared to C1 and C2 throughout gall development. The presence of calcium-binding proteins in aphid saliva (Will and Van Bel 2006) could be related to the decrease of free Ca in galls. In conclusion, preliminary analysis of the time course of water-soluble sugars, starch and mineral elements in galled and ungalled tissues suggest that the galls induced by *G. utricularia* in *P. therebinthus* behave as a physiological sink, the strength of which increases as they mature.

4 Perspectives in Structural and Functional Studies

Galls induced by aphids were proven as excellent study models for understanding the modulation of plant development by gall-inducing organisms. Host plants typically harbor multiple species of galling aphids, making these galls valuable for investigating the structural details that shed light on the roles of plant features in dermal, ground, and vascular tissues. Additionally, exploring their adaptive values to gall inducers and understanding their phylogenetic relationships adds further depth to the study. Aphid galls on superhosts offer valuable insights into evaluating the comparative effects of distinct stimuli on the physiological and histological features of host plant organs. This exploration can lead to a better understanding of alterations in source-sink relationships and mineral contents. Galls inhabited by other ecological guilds also present intriguing models, providing opportunities to assess the effectiveness of presumed protective features.

The detailed study shown here on aphid-induced galls leaves some questions and possibilities of future studies. Gall-inducing aphids determine the formation of galls from a single individual: the fundatrix. In most species, the fundatrix induces the gall in which winged individuals will eventually develop, leaving the gall to continue their life cycle. Nature provides an interesting experimental situation: there are fundatrices that are parasitized and do not give birth to individuals, which would continue giving parthenogenetic generations until winged individuals. Field observation shows that these parasitized fundatrices induce galls that do not reach the size of their sister galls. It arises then, if the intervention of the descendants is necessary so that the gall continues to grow from a starting point established by the fundatrix. The histological and physiological study of these galls may perhaps provide information in this regard. Other situation may be used as a study model: in some species of gall-inducing aphids, the fundatrix induces the formation of a first gall (temporary gall) in which individuals (the fundatrigeniae) give birth to a second generation, which, when emerging from that gall, will induce the definitive galls, which will finally house the winged individuals. It would be appropriate to carry out the study of these temporary galls.

The mature galls finally open to allow the exit of the winged individuals. Sometimes the closure zone of the gall opens, leaving space for the release of the individuals. In the case of closed galls, there comes a time when they partially explode, thus leaving a clear exit for the winged individuals. Both opening mechanisms invite to carry out studies in this regard. And, although it is not a histological question, the fact that the phloem or xylem may be the vascular tissue closest to the chamber on galls induced by aphids, invites entomological studies on the structure of the rostrum and/or the length of the stylets of the aphids that inhabit one or the other galls. At the end, the rates of predation or parasitism of the galls induced by aphids, the study of the organisms involved in the predation or parasitism, as well as the histological and physiological effects that these individuals determine in the galls, are also future interesting possibilities of study.

References

Álvarez R (2011) Initial stages in the formation of galls induced by *Geoica utricularia* in *Pistacia terebinthus* leaflets: origin of the two vascular bundles which characterize the wall of the galls. Am J Plant Sci 2:175–179

Álvarez R (2012) Microscopic study of the walls of galls induced by *Geoica utricularia* and *Baizongia pistaciae* in *Pistacia terebinthus*: a contribution to the phylogeny of Fordini. Arthropod-Plant Interact 6(1):137–145

Álvarez R, Encina A, Pérez Hidalgo N (2008) *Pistacia terebinthus* L. leaflets: an anatomical study. Plant Syst Evol 272:107–118

Álvarez R, Encina A, Pérez Hidalgo N (2009) Histological aspects of three *Pistacia terebinthus* galls induced by three different aphids: *Paracletus cimiciformis*, *Forda marginata* and *Forda formicaria*. Plant Sci 176:303–314

Álvarez R, González S, Candelas A, Martinez JJ (2013) Histological study of galls induced by aphids on leaves of *Ulmus minor*: *Tetraneura ulmi* induces globose galls and *Eriosoma ulmi* induces pseudogalls. Arthropod-Plant Interact 7(6):643–650

Álvarez R, Molist P, González–Sierra S, Martinez JJ, Nieto Nafría JM (2014) The histo structure of galls induced by insects as a useful taxonomic character: the case of *Rectinasus* (Hemiptera, Aphididae, Eriosomatinae). Zootaxa 3861(5):487–492

Álvarez R, Martinez JJ, Muñoz-Viveros AL, Molist P, Abad-González J, Nieto Nafría JM (2016) Contribution of gall microscopic structure to taxonomy of gallicolous aphids on Pistacia. Plant Biol 18:868–875

Álvarez R, Moreno-González V, Martinez JJ, Ferreira BG, Pérez Hidalgo N (2020) Microscopic study of nine galls induced in *Populus nigra* by aphids of the Iberian Peninsula. Arthropod-Plant Interact 14:799–809

Álvarez R, García-Ferreira B, López B, Martínez JJ, Boso S, Martínez MC (2021) Histological study of leaf galls induced by Phylloxera in *Vitis* (Vitaceae) leaves. Proc Natl Acad Sci India Sect B Biol Sci 91(1):117–122

Álvarez R, Ferreira BG, Moreno-González V, Alonso-Redondo R, Penas Á, del Río S (2022) Leaf anatomy of varieties of *Vitis vinifera* from DO León (Spain) and its relationship to the susceptibility to *Plasmopara viticola*. Flora 292:152077

Aoyama T, Akimoto SI, Hasegawa E (2012) Gall distribution as a compromise between the optimal gall–site selection and the synchrony to host–plant phenology in the aphid *Kaltenbachiella japonica*. Arthropod–Plant Interact 6:461–469
Arduin M, Kraus JE (1995) Anatomia e ontogenia de galhas foliares de *Piptadenia gonoacantha* (Fabales, Mimosaceae). Boletim de Botânica da Universidade de São Paulo 14:109–130
Arduin M, Fernandes GW, Kraus JE (2005) Morphogenesis of galls induced by *Baccharopelma dracunculifoliae* (Hemiptera: Psyllidae) on *Baccharis dracunculifolia* (Asteraceae) leaves. Braz J Biol 65(4):559–571
Benheim D, Rochfort S, Robertson E, Potter ID, Powell KS (2012) Grape phylloxera (*Daktulosphaira vitifoliae*) – a review of potential detection and alternative management options. Ann Appl Biol 161:91–115
Blackman R, Eastop V (2020) Aphids on the world's plants. Avalaible on http://www.aphidsonworldsplants.info
Burstein M, Wool D, Eshel A (1994) Sink strength and clone size of sympatric, gall-forming aphids. Eur J Entomol 91:57–61
Carneiro RG, Castro AC, Isaias RM (2014) Unique histochemical gradients in a photosynthesis-deficient plant gall. S Afr J Bot 92:97–104
Castro AC, Oliveira DC, Moreira ASF, Lemos-Filho J, Isaias R (2012) Source–sink relationship and photosynthesis in the horn-shaped gall and its host plant *Copaifera langsdorffii* Desf. (Fabaceae). S Afr J Bot 83:121–126
Chen J, Qiao G (2009) A study on diversity of aphid's galls of Hormaphidinae. Acta Zootaxon Sin 34(2):269–276
Chen J, Qiao GX (2012) Galling aphids (Hemiptera: Aphidoidea) in China: diversity and host specificity. Psyche 2012:1–11
Chen J, Jiang LY, Qiao GX (2014) A total-evidence phylogenetic analysis of Hormaphidinae (Hemiptera: Aphididae), with comments on the evolution of galls. Cladistics 30:26–66
Chen X, Yang Z, Chen H, Qi Q, Liu J, Wang C, Shao S, Lu Q, Li Y, Wu H, King-Jones K, Chen MS (2020) A complex nutrient exchange between a gall-forming aphid and its plant host. Front Plant Sci 11:811
Compson ZG, Larson KC, Zinkgraf MS, Whitham TG (2011) A genetic basis for the manipulation of sink–source relationships by the galling aphid *Pemphigus betae*. Oecologia 167:711–721
Dawkins R (1982) The extended phenotype: the gene as the unit of selection. Oxford University Press, Oxford
Douglas AE (2003) The nutritional physiology of aphids. Adv Insect Physiol 31(31):73–140
Dsouza MR, Ravishankar BE (2014) Nutritional sink formation in galls of *Ficus glomerata* Roxb. (Moraceae) by the insect *Pauropsylla depressa* (Psyllidae, Hemiptera). Trop Ecol 55:129–136
Evert RF (2006) Esau's plant anatomy: meristems, cells, and tissues of the plant body: their structure, function, and development, 3rd edn. Wiley
Fahn A (1990) Plant anatomy, 4th edn. Pergamon
Fernandes GM, Preszler RW, Grim JN (1989) The occurrence of crystals in a cynipid leaf gall on *Quercus turbinella*. Beiträgezur Biologie der Pflanzen 65:377–383
Ferreira BG, Álvarez R, Avritzer SC, Isaias RMS (2017) Revisiting the histological patterns of storage tissues: beyond the limits of gall–inducing taxa. Botany 95:173–184
Ferreira B, Álvarez R, Bragança G, Alvarenga D, Pérez-Hidalgo N, Isaias R (2019) Feeding and other gall facets: patterns and determinants in gall structure. Bot Rev 85:78–106
Garrido Briñón E, Álvarez Nogal R, Hidalgo P (2009) Incidencia de la depredación de *Alophia combustella* (Piralidae) sobre las poblaciones gallícolas de pulgones (Aphididae) en *Pistacia terebinthus* (Anacardiaceae). XXVI Jornadas de la Asociación Española de Entomología
Giron D, Huguet E, Stone GN, Body M (2016) Insect-induced effects on plants and possible effectors used by galling and leaf-mining insects to manipulate their host-plant. J Insect Physiol 84:70-89. https://doi.org/10.1016/j.jinsphys.2015.12.009
Hałaj R, Osiadacz B (2017) Woolly ash aphid – is the alien bug posing a threat to European ash trees? – a review. Plant Prot Sci 53:127–133

Hartley SE (1998) The chemical composition of plant galls: are levels of nutrients and secondary compounds controlled by the gall-former? Oecologia 113:492–501
Hartley SE, Lawton JH (1992) Host–plant manipulation by gall insects: a test of the nutrition hypothesis. J Anim Ecol 61:113–119
Heie OE (1987) Morphological structures and adaptations. In: Minks AH, Harrewijn P (eds) Aphids. Their biology, natural enemies and control, vol A. Elsevier, Amsterdam, pp 393–400
Hemmam S, Benhassaini H, Alvarez R, Bennaoum Z (2022) Morphological changes in Atlas pistachio leaves *Pistacia atlantica* Desf., caused by the galling aphid *Forda riccobonii* (Stephani 1899) in western Algeria. Int J Environ Stud 80:733–741
Hessler M (2017) World plants: synonymic checklists of the vascular plants of the world (version Feb 2017). In: Roskov Y, Abucay L, Orrell T, Nicolson D, Bailly N, Kirk PM, Bourgoin T, RE DW, Decock W, De Wever A, van Nieukerken E, Zarucchi J, Penev L (eds) Species 2000 & IT IS catalogue of life, 2017 annual checklist
Huang M-Y, Huang WD, Chou HM, Lin KH, Chen CC, Chen PJ, Chang YT, Yang CM (2014) Leaf-derived cecidomyiid galls are sinks in *Machilus thunbergii* (Lauraceae) leaves. Physiol Plant 152:475–485
Huang M-Y, Huang W-D, Chou H-M, Chen C-C, Chen P-J, Chang Y-T, Yang C-M (2015) Structural, biochemical, and physiological characterization of photosynthesis in leaf-derived cup-shaped galls on *Litsea acuminata*. BMC Plant Biol 15:61
Inbar M, Mayer RT, Doostdar H (2003) Induced activity of pathogenesis related (PR) proteins in aphid galls. Symbiosis 34:293–300
Inbar M, Wink M, Wool D (2004) The evolution of host plant manipulation by insects: molecular and ecological evidence from gall-forming aphids on *Pistacia*. Mol Phylogenet Evol 32:504–511
Isaias RMS, Carneiro RGS, Santos JC, Oliveira DC (2014) Gall morphotypes in the Neotropics and the need to standardize them. In: Fernandes GW, Santos JC (Orgs) Neotropical insect galls. Springer, Dordrecht pp 51–67
Kingston KB, Powell KS, Cooper PD (2007) Grape phylloxera external morphology: observations under scanning electron microscopy. Acta Hortic 733:107–114
Kmiec K, Złotek U, Jakubczyk A, Karas M (2018) Biochemical alterations in *Ulmus pumila* L. leaves induced by galling aphid *Tetraneura ulmi* L. Acta Sci Pol Hortorum Cultus 17:175–183
Kot I, Kmiec K (2020) Poplar tree response to feeding by the petiole gall aphid *Pemphigus spyrothecae* pass. Insects 1:271
Koyama Y, Yao I, Akimoto SI (2004) Aphid galls accumulate high concentrations of amino acids: a support for the nutrition hypothesis for gall formation. Entomol Exp Appl 113:35–44
Kurzfeld-Zexer L, Lev-Yadun S, Inbar I (2015) One aphid species induces three gall types on a single plant: comparative histology of one genotype and multiple extended phenotypes. Flora 210:19–30
Küster E (1911) Die gallen der pflanzen. S. Hirzel, Leipzig
Kutsukake M, Meng XY, Katayama N, Nikoh N, Shibao H, Fukatsu T (2012) An insect–induced novel plant phenotype for sustaining social life in a closed system. Nat Commun 3:1187
Kutsukake M, Uematsu K, Fukatsu T (2019) Plant manipulation by gall–forming social aphids for waste management. Front Plant Sci 10:933
Laakso K, Sullivan JH, Huttunen S (2000) The effects of UV–B radiation on epidermal anatomy in loblolly pine (*Pinus taeda* L.) and Scots pine (*Pinus sylvestris* L.). Plant Cell Environ 23:461–472
Larson KC (1998) The impact of two gall-forming arthropods on the photosynthetic rates of their hosts. Oecologia 115:161–166
Larson KC, Whitham TG (1991) Manipulation of food resources by a gall-forming aphid: the physiology of sink-source interactions. Oecologia 88:15–21
Mani MS (1964) Ecology of plant galls. Dr. Junk Publisher, The Hague
Martinez JJ (2010) Anti–insect effects of the gall wall of *Baizongia pistaciae* [L.], a gall–inducing aphid on *Pistacia palaestina* Boiss. Arthropod-Plant Interact 4:29–34

Martinez JJ, Moreno-González V, Jonas-Levi A, Alvarez R (2018) Quantitative differences detected in the histology of galls induced by the same aphid species in different varieties of the same host. Plant Biol 20:516–524

Martinez JI, Jonas-Levi A, Gonzalez-Karlsson A, Moreno-González V, Álvarez R (2021) Experimental validation of the soil fertility hypothesis in a natural Mediterranean tri-trophic system. Entomol Exp Appl 169:eea.13041

Mellah R, Benhassaïni H, Álvarez R (2016) Histo–anatomy of *Pistacia terebinthus* L. leaflets galls induced by *Paracletus cimiciformis* von Heyden and *Geoïca utricularia* Passarini. Two aphids in western Algerian region. Int J Geobot Res 6:1–8

Meyer J (1987) Plant galls and gall inducers. Gebru¨ der Borntraeger, Berlin

Meyer J, Maresquelle J (2003) Anatomie des Galles. Ggebrüder Borntrager, Berlin. 1983

Miles PW (1999) Aphid saliva. Biol Rev 74:41–85

Miller DG, Ivey CT, Shedd JD (2009) Support for the microenvironment hypothesis for adaptive value of gall induction in the California gall wasp, *Andricus quercuscalifornicus*. Entomol Exp Appl 132:126–133

Moran NA (1992) The evolution of aphid life cycles. Annu Rev Entomol 37:321–348

Muñoz-Viveros AL, Martinez JJ, Molist P, González-Sierra S, González Julián P, Álvarez R (2014) Microscopic study of galls induced by three species of *Geopemphigus* (Hemiptera: Aphididae: Eriosomatinae) on *Pistacia mexicana*. Arthropod-Plant Interact 8:531–538

Nieto Nafría JM, Mier Durante MP (1998) Hemiptera, Aphididae I. Museo Nacional de Ciencias Naturales, Madrid

Nováková E, Hypša V, Klein J, Foottit RG, von Dohlen CD, Moran NA (2013) Reconstructing the phylogeny of aphids (Hemiptera: Aphididae) using DNA of the obligate symbiont Buchnera aphidicola. Mol Phylogenet Evol 68(1):42–54

Nyman T, Julkunen-Tiitto R (2000) Manipulation of the phenolic chemistry ofwillows by gall–inducing sawflies. Proc Natl Acad Sci USA 97(24):13184–13187

Oliveira DC, Isaias RMS (2010) Cytological and histochemical gradients induced by a sucking insect in galls of *Aspidosperma australe* Arg. Muell (Apocynaceae). Plant Sci 178:350–358

Oliveira DC, Moreira ASF, Isaias RM, Martini V, Rezende UC (2017) Sink status and photosynthetic rate of the leaflet galls induced by *Bystracoccus mataybae* (Eriococcidae) on *Matayba guianensis* (Sapindaceae). Front Plant Sci 8:1249

Ortiz-Rivas B, Martínez-Torres D (2010) Combination of molecular data support the existence of three main lineages in the phylogeny of aphids (Hemiptera: Aphididae) and the basal position of the subfamily Lachninae. Mol Phylogenet Evol 55:305–317

Pérez Hidalgo N, Nieto Nafría JM (2003) Sufamilia Eriosomatinae Kirkaldy, 1905 (1843). En: Hemiptera, Aphididae II. Nieto Nafría et al. In: R. MA (ed) Fauna Ibérica vol 19. Naturales. CSIC, Madrid, pp 141–275

Pike N, Richard D, Foster W, Mahadevan F (2002) How aphids lose their marbles. Proc R Soc Lond Ser B Biol Sci 269:1211–1215

Price PW, Waring GL, Fernandes GW (1986) Hypotheses on the adaptive nature of galls. Proc Entomol Soc Wash 88:361–363

Price PW, Fernandes GW, Waring GL (1987) Adaptive nature of insect galls. Environ Entomol 16:15–24

Raman A, Beiderbeck R, Herth W (2009) Early subcellular responses of susceptible and resistant *Vitis* taxa to feeding by grape phylloxera *Daktulosphaira vitifoliae*. Bot Helv 119:31–39

Richardson RA, Body M, Warmund MR, Schultz JC, Appel HM (2017) Morphometric analysis of young petiole galls on the narrow-leaf cottonwood, *Populus angustifolia*, by the sugarbeet root aphid, *Pemphigus betae*. Protoplasma 254:203–216

Sano M, Akimoto SI (2011) Morphological phylogeny of gall-forming aphids of the tribe Eriosomatini (Aphididae: Eriosomatinae). Syst Entomol 36:607–627

Simmons AT, Gurr GM, McGrath D, Nicol HI, Martin PM (2003) Trichomes of *Lycopersicon* spp. and their effect on *Myzus persicae* (Sulzer) Hemiptera: Aphididae. Aust J Entomol 42:373–378

Spiller HJ, Kimmins FM, Llewellyn M (1985) Fine structure of aphid stylet pathways and its use in host plant resistance studies. Entomol Exp Appl 38(3):293–295
Stern DL (1995) Phylogenetic evidence that aphids, rather than plants, determine gall morphology. Proc R Soc Lond B Biol Sci 260:85–89
Stone GN, Schönrogge K (2003) The adaptive significance of insect gall morphology. Trends Ecol Evol 18:512–522
Tjallingii WF, Hogen-Esch T (1993) Fine structure of aphid stylet routes in plant tissues in correlation with EPG signals. Physiol Entomol 18(3):317–328
Uematsu K, Kutsukake M, Fukatsu T (2018) Water–repellent plant surface structure induced by gall–forming insects for waste management. Biol Lett 14:20180470
Watson L, Dallwitz MJ (1992) The families of flowering plants: descriptions, illustrations, identification, and information retrieval. Version: 27th May 2023. delta–intkey.com
Will T, Van Bel AJ (2006) Physical and chemical interactions between aphids and plants. J Exp Bot 57:729–737
Wool D (2004) Galling aphids: specialization, biological complexity, and variation. Annu Rev Entomol 49(1):175–192
Wool D (2005) Gall–inducing aphids: biology, ecology and evolution. In: Raman R, Schaefer CW, Withers TM (eds) Biology, ecology, and evolution of gall–inducing arthropods. CRC Press, Boca Raton, pp 73–132
Wool D, Ben Zvi O (1998) Population ecology and clone dynamics of the galling aphid *Geoica wertheimae* Brown and Blackman (Pemphigidae: Fordinae). Eur J Entomol 75:509–518
Zhang H, Qiao G (2007) Systematic status of the genus *Formosaphis* Takahashi and the evolution of galls based on the molecular phylogeny of Pemphigini (Hemiptera: Aphididae: Eriosomatinae). Syst Entomol 32(4):690–699

Eriophyoid Mite Galls: From Structural Complexity to Economic Impact

Bruno Garcia Ferreira, Filipe Rodrigues Valeriano, Leticia Ponticel Nobrega, Rayssa Rosa Marquesine, and Lubia María Guedes

1 Introduction

Mites (Acarina), along with insects and nematodes, constitute the most prominent groups of gall-inducing organisms (Mani 1964; Ferreira et al. 2019). Currently, the "gall mites", are represented by the superfamily Eriophyoidea (Chetverikov et al. 2021). All eriophyoids are obligatory plant-feeding mites. Although they are popularly known as gall mites (Lindquist 1996; Klimov et al. 2018), several of them are free-living organisms (Petanović and Kielkiewicz 2010a). The exact numbers are unknown, but it is supposed that maybe half of the eriophyoids are free-living (vagrant) organisms, and the other half is composed of organisms that can induce simple (such as erinea and bud galls) or complex galls (mainly blister and pouch galls) (Petanović and Kielkiewicz 2010a).

Eriophyoids consist of three extant families: Phytoptidae, Eriophyidae, and Diptilomiopidae, comprising nearly 5000 described species (Zhang 2017; Klimov et al. 2018; Chetverikov et al. 2015a, 2021). The largest diversity of Eriophyoidea is present within the family Eriophyidae (Craemer 2010). Phytoptidae and Eriophyidae encompass organisms that modify the development of host plant cells and tissues, being considered gall-inducing organisms. Most of gall-inducers belong to the Eriophyinae (Eriophyidae), although much is unknown about the extant species. Phytoptidae is a smaller and more restricted group of mites, and the gall-inducing phytoptids are reported to parasite eudicots from the clade superrosids,

B. G. Ferreira (✉) · F. R. Valeriano · L. P. Nobrega · R. R. Marquesine
Department of Botany, Universidade Federal do Rio de Janeiro (UFRJ), Rio de Janeiro, Brazil
e-mail: bgf@biologia.ufrj.br

L. M. Guedes
Facultad de Ciencias Forestales, Laboratorio de Semioquímica Aplicada, Universidad de Concepción (UdeC), Concepción, Chile

D. C. de Oliveira, R. M. dos Santos Isaias (eds.), *Plant Galls*,
https://doi.org/10.1007/978-3-031-80064-1_12

with a single exception, *Phytoptus dentatae* Smith Meyer, 1991, inducing galls on superasterids (Chetverikov et al. 2021).

Several works on the gall anatomy does not identify the species of the inducer Eriophyidae, since there is a lack in the knowledge of the diversity and taxonomy of eriophyoid species, especially in the Tropics. Additionally, the systematics of Acari, including the superfamily Eriophyoidea, is far to be consolidated. A significant, yet undiscovered diversity is estimated to range from 500,000 to over one million species, the majority of which remain undescribed (Craemer 2010; Chetverikov et al. 2021). Differences among eriophyoid species appear to be linked to their specialization in parasitizing specific plant groups. Some eriophyoid groups are confined to gymnosperms, while others are associated with certain orders of angiosperms, and still others with ferns (*see* Chetverikov et al. 2015a and references therein). Economically, the eriophyoid mites are very important, since several of them can transmit pathogenic viruses among host plants, being associated to diseases in several crops (De Lillo and Skoracka 2010) (Table 1), and several species are associated to gall formation (Klimov et al. 2018).

The impacts of Eriophyoidea mites in plants are evident and the changes provoked by them reveal fine control in host plant metabolism and development. The structural features of their galls and the diversity of cytological, physiological and chemical manipulation are herein explored, revealing their functional aspects and the gaps in the comprehension of these widespread and diverse interactions.

2 General Morphology and Life Cycle of Gall-Inducing Mites

The eriophyoids are distinct from other arachnids by having only two pairs of five-segmented legs—giving them a vernacular name, "four-legged mites"—and an elongated opisthosoma with annuli, a reduced number of setae, ending in retractile caudal lobes (Mani 1964; Chetverikov et al. 2015a; Klimov et al. 2018). They are tiny, often microscopic, elongated, and cylindrical mites characterized by unpaired genital organs. Their size may range from 80 to nearly 500 μm, but they are 100–200 μm long in average (Lindquist 1996; Chetverikov et al. 2015b; Propistsova et al. 2023). They lack respiratory and circulatory organs as well as eyes. Typically, they exhibit colors ranging from white and yellowish-white to yellowish-red or brown (Mani 1964; Klimov et al. 2018; Desnitskiy and Chetverikov 2022).

The eriophyid mites have mouthparts with a needle-shaped mandible, capable of boring plant tissues and sucking the liquid cellular contents (Klimov et al. 2018). The feeding organs are also called gnathosoma, and they consist of in a prominent subcapitulum or rostrum, enclosing the cheliceral stylets, which are slightly protruded during feeding. Two palps form a suction cup terminally (Westphal and Manson 1996). The cheliceral stylets are rigid and do not form a food channel.

Table 1 Eriophyoids capable of transmitting viruses to plants of agricultural importance

Family	Plant species	Common name	Eriophyid species	Vagrant or gall-inducer	Affected organ	Virus type	Damage caused	References
Amaryllidaceae	*Allium sativum* L.	Garlic	*Aceria tulipae* Keifer	Vagrant	Bulb	GarV-B1	Distorted leaves with strips of yellow mosaic.	De Lillo and Skoracka et al. (2010)
Fabaceae	*Cajanus cajan* L. Millsp.	Pegeon pea	*Aceria cajani* Channabasavanna	Vagrant	Leaf	Pigeon pea sterility mosaic virus (PPSMV)	Plants remain in vegetative state, without flower production.	Kulkarni et al. (2002)
Grossulariaceae	*Ribes nigrum* L.	Blackcurrant	*Cecidophyopsis ribis* Westwood	Gall-inducer	Branches	Blackcurrant Reversion Virus (BRV)	Production of smaller and distorted leaves in a few numbers, decreased hairiness of the floral buds and flowers, darkening of the flower from dark red to violet, abnormal and smaller fruits.	Lemmetty et al. (2004)
Moraceae	*Ficus carica* L.	Fig	*Aceria ficus* Cotte	Vagrant	Leaf	Fig Mosaic Virus (FMV)	Mosaic appearance, vein clearing, and small yellowish spots.	Caglayan et al. (2012)
Poaceae	*Lolium multiflorum* L. and *Agropyron repens* L.	Annual ryegrass and couch grass	*Abacarus hystrix* Nalepa	Vagrant	Leaf	Ryegrass mosaic virus (RyMV) and Agropyron Mosaic Virus (AMV)	Reduction of dry matter production.	De Lillo and Skoracka (2010)
	Triticum aestivum L.	Bread wheat	*Aceria tosichella* Keifer	Vagrant	Leaf	Wheat streak mosaic virus (WSMV), Wheat mosaic virus (WMoV), Brome streak mosaic virus (BrSMV) and Triticum mosaic virus (TriMV)	Mosaic pattern with chlorotic streaks, stunted plants, and loss of productivity.	Navia et al. (2013), De Lillo et al. (2018)

(continued)

Table 1 (continued)

Family	Plant species	Common name	Eriophyid species	Vagrant or gall-inducer	Affected organ	Virus type	Damage caused	References
Rosaceae	*Malus domestica* Borkh and *Pyrus communis* L.	Apple and pear	*Eriophyes pyri* Pagenstecher	Gall-inducer	Young leaves	Emaravirus (EMARaV)	Chlorotic spots, necrotic and black blisters, reduced photosynthesis, partial leaf fall, deformation, and premature fruit fall.	von Bargen et al. (2019)
	Prunus emarginata Douglas	Bitter cherry	*Eriophyes inaequalis* Wilson & Oldfield	Gall-inducer	Lateral buds	Cherry mottle leaf virus (ChMLV)	Irregular chlorotic mottling, edge tattering, reduction in the leaf size, fruits with little flavor and late ripening, shortening of internodes.	De Lillo et al. (2021)
	Prunus persica L. Batsch	Peach	*Eriophyes insidiosus* Keifer & Wilson	Gall-inducer	Buds	Peach mosaic virus (PcMV)	Stunted leaves, dwarfing of the twigs, deformation of petals, and fruit deformity.	Gispert et al. (1997, 1998)
	Rosa multiflora Thunb.	Rose	*Phyllocoptes fructiphilus* Keifer	Vagrant	Flowers	Rose virus 1 (RSV-1)	Deformation of flowers, normal growth and development of rose bushes, increased density of thorns, redness of the stems and death.	De Lillo and Skoracka (2010), Monterrosa et al. (2021)
Vitaceae	*Vitis vinifera* L.	Grape	*Colomerus vitis* Pagenstecher	Gall-inducer (erinea)	Leaf	Grapevine Pinot Gray virus	Mottling and deformation of leaves.	Malagnini et al. (2016)

These specialized mouthparts reflect their extreme adaptations as plant feeders (De Lillo and Skoracka 2010; Klimov et al. 2018).

There is some divergence among acarologists when considering the two stages between eggs and adults. Some of them may consider as first and second-stage nymphae, and others as larvae and nymphae, respectively (Lindquist 1996). Larvae and nymphs closely resemble adults, possessing two pairs of legs, but are significantly smaller and do not possess genital apparatus (Lindquist 1996). Generally, adult females are larger than males, and the dispersive form is the deutogyne females, which are distinct from the protogynes (see below). The small size of gall-inducing species, variations in their life cycles among individuals, and the challenges associated with collecting and observing inducers contribute to the difficulty of identifying gall inducers at the species level.

The eriophyoid females are usually more abundant than the males (Mani 1964; Manson and Oldfield 1996), even though there is no obvious sexual dimorphism and an aedeagus for sperm transfer is absent in males (Helle and Wysoki 1996). Only females are usually present in hibernation or aestivation conditions. The overwinter females are deutogynes, while the active generation (males, females and immatures) are protogynes (Keifer et al. 1982). The deutogyne females are morphologically distinct from the protogynes, adding confusion in species description and identification (Lindquist and Oldfield 1996).

Helle and Wysoki (1996) reinforced that parthenogenesis in eriophyoids was erroneously reported to occur based on the evidence of the "absence" of males. The eriophyoids commonly reproduce by parthenogenesis (Mani 1964; Navia et al. 2010), but not the thelytokous type (= female producing) (Helle and Wysoki 1996). The only type of parthenogenesis observed in eriophyoids is arrhenotokous parthenogenesis, with fertilized eggs originating females and unfertilized eggs becoming males, a mechanism that favor the colonization of novel habitats by a population or even by a single eriophyoid female (Helle and Wysoki 1996; Navia et al. 2010).

The sexual behavior and dispersion of eriophyoids was revised by Michalska et al. (2010), and several of these organisms demonstrate to have an independent sexual reproduction, without direct contact between males and females. Spermatophores are deposited on the plant surface (Michalska et al. 2010) and are responsible to transfer sperms to the females (Helle and Wysoki 1996). However, several variations in reproductive cycle may be observed inside the group, which deserve more studies (Michalska et al. 2010). Several reports strongly support that most of eriophyoids are bisexual and therefore females always have spermathecae (Helle and Wysoki 1996).

In gall-inducing eriophyoids, deutogynes initiate feeding on host plant organs, typically targeting meristematic tissues, and/or young leaves, stems, flowers, or, thereby inducing the formation of galls that will accommodate their offspring (Ferreira et al. 2019). Only in one case, was reported that nymphs may induce galls, although less effective than fertile females (Westphal et al. 1990; Westphal and Manson 1996). The gall-inducing eriophyoids are mostly species-specific (De Lillo and Skoracka 2010; Skoracka et al. 2010), although in some cases closely related plant species may host the same gall mite species (Westphal et al. 1990; Araújo and

Kollár 2019). Some eriophyoid species are non-monophagous, but symptoms and responses of host plants may vary and the performance of the eriophyoids is expected to be higher in some of the closely related host plants. Recent case studies on several species of Eriophyidae and Phytoptidae demonstrated high specificity among gall-inducing organisms and their host plants. While gall-inducing mites might inhabit or feed on other host plant organs or species, a specific host plant species and/or organ may be essential for completing their lifecycle.

3 Gall-Inducing Feeding Behavior and Early Cellular Changes on Host Plants

The feeding behavior is a crucial mechanism in gall induction, including in eriophyoid galls (Ferreira et al. 2019). In this context, fertile females, referred to as foundatrices (singular: fundatrix), feed on the plant's surface, ultimately triggering gall formation. Subsequently, they oviposit, and upon hatching, the nymphs commence feeding on gall tissues, perpetuating the alterations and, in certain instances, stimulating the development of new nutritive cells. Feeding stimulus, as occur in insect galls, is important to maintenance of gall growth and metabolism (Yang et al. 2023). After adults' dispersion from the gall, the mites may hibernate whether the host plants are deciduous, and a resistant form – the deutogyne females – overwinter in other plant organs or unknown places (Keifer et al. 1982). The environmental temperatures seem to be an important factor to allow their emergence (De Lillo and Skoracka 2010). Deutogynes may hibernate inside the apical or axillary buds, below the protective scales, or in the angles of leaves and branches (Mani 1964). They move actively and usually infect young parts of the plants (De Lillo and Skoracka 2010). The mites resistant to unfavorable conditions are commonly deutogyne females (secondary females), which are dormant mites, also considered the dispersive generation in unfavorable conditions (Manson and Oldfield 1996; Sabelis and Bruin 1996; Westphal and Manson 1996). For some species, it was suggested that deutogynes are responsible to induce new galls (Michalska et al. 2010), reinforcing the importance of females in colonization of new host plants or habitats.

The mite feeding behavior has been observed in detail only in a few works, and in artificial conditions (Westphal et al. 1981; Westphal and Manson 1996). In the case of galling eriophyoids, the foundatrices are responsible to feed and induce the gall formation before the oviposition inside the galls (Oldfield 2005). The galling mites move randomly on the host plant surface to find a suitable site to feeding. They anchor the rostrum to the host surface and contracts the telescopic palpal segments, allowing the protrusion of the stylets for a short distance—only about 2 μm (Westphal and Manson 1996). Due to their small sizes, most eriophyids can pierce single cells and suck its contents (Mani 1964). According to Westphal and Manson (1996), it is unlikely that saliva deposited on host surface acts in the dissolution of cell walls, and the time of the stylet insertion is short after probing. The perforation is small and seems to affect the cell wall only in a small puncture. The saliva is poured into the plant cell through the wound at the time of sucking, which is important to induce galls and maintain their tissues active (Mani 1964). After cell wall perforation, the mite withdraws its cheliceral stylets while the cupping action of the palps facilitates food suction. Westphal et al. (1981, 1990) observed that mite feeding duration may vary from 1 min to 1 h.

As an example, *Eriophyes cladophthirus* Nalepa on *Solanum dulcamara* L. may last 45 s to 2 min to insert the stylets into plant cells, and after perforation, the pharyngeal pump occurs from 4 to 20 min, and then the mite may remain motionless from 30 s to 40 min before locating other feeding sites (Westphal et al. 1981). In resistant plants, necrosis occurs after 1 h of infestation, and tiny areas of collapsed cells became perceptible and enlarge to 100 μm of diameter, becoming into brown points. Polyphenols were detected after 4 h, and callose is deposited in lateral walls of the living cells lining the necrotic area after 3 or 4 days. In susceptible plants, injured cells produce a wall thickening of callose (cone-shaped callose deposits) around the puncture after 15–20 min. In the first hour, there is an alkalinization of vacuoles, the nuclei of affected cells enlarge, cytoplasm become abundant, vacuoles reduce in size, and nucleoli hypertrophy, leading to the differentiation of typical nutritive cells (Westphal et al. 1981, 1990; Westphal and Manson 1996; Ferreira et al. 2017). Chitosan, derived molecules from chitin, are thought to act in DNA changes of plant cells and in resistance mechanisms. Cell divisions started at 2 h, with modifications extending to adjacent cells (Westphal et al. 1981). Rust eriophyoid mites usually cause necrosis and move to other plant parts to continue to feed, causing massive destruction of epidermal surface, lignification or formation of phellogen-like tissues (Petanović and Kielkiewicz 2010b). On the other hand, galling eriophyoid may induce the formation of nutritive cells in susceptible plants and the production of nutritive adjacent cells, after the collapse and death of the cell used as food source (Westphal and Manson 1996; Petanović and Kielkiewicz 2010b). A constant activity of puncturing distinct cells would lead to the formation, gall maintenance and nutritive tissue differentiation (Westphal and Manson 1996, Petanović and Kielkiewicz 2010b).

The eriophyoid foundatrix—the fertile female—moves to localize the suitable place for the gall induction through feeding and deposit its eggs inside the gall. When the foundatrix stops feeding, the epidermal nutritive cells die and subepidermal cells redifferentiate into nutritive cells after egg hatching, when the nymphs begin to feed (Petanović and Kielkiewicz 2010b). The plant cell capabilities of differentiate into distinct cell types on response to external agents are elegantly demonstrated by observing eriophyoid life cycle on host plants, and it is likely that the formation of nutritive cells in mite galls is a defense response of the host plants (Ferreira et al. 2019). If the galling stimuli ceases, such cells, with meristematic features, will recover the typical plant tissues, forming including novel chlorophyllian tissues and stomata (Petanović and Kielkiewicz 2010b).

4 Structure of Eriophyidae Galls

4.1 Simple Galls: Erinea or Hairy Galls and Leaf Rolling

The simplest mite galls are those called as erineum (plural erinea, also called erinoses), hairy galls or filzgalls (Mani 1964; Westphal and Manson 1996; Ferreira et al. 2017, 2019). They are products of changes in indumentum density or type (Ferreira et al. 2017, 2019; Freitas et al. 2023; Nobrega et al. 2023) (Fig. 1), although other symptoms occur, as leaf bulging or leaf distortion (Figs. 1a, b and 2a), when infecting the whole or significant part of the leaf (Westphal and Manson 1996; Ferreira et al. 2017; Freitas et al. 2023; Nobrega et al. 2023; Guedes et al. 2024a). The leaf bulging (Figs. 1a–d and 2a), as observed in some erinea, are sometimes called by "blister gall", but we avoid this term here because we considered the concept of blister galls which involve the eriophyoid entrance inside leaf mesophyll (see below) (Mani 1964; Westphal and Manson 1996).

Not only dense hairs characterize erinea. Some galls may have increased density of emergences, which are plant outgrowths or appendices formed not only by epidermal cells (i.e., of protodermal origin), but also by parenchyma cells (originated from ground meristem cells). Examples of this peculiarity are *Miconia ibaguensis* Bonpl. (Fig. 1c, d) and *M. notabilis* Triana (Melastomataceae) galls (Ferreira et al. 2017, 2019; Freitas et al. 2023). The changes in trichome types usually occur, as demonstrated in *Avicennia schaueriana* Stapf & Leechm. ex Moldenke (Acanthaceae) (black mangrove) galls (Fig. 1e–g), which change the host plant pattern from short pyriform trichomes to elongated filiform trichomes in galls (Fig. 1h–j) (Nobrega et al. 2023). Additionally, salt glands appear projected in galled surface (Fig. 1j), and sunken in the adaxial surface (Fig. 1i), when compared to the non-galled leaves (Fig. 1h). The trichome density is not altered, but the change in trichome size could lead to additional protection against desiccation, high temperatures, and predators (Nobrega et al. 2023).

The hairs and emergences of erinea are usually described as nutritive hairs (Westphal and Manson 1996), where reducing sugars and proteins are usually accumulated (Ferreira et al. 2017, Freitas et al. 2023, Nobrega et al. 2023, Guedes et al. 2024a). However, the emergence epidermal cells on erinoses of *M. ibaguensis*, food sources of eriophyids, do not have features of typical nutritive cells such as dense cytoplasm and prominent nuclei. Such cells are vacuolated, with small plastids, and multivesicular bodies (Ferreira et al. 2020).

The erinea are expected to be concentrated alterations in epidermal cells, due to the feeding activity of the mites (Nobrega et al. 2023). In galls induced by unidentified eriophyids on *A. schaueriana* leaves, the mesophyll is maintained dorsiventral as in non-galled leaves, although smaller intercellular spaces and spongy parenchyma cells were detected in galls when compared to non-galled leaves (Fig. 1h–j). Additionally, other epidermal changes than trichome alterations were observed in abaxial epidermis (Fig. 1j), such as stomatal malformation and alteration of stomatal types, which could lead to reduced photosynthetic rates.

Fig. 1 Erinea (hairy galls) induced by eriophyoids. (**a, b**) Erinea on young (YL) and mature leaves (ML) of *Miconia ibaguensis*. Arrows show the abaxial regions of the leaves with erinea, where these eriophyoids usually occur and feed. The asterisks indicate the bulging of the leaf, causing sometimes leaf distortion. (**c, d**) Transversal section of galled leaves of *M. ibaguensis*. Arrows point the emergences in the abaxial surface. (**c**) Galled portion (GP) and a non-galled portion (NGP) and the midrib (Mi). (**d**) Detail of the gall mesophyll and emergences where mites feed. (**e–j**) Erinea on *Avicennia schaueriana* leaves. (**e**) Leaf bulging on adaxial side (arrows). (**f**) Dense indumentum of an erineum (arrows), seen on abaxial side. (**g**) Eriophyoid found among the neo-formed trichomes. (**h–j**) Transversal sections. (**h**) Non-galled portion, with indumentum (arrow) formed by pyriform trichomes. (**i**) Galled portion with filiform trichomes (arrow). (**j**) Detail of abaxial surface of galled portion, showing projected salt glands (asterisks) and filiform trichomes (arrow). Abbreviations: AdE adaxial epidermis, AbE abaxial epidermis, Dr druses, GL galled leaf, HP homogeneous parenchyma, NGL non-galled leaf, MB minor vascular bundle, PP palisade parenchyma, SP spongy parenchyma, WP water-storage parenchyma, Scale bars: (**a, b, e, f**) 1 cm. (**c, g–i**) 200 μm. (**d**) 50 μm

Reduced intercellular spaces in a homogenized mesophyll were observed in erinea of *M. ibaguensis* and *M. notabilis* (Freitas et al. 2023). There is no hyperplasia of mesophyll parenchyma, whose number of cell layers is determined by host plant features in response to abiotic factors (Freitas et al. 2023). The homogeneous mesophyll of *M. ibaguensis* galls (Fig. 1c, d) was interpreted as a common storage parenchyma, with well-developed chloroplasts, thylakoid lamellation and starch grain

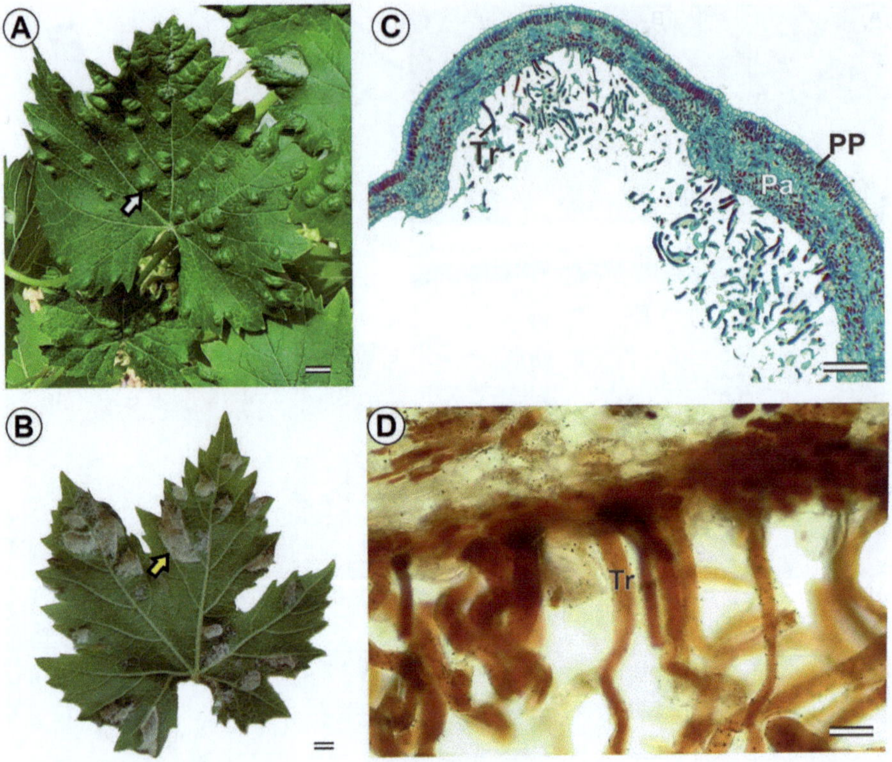

Fig. 2 Erinea (hairy galls) induced by *Colomerus vitis* on *Vitis vinifera* leaf. (**a**) Adaxial leaf surface with numerous erinea (white arrow). (**b**) Abaxial leaf surface with numerous erinea (yellow arrow). (**c**) Transversal section of the erineum. (**d**) Accumulation of reducing sugars (brown color after Fehling's test) in the abaxial epidermis and in the trichomes. Abbreviations: PP palisade parenchyma, Pa parenchyma, Tr trichomes. Scale bars: (**a, b**) 1 cm; (**c**) 250 μm; (**d**) = 100 μm

accumulation (Ferreira et al. 2020). Such galls, however, presented reduced chlorophyll a and b concentrations, although carotenoids remained similar in galls when compared to non-galled leaves (Ferreira et al. 2018). Such changes explain the changes in coloration of galls compared to non-galled portions of leaves (Fig. 1b). In these galls, the concentration of phenolics in the protoplast and cell walls increased 37% and 30%, respectively, which seems to avoid the uncontrolled increase in reactive oxygen species (Ferreira et al. 2018).

Another example of blister-like galls are those induced by *Colomerus vitis* Pagenstecher on *Vitis vinifera* L. leaves (Vitaceae) (Guedes et al. 2024a). The mite infection caused morphological and anatomical changes on *V. vinifera* leaves, affecting both abaxial and adaxial surfaces and inducing alterations in the leaf mesophyll (Fig. 2). On the adaxial side of the leaf, erineum appears as a green blister (Fig. 2a) affecting approximately 6% of the leaf area, while on the abaxial side, it appears whitish due to the accumulation of trichomes and emergences (Fig. 2b).

The mesophyll thickness increased significantly in the erineum formation, which consisted of up to 4 layers of palisade parenchyma and up to 11 layers of compact parenchyma cells (Fig. 2c). In some erineum areas, mesophyll homogenization occurs. The mesophyll layers functioned as storage tissue, accumulating starch and proteins. The abaxial epidermis became hyperplasic, with numerous long trichomes and emergences that accumulated reducing sugars, providing a feeding site for the mite colony (Fig. 2b, c, d).

Leaf rolling galls may also be produced by eriophyids and are considered simple galls. Histometric and cytometric changes are generally observed in these galls, leading to the leaf blade rolling towards abaxial surface (Rančić and Petanović 2002). Differential cell hypertrophy and hyperplasia occur, in which the region of the abaxial side of the affected leaves experiment a decrease in cell size and an increase in the number of tissue layers, and minute intercellular spaces can be observed, although spongy and palisade parenchymas are also observed in galls (Rančić and Petanović 2002). In some cases, anatomical alterations are very diminutive—such as deceasing of intercellular space area or cell size—or not observed, and the leaf rolling is only consequence of disruption of epidermal cell in the sites of mite feeding. These alterations are not usually considered galls, due to absence of anatomical changes besides cell disruption, and are known as toxemias (Rančić and Petanović 2002).

4.2 Complex Galls: Blister, Pouch, and Other Gall Formation Types

Complex galls are defined according to the duration of their development and the severity of anatomical, cytological, histometrical, and histochemical changes induced by mites (Moura et al. 2009; Formiga et al. 2015; Ferreira et al. 2019). Similar to simple galls, the mite stimuli occur on epidermal cells or in the cells in direct contact with the mites. However, more complex changes besides erinea and leaf rolling galls, occur in mite-induced galls, such as blister (or pocket), pouch, bud, and stem galls (Mani 1964; Westphal and Manson 1996; Nasareen et al. 2012; Paponova et al. 2018; De Lillo et al. 2020; Anand and Ramani 2021; Guedes et al. 2023) (Table 2).

Pocket galls are also called blister galls and constitute usually lenticular biconvex galls in which mites may enter through stomata and feed on parenchyma cells, creating a nymphal chamber inside the mesophyll (Mani 1964; Westphal and Manson 1996). An example is the gall induced by *Aceria tristriata* Nalepa on leaves of *Juglans regia* L. (Juglandaceae) (Fig. 3a), as studied by Ferreira et al. (2017). In this gall, mite feeding activity leads to the formation of typical nutritive cells (Fig. 3b), characterized by granular cytoplasm, prominent nucleus, and protein accumulation. The feeding activity of *A. tristriata* lead to local cell death, and the adjacent cells become nutritive as a reaction of the plant tissues. The biconvex shape

Table 2 Eriophyid galls of distinct levels of complexity

Family	Host plant	Inducer	Site	Morphotype	Complexity	Color	Ornamentation	References
Acanthaceae	*Avicennia schaueriana* Stapf & Leechm ex Moldenke	Unidentified Eriophyidae	Leaves	Erinea (leaf bulging)	Simple	Brownish	Trichomes	Nobrega et al. (2023)
Asteraceae	*Cirsium arvense* (L.) Scop	*Aceria anthocoptes* (Nal.)	Young leaves	Marginal roll	Simple	Brownish	Russeting of the leaves	Rančić and Petanović (2002), Rančić et al. (2006)
Fabaceae	*Dichrostachys cinerea* W & A	*Eriophyes dichrostachia* Tuck.	Leaflets	Cylindrical	Complex	Sometimes become pink	Trichomes	Jayaraman (1987)
Fabaceae	*Pongamia pinnata* L.	*Aceria pongamiae* Keifer 1966	Leaves	Clavate, pouch galls	Complex	Green	Glandular trichomes	Nasareen et al. (2012)
Juglandaceae	*Juglans regia* (L.)	*Aceria tristriata* (Nalepa, 1890)	Leaves	Lenticular (blister)	Simple	Brownish	Abaxial aperture	Ferreira et al. (2017)
Lygodiaceae	*Lygodium microphyllum* (Cav.) R. Br.	*Floracarus perrepae* Knihinicki & Boczek	Leaves (subpinna)	Marginal roll	Simple	Green or brownish	–	Freeman et al. (2005)
Malvaceae	*Sphaeralcea grossulariae folia* (Hook. & Arn.) Rydb.	*Acalitus sphaeralceae* Keifer	Branch tips and young leaves.	Bud gall	Complex	Ashy gray and reddish brown	Crumpled	Hoover and Shields (1972)

(continued)

Table 2 (continued)

Family	Host plant	Inducer	Site	Morphotype	Complexity	Color	Ornamentation	References
Malvaceae	*Tilia platyphyllos* Scop	*Eryophides tiliae* (Pagenstecher 1857)	Leaves	Nail-shaped pouch gall	Complex	Red	Trichomes	Guedes et al. (2023)
Melastomataceae	*Miconia ibaguensis* (Bonpl.) Triana; *M. notabilis* Triana	Unidentified Eriophyidae	Leaves	Erinea (leaf bulging)	Simple	Brownish	Trichomes and emergences	Ferreira et al. (2020), Freitas et al. (2023)
Rosaceae	*Fragaria viridis*	*Fragariocoptes setiger*	Leaves	Globoid pocket-shaped gall	Complex	Red	Trichomes	Paponova et al. (2018)
Verbenaceae	*Lantana camara* L.	*Aceria lantanae* Cook	Leaves	Globoid, pouch galls	Complex	Whitish	Trichomes	Moura et al. (2008, 2009)
Verbenaceae	*Vitex agnus-castus* L.	*Aceria massalongoi* (Canestrini, 1890)	Leaflet, petioles and receptacle	Globoid	Complex	Green, Pale red	–	De Lillo et al. (2020)

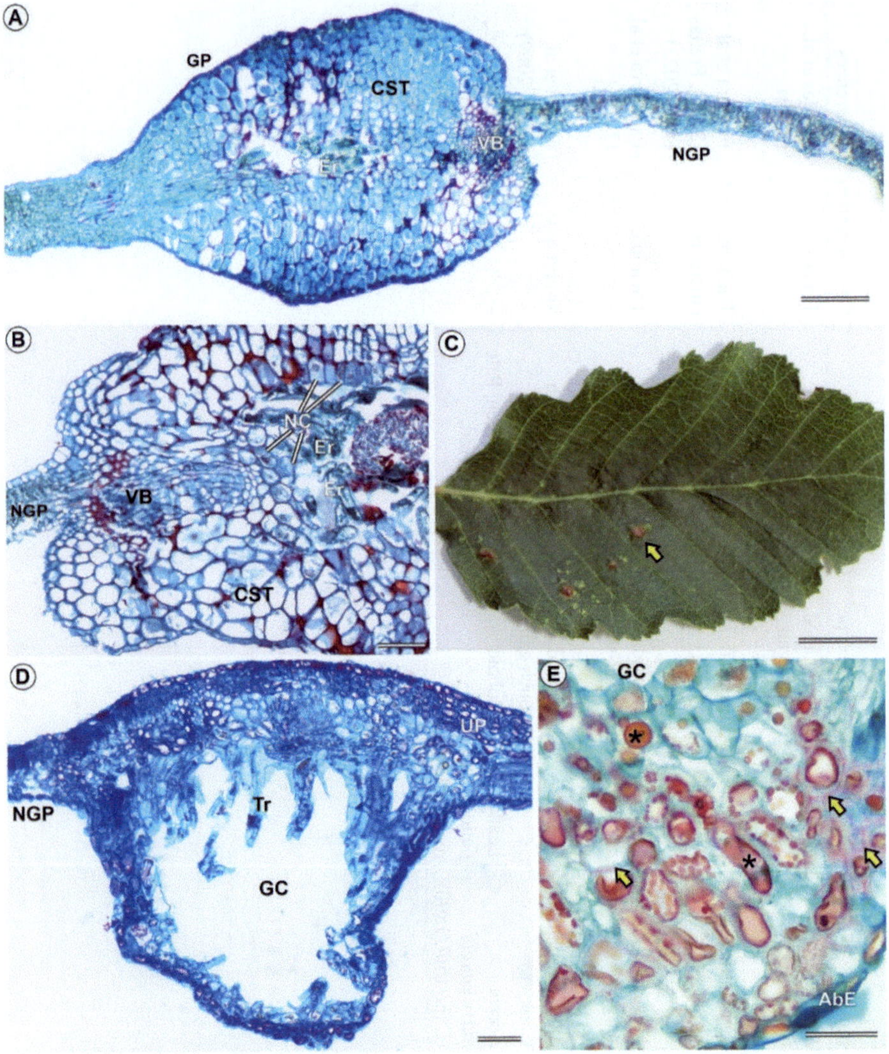

Fig. 3 Blister galls induced by eriophyoids. (**a**, **b**) Blister gall induced by *Aceria tristriata* on *Juglans regia*. The eriophyids (Er) occupy and feed on mesophyll cells. The immediate cells in contact with the eriophyids and the adjacent cells redifferentiate into nutritive cells (NC). The outer layers have hypertrophied and vacuolated cells of the common storage tissue (CST). (**c**–**e**) Red blister galls on *Nothofagus obliqua* leaves. (**c**) Macroscopic images of galls (arrow). (**d**, **e**) Transversal sections of the gall. (**d**) Gall chamber (GC) with neoformed tricomes (Tr), and an upper homogeneous parenchyma (UP), correspondent to the palisade parenchyma of non-galled leaf portion. (**e**) Abaxial portion of the gall, showing lignified parenchyma cells (arrows) and tannins (asterisks). Abbreviations: AbE abaxial epidermis, GP galled portion, NGP non-galled portion, VB vascular bundle. Scale bars: (**a**) 200 μm. (**b**, **d**) 100 μm. (**c**) 1 cm. (**e**) 50 μm

is related to the parenchyma homogenization, cell hypertrophy and hyperplasia of mesophyll. The cells more distant from the eriophyids are vacuolated and hypertrophied, accumulating starch, and functioning as a common storage parenchyma, an energetical reserve that maintains gall functioning.

Another examples of pocket galls are those induced by an unidentified species of Eriophyoidea on *Nothofagus obliqua* (Mirb.) Oerst. (Nothofagaceae) leaves (Fig. 3c). These typical pocket galls are mostly red on their adaxial projection and whitish on their abaxial side. The mite induces the homogenization of the mesophyll, and the formation of numerous modified hairs that project towards the gall chamber (Fig. 3d). The parenchymatic and epidermal cells that surround the closed nymphal chamber have large vacuoles with abundant contents, strongly stained with safranin, which seems to be tannins (Fig. 3d, e). Other cells have thick and granular cytoplasm. Both the abaxial and adaxial epidermis, as well as the parenchyma cells and the hairs or projections, become lignified (Fig. 3e).

Pouch galls form due to the differential growth—hyperplasia and hypertrophy—of the outer layers of the galls compared to the inner layers. Along with the hypertrophic and hyperplasic processes, several periclinal and anticlinal cell divisions occur in the ground meristem, allowing the invagination of leaf blade, usually protruded to adaxial surface (Fig. 4). The invagination of the leaf blade leads to the formation of an open to a relatively closed gall chamber, in which there is an ostiole in the opposite side of the invagination, closed by emergences originated from the leaf mesophyll and dense trichomes (Moura et al. 2008, 2009). The common sites

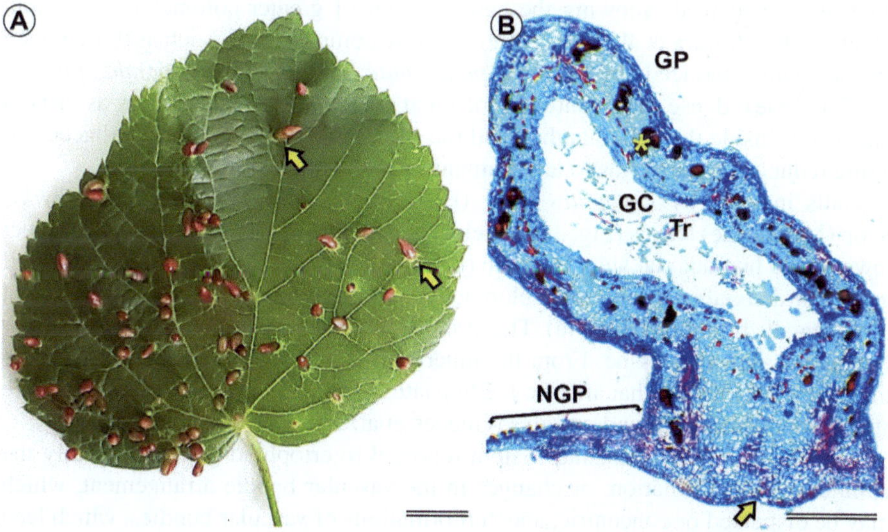

Fig. 4 Pouch galls induced by *Eriophyies tiliae* on *Tilia platyphyllos* leaves. (**a**) Red nail-shaped pouch galls (arrows) projected towards adaxial surface. (**b**) Tangential section of gall (GP), transversal to non-galled portion of leaf (NGP). Arrow indicate the ostiole. Note trichomes (Tr) projected towards the gall chamber (GC) and mucilaginous cells in gall parenchyma (asterisks). Scale bars: (**a**) 1 cm. (**b**) 500 μm

of gall induction are the abaxial regions of the leaves (Ferreira et al. 2017; Paponova et al. 2018; Guedes et al. 2023), or often both leaf surfaces (Jayaraman 1987; Nasareen et al. 2012; Anand and Ramani 2021). The induction in the leaf abaxial surface cause the formation of leaf blade invagination (pouch) towards the adaxial surface. Therefore, the ostioles in these cases are positioned toward abaxial surface. Pouch galls may have distinct shapes as nail (Guedes et al. 2023), horn-shaped, clavate (Nasareen et al. 2012), and globoid galls (Moura et al. 2008, 2009).

Epidermal cells of the outer gall surface are simple, sometimes with the presence of an indument that helps protect the eriophyids—maintaining the host plant organ potentials in the galls (Moura et al. 2008, 2009; Nasareen et al. 2012; Anand and Ramani 2021). Sometimes, the adaxial epidermal cells accumulate pigments, such as anthocyanins, promoting the pigmentation of the galls (Hoover and Shields 1972; Jayaraman 1987; Moura et al. 2008, 2009; Nasareen et al. 2012; Paponova et al. 2018; De Lillo et al. 2020; Anand and Ramani 2021) (Table 2), and provide a certain type of protection, especially in high UV environments (Guedes et al. 2023, 2024b).

In some galls, the epidermal cells surrounding gall chamber—inner epidermal cells—become nutritive tissue cells, important for the direct feeding by eriophyids (Hoover and Shields 1972; Guedes et al. 2023). These epidermal cells may be hypertrophied, some with a radially elongated palisade shape, with a dense granular cytoplasm and a hypertrophied nucleus (Jayaraman 1987; Moura et al. 2008, 2009) (see below). In some galls, processes of hyperplasia and cell hypertrophy of dermal and ground tissues surrounding gall chambers lead to the appearance of clefts or emergences, which increase the gall inner surface, in which the eriophyoids protect themselves and feed, allowing the better shelter of greater colonies (Hoover and Shields 1972; Moura et al. 2008, 2009). In more complex galls, such as those pouch clavate galls induced by *Aceria pongamiae* Keifer on *Pongamia pinnata* L. (Fabaceae), dense indumentum neoformation (called by the authors as erinea) may occur inside the gall chamber, and the dense trichomes are nutritive, accumulating reducing sugars (Anand and Ramani 2021).

Galls induced by *Eryophies tiliae* (H.A. Pagenstecher) on *Tilia platyphyllos* Scop (Malvaceae) leaves (Fig. 4) is an elegant example of complex galls that project towards the leaf adaxial side. Because of their shape, these galls are known as nail galls. The adaxial epidermis and chlorenchyma layers accumulate anthocyanins that turn the galls bright red (Fig. 4a). The inner layers that border the nymphal chamber function as nutritive tissue. From the innermost layer, trichomes and long projections emerge (Fig. 4b) that are also feeding sites for the mite colony and accumulate starch, reduced sugars, and proteins (Guedes et al. 2023).

The changes in the vascular system reported to eriophyid galls are usually the maintenance, deformation, or changes in the vascular bundle arrangement, which may be dispersed or concentric, and neoformations of vascular bundles, which lead to greater supply of water and nutrients for galls (Moura et al. 2008, 2009; Paponova et al. 2018; Anand and Ramani 2021; Guedes et al. 2023, 2024a). Some more complex galls have also a neoformed sclerenchyma sheath (mechanical tissue), as occur in some insect galls. Neoformed parenchyma cells in galls induced by *A. pongamiae* on *P. pinnata* redifferentiate into hypertrophied sclerenchyma cells, sparsely distributed in the outermost layer of parenchyma cells (Anand and Ramani 2021).

Other complex gall morphotypes induced by mites, such as bud, stem, and flower galls are reported in the literature, with similar alterations when compared to those induced on leaves (Westphal and Manson 1996). Although very common in leaf blades, eriophyid erineum galls on *M. ibaguensis* were also found in petioles and stems (Ferreira et al. 2017). It similarly occurs in *Acalitus sphaeralceae* Keifer on *Sphaeralcea grossulariaefolia* (Hook. and Arn.) Rydb (Malvaceae), where bud galls remain in the stems even after leaf fall during autumn, and therefore such galls are important to the survivorship of mites in this season (Hoover and Shields 1972). In bud galls, the mites are lodged among leaf primordia and may induce changes as cell hypertrophy, hyperplasia, and differentiation of nutritive cells (Westphal and Manson 1996). Stem galls reported to Eriophyidae commonly are products of outgrowths in the host organs induced by feeding activity of the inducer, forming coverships (emergences), where the mites install and feed (Mani 1964; Westphal and Manson 1996).

4.3 Nutritive Cells and Common Storage Cells in Mite Galls

In pouch galls, the larval chamber is lined by epidermal cells and adjacent layers of parenchyma cells. The cells lining the larval chamber are under stimuli of mites feeding activity, and then it appears to increase in size and change their cytoplasm density and chemical composition. These redifferentiated cells form a typical nutritive tissue (Ferreira et al. 2017), that constitutes the mites' food source (Hoover and Shields 1972; Freeman et al. 2005; Moura et al. 2008; Moura et al. 2009; Paponova et al. 2018). Nutritive cells have enlarged nuclei and nucleoli, granular cytoplasm, and small vacuoles (Jayaraman 1987; Westphal and Manson 1996; Guedes et al. 2023), and they can accumulate starch, reducing sugars, lipids, phenolic compounds, flavonoids, tannins, and proteins (Guedes et al. 2023).

In galls induced by *A. lantanae*, the epidermal and ground system cells that make up the nutritive tissue undergo hyperplasia, forming emergences that facilitate the mites' feeding (Moura et al. 2008). The emergence also increases the surface/volume rate, which allows the gall to shelter a greater population of eriophyids inside the gall chamber. In the case of simpler galls such as erinea, cell hypertrophy usually is not involved, and vacuolated epidermal cells with reducing sugars serve as feeding sites for eriophyid mites, as in *M. ibaguensis* and *M. notabilis* leaf galls (Ferreira et al. 2017; Freitas et al. 2023). The neoformed filiform trichomes are also sites of mite feeding in erinoses induced on *A. schaueriana* and *V. vinifera* leaves. Such trichomes accumulate proteins and reducing sugars as the main food sources for the eriophyids (Nobrega et al. 2023; Guedes et al. 2024a).

Besides typical nutritive tissues or nutritive hairs, the galls induced by eriophyoids also have common storage tissues in the outer layers of the galls—in similarity with insect galls (Ferreira et al. 2019). Common storage cells are usually parenchymatic and are more distant from the gall chamber or eriophyoid feeding sites than nutritive cells. Such cells are usually more hypertrophied and vacuolated than nutritive cells and accumulate starch and/or lipids—depending on the host plant potentialities –, that are considered important to gall tissue growth and maintenance (Ferreira et al. 2017; Guedes et al. 2023, 2024a). Phenolics and other secondary metabolites also accumulate in outer layers, which may be important to antioxidant and photoprotective defenses to gall tissues and to the gall inducers (Ferreira et al. 2018; Guedes et al. 2023, 2024b).

4.4 Cell Wall Changes

Cell wall changes are not commonly studied in mite galls. In those eriophyid galls induced on *M. ibaguensis* leaves, few changes in cell wall composition of ground tissues and in emergences that compose the erinea were found in comparison to non-galled leaves, which may be associated with the simplest changes promoted in the formation of erinoses (Ferreira et al. 2020). A more intense accumulation of galactans in cell walls of the gall mesophyll guarantees the reinforcement of cell walls during cell hypertrophy. In the storage tissue and in the feeding cells of the emergences of this gall, a combination of pectins with arabinans and galactans may provide a greater porosity to the cell walls, allowing an increase in water and sugar flow from the storage tissues to nutritive cells. Lignification is another change induced by some eriophyoids on leaf affected area, as previously described for *N. obliqua* (Nothofagaceae) leaf galls.

Other common changes in mite galls are the deposition of suberin as a form of healing the injuries caused by feeding, as observed in pouch galls induced by *A. lantanae* on *L. camara* (Moura et al. 2009) and in subepidermal layers of *A. schaueriana* erinea, where a phellogen-like tissue differentiates (Nobrega et al. 2023).

5 Multitrophic Relationships in Eriophyoid Galls: Ecological Apects

Several associations and natural enemies have been documented concerning galls induced by eriophyoids. Eriophyoids, being smaller than most predatory mites, are particularly susceptible to predation (Sabelis and Bruin 1996). Host plants serve as refuges, allowing eriophyoids to evade their predators. These refuges can be found in confined spaces where predatory mites cannot easily access, such as within modified structures like erinea, leaf edge rolling galls, blister and pouch galls enclosed by trichomes at the ostiole, as well as in the gaps between scales and within buds. These galls not only provide protection against desiccation, due to their closed or densely covered surfaces, but also serve as protective shelters against possible predators (Sabelis and Bruin 1996; Michalska et al. 2010).

The eriophyoids may use the plant structure and their developmental changes induced by gall formation as protective apparatuses against the invasion of the galls and attack by predators and opportunistic herbivores. The plants may also attract predators to control the population of galling eriophyoids. Eriophyoid mites are food sources for predatory mites (Phytoseiidae, Ascidae, Stigmaeidae, Tydeidae, Anystidae) and insects (Heteroptera, Lepidoptera, and Diptera) that may be used as biological control agents against other targets (*see* De Lillo and Skoracka 2010 and references therein). Stigmateids and phytoseiids (Acari) are the main predators of eriophyoids. Stigmateids can detect them at a distance by the volatiles emitted by the host plants and are smaller than the phytoseiids. Because of their sizes, stigmateids may penetrate hiding places inaccessible for the phytoseiids (Sabelis 1996; Michalska et al. 2010). The phytoseiids may easily and quickly capture the eriophyoids, which seems to be their preferential food source (Thistlewood et al. 1996), and they usually feed on distinct eriophyoid species and on all developmental stages (Sabelis 1996). They may invade old galls, with drying tissues and opening ostiole, and eat the mites. The erinea are also often visited by predatory mites.

Besides Eriophyoidea, Tarsonemidae mites were considered by Mani (1964) as another group of mites able to induce galls. We have not found works in the literature that demonstrate the gall-inducing capability of the Tarsonemidae to make comparisons with Eriophyoid galls. However, tarsonemid mites were observed inhabiting galls induced by Eriophyidae, such as those induced by the maple spindle gall mite *Vasates aceriscrumena* Riley & Vasey in the sugar maple *Acer saccharum* Marsh (Sapidaceae) (Patankar et al. 2012). In this case, there is evidence that the tarsonemid *Tarsonemus acerbilis* not only occupies the galls, especially those with more open ostioles, but they also eat the eggs of galling eriophyids (Patankar et al. 2012). This work demonstrates that the more enclosed galls are the best protective structures against the predation of the eriophyoids by natural enemies.

Despite of the observations of Patankar et al. (2012), the tarsonemids are usually reported as non-predatory invaders of eriophyid gall, as reported by Beer (1963) and Perring and McMurtry (1996), who observed the presence of tarsonemid in galls induced by distinct species of eriophyids. In these cases, the tarsonemids did not cause changes in host plant cells, but they seem to use the galled cells as food—although non-galled host plant cells were not observed as food sources of these predators.

Beer (1963) and Patankar et al. (2012) also observed the invasion of Eriophyidae galls by other arthropods, such as Cecidomyiidae larvae and mites of other families. Cecidomyiidae larvae were reported as predators occurring inside eriophyoid galls (Beer 1963; Perring and McMurtry 1996), as also coleopterans, as some species of Coccinelidae, insects of other orders, and mites of distinct families (Perring and McMurtry 1996). The protection of the closely packed trichomes in gall ostioles is important, although they may not completely impair the action of the invaders (Patankar et al. 2012). Cecidomyiidae larvae may open the gall ostioles by mechanical force, allowing the entrance of mite predators. Beer (1963) also observed that erinea difficult the action of predatory and invader mites, since they are usually larger than the tiny gall-inducing eriophyids.

Recently, the braconid larvae of *Bracon predatorius* Ranjith & Quicke (Braconidae: Hymenoptera) was reported as a predator solely of the eriophyoids *Aceria doctersi* Nalepa on *Cinnamomum verum* (Lauraceae) J. Presl, an important crop (Ranjith et al. 2022). This feeding behavior is novel among the braconids, which had already reported parasitoid species, gall inducers, entomophagous, and egg predators. *Bracon predatorius* adults were observed ovipositing inside the eriophyid galls (Ranjith et al. 2022). Besides predators, pathogens may attack eriophyoids. However, according to McCoy (1996), virus, bacteria, and protozoa were not observed in eriophyoids. Some pathogenic fungi were reported in eriophyoids, although less than in other mite lineages. The wet and dark habitat provided by the galls could facilitate the infection of the galls by fungi.

Taking all interactions into account, galls offer mites and their offspring a good refuge from natural enemies, generally predators and abiotic stresses, as previously described for galling insects. However, the wet environment may facilitate the presence of fungal pathogens in mite galls. Even with these physical protections against natural enemies, several predators and invaders are reported to affect the mite galls. As proposed by Stone and Schönrogge (2003) for insect galls, the eriophyoid gall structure may be under constant action of negative pressures, and then the most rigid and thicker gall walls—or longer indumentum in erinea—could best avoid predators, which deserves to be studied.

6 Physiology of Eriophyid Galls

The biochemistry and physiology of the host plant is clearly altered during gall-inducing eriophyid infestation (Desnitskiy et al. 2023; Nobrega et al. 2023; Guedes et al. 2024a, b). These alterations trigger significant responses in the physiological processes of host plants, such as photosynthesis, primary and secundary metabolite synthesis, transpiration, stomatal conductance, efficient water use, and gas exchange (Patankar et al. 2011; Jiang et al. 2020; Guedes et al. 2024a, b). However, limited available data suggest that plant physiological responses triggered by gall-inducing mites can range from compensatory gains (Fay et al. 1993; Dorchin et al. 2006; Tung and Martin 2017) to significant reductions in plant productivity (Patankar et al. 2011; Daud et al. 2012; Takeda et al. 2019). Such physiological parameters vary depending on the type of gall-inducer and the host plant, demanding research to elucidate the determinants of complexity levels of eriophyid gall impacts on host plants (Welter 1989; Tung and Martin 2017).

The modifications caused by galling eriophyoids generally involve changes in primary and secondary metabolism, nutrient contents, and in the balance of plant hormones (De Lillo and Monfreda 2004; Chetverikov et al. 2015a, b; Jiang et al. 2020), as for example in *V. vinifera* - *Co. vitis* system (Guedes et al. 2024a). *Co. vitis* induces an increase in anthocyanin pigments and total soluble sugars in corinto cultivar leaves, while starch content decreases. Polyphenol content remains unaffected by the mite, but both the profile and concentration of detected polyphenols were also affected.

As eriophyid mites cause damage in the first cell layers of host leaves, these injuries can result in potential impacts on gas exchange and, consequently, on photosynthesis (Patankar et al. 2011; Jiang et al. 2020). The effects of the saliva of eriophyoids on plant tissues may be similar to those produced by phytohormone imbalance (De Lillo and Monfreda 2004). Their saliva may contain auxins and cytokinins, as in galling insects (Tooker and Helms 2014; Andreas et al. 2020; Ponce et al. 2021). The difficulty in collecting enough eriophyid saliva is a limiting factor to the advances in studies on composition and effects of their secretions in plant metabolism (Desnitskiy et al. 2023). In the gall induction process by *Aceria pallida* Keifer on *Lycium barbarum* L. (Solanaceae), expression of genes involved in indolacetic acid (IAA) and cytokinins (CK) are upregulated (Yang et al. 2023). Additionally, a profound impact on the chlorophyll synthesis and photosynthetic capacity of wolfberry was observed, related to the dowregulation of biosynthesis of chlorophyll and carotenoids. Given the strong suppression, there was a need for assimilates in tissues adjacent to the galls. To meet this demand, genes related to transmembrane transport were significantly upregulated in the tissues affected by the galls, accumulating nutrients in galls through transmembrane transportation. This enrichment was evidenced by the substantial upregulation of genes associated with the tricarboxylic acid (TCA) cycle, oxidative phosphorylation process, transmembrane transportation, and carbohydrate and amino acid synthesis and accumulation (Yang et al. 2023).

Reduction of photosynthesis rates and stomatal conductivity are reported on *Vasates aceriscrumena* (Riley) galls on *Acer saccharum* Marsh. (Patankar et al. 2011) and *A. erinea* on *J. regia* leaves (Popitanu et al. 2021). These effects were accompanied by lower water use efficiency (WUE) in *A. saccharum* (Patankar et al. 2011) and a decrease in photosynthetic pigment contents (chlorophyll and β-carotene), along with an increase in the production of volatile compounds on *J. regia* leaves (Popitanu et al. 2021). The decrease in stomatal conductivity could be a potential mechanism to avoid non-stomatal water loss within and around the gall (Patankar et al. 2011). Additionally, the *V. aceriscrumena* galls have shown reduced nitrogen content and increased C:N ratios in galled leaves.

Eriophyoids also can trigger a significant source of plant resistance through the production of secondary metabolites. For example, in grapes, elevated levels of anthocyanins and total phenols, along with the presence of the phenol pinocembrin in the corinto cultivar, suggest potential resistance to *Co. vitis* attacks (Guedes et al. 2024a). In another study, infestation of *Aceria schlechtendali* Nalepa at a density of 22–27 specimens per cm^2 of leaf caused an increase in the ratio of total phenolic compounds to carbohydrates, shifting from a 3:1 ratio in the control treatment to 5:1 in mite-infested leaves of a susceptible apple cultivar. This alteration affected the palatability of the host plant, reducing it for the mites (Kozlowski and Kozlowska 1998). Similar results were observed in field-cultivated blackberry cultivars infested with the blackberry rust mite, *Epitrimerus gibbosus* (Nalepa) (Shi and Tomczyk 2001). When infested, host plants initiate various defense responses against gall inducers, involving both direct defenses such as the accumulation of secondary metabolites and indirect defenses like the release of volatile compounds, such responses may even lead to the death of gall-inducing organisms (Jiang et al. 2020; Popitanu et al. 2021; Guedes et al. 2023; Yang et al. 2023). As a defense mechanism against the gall mite, the wolfberry (*L. barbarum*) triggered a defensive response, leading to the accumulation of secondary metabolites. This was supported by a significant positive regulation of most genes involved in the defensive response process and the phenylpropanoid (PPP) pathway, along with a noticeable increase in the relative abundance of flavonoids and coumarins in the gall-affected tissues (Yang et al. 2023).

In general, tissues outside the galls exhibit an abundance of phenols, flavonoids, and other secondary metabolites (Kuster et al. 2019). Additionally, other responses are modulated during infestation, including an increase in the production of oxidative enzymes such as superoxide dismutase, catalase, peroxidase, and polyphenol oxidase, as well as defense-related phytohormones like salicylic acid and jasmonic acid (Zhu et al. 2011; Soika et al. 2017; Dorleku and Gbewonyo 2021; Yang et al. 2023).

7 Impacts in Productivity and Examples of Economic Interest

As Eriophyoids are distributed worldwide and cause damages by gall induction and/or transmitting phytopathogenic viruses (Table 1), they affect a wide variety of commercial interest plants. Although the study of economically important galling mites has been increasing in recent decades, the magnitude of the damage caused and yield losses in affected crops still requires detailed studies. These studies will allow us to establish prevention and control strategies. Ornamental, medicinal and food plants are affected by galling mites, mainly of the Eriophyidae family. Some species that are affected by galling eriophyoids are alder, aspen, *Baccharis*, beech, elm, grape, linden, maple, and willow (Table 1).

Among ornamental crops, the aloe eriophyoid mite *Eriophyes aloinis* Keifer (= *Aceria aloinis*) is known to induce cauliflower-shaped malformations in leaves, stems, and inflorescences of *Aloe* spp. (Asphodelaceae) (Vachajitan and Harvey 2021). The aloe producers known this condition as “aloe cancer”, affecting the plant aesthetics and its commercialization. The genus *Fuchsia* (Onagraceae, Myrtales) is also affected by the gall-inducing eriophyoid mite *Aculops fuchsia* Keifer. This mite probably originated in South America but has spread throughout North America and Europe (Streito et al. 2004; Salisbury et al. 2014). The mite induces galls on the leaves and flowers of fuchsias, beginning as a noticeable redness at the tips of the shoots that produces thickening and deformation of the leaves and flowers (Salisbury et al. 2014). Terminal shoots can be affected, suppressing new growth, and the infestation of flower buds, also causes abnormal flowering (Koehler et al. 1985). *Aculops fuchsia* is considered one of the main pests of fuchsia crops, impacting the production of plants and flowers of this species in the countries where it has been introduced. For example, in California, USA, producers have abandoned the cultivation of this ornamental species due to the strong impact of the galling mite (Koehler et al. 1985). In Europe, it is a 40-year pest, included on the alert list of the European and Mediterranean Plant Protection Organization (EPPO) (EPPO datasheet 2022; Picard et al. 2018).

Eriophyoids affect several crops worldwide (Lindquist 1996). The impacts of eriophyids in plant productivity are low when compared with other herbivores since they have short stylets and usually affect superficial cells or cells in direct contact with them (Sabelis and Bruin 1996). However, some cases of relevant impact on the host plants are reported in the literature, such as the erinea/leaf edge rolling induced by *Cecidophyes caroliniani* Chandrapatya & Baker on wild geranium, massive defoliation caused by *Phyllocoptruta oleivora* Ashmead on citrus trees, and wilting and death of tomato plants caused by *Aculops lycopersici* A. M. Massee. The two latter cases are product of a collapse of mesophyll cells and destruction of epidermal cells (Sabelis and Bruin 1996). In the case of *A. lycopersici,* there are native closely related species of *Solanum* which are not affected as tomato plants.

Tilia species are known for their medicinal and ornamental use (Shikov et al. 2021). On the leaves of these species, at least two eriophyoid species induce galls, *Eriophyes tiliae* Pagenstecher causing nail galls and *E. exilis* Nalepa causing erinea. Not many studies have addressed the implications of the establishment of eriophyoid mites on the medicinal properties of *Tilia* spp. However, recent studies demonstrated that *E. tiliae* induces alterations in the concentrations of total phenols and anthocyanin that are reflected in an increase in the antioxidant potential of *T. platyphyllos* Scop (Guedes et al. 2023, 2024b). This evidence may imply possible changes in the composition of the phenolic compounds of *T. platyphyllos* leaves that modify the medicinal properties of this species and other *Tilia* spp. hosts of eriophyoid galling mites.

Many fruit species are also affected by eriophyoid gall-inducing mites, including pear (*Pyrus communis* L., Rosaceae), black currant (*Ribes nigrum* L., Grossulariaceae), and grape (*V. vinifera*). *Cecidophyopsis ribis* Westwood is considered the most serious pest of blackcurrant in European countries. The mite migrates from dry buds to newly formed buds in the spring, during flowering, where it feeds on new tissues (Duso and De Lillo 1996). The establishment of mites and their colonies induces delayed bud growth, deformation of leaves and fruits, blistering, necrosis, and leaf discoloration, drying of buds, and loss of fruit yield (Duso and de Lillo 1996; Morocko-Bicevska et al. 2022). In addition to the described damage associated with gall formation by *Ce. ribis*, this mite species is the vector of the Blackcurrant Reversion Virus (BRV) (Lemmetty et al. 2004). In Europe, most blackcurrant cultivars are susceptible to this mite (Brennan 1996). The BRV induces a reversion of the plant to its wild phenotype, producing smaller, distorted leaves in a few numbers, decreased hairiness of the floral buds and flowers, darkening of the flower from dark red to violet, abnormal and smaller fruits that fall before ripening (Adams and Thresh 1987). The combination of the mite and BRV in susceptible cultivars can cause between 70% and 100% yield losses (Pluta et al. 2000; Łabanowska and Pluta 2010). Currently, the virus has spread throughout Europe, New Zealand and recently Canada and the United States (Zurn et al. 2019).

The pear leaf blister mite, *Eriophyes pyri* Pagenstecher, is considered an important pest affecting fruit trees worldwide (Easterbrook 1996). The mite induces blisters (erinea) on the new leaves of pear and apple trees. The blisters begin as discolored spots on new leaves that turn red and at the end of their development become necrotic, inducing a decrease in photosynthesis and eventual leaf fall (Easterbrook 1996). Blisters also form on the fruit calyx, causing deformation and fall. The ability of *E. pyri* to transmit the European mountain ash ringspot-associated emaravirus (EMARaV) has also been reported (von Bargen et al. 2019), causing chlorotic ring spots, mottling, and line patterns on leaves (Roßbach et al. 2015). Although the virus was initially reported in *Sorbus aucuparia* L. (European mountain ash), it is currently known that it can affect other species of the genus *Sorbus* (Grimová et al. 2015; von Bargen et al. 2019).

Probably the most studied gall-inducing eriophyid is *Co. vitis*, due to its implications in the grapevine cultivation (*V. vinifera*), although many of the studies have been carried out in a very restricted geographical area (Iran, Switzerland, and Italy). *Co. vitis* induces leaf erinea on different table grape cultivars throughout the world (Guedes et al. 2024a). It has been reported that this mite induces zigzag sprouts and, in smaller numbers, a decrease in yield, in the quality, and in the quantity of grapes produced (Javadi et al. 2014, 2018a, b). Like other galling eriophyoids, *Co. vitis* is considered a vector transmitting the grapevine Pinot Gray virus (Malagnini et al. 2016) that causes mottling and deformation of grapevine leaves (Giampetruzzi et al. 2012).

The eriophyoid galling species overwinter under bud scales of their host plants, which also involves damage during this stage. For example, during winter, high infestations of mite colonies of the dormant form of *Co. vitis*, can cause total or partial necrosis of the primary bud, which does not sprout or generates defective shoots, with short, thin, and zigzag internodes, without fruit. In many grape cultivars, the secondary buds are mostly infertile, which directly affects the productive yield in these cultivars (Maboeei 2022).

Some eriophyoids cause damages in several crop species—such as *Aceria tosichella* Keifer, the wheat curl mite, which infects a wide range of cultivated and wild Poaceae, affecting the yield of grass crops by inducing curled, looped and trapped leaves. As *A. tosichella,* several eriophyoid are also important viruses' vectors which severely affect economically important plants worldwide (Lindquist 1996; De Lillo and Skoracka 2010). Other important crops such as litchi and mango are also affected by eriophyoid mites, which can be important fungal vectors (De Lillo et al. 2018). Several ornamental and forest are also affected by eriophyoids, and distinctive approaches may be proposed to their control (Castagnoli et al. 2010).

While the examination of economically significant galling mites has seen an increase in recent decades, a comprehensive understanding of the extent of damage and associated yield losses in affected crops necessitates more in-depth investigations. These studies are crucial for developing effective prevention and control strategies. Despite inducing seemingly subtle alterations in leaf anatomy compared to other gall inducers, certain eriophyoids can extensively infect host plants, resulting in leaf distortion. It is imperative to thoroughly assess these diminutive anatomical effects, particularly on the epidermis, as changes in epidermal development—such as trichome neoformation, stomatal malformations, reduced photosynthesis rates, and drainage of primary metabolites—may lead to decreased yields in both crop and native plants. An integrated study of the anatomy and physiology of eriophyoid galls is essential for comprehending and mitigating the damages inflicted by eriophyoids on plant productivity.

8 Conclusions and Perspectives

While galling eriophyoids are globally distributed and infect various plant groups, their small size poses challenges for accurately understanding their taxonomy, behavior, and associations. These parasitic organisms establish antagonistic relationships with host plants, potentially causing severe infestations on plant performance. Gall inducers compromise plant development and vigor by dispersing and covering significant leaf areas. Research on eriophyid mite-induced galls has focused on understanding gall formation mechanisms, including mechanical stimulation and fluid deposition during feeding, leading to reprogramming of host tissue development. This results in the formation of nutritive tissue, providing a favorable environment for mites and protection from predators and abiotic factors. Efforts in recent decades have aimed to understand the ecological and evolutionary aspects of gall-inducing organisms and develop methods for their control. Eriophyoid galls, while appearing simple, reveal complex control mechanisms exerted by mites on host tissues, leading to physiological effects on host plants and potential reductions in productivity and susceptibility to pathogens. Integrating taxonomical and behavioral studies of mites with gall anatomy, cytology, and physiology can enhance our understanding of their effects and control mechanisms.

References

Adams AN, Thresh JM (1987) Virus and viruslike diseases of black currant. Mite–borne diseases. In: Converse RH (ed) Virus diseases of small fruits. US Government Printing Office, Washington, pp 133–166

Anand PP, Ramani N (2021) Dynamics of limited neoplastic growth on *Pongamia pinnata* (L.) (Fabaceae) leaf, induced by *Aceria pongamiae* (Acari: Eriophyidae). BMC Plant Biol 21. https://doi.org/10.1186/s12870-020-02777-7

Andreas P, Kisiala A, Emery RJN, De Clerck-Floate R, Tooker JF, Price PW, Miller DG III, Chen MS, Connor EF (2020) Cytokinins are abundant and widespread among insect species. Plants 9:208. https://doi.org/10.3390/plants9020208

Araújo WS, Kollár J (2019) First characterization of a highly specialized ecological network composed by gall-inducing mites and their host plants. Int J Acarol 45:223–226. https://doi.org/10.1080/01647954.2019.1591505

Araújo WS, Freitas ÉVD, Kollár J, Pessoa RO, Corgosinho PHC, Valério HM, Falcão LAD, Fagundes M, Pimenta MAS, Faria ML, Martins WP, Borges MAS (2019) Host specialization in plant-galling interactions: contrasting mites and insects. Diversity 11:180. https://doi.org/10.3390/d11100180

Beer RE (1963) Social parasitism in the Tarsonemidae, with description of a new species of Tarsonemid mite involved. Ann Entomol Soc Am 56:153–160

Brennan RM (1996) Currants and gooseberries. In: Janick J, Moore JN (eds) Fruit breeding, Vine and small fruits crops, vol II. Wiley, New York, pp 191–295

Caglayan K, Elci E, Serce CU, Kaya K, Gazel M, Medina V (2012) Detection of fig mosaic virus in viruliferous eriophyid mite *Aceria ficus*. J Plant Pathol 94:629–634

Castagnoli M, Lewandowski M, Łabanowski GS, Simoni S, Soika GM (2010) An insight into some relevant aspects concerning eriophyoid mites inhabiting forests, ornamental trees and shrubs. Exp Appl Acarol 51:169–189. https://doi.org/10.1007/s10493-009-9313-8

Chetverikov PE, Cvrković T, Makunin A et al (2015a) Basal divergence of Eriophyoidea (Acariformes, Eupodina) inferred from combined partial COI and 28S gene sequences and CLSM genital anatomy. Exp Appl Acarol 67:219–245. https://doi.org/10.1007/s10493-015-9945-9

Chetverikov PE, Vishyakov AE, Dodueva IT, Osipova MA, Sukhareva SI, Shavardab AL (2015b) Gallogenesis induced by Eryophyoid mites (Acariformes: Eriophyoidea). Entomol Rev 95:1137–1143

Chetverikov PE, Craemer C, Cvrković T et al (2021) Molecular phylogeny of the phytoparasitic mite family Phytoptidae (Acariformes: Eriophyoidea) identified the female genitalic anatomy as a major macroevolutionary factor and revealed multiple origins of gall induction. Exp Appl Acarol 83:31–68. https://doi.org/10.1007/s10493-020-00571-6

Craemer C (2010) A systematic appraisal of the Eriophyoidea (Acari: Prostigmata). PhD Thesis (Zoology), University of Pretoria

Daud RD, Conforto EC, Feres RJF (2012) Changes in leaf physiology caused by Calacarus heveae (Acari, Eriophyidae) on rubber tree. Exp Appl Acarol 57:127–137. https://doi.org/10.1007/s10493-012-9552-y

De Lillo E, Monfreda R (2004) 'Salivary secretions' of eriophyoids (Acari: Eriophyoidea): first results of an experimental model. Exp Appl Acarol 34:291–306. https://doi.org/10.1007/s10493-004-0267-6

De Lillo E, Skoracka A (2010) What's "cool" on eriophyoid mites? Exp Appl Acarol 51:3–30. https://doi.org/10.1007/s10493-009-9297-4

De Lillo E, Pozzebon A, Valenzano D, Duso C (2018) An intimate relationship between Eriophyoid mites and their host plants—a review. Front Plant Sci 9:1786. https://doi.org/10.3389/fpls.2018.01786

De Lillo E, Fanelli E, Valenzano D, Monfreda R, Troccoli A, Vovlas A, De-Luca F (2020) Characterisation of *Aceria massalongoi* and a histopathological study of the leaf galls induced on chaste trees. Exp App Acarol 82:33–57

De Lillo E, Freitas-Astúa J, Watanabe Kitajima E, Ramos-González PL, Simoni S, Tassi AD, Valenzano D (2021) Phytophagous mites transmitting plant viruses: update and perspectives. Entomol Gen 41:439–462. https://doi.org/10.1127/entomologia/2021/1283

Desnitskiy AG, Chetverikov PE (2022) Induction of leaf galls by four-legged mites (Eriophyoidea) as a problem of developmental biology. Russ J Dev Biol 53:6–14. https://doi.org/10.1134/S1062360422010039

Desnitskiy AG, Chetverikov PE, Ivanova LA, Kuzmin IV, Ozman-Sullivan SK, Sukhareva SI (2023) Molecular aspects of gall formation induced by mites and insects. Life 13:1–14. https://doi.org/10.3390/life13061347

Dorchin N, Cramer MD, HoVmann JH (2006) Photosynthesis and sink activity of wasp-induced galls in *Acacia pycnantha*. Ecology 87:1781–1791. https://doi.org/10.1890/0012-9658(2006)87[1781:pasaow]2.0.co;2

Dorleku WP, Gbewonyo WSK (2021) Levels of polyphenol oxidase activity in leaves of Milicia seedlings is indicative of the resistance to Phytolyma lata-induced gall disease. Agric For Entomol 23:518–526

Duso C, de Lillo E (1996) Damage and control of Eriophyoid mites in crops: 3.2.5 grape. In: Lindquist EE, Sabelis MW, Bruin J (eds) Eriophyoid mites – their biology, natural enemies and control, World Crop Pests, vol 6. Elsevier Science Publishing, Amsterdam, pp 571–582

Easterbrook MA (1996) Damage and control of eriophyoid mites in apple and pear. In: Lindquist EE, Sabelis MW, Bruin J (eds) Eriophyoid mites: their biology, natural enemies and control. Elsevier, pp 527–541

EPPO datasheet (2022) Aculops fuchsiae. Disponible en gd.eppo.int/taxon/ACUPFU/download/datasheet_pdf

Fay PA, Hartnett DC, Knapp AK (1993) Increased photosynthesis and water potentials in *Silphium integrifolium* galled by cynipid wasps. Oecologia 93:114–120. https://doi.org/10.1007/BF00321200

Ferreira BG, Álvarez R, Avritzer SC, Isaias RMS (2017) Revisiting the histological patterns of storage tissues: beyond the limits of gall-inducing taxa. Botany 95:173–184. https://doi.org/10.1139/cjb-2016-0189

Ferreira BG, Oliveira DC, Moreira ASFP, Faria AP, Guedes LM, França MGC, Álvarez R, Isaias RMS (2018) Antioxidant metabolism in galls due to the extended phenotypes of the associated organisms. PLOS One 13:e0205364. https://doi.org/10.1371/journal.pone.0205364

Ferreira BG, Álvarez R, Bragança GP, Alvarenga DR, Pérez-Hidalgo N, Isaias RMS (2019) Feeding and other gall facets: patterns and determinants in gall structure. Bot Rev 85:78–106. https://doi.org/10.1007/s12229-019-09207-w

Ferreira BG, Bragança GP, Isaias RMS (2020) Cytological attributes of storage tissues in nematode and eriophyid galls: pectin and hemicellulose functional insights. Protoplasma 257:229–244. https://doi.org/10.1007/s00709-019-01431-w

Formiga AT, Silveira FAO, Fernandes GW, Isaias RMS (2015) Phenotypic plasticity and similarity among gall morphotypes on a superhost, *Baccharis reticularia* (Asteraceae). Plant Biol 17:512–521

Freeman TP, Goolsby JA, Ozman SK, Nelson DR (2005) An ultrastructural study of the relationship between the mite *Floracarus perrepae* Knihinicki & Boczek (Acariformes: Eriophyidae) and the fern *Lygodium microphyllum* (Lygodiaceae). Aust J Entomol 44:57–61

Freitas MSC, Ferreira BG, Bragança GPP, Boanares D, Isaias RMS (2023) Can the galling Eriophyidae (Trombidiformes) manipulate leaf structural and histochemical profiles over environmental stressors? Aust J Bot 71:146–156. https://doi.org/10.1071/BT22091

Giampetruzzi A, Roumi V, Roberto R, MalossSSini U, Yoshikawa N, La Notte P, Terlizzi F, Credi L, Saldarelli P (2012) A new grapevine virus discovered by deep sequencing of virus- and viroid-derived small RNAs in cv "Pinot gris". Virus Res 163:262–268

Gispert C, Perring TM, Oldfield GN (1997) Rearing *Eriophyes insidiosus* Keifer and Wilson (Acari: Eriophyoidea), a fastidiosus bud mite. Int J Acarol 23:227–231. https://doi.org/10.1080/01647959708683570

Gispert C, Oldfield GN, Perring TM, Creamer R (1998) Biology of the transmission of peach mosaic virus by *Eriophyes insidiosus* (Acari: Eriophyidae). Plant Dis 82:1371–1374. https://doi.org/10.1094/PDIS.1998.82.12.1371

Grimová L, Marek M, Konrady M, Ryšánek P (2015) Newly identified host range of European mountain ash ringspot-associated virus (EMARaV) and its distribution in The Czech Republic. For Pathol 45:177–189

Guedes LM, Sanhueza C, Torres S, Figueroa C, Gavilán E, Pérez CI, Aguilera N (2023) Gall-inducing *Eriophyes tiliae* stimulates the metabolism of *Tilia platyphyllos* leaves towards oxidative protection. Plant Physiol Biochem 195:25–36

Guedes LM, Henríquez IA, Sanhueza C, Rodríguez-Cerda L, Figueroa C, Gavilán E, Aguilera N (2024a) Alterations induced by *Colomerus vitis* on the structural and physiological leaf features of two grape cultivars. Exp Appl Acarol 92:1–19

Guedes LM, Rodriguez-Cerda L, Gavilan E, Aguilera N (2024b) Can the galling mite *Eriophyes tiliae* alter the phenolic profile and antioxidant capacity of *Tilia platyphyllos* in a high UV radiation environment? Entomol Exp Appl 172:777–784. https://doi.org/10.1111/eea.13462

Helle W, Wysoki M (1996) Arrhenotokous parthenogenesis. In: Lindquist EE, Sabelis MW, Bruin J (eds) Eriophyoid mites – their biology, natural enemies and control, World crop pests, vol 6. Elsevier Science, pp 169–172

Hoover JB, Shields LM (1972) Mite-induced bud gall in globemallow. Southwest Nat 16:413–418

Javadi KS, Khanjani M, Bahman Asali F (2014) Resistance of three grapevine cultivars to Grape Erineum Mite, *Colomerus vitis* (Acari : Eriophyidae), in field conditions. Persian J Acarol 3:63–75

Javadi KS, Mohammad K, Mansur K (2018a) Impact of the erineum strain of *Colomerus vitis* (Acari: Eriophyidae) on the development of plants of grapevine cultivars of Iran. Exp Appl Acarol 74:347–363

Javadi KS, Mohammad K, Mansur K (2018b) Influence of the erineum strain of *Colomerus vitis* (Acari: Eriophyidae) on grape (*Vitis vinifera*) defense mechanisms. Exp Appl Acarol 75:1–24

Jayaraman P (1987) Anatomical study on mite induced gall on *Dichrostachys cinera*. Ind Bot Soc 67:145–148

Jiang Y, Ye J, Veromann-Jürgenson LL, Niinemets U (2020) Gall – and – erineum forming *Eriophyes* mites alter photosynthesis and volatile emissions in an infection severity-dependent manner in broad-leaved trees *Alnus glutinosa* and *Tilia cordata*. Tree Physiol 41:1122–1142. https://doi.org/10.1093/treephys/tpaa173

Keifer HH, Baker EW, Kono T, Delfinado M, Styer WE (1982) An illustrated guide to plant abnormalities caused by eriophyid mites in North America, No. 573. US Department of Agriculture, Agricultural Research Service

Klimov PB, O'Connor BM, Chetverikov PE, Bolton SJ, Pepato AR, Mortazavi AL, Tolstikov AV, Bauchan GR, Ochoa R (2018) Comprehensive phylogeny of acariform mites (Acariformes) provides insights on the origin of the four-legged mites (Eriophyoidea), a long branch. Mol Phylogen Evol 119:105–117. https://doi.org/10.1016/j.ympev.2017.10.017

Koehler CS, Allen WW, Costello LR (1985) Fuchsia gall mite management. Calif Agric 39:10–12

Kozlowski J, Kozlowska M (1998) Effect of feeding of the apple rust mite *Aculus schlechtendali* [Nal.][Acari, Eriophyidae] on apple shoots elongation. Zeszyty Naukowe Akademii Techniczno-Rolniczej w Bydgoszczy. Ochrona Środowiska 2:49–54

Kulkarni NK, Kumar PL, Muniyappa V, Jones AT, Reddy DVR (2002) Transmission of pigeon pea sterility mosaic virus by the eriophyid mite, *Aceria cajani* (Acari: Arthropoda). Plant Dis 86:1297–1302. https://doi.org/10.1094/PDIS.2002.86.12.1297

Kuster VC, Rezende UC, Cardoso JCF, Isaias RMS, Oliveira DC (2019) How galling organisms manipulate the secondary metabolites in the host plant tissues? A histochemical overview in Neotropical gall systems. In: Merillon J, Ramawat KG (eds) Co-evolution of secondary metabolites. Springer, Heidelberg, pp 1–20

Łabanowska BH, Pluta S (2010) Assessment of big bud mite (*Cecidophyopsis ribis* Westw.) infestation level of blackcurrant genotypes in the field. J Fruit Ornam Plant Res 18:283–295

Lemmetty A, Tikkanen M, Tuovinen T, Lehto K (2004) Identification of different *Cecidophyopsis* mites on *Ribes* in Finland. Acta Hortic 656:115–118

Lindquist EE (1996) External anatomy and notation of structures. In: Lindquist EE, Sabelis MW, Bruin J (eds) Eriophyoid mites – their biology, natural enemies and control, World crop pests, vol 6. Elsevier Science, pp 3–31

Lindquist EE, Oldfield GN (1996) Evolution of eriophyoid mites in relation to their host plants. In: Lindquist EE, Sabelis MW, Bruin J (eds) Eriophyoid mites – their biology, natural enemies and control, World crop pests, Elsevier, pp. 277–300. https://doi.org/10.1016/S1572-4379(96)80018-2

Maboeei PM (2022) Solving the "*Colomerus vitis* conundrum ": genetic evidence reveals a complex of highly diverged groups with little morphological differentiation. Thesis presented in partial fulfilment of the requirements for the degree of Master of Science in the Faculty of Sciences. Stellenbosch University

Malagnini V, De Lillo E, Saldarelli P, Beber R, Duso C, Raiola A, Zanotelli L, Valenzano D (2016) Transmission of grapevine "Pinot gris" virus by *Colomerus vitis* (Acari: Eriophyidae) to grapevine. Arch Virol 161:2595–2599

Mani MS (1964) Ecology of plant galls. Dr. W. Junk Publishers, The Hague

Manson DCM, Oldfield GN (1996) Life forms, deuterogyny, diapause and seasonal development. In: Lindquist EE, Sabelis MW, Bruin J (eds) Eriophyoid mites – their biology, natural enemies and control, World crop pests, vol 6. Elsevier Science, pp 173–183

McCoy CW (1996) Pathogens of eriophyoid mites. In: Lindquist EE, Sabelis MW, Bruin J (eds) Eriophyoid mites – their biology, natural enemies and control, World crop pests, Elsevier, pp. 481–490

Michalska K, Skoracka A, Navia D (2010) Behavioural studies on eriophyoid mites: an overview. Exp Appl Acarol 51:31–59. https://doi.org/10.1007/s10493-009-9319-2

Monterrosa A, Iriarte FB, Paret ML, Joseph SV (2021) Effects of relative humidity on the vector of rose rosette disease, *Phyllocoptes fructiphilus* (Eriophyidae), and incidence of disease symptoms. Fla Entomol 104:173–177. https://doi.org/10.1653/024.104.0305

Morocko-Bicevska I, Stalažs A, Lacis G, Laugale V, Baļķe I, Zuļģe N, Strautin S (2022) Cecidophyopsis mites and blackcurrant reversion virus on *Ribes* hosts: current scientific progress and knowledge gaps. Ann Appl Biol 180(1):26–43. https://doi.org/10.1111/aab.12720

Moura MZD, Soares GLG, Isaias RMS (2008) Species-specific changes in tissue morphogenesis induced by two arthropod leaf gallers in *Lantana camara* L. (Verbenaceae). Aust J Bot 56:153–160

Moura MZD, Soares GLG, Isaias RMS (2009) Ontogenesis of the leaf and leaf galls induced by *Aceria lantanae* Cook (Acarina: Eriophyidae) in *Lantana camara* L. (Verbenaceae). Braz J Bot 32:271–282

Nasareen PNM, Vardhanan YS, Ramani N (2012) Damage assessment of the Gall Mite *Aceria pongamiae* Keifer 1966 (Acari: Eriophyidae) on *Pongamia pinnata* (L.) Pierre. In: Sabu A, Augustine A (eds) Prospects in bioscience: addressing the issues. Springer. https://doi.org/10.1007/978-81-322-0810-5_38

Navia D, Ochoa R, Welbourn C, Ferragut F (2010) Adventive eriophyoid mites: a global review of their impact, pathways, prevention and challenges. Exp Appl Acarol 51:225–255. https://doi.org/10.1007/s10493-009-9327-2

Navia D, Mendonça RS, Skoracka A, Szydło W, Knihinicki D, Hein GL, Pereira PRVS, Truol G, Lau D (2013) Wheat curl mite, *Aceria tosichella*, and transmitted viruses: an expanding pest complex affecting cereal crops. Exp Appl Acarol 59:95–143. https://doi.org/10.1007/s10493-012-9633-y

Nobrega LP, Sá-Haiad B, Ferreira BG (2023) Epidermal and subepidermal changes during the formation of hairy galls induced by Eriophyidae on *Avicennia schaueriana* leaves. Sci Nat (Naturwissenschaften) 110(5):49. https://doi.org/10.1007/s00114-023-01876-3

Oldfield GN (2005) Biology of gall-inducing Acari. In: Raman A, Schaefer CW, Withers TM (eds) Biology, ecology and evolution of gall-inducing arthropods. Science Publishers Inc, Enfield, pp 35–58

Paponova SS, Chetverikov PE, Pautov AA, Yakovleva OV, Zukoff SN, Vishnyakov AE, Lutova LA (2018) Gall mite *Fragariocoptes setiger* (Eriophyoidea) changes leaf developmental program and regulates gene expression in the leaf tissues of *Fragaria viridis* (Rosaceae). Ann Appl Biol 172:33–46

Patankar R, Thomas SC, Smith SM (2011) A gall-inducing arthropod drives declines in canopy tree photosynthesis. Oecologia 167:701–709. https://doi.org/10.1007/s00442-011-2019-8

Patankar R, Beaulieu F, Smith SM, Thomas SC (2012) The life history of a gall-inducing mite: summer phenology, predation and influence of gall morphology in a sugar maple canopy. Agric For Entomol 14:251–259. https://doi.org/10.1111/j.1461-9563.2011.00563.x

Perring TM, McMurtry JA (1996) Other predatory arthropods. In: Lindquist EE, Sabelis MW, Bruin J (eds) Eriophyoid mites – their biology, natural enemies and control, World crop pests, vol 6. Elsevier Science, pp 471–479

Petanović R, Kielkiewicz M (2010a) Plant–eriophyoid mite interactions: specific and unspecific morphological alterations. Part II. Exp Appl Acarol 51:81–91. https://doi.org/10.1007/s10493-009-9328-1

Petanović R, Kielkiewicz M (2010b) Plant–eriophyoid mite interactions: cellular biochemistry and metabolic responses induced in mite-injured plants. Part I. Exp Appl Acarol 51:61–80. https://doi.org/10.1007/s10493-010-9351-2

Picard C, Afonso T, Benko-Beloglavec A, Karadjova O, Matthews-Berry S, Paunovic SA, Ward M (2018) Recommended regulated non-quarantine pests (RNQP s), associated thresholds and risk management measures in the European and Mediterranean region. EPPO Bull 48(3):552–568

Pluta S, Zurawicz E, Malinowski T, Gajek D (2000) Breeding of black currant (*Ribes nigrum* L.) resistant to gall mite and reversion virus. Acta Hortic 538:463–468

Ponce GE, Fuse M, Chan A, Connor EF (2021) The localization of phytohormones within the gall-inducing insect *Eurosta solidaginis* (Diptera: Tephritidae). Arthropod Plant Interact 15:375–385. https://doi.org/10.1007/s11829-021-09817-5

Popitanu C, Lupitu A, Copolovici L, Bungău S, Niinemets U, Copolovici DM (2021) Induced volatile emissions, photosynthetic characteristics, and pigment content in Juglans regia leaves infected with the Erineum-forming mite Aceria erinea. Forests 12:1–12. https://doi.org/10.3390/f12070920

Propistsova EA, Makarova AA, Chetverikov PE, Polilov AA (2023) Arthropod Struct Dev 72:101235. https://doi.org/10.1016/j.asd.2023.101235

Rančić D, Petanović R (2002) Anatomical alterations of *Galium mollugo* L. leaves caused by eriophyoid mite *Aculus anthobius* (Nal.). Acta Entomol Serb 7:119–128

Rančić D, Stevanovic B, Petanović R, Magud B, Tosevski I, Gassmann A (2006) Anatomical injury induced by the eriophyid mite *Aceria anthocoptes* on the leaves of *Cirsium arvense*. Exp Appl Acarol 38:243–253. https://doi.org/10.1007/s10493-006-0013-3

Ranjith AP, Quicke DLJ, Manjusha K, Butcher BA, Nasser M (2022) Completely predatory development is described in a braconid wasp. Sci Rep 12:1747. https://doi.org/10.1038/s41598-022-05705-x

Roßbach J, Dieckmann HL, Büttner T, Mühlbach HP, von Bargen S, Büttner C (2015) Genetic variability and phylogeny of European mountain ash ringspot-associated virus RNA3 and RNA4. Forests 6:4072–4087

Sabelis MW (1996) Phytoseiidae. In: Lindquist EE, Sabelis MW, Bruin J (eds) Eriophyoid mites – their biology, natural enemies and control, World crop pests, vol 6. Elsevier Science, pp 427–456

Sabelis MW, Bruin J (1996) Evolutionary ecology: life history patterns, food plant choice and dispersal. In: Lindquist EE, Sabelis MW, Bruin J (eds) Eriophyoid mites – their biology, natural enemies and control, World crop pests, vol 6. Elsevier Science, pp 329–366

Salisbury A, Ostoja-Starzewski J, Halstead AJ (2014) The establishment of Fuchsia gall mite, *Aculops fuchsiae* (Acari: Eriophyidae) in England, a serious pest of Fuchsia. Br J Entomol Nat Hist 27(3):145–150

Shi A, Tomczyk A (2001) Impact of feeding of eriophyid mite *Epitrimerus gibbosus* (Nalepa) (Acari: Eriophyoidea) on some biochemical components of blackberry (*Rubus* spp.). Bull Polish Acad Sci Biol Sci 49:41–47

Shikov AN, Narkevich IA, Flisyuk EV, Luzhanin VG, Pozharitskaya ON (2021) Medicinal plants from the 14th edition of the Russian Pharmacopoeia, recent updates. J Ethnopharmacol 268:113685. https://doi.org/10.1016/j.jep.2020.113685

Skoracka A, Smith L, Oldfield G, Cristofaro M, Amrine JW (2010) Host-plant specificity and specialization in eriophyoid mites and their importance for the use of eriophyoid mites as biocontrol agents of weeds. Exp Appl Acarol 51:93–113. https://doi.org/10.1007/s10493-009-9323-6

Soika G, Kozak M (2012) Population dynamics of gall-inducing eriophyoid mites and seasonal variations in the gall density on linden tres. Int J Acarol 38:715–725. https://doi.org/10.1080/01647954.2012.726275

Soika G, Kozak M (2013) Eriophyes species (Acari: Eriophyoidea) inhabiting lime trees (*Tilia spp.*: Tiliaceae)—supplementary description and morphological variability related to host plants and female forms. Zootaxa 3646:349–385

Soika G, Tomczyk A, Kozak M (2017) Biochemical reaction of *Tilia* leaves on infestation by some species of gall-inducing eriophyoid mites. Int J Acarol 43:16–21

Stone GN, Schönrogge K (2003) The adaptive significance of insect gall morphology. Trends Ecol Evol 18:512–522. https://doi.org/10.1016/S0169-5347(03)00247-7

Streito JC, Coroller MH, Kreiter S, Flechtmann CHW (2004) A new gall mite of Fuchsia: discovery in France of *Aculops fuchsiae*, first record in Europe. Phytoma 572:32–34

Takeda S, Yoza M, Amano T, Ohshima I, Hirano T, Sato MH, et al. (2019) Comparative transcriptome analysis of galls from four different host plants suggests the molecular mechanism of gall development. PLoS ONE 14(10):e0223686. https://doi.org/10.1371/journal.pone.0223686

Thistlewood HMA, Clements DR, Harmsen R (1996) Stigmaeidae. In: Lindquist EE, Sabelis MW, Bruin J (eds) Eriophyoid mites – their biology, natural enemies and control, World crop pests, vol 6. Elsevier Science, pp 457–470

Tooker JF, Helms AM (2014) Phytohormone dynamics associated with gall insects, and their potential role in the evolution of the gall-inducing habit. J Chem Ecol 40:742–753. https://doi.org/10.1007/s10886-014-0457-6

Tung G-S, Martin CE (2017) Physiological effects of four gall types on leaves of hackberry (Celtis occidentalis) in Northeastern Kansas, U.S.A. J Kans Entomol Soc 90:304–312. https://doi.org/10.2317/141204.1

Vachajitan P, Harvey T (2021) A new pest of aloes. Cact Succ J 93:215–217

von Bargen S, Dieckmann HL, Candresse T, Mühlbach HP, Roßbach J, Büttner C (2019) Determination of the complete genome sequence of European mountain ash ringspot-associated emaravirus from Sorbus intermedia reveals two additional genome segments. Arch Virol 164:1937–1941

Welter SC (1989) Arthropod impact on plant gas exchange. In: Bernays EA (ed) Insect–plant interactions. CRC, Boca Raton, pp 135–150

Westphal E, Manson DCM (1996) Feeding effects on host plants: gall formation and other distortions. In: Lindquist EE, Sabelis MW, Bruin J (eds) Eriophyoid mites – their biology, natural enemies and control, World crop pests, vol 6. Elsevier Science, pp 231–242

Westphal E, Bronner R, Le Ret M (1981) Changes in leaves of susceptible and resistant *Solanum dulcamara* infested by the gall mite *Eriophyes cladophthirus* (Acarina, Eriophyoidea). Can J Bot 59:875–882. https://doi.org/10.1139/b81-122

Westphal E, Dreger F, Bronner R (1990) The gall mite *Aceria cladophthirus*. I. Life-cycle, survival outside the gall and symptoms' expression on susceptible or resistant *Solanum dulcamara* plants. Exp Appl Acarol 9:183–200

Yang M, Li H, Qiao H, Guo K, Xu R, Wei H, Wei J, Liu S, Xu C (2023) Integrated transcriptome and metabolome dynamic analysis of galls induced by the gall mite *Aceria pallida* on *Lycium barbarum* reveals the molecular mechanism underlying gall formation and development. Int J Mol Sci 24:1–21. https://doi.org/10.3390/ijms24129839

Youngman RR, Jones VP, Welter SC, Barnes MM (1986) Comparison of feeding damage caused by four tetranychid mite species on gas-exchange rates of almond leaves. Environ Entomol 15:190–193

Zhang Z-Q (2017) Eriophyoidea and allies: where do they belong? Syst Appl Acarol 22:1091–1095

Zhu L, Chen MS, Liu X (2011) Changes in phytohormones and fatty acids in wheat and rice seedlings in response to Hessian fly (Diptera: Cecidomyiidae) infestation. J Econ Entomol 104:1384–1392

Zurn JD, Ho T, Li R, Bassil NV, Tzanetakis IE, Martin RR, Postman JD (2019) First report of blackcurrant reversion virus in *Ribes nigrum* germplasm in the United States. Plant Dis 103:1051. https://doi.org/10.1094/PDIS-03-18-0526-PDN

Structural and Functional Diversity of Neotropical Psyllid Galls (Hemiptera: Psylloidea) Refute the "Simplicity" Hypothesis

René Gonçalves da Silva Carneiro, Dalva Luiz de Queiroz, Daniel Burckhardt, and Lubia Maria Guedes

1 Introduction

Psylloidea constitutes a superfamily of mostly phloem-feeding Sternorrhycha. The slightly over 4000 described species worldwide are classified into seven families (Aphalaridae, Calophyidae, Carsidaridae, Liviidae, Mastigimatidae, Psyllidae and Triozidae) (Burckhardt et al. 2021). Most species are host specific, and often related psyllid species develop on related plant taxa. Most psyllids are associated with eudicots and magnoliids, a few also with monocots and less than a handful with conifers (Burckhardt 2005; Burckhardt et al. 2014; Ouvrard et al. 2015). Gall-induction has been reported from all psyllid families except for the species-poor Mastigimatidae, but the feature is not evenly distributed across the superfamily. In some taxa (e.g. *Baccharopelma* or Pachypsyllinae) all, in others (*Calophya* or *Pseudophacopteron*) some and still others none of the species (Carsidarinae) are gall-inducing (Burckhardt 2005). Gall-inducing psyllids appear more numerous in the tropics but this may be an artefact of the unequal knowledge on the faunas of the different biogeographic regions (Burckhardt et al. 2018).

Psyllid galls from South America are known since the nineteenth century. Most of the early records are of a descriptive nature. They are summarised by Houard

R. G. S. Carneiro (✉)
Department of Botany, Universidade Federal de Goiás (UFG), Goiânia, Goiás, Brazil
e-mail: renecarneiro@ufg.br

D. L. de Queiroz
Embrapa Florestas, Colombo, Paraná, Brazil

D. Burckhardt
Naturhistorisches Museum, Basel, Switzerland

L. M. Guedes
Facultad de Ciencias Forestales, Laboratorio de Semioquímica Aplicada, Universidad de Concepción (UdeC), Concepción, Chile

D. C. de Oliveira, R. M. dos Santos Isaias (eds.), *Plant Galls*,
https://doi.org/10.1007/978-3-031-80064-1_13

(1933). More detailed information on the structure of neotropical psyllid galls (Table 1) were published since the 1940s of species from Argentina (Lizer and Molle 1945; Agudelo et al. 2018), Chile (Guedes et al. 2016, 2018a, b, 2019) and, especially, Brazil (Arduin et al. 2005; Carneiro et al. 2014a, b, 2015; Carneiro and Isaias, 2015a, b; Dias et al. 2013; Formiga et al. 2015; Isaias et al. 2011; Kuster et al. 2022; Martini et al. 2019, 2020; Oliveira et al. 2006, 2014, 2016, 2019; Oliveira and Isaias 2010; Picanço et al. 2022; Rosa et al. 2024).

The literature on psyllid galls is quite extensive, but a modern synthesis on morphological and structural-functional aspects of these galls is lacking. The recognition of the most common gall morphotypes is important for inventories and estimates of gall diversity (Isaias et al. 2013) particularly considering that the neotropical fauna is probably the richest compared to the psyllid fauna of other biogeographic realms (Burckhardt and Queiroz 2012). Also, the structural-functional simplicity of psyllid galls, advocated by some authors, has been questioned by others (Mani 1964; Bronner 1992; Carneiro et al. 2015; Ferreira et al. 2017, 2019), based on different views on the "non-nutritive" nature of their tissues.

Here we review the existing literature and propose a new hypothesis documenting psyllid galls as complex and structurally-functionally diverse organs.

2 Morphology

Psyllids induce galls on virtually all plant organs: leaves (on both the lamina and petioles), stems, buds, flowers and roots, with leaf galls being the most commonly reported (Hodkinson 1984; Burckhardt 2005; Sharma and Raman 2022). Many psyllid galls are morphologically simple, ranging from small open pits to leaf rolls that affect larger areas of the leaf lamina. Pocket-shaped galls contain a single immature psyllid, while leaf rolls typically shelter several immatures forming a colony (Hodkinson 1984). Some psyllids also induce closed leaf galls, unilocular with a single or several psyllids or multilocular (Hodkinson 1984, D. Burckhardt pers. obs.). The closed galls can be disk or nipple shaped, spherical or hemispherical, as exemplified by galls on leaves and stems of *Schinus* spp. induced by *Calophya* spp., which also induce pit galls and galls on the flowers (Burckhardt and Basset 2000). Here, terminology regarding the external morphology of galls will follow Isaias et al. (2013) for standardized gall morphotypes. Galls originally described as "pit" or "pouch" are referred as pocket-shaped galls; galls derived from the rolling or folding of the leaf lamina are referred as leaf-roll or leaf-fold galls, respectively; flattened galls described as "disk" or "nipple" as referred as lenticular galls; round-shaped galls described as "spherical", "hemispherical" or other terminology are referred as globoid galls.

The neotropical psyllid galls studied anatomically reflect the morphological aspects observed worldwide, varying from simpler structures like pocket-shaped (Fig. 1a), leaf-roll (Fig. 1b) and leaf-fold (Fig. 1c) galls, to more complex ones such as lenticular (Fig. 1d), conical (Fig. 1e), bivalve-shaped (Fig. 1f) and globoid

Table 1 Published information on structural-functional aspects of neotropical psyllid galls organized chronologically. The names of the psyllids and host plants are those currently valid with names listed in the reference in brackets, if different

References	Inducer—Species (Family)	Host-plant—Species (Family)	Host-plant—Species (Family)	Study approach
Lizer and Molle (1945)	*Calophya duvauae* (Scott, 1882) [*Psylla duvauae*] (Calophyidae)	*Schinus longifolia* (Lindl.) Speg. [*Schinus polygamus;* see Burckhardt and Basset 2000, Burckhardt and Queiroz 2017] (Anacardiaceae)	Globoid (leaves)	Biology (life history, distribution), descriptive gall morphology and developmental anatomy (LM)
	Calophya sp. nov. [*Calophya gallifex* (Kieffer and Jörgensen 1910)] (Calophyidae)		Conical (leaves)	
	Tainarys schini Brèthes, 1920 (Aphalaridae)	*Schinus longifolia* (Lindl.) Speg. [*Schinus polygamus;* see Burckhardt and Queiroz 2017) (Anacardiaceae)	Leaf-roll (leaves)	
	Trioza ocoteae Houard, 1933 (Triozidae)	*Ocotea acutifolia* (Nees) Mez (Lauraceae)	Lenticular (leaves)	
Arduin et al. (2005)	*Baccharopelma dracunculifoliae* Burckhardt et al., 2004 (Psyllidae)	*Baccharis dracunculifolia* DC. (Asteraceae)	Leaf-fold (leaves)	Developmental morphology and anatomy (LM) and histochemical analysis (LM; lipids)
Oliveira et al. (2006)	*Euphalerus ostreoides* Crawford, 1925 (Psyllidae)	*Lonchocarpus muehlbergianus* Hassl. (Fabaceae)	Bivalve-shaped (leaves)	Morphological, anatomical (LM, SEM) and histochemical analyses (LM; primary and secondary metabolites)
Oliveira and Isaias (2010)	*Pseudophacopteron aspidospermi* Malenovský et al., 2015 [*Pseudophacopteron* sp.] (Aphalaridae)	*Aspidosperma australe* Müll. Arg. (Apocynaceae)	Lenticular (leaves)	Cytological (TEM) and histochemical analyses (LM; nucleic acids, primary and secondary metabolites, ROS, enzyme activity)

(continued)

Table 1 (continued)

References	Inducer—Species (Family)	Host-plant—Species (Family)	Host-plant—Species (Family)	Study approach
Isaias et al. (2011)	*Euphalerus ostreoides* Crawford, 1925 (Psyllidae)	*Lonchocarpus muehlbergianus* Hassl. (Fabaceae)	Bivalve-shaped (leaves)	Developmental morphology, anatomy, and cytology (LM, SEM, TEM), and histochemical analyses (phenolics, ROS; enzyme activity)
Carneiro et al. (2013)	*Nothotrioza cattleiani* Burckhardt, 2013 (Triozidae)	*Psidium cattleyanum* Sabine (Myrtaceae)	Globoid (leaves)	Biology (phenology, gall morphology and morphometry, host-plant association, distribution, life cycles, attack by parasitoids), taxonomy and systematics
	Nothotrioza myrtoidis Burckhardt, 2013 (Triozidae)	*Psidium myrtoides* O.Berg (Myrtaceae)		
	Nothotrioza tavaresi (Crawford, 1925) (Triozidae)	*Psidium* sp. (Myrtaceae) [unidentified Malpighiaceae; see Burckhardt et al. 2022)		
Dias et al. (2013)	*Calophya duvauae* (Scott, 1882) (Calophyidae)	*Schinus longifolia* (Lindl.) Speg. [*Schinus polygama* (Cav.) Cabrera] (Anacardiaceae)	Globoid (leaves)	Developmental morphology and anatomy (LM, SEM; cytometry and histometry), histochemical analyses (crystals and phenolics)
Oliveira et al. (2014)	*Baccharopelma dracunculifoliae* Burckhardt et al., 2004 (Psyllidae)	*Baccharis dracunculifolia* DC. (Asteraceae)	Leaf-fold (leaves)	Histochemical analyses (LM and FM; pectins) and immunocytochemistry of cell walls (FM; pectins and proteins)
Carneiro et al. (2014a)	*Nothotrioza myrtoidis* Burckhardt, 2013 (Triozidae)	*Psidium myrtoides* O.Berg (Myrtaceae)	Globoid (leaves)	Developmental anatomy (LM; cytometry and histometry) and immunocytochemistry of cell walls (FM; pectins and proteins)
Carneiro et al. (2014b)	*Nothotrioza myrtoidis* Burckhardt, 2013 (Triozidae)	*Psidium myrtoides* O.Berg (Myrtaceae)	Globoid (leaves)	Histochemical (LM; primary and secondary metabolites, ROS, enzyme activity), biochemical (quantification of nitrogen, carbohydrates, and of photosynthetic pigments using SP)

(continued)

Table 1 (continued)

References	Inducer—Species (Family)	Host-plant—Species (Family)	Host-plant—Species (Family)	Study approach
Carneiro et al. (2015)	*Nothotrioza cattleiani* Burckhardt, 2013 (Triozidae)	*Psidium cattleyanum* Sabine (Myrtaceae)	Globoid (leaves)	Developmental anatomy (LM; cytometry and histometry) and immunocytochemistry of cell walls (FM; pectins and proteins)
Formiga et al. (2015)	Psylloidea? [sap-sucking Hemiptera]	*Baccharis reticularia DC.*	Leaf-fold and leaf-roll (leaves)	Morphological and anatomical analyses (LM; cytometry and histometry), assessing phenotypic plasticity
Malenovský et al. (2015)	*Pseudophacopteron aspidospermi* Malenovský et al., 2015 (Aphalaridae)	*Aspidosperma australe* Müll. Arg. (Apocynaceae)	Lenticular (leaves)	Biology (gall morphology, host-plant association, distribution, life cycles) and taxonomy
	Pseudophacopteron longicaudatum Malenovský et al., 2015 (Aphalaridae)	*Aspidosperma macrocarpon* Mart. (Apocynaceae)		
		Aspidosperma tomentosum Mart. (Apocynaceae)		
Carneiro and Isaias (2015a)	*Nothotrioza myrtoidis* Burckhardt, 2013 (Triozidae)	*Psidium myrtoides* O.Berg (Myrtaceae)	Globoid (leaves)	Developmental anatomy (LM) and cytology (TEM)
Carneiro and Isaias (2015b)	*Nothotrioza cattleiani* Burckhardt, 2013 (Triozidae)	*Psidium cattleyanum* Sabine (Myrtaceae)	Globoid (leaves)	Developmental anatomy (LM) and cytology (TEM), histochemical analyses (LM; primary and secondary metabolites, ROS; enzyme activity)
Guedes et al. (2016)	*Calophya rubra* (Blanchard, 1852) (Calophyidae)	*Schinus polygama* (Cav.) Cabrera (Anacardiaceae)	Conical (stems)	Anatomical (LM, SEM) and biochemical analyses (identification of secondary metabolites using CG-MS)
	Calophya mammifex Burckhardt and Basset, 2000 (Calophyidae)		Globoid (leaves)	
Agudelo et al. (2018)	*Baccharopelma* spp. (Psyllidae)	*Baccharis spicata* (Lam.) Baill. (Asteraceae)	Leaf-fold (Leaves)	Anatomical (LM, SEM), histochemical (TLC and FM; phenolics), and biochemical analyses (quantification of secondary metabolites)

(continued)

Table 1 (continued)

References	Inducer—Species (Family)	Host-plant—Species (Family)	Host-plant—Species (Family)	Study approach
Guedes et al. (2018a)	*Calophya rubra* (Blanchard, 1852) (Calophyidae)	*Schinus polygama* (Cav.) Cabrera (Anacardiaceae)	Conical (stems)	Phenology, developmental anatomy (LM; cytometry and histometry), histochemical analyses (primary and secondary metabolites)
Guedes et al. (2018b)	*Calophya mammifex* Burckhardt and Basset, 2000 (Calophyidae)	*Schinus polygama* (Cav.) Cabrera (Anacardiaceae)	Globoid (leaves)	Phenology, developmental anatomy (LM; cytometry and histometry), histochemical analyses (primary metabolites)
Guedes et al. (2019)	*Calophya rubra* (Blanchard, 1852) (Calophyidae)	*Schinus polygama* (Cav.) Cabrera (Anacardiaceae)	Conical (stems)	Histochemical (phenolics and lignins) and biochemical analyses (quantification of polyphenols by spectrophotometric method)
	Calophya mammifex Burckhardt and Basset, 2000 (Calophyidae)		Globoid (leaves)	
Martini et al. (2019)	*Pseudophacopteron longicaudatum* Malenovský et al., 2015 (Phacopteronidae)	*Aspidosperma tomentosum* Mart. (Apocynaceae)	Lenticular (leaves)	Developmental anatomy (LM) and immunocytochemistry of cell walls (FM; pectins and arabinogalactan-proteins)
Oliveira et al. (2019)	*Ceropsylla pouteriae* Burckhardt, 2019 (Triozidae)	*Pouteria ramiflora* (Mart.) Radlk. (Sapotaceae)	Pocket-shaped (leaves)	Biology (gall morphology, host-plant association, distribution), taxonomy, anatomical (LM) and histochemical analyses (pectins, primary and secondary metabolites, and ROS)
Martini et al. (2020)	*Pseudophacopteron longicaudatum* Malenovský et al., 2015 (Phacopteronidae)	*Aspidosperma tomentosum* Mart. (Apocynaceae)	Lenticular (leaves)	Developmental anatomy (LM; histometry) and cytology (TEM); histochemical analyses (primary metabolites and ROS)

(continued)

Table 1 (continued)

References	Inducer—Species (Family)	Host-plant—Species (Family)	Host-plant—Species (Family)	Study approach
Burckhardt et al. (2022)	*Nothotrioza acuminata* Burckhardt, 2022 (Triozidae)	*Psidium laruotteanum* Cambess. (Myrtaceae)	Globoid (leaves)	Biology (gall morphology, host-plant association, distribution, life cycles, parasitoid attack) and taxonomy
	Nothotrioza camptyla Burckhardt, 2022 (Triozidae)	*Psidium rufum* Mart. ex DC. (Myrtaceae)		
	Nothotrioza cattleiani Burckhardt, 2013 (Triozidae)	*Psidium cattleyanum* Sabine (Myrtaceae)		
	Nothotrioza myrtoidis Burckhardt, 2013 (Triozidae)	*Psidium myrtoides* O.Berg (Myrtaceae)		
	Nothotrioza tavaresi (Crawford, 1925) (Triozidae)	*Psidium* sp. (Myrtaceae)		
	Nothotrioza sp. nov.	*Psidium* sp. (Myrtaceae)		
Kuster et al. (2022)	*Pseudophacopteron aspidospermi* Malenovský et al., 2015 (Phacopteronidae)	*Aspidosperma australe* Müll. Arg. (Apocynaceae)	Lenticular (leaves)	Anatomy (LM); histochemical (phenolics and ROS), biochemical (quantification of phenolics and lipid peroxidation, and photosynthetic parameters)
	Pseudophacopteron longicaudatum Malenovský et al., 2015 (Phacopteronidae)	*Aspidosperma macrocarpon* Mart. (Apocynaceae)		
		Aspidosperma tomentosum Mart. (Apocynaceae)		
Picanço et al. (2022)	*Triozoida limbata* (Enderlein, 1918) (Triozidae)	*Psidium guajava* L. (Myrtaceae)	Leaf-roll (leaves)	Developmental morphology and anatomy (LM); biochemical analyses (hormone quantification)
Rosa et al. (2024)	*Neolithus fasciatus* Scott, 1882 (Triozidae)	*Sapium glandulosum* (L.) Morong (Euphorbiaceae)	Globoid (leaves)	Developmental morphology, anatomy (LM) and cytology (TEM); histochemical (primary and secondary metabolites) and biochemical analyses (primary metabolites)

Abbreviations: *FM* Fluorescence microscopy, *GC-MS* Gas chromatography-mass spectrometry, *LM* Light microscopy, *PS* Photospectrometry, *ROS* Reactive oxygen species, *SEM* Scanning electron microscopy, *TEM* Transmission electron microscopy, *TLC* Thin layer chromatography

Fig. 1 Representative morphologies of galls induced by Neotropical pysllids, named according to standardized nomenclature of gall morphotypes (Isaias et al. 2013). (**a**) Pocket-shaped gall induced by *Calophya schini* (Calophydae) on *Schinus molle* (Anacardiaceae), seen from the adaxial leaf surface. The immature is absent, showing the invagination of the leaf lamina. (**b**) Leaf-roll gall induced by *Triozoida limbata* (Triozidae) on *Psidium guajava* (Myrtaceae) seen from the adaxial leaf surface. Immature psyllids feeding lead the leaf margins to roll upwards. (**c**) Leaf-fold gall **gall** induced by *Baccharopelma dracunculifoliae* (Psyllidae) on *Baccharis dracunculifolia* (Asteraceae). The leaf folds upwards along the midrib, forming a longitudinal slit (arrows) delimited by the leaf margins. Note white waxes produced by psyllid immatures inside the gall. (**d**) Lenticular gall induced by *Pseudophacopteron aspidospermi* (Aphalaridae) on *Aspidosperma australe* (Apocynaceae), viewed from the adaxial leaf surface. Upper box shows a lateral view of the gall. (**e**) Conical gall induced by *Calophya rubra* (Calophyidae) on the stems of *Schinus polygama* (Anacardiaceae). Note the white trichomes differentiated on the gall ostiole (arrows). (**f**) Bivalve-shaped gall induced by *Euphalerus ostreoides* (Psyllidae) on *Lonchocarpus muehlbergianus* (Fabaceae) seen from the adaxial leaf surface, resembling a bivalve mollusk's shell. (**g**) Globoid galls induced by *Nothotrioza acuminata* (Triozidae) on *Psidium laruotteanum* (Myrtaceae) seen from the abaxial leaf surface. These galls are induced on the adaxial leaf surface and grow downwards. (**a, c, e**) Courtesy of Rosa Nejapa, Thiago Magalhães, and Lubia Guedes, respectively. (**b, d–g**). Authors' archive. Bars = 0.5 cm

(Fig. 1g) galls. Pocket-shaped galls are understood as the simplest gall in terms of morphology, with the lowest levels of distortion of the host organs and a very localized cecidogenetic field (Mani 1964), which means that the area affected by the gall inducer is restricted to the tissues around the feeding site of the gall inducer (Fig. 2a). This open gall morphotype is paralleled with morphological adaptations of the psyllid inducer: the dorsal body face of the immature is more or less strongly sclerotised while the ventral face (in the gall) in often membranous (Hodkinson 1984). Other taxa of galling organisms commonly induce pocket-shaped galls, such as the Eriophyidae (Acarina), but psyllid galls lack the typical erinea and/or emergences commonly observed in mite galls (Mani 1964; Ferreira et al. 2019).

Together with pocket-shaped galls, leaf-rolls are simple gall morphotypes. Leaf rolls are also known to be induced by Acarina, Thysanoptera, Diptera and Hymenoptera (Mani 1964) but those by psyllids are usually much larger and more complex due to the greater body size and the feeding habit, which involves extracting nutrients from tissues deep inside the plant body. This results in the establishment of a far more intense cecidogenetic field on the galled leaves (Fig. 2b), compared to Acarina and Thysanoptera galls (Mani 1964). A characteristic leaf roll gall on the leaves of guava (*Psidium guajava*—Myrtaceae) reported from the neotropics is induced by *Triozoida limbata* (Triozidae) (Table 1; Fig. 1b), a pest of guava orchards that causes major productivity losses due to profound alterations to plant metabolism (Picanço et al. 2022).

A gall morphotype with intermediary structural complexity is the leaf-fold, in which an entire leaf turns fleshy by the thickening of the mesophyll and folds upwards along the midrib. A longitudinal, slightly obliterated ostiole is delimited by the leaf margins. This process involves the uneven growth of the leaf surface, when the abaxial surface grows more than the adaxial surface, as the cecidogenetic field extends throughout the whole leaf (Fig. 2c). This gall morphotype has been anatomically studied in the system of *Baccharopelma dracunculifoliae* (Psyllidae) and *Baccharis dracunculifolia* (Asteraceae) (Arduin et al. 2005) (Table 1). Additionally, leaf fold galls were studied in *Baccharis reticularia*, whose galls are induced by an unidentified Hemiptera (most probably a psyllid) (Formiga et al. 2015). The latter galls are referred to as "kidney-shaped" and can be classified as leaf-fold galls (Isaias et al. 2013).

More complex gall morphotypes induced by neotropical psyllids are the lenticular, conical, bivalve-shaped and globoid galls (Table 1). Based on morphological and anatomical analyses, most of these galls fit the "pouch galls" of Mani (1964). These are regularly-shaped, thick-walled and fleshy with a completely closed or obliterated ostiole. Additionally, they exhibit a dehiscence mechanism for gall opening during the transition from the mature to the senescent developmental stage of the gall. The term "pouch-gall" refers above all to a developmental mechanism by which they originate, i.e., the formation of invaginations and/or emergences on the plant tissues around the galling organisms, which usually end up enclosed by the plant tissues (Mani 1964; Hodkinson 1984). The process of tissue growth around the gall inducer is well documented anatomically for neotropical psyllid galls, but does not determine the final gall morphotype (Isaias et al. 2011; Dias et al. 2013;

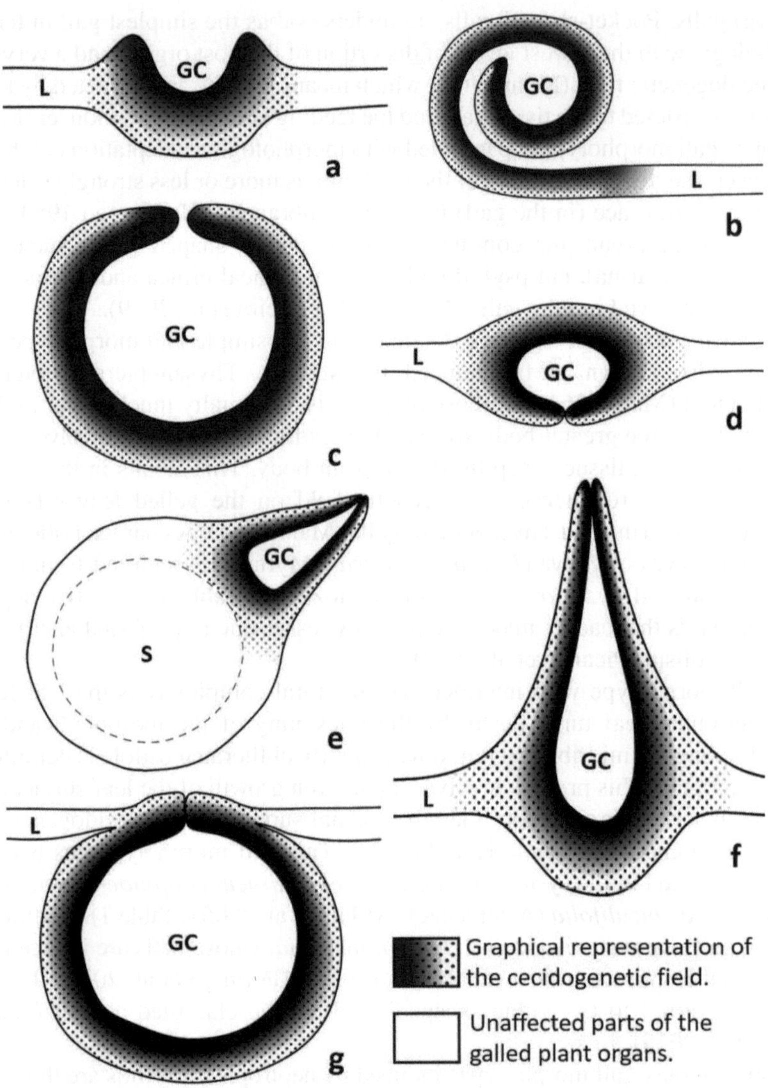

Fig. 2 Schematic representations of cross-sectioned Psylloidea-induced galls, evidencing the extension of distortions caused on galled plant organs by the establishment of a cecidogenetic field (*sensu* Mani 1964). Black gradients represent the stimuli for the establishment of the cecidogenetic field in gall tissues, and dotted patterns indicate the extension of the cecidogenetic field. As the stimuli come from the gall inducers inside the gall chamber (GC), the cecidogenetic field is more intense in the innermost gall tissues, and less intense in the outermost gall tissues. At the periphery of the cecidogenetic field, there is a transition between galls and the unaffected parts of the galled organs, either leaves (L) or stems (S), forming a *continuum*. (**a**) pocket-shaped leaf gall, which is open and morphologically simple, with restricted cecidogenetic field. (**b**) leaf-roll gall, which is open and morphologically simple, with cecidogenetic field extended to a larger area. (**c**) leaf-fold gall, which is open and slightly more complex morphologically; cecidogenetic field is extended to the whole leaf. (**d**) lenticular leaf gall, which is functionally closed, morphologically complex,

Carneiro et al. 2014a, 2015; Carneiro and Isaias 2015a; Guedes et al. 2016, 2018a, b; Martini et al. 2020). Instead, the recruitment and stimulation for the differential growth of specific cell lineages within the cecidogenetic field and, thus, of gall parts, are crucial for the determination of the gall morphotypes. The lenticular gall morphotype arises when both the emergences that grow over the inducer's body and the invagination formed around it grow evenly (Fig. 2d). When the growth in uneven, the development of the emergences is detrimental to that of the invaginations, or vice-versa. In cases where the emergences develop more than invaginations, conical and bivalve-shaped gall morphotypes are determined (Fig. 2e, f), while globoid galls will be determined by major development of the invaginations (Fig. 2g). These dynamics will also determine whether the gall will project toward the adaxial or abaxial leaf surface. Most psyllid galls studied so far on a structural basis show that induction occurs on the abaxial leaf surface and the galls project upwards (Mani 1964; Hodkinson 1984; Dias et al. 2013; Guedes et al. 2018a, b), while others are induced on the adaxial leaf surface and project downwards (Carneiro et al. 2013; Burckhardt et al. 2022). The case of the bivalve-shaped galls of *Euphalerus ostreoides* (Psyllidae) on *Lonchocarpus muehlbergianus* (Fabaceae) is an exception, as the galls are induced on the adaxial leaf surface and most of the gall volume (the valves) develops upwards (Oliveira et al. 2006; Isaias et al. 2011).

The growth and development of galls naturally involve increase in size, due to intense cellular activity, whose development relies on the constant stimulation of plant tissues by the inducer (Rohfritsch 1992). In the case of psyllids, the life cycle consists of an egg stage and five immature stages (Hodkinson 1984), and mature galls reach maximum size when immatures are at the 5th instar. While the pocket-shaped and lenticular galls have a more reduced size, the leaf-roll, leaf-fold, bivalve-shaped, conical, and globoid galls reach higher volumes, as documented for mature globoid galls of *Nothotrioza myrtoidis* (Triozidae) on *Psidium myrtoides* (Myrtaceae) that reach a volume of up to 1.5 cm^3 (Carneiro et al. 2013). Considering that life cycles and adaptive strategies of psyllids are highly variable (Hodkinson 2009) and that the neotropical flora is among the richest in the world (Brazil Flora Group et al. 2022), much is yet to be documented about how the association of psyllids and plants may result in different gall morphologies.

Fig. 2 (continued) with evenly developed emergences and invagination, and restricted cecidogenetic field. (**e, f**) conical stem gall and bivalve-shaped leaf gall, respectively, which are functionally closed, morphologically complex, with well-developed emergences, and restricted cecidogenetic fields. (**g**) globoid leaf gall, which is closed, morphologically complex, with well-developed invagination, and restricted cecidogenetic field. Schemes are not to scale; please refer to Fig. 1 for actual gall sizes or to the original publications in which they are featured ((**a**) Oliveira et al. 2019; (**b**) Picanço et al. 2022; (**c**) Arduin et al. 2005; (**d**) Kuster et al. 2022; (**e**) Guedes et al. 2018a; (**f**) Isaias et al. 2011; (**g**) Carneiro et al. 2015)). Figure executed by Marcelo Kubo

3 Structure and a Proposed Developmental Model for Psyllid Galls

Anatomical alterations caused by gall inducers in plant tissues are inherently correlated with the feeding behaviour of the galling organisms, which can be piercing-sucking, as in Acarina, Thysanoptera and Hemiptera, chewing, as in Lepidoptera, Coleoptera and Hymenoptera, and scraping-sucking as in Diptera (Mani 1964; Oliveira et al. 2016). To a far extent, evidence of gall anatomy reinforces that premise, being the galls induced by the Diptera (Cecidomyiidae) and Hymenoptera (Cynipidae) classically understood as the most complex ones, with concentrical tissue layers organized around the larval chamber—innermost layers composed of nutritive tissues, followed by a sclerenchymatous sheath (frequently; not always), a vascularized parenchymatous cortex and a dermal system with different degrees of alterations (with or without indumentum) (Rohfritsch 1992; Ferreira et al. 2019). The same tissue topology (maybe except for the presence of a sclerenchymatous sheath) applies to Lepidoptera galls, which have been increasingly studied in the neotropics and exhibit a similar degree of complexity (Ferreira et al. 2022; Guedes et al. 2023; Martins et al. 2023; Rezende et al. 2023), usually with greater size when compared to Diptera and Hymenoptera galls. Galls induced by Coleoptera remain poorly known from an anatomical point of view (Korotyaev et al. 2005).

Amongst the piercing-sucking gall inducers, the Acarina and Thysanoptera, which mainly feed on the epidermal cells, induce the formation of less complex galls, while the Hemiptera, whose mouthparts are bigger and feed mostly from ground and vascular tissues deeper inside the plant body (Mani 1964; Hodkinson 1984), induce more complex galls. The stylets of psyllids serve to inject saliva that contains cell-degrading enzymes, amylases, proteases and lipases and to ingest the predigested plant sap (Sharma and Raman 2022). This results in structural alterations in all tissue systems—dermal, ground and vascular—that vary according to the galling species. Anatomical studies from the beginning to the mid-twentieth century describe such alterations in the gall systems: *Tainarys schini*, *Calophya duvauae* [as *Psylla duvauae*], and *Calophya* sp. nov. [as *Calophya gallifex*] developing in *Schinus longifolia* [as *Schinus polygama*]; and *Trioza ocotae* associated to *Ocotea acutifolia* (Lizer and Molle 1945; Table 1). Although these galls develop into different morphotypes, some cellular profiles are similar, such as little alteration in epidermal cells (generally hypertrophic); hypertrophy and hyperplasia of both ground and vascular tissues, where the insects primarily feed (Mani 1964).

As far as the dermal system is concerned, epidermal cells tend to be more altered around the ostiole, as most galls are completely or slightly closed (Mani 1964) and such a closure may be conferred by the overproduction of trichomes (Arduin et al. 2005; Dias et al. 2013; Guedes et al. 2016, 2018a, b; Martini et al. 2020; Kuster et al. 2022), deposition of calcium oxalate crystals (Dias et al. 2013) and lignification (Carneiro et al. 2014a). In galls that develop as leaf invaginations, the inner epidermis may be discontinuous and lack stomata (Carneiro et al. 2015) as a sign of function loss. In fact, normal-functioning stomata are rarely observed in galls (Mani

1964), which also seems to be true for the psyllid galls. Alterations on the ground tissues follow the standard rules of gall anatomy, i.e., hyperplasia and hypertrophy are common processes through which parenchyma undergoes to originate a somewhat homogenous cortex (Mani 1964; Ferreira et al. 2019), usually chlorophyllous. Sclerenchyma occurs as isolated to clustered cells (Sharma et al. 2015; Sharma and Raman 2022), sometimes forming a sheath (Carneiro et al. 2014a, 2015). Psyllid galls do not form typical nutritive tissues (Bronner 1992; Ferreira et al. 2019), but the innermost parenchymatous layers lining the gall chamber and sheathing the vascular bundles may have peculiar cytological features. The vasculature of the galled organ seems to be either maintained in gall structure or neoformed vascular bundles differentiate among the parenchyma cells nearest to the gall chamber, which is well documented in neotropical galls (see Carneiro et al. 2014a, 2015; Dias et al. 2013; Guedes et al., 2016, 2018a, b; Isaias et al. 2011; Martini et al. 2020). The composition of vascular tissues in psyllid galls is frequently altered when compared to non-galled organs, as vascular bundles within the cecidogenetic field usually turn hyperplasic and hypertrophic, with increased relative abundance of vascular parenchyma; phloem-rich vascular bundles with poorly-developed xylem may also occur (Isaias et al. 2011; Carneiro et al. 2014a, 2015; Agudelo et al. 2018; Guedes et al. 2018a, b).

The set of cell alterations observed in psyllid galls are somewhat similar, but differences on the processes of cell division, expansion, and differentiation during the ontogenesis are unique for each gall. Such differences are observed as a function of the high variety of host plant-galling psyllid systems studied so far (Table 1), and are also important for the recognition of the gall phenotype (Carneiro et al. 2015). The processes of gall initiation during the induction stage (Rohfritsch 1992) are generally poorly documented for psyllid galls from across the world. There is evidence that the feeding activity of psyllids involves early events of cell hypertrophy, both in the epidermis and in the subjacent parenchyma, leading to the breakdown of mesophyll cells, where some species primarily feed. Hyperplasia is evident around the apex of the stylet (Hodkinson 1984), most probably due to the protein-rich saliva injected by the immatures that affect the levels of plant growth regulators (Sharma and Raman 2022). Evidence also shows that psyllid immatures produce hormones such as cytokinins which they inject into plant tissue (Straka et al. 2010; Andreas et al. 2020), which directly affects the plant cell cycle by increasing cell division, thus leading to tissue hyperplasia. Cytologically, reports suggest signs of cell digestion by autophagic vacuoles (Sharma and Raman 2022), but detailed information on cytological profiles is lacking.

As far as neotropical psyllid galls are concerned, only a few studies describe the induction stage of psyllid galls. These studies provide information on how tissue lineages change during gall development, as well as on the role of the changes for the functioning of gall structure (Carneiro and Isaias 2015a; Carneiro et al. 2015). In general, psyllid galls have metabolically-active tissues, especially when ground tissues are analysed along the stages of gall induction, growth and development, and early maturation (Oliveira and Isaias 2010; Isaias et al. 2011; Sharma et al. 2015; Carneiro and Isaias 2015a, b; Martini et al. 2020; Rosa et al. 2024). Evidence of a high metabolism include the register of inner cortical cells with large nuclei and conspicuous nucleoli, electron-dense cytoplasm with ribosomes, rough endoplasmic reticulum, numerous mitochondria, and plastids with poorly to well-developed thylakoid and starch grains, and fragmented vacuole (vacuome) (Oliveira and Isaias 2010; Isaias et al. 2011; Sharma et al. 2015; Carneiro and Isaias 2015a, b; Rosa et al. 2024). Sinuous and thin to thick-walled cells are reported as a sign of hyperplasia throughout gall cortex; often a Golgi apparatus is observed in such cells. The outermost cell layers typically exhibit distinct characteristics, often featuring large central vacuoles and peripheral cytoplasm. This cytoplasm may contain well-developed chloroplasts, occasionally accompanied by starch grains or amyloplasts, mitochondria, and normal-sized nuclei (Oliveira and Isaias 2010; Isaias et al. 2011; Carneiro and Isaias 2015a, b; Martini et al. 2020; Rosa et al. 2024).

Antioxidant organelles are often observed in the green, photosynthesizing galls, with the formation of electron-dense plastoglobules in plastids, and of lomasomes, multivescular bodies, and lamellar bodies associated to the plasma membrane (Oliveira and Isaias 2010; Carneiro and Isaias 2015a, b; Martini et al. 2020; Rosa et al. 2024). On one hand, plastoglobules act on the detoxification of excessive reactive oxygen species (ROS), common in galls (Isaias et al. 2014, 2015), as they accumulate ROS-scavenging substances (Oliveira et al. 2011). On the other hand, lomasomes, multivescular bodies and lamellar bodies act on the restoration of the membrane systems where ROS accumulation affects their functional integrity (Carneiro and Isaias 2015a, b; Martini et al. 2020). In the photosynthesis-deficient gall induced by *Nothotrioza myrtoidis* (Triozidae) on *Psidium myrtoides* (Myrtaceae) (Carneiro et al. 2014b), antioxidant organelles are absent, even though they are present in ungalled leaves; this has been attributed to the low gall metabolism (Carneiro and Isaias 2015a). On the contrary, in the association of *Pseudophacopteron aspidospermi* (Aphalaridae) to *Aspidosperma australe* (Apocynaceae), plastoglobules are present in galls and absent in non-galled leaves, which is attributed to increased oxidative stress in galls. Still, the galls can photosynthesize at levels similar to those of leaves, corroborating the beneficial role of antioxidant organelles for gall metabolism (Oliveira et al. 2011) and development (Carneiro and Isaias 2015a, b). A summary of cytological profiles and gradients associated with gall development and cell metabolism induced by the feeding activity of psyllids is presented in Fig. 3.

By compiling anatomical and cytological characteristics of gall tissues, together with information regarding processes of cell division and expansion, it is possible to trace the origin and final destinations of tissue lineages and explain how galls and their peculiar morphologies arise from the plant body by the alteration of standard plant organogenesis. In terms of gall morphogenesis, most authors report the same initial process: the tissues underneath the feeding site of the gall inducer arches downwards to form an invagination, while the tissues surrounding the immatures grow upwards to form emergences that enclose the insects (Lizer and Molle 1945; Hodkinson 1984; Carneiro et al. 2014a, 2015; Sharma and Raman 2022). However, some gall morphologies are completely open, which is the case of leaf-roll and leaf-fold galls. The morphogenesis of leaf rolls is perhaps the most different and simple, as it involves uneven growth of the adaxial and abaxial domains of the leaf lamina, as demonstrated for Thysanoptera-induced galls (Jorge et al. 2018). To roll upwards or downwards, the abaxial or the adaxial leaf domains, respectively, must express higher degrees of cell hypertrophy, with periclinal cell elongation, and of tissue hyperplasia, with anticlinal cell divisions, in comparison to the opposite leaf side (Fig. 4a"). This seems to be the case of the galls induced by *Tainarys schini* (Aphalaridae) on *Schinus longifolia* [as *S. polygama*] (Anacardiaceae) (Lizer and Molle 1945), and by *Triozoida limbata* (Triozidae) on *Psidium guajava* (Myrtaceae) (Picanço et al. 2022). In the leaf-fold galls, the morphogenesis involves similar processes, except that the entire adaxial leaf surface grows less than the abaxial leaf surface, with anticlinal cell divisions and periclinal cell elongation, which causes the whole leaf to fold upwards. Additionally, the gall cortex is thickened by additional processes of cell hypertrophy, with anticlinal elongation, and tissue hyperplasia, with periclinal cell divisions (Fig. 4b"). Such processes are well documented for the system *Baccharopelma dracunculifoliae* (Psyllidae)—*Baccharis dracunculifolia* (Asteraceae) (Arduin et al. 2005). Agudelo et al. (2018) briefly comment on anatomical features of *Baccharopelma* spp. (Psyllidae) galls on *Baccharis spicata* (Asteraceae), which are similar to the former species.

Another completely open gall morphotype, the pocket-shaped galls are still poorly studied on the anatomical basis; they develop by an invagination of the leaf lamina by uneven growth of the adaxial and abaxial domains (as described above), but they also have emergences that grow around the insect body, without covering it. Such emergences have limited growth and do not enclose the insect. Although pocket-shaped galls are considered morphologically simpler than the leaf-roll and leaf-fold galls, their morphogenesis is slightly more complex, as it may involve cell hypertrophy with periclinal elongation, and tissue hyperplasia with anticlinal cell divisions to form the invagination, as well as anticlinal cell elongation and periclinal cell divisions to form the underdeveloped emergences (Fig. 5a). This seems to be the case of the leaf galls induced by *Trioza ocoteae* on *Ocotea acutifolia* (Lizer and Molle 1945). In the case of the *Ceropsylla pouteriae* (Triozidae) galls on *Pouteria ramiflora* (Mart.) Radlk. (Sapotaceae), only anticlinal cell elongation and periclinal cell divisions seem to form both the invagination and the emergence (see Oliveira et al. 2019). Very similar processes are observed in the morphogenesis of lenticular galls, with the difference that the emergences growing around the insect body develop further, with anticlinal cell elongation and additional periclinal cell divisions, and fully enclose the gall inducer inside the gall chamber. In lenticular galls, both the invagination and the emergency develop somewhat evenly (Fig. 5b). This is the case of lenticular galls induced by and *C. mammifex* on *Schinus polygama* (Guedes et al. 2018b), and on the association of *Aspidosperma* spp. (Apocynaceae) to *Pseudophacopteron* spp. (Aphalaridae). *Pseudophacoteron aspidospermi* induces galls on *A. australe* Müll. Arg. (Oliveira and Isaias 2010; Malenovský et al. 2015), while *P. longicaudatum* induces galls both on *A. macrocarpon* and *A. tomentosum* (Malenovský et al. 2015; Martini et al. 2020). It is worth mentioning that the galls of *P. longicaudatum* on *A. macrocarpon* are phenotypically distinct from the ones on *A. tomentosum*, confirming that gall phenotype also depends on the plant, not

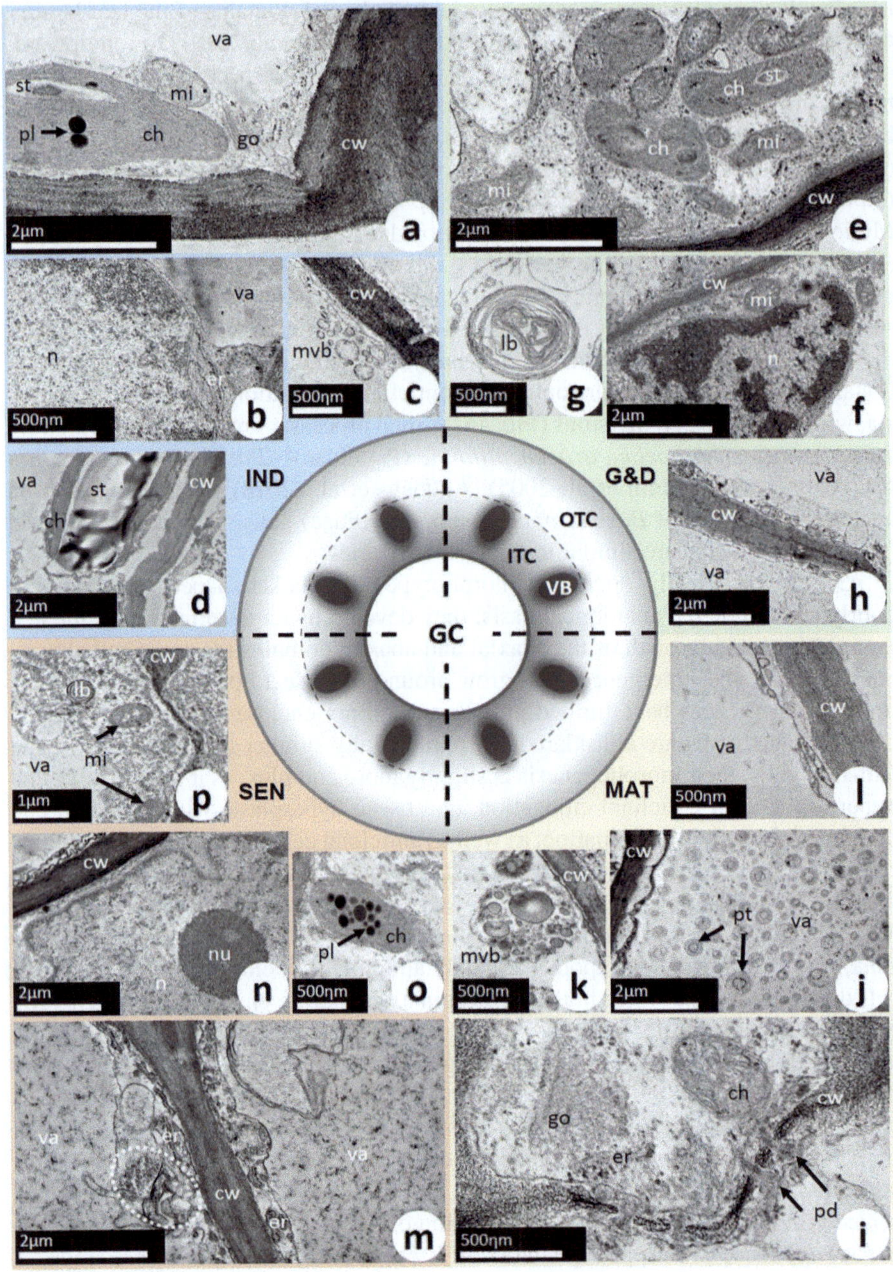

Fig. 3 General aspects of the cytological ontogenesis of Psylloidea galls based on ultrastructural evidences of the system *Neolithus fasciatus* (Triozidae)—*Sapium glandulosum* (Euphorbiaceae) [see Rosa et al. 2024 for further information]. At the center, a scheme of a hypothetical gall is presented, showing a gall chamber (GC) surrounded by the gall cortex, divided into outer and inner tissue compartments (OTC and ITC, respectively), separated by a dotted line. The OTC, composed

only the inducer. This is also observed on the association of *C. mammifex* to different species of *Schinus* (D. Burckhardt pers. obs.). The basic morphogenetic steps of pocket-shaped and lenticular galls are slightly similar to that of other gall morphotypes induced by psyllids (see Fig. 5, upper part), especially when observing the induction stage (see Isaias et al. 2011; Dias et al. 2013; Carneiro et al. 2014a, 2015). This can either be a convergent trait, as the galling habit is widely homoplastic (Harris and Pitzschke 2020), or it can be a conserved trait of all galling Psyllids. Evolutionary studies, especially those focused on the effector mechanisms employed by galling organisms to induce galls, should set further light into this question, which is long in need for answers.

Progressing toward the more complex galls, the conical, the bivalve-shaped and the globoid ones, additional morphogenetical processes are observed. Such processes follow the previously described basic steps, which are common to the development of galls induced by Psyllids. The morphogenesis of these gall morphotypes best fit the description of a "covering gall" (Mani 1964), in which plant tissues grow

Fig. 3 (continued) of an outer epidermis and outermost ground tissue layers, is indicated by lighter shades of grey, representing less intense cytological gradients and cell metabolism. The ITC, consisting of vascular bundles (VB), innermost ground tissue layers, and an inner epidermis, is shown in darker shades of grey, indicating more intense cytological gradients and cell metabolism toward the gall chamber. (**a–d**) Induction stage (IND), showing intense cell metabolism both at the ITC (**a–c**) and at the OTC (**d**); (**e–h**) Growth and development stage (G&D); intense cell metabolism maintained at the ITC (**e-g**) but not at the OTC (**h**); (**i–l**) Maturation stage (MAT), with intense cell metabolism and accumulation of metabolites at the ITC (**i–k**) and cells with low metabolism at the OTC (**l**); (**m–p**) Senescent stage (SEN), showing signs of cell degradation but still with intense metabolism at the ITC (**m–o**); OTC cells showing advanced signs of cell degradation (**p**). ITC cells display mechanisms of oxidative stress control (**c, g, k, o**), responsible for maintaining cell functional integrity against the action of reactive oxygen species (ROS) along gall development. (**a**) ITC cell with irregularly thickened primary wall, large central vacuole, chloroplast with few plastoglobuli and starch grain, and Golgi apparatus. (**b**) ITC cell with large nucleus, rough endoplasmic reticulum, and central vacuole. (**c**) detail of ITC cell with ROS-scavenging multivesicular bodies close to the plasma membrane. (**d**) OTC cell with irregularly shaped wall, chloroplast with large starch grain and large central vacuole. (**e**) ITC cell rich in mitochondria and chloroplasts with starch grains, dense cytoplasm and electrondense primary cell wall. (**f**) ITC cell with dense cytoplasm, large nucleus with condensed heterochromatin, numerous mitochondria, and thin primary walls. (**g**) detail of lamellar body next to the plasma membrane. (**h**) Detail of OTC cells with few organelles and large central vacuoles. (**i**) ITC cells with dense cytoplasm, poorly differentiated chloroplasts, Golgi apparatus, rough endoplasmic reticulum, and abundant plasmodesmata. (**j**) Detail of ITC cells accumulating protein crystals. (**k**) Detail of ITC cell with electrondense multivesicular bodies. (**l**) detail of OTC cells with few organelles and large central vacuoles. (**m**) ITC cells showing initial signs of cell degradation, with encapsulated organelles (dotted circle), accumulation of phenolics in the vacuoles, and increased proteosynthesis (see rough endoplasmic reticulum). (**n**) Detail of metabolically active ITC cell with large lobed nucleus, evident nucleolus, and electrondense wall. (**o**) Detail of ITC cell with chloroplast with numerous plastoglobuli. (**p**) Detail of OTC cells with signs of degradation, displaying irregular cell wall, dense cytoplasm, numerous mitochondria and lamellar body. Abbreviations: ch chloroplast, cw cell wall, er rough endoplasmic reticulum, go Golgi apparatus, lb. lamellar bodies, mi mitochondrion, mvb multivesicular bodies, n nucleus, nu nucleolus, pd. plasmodesmata, pl plastoglobulus, pt. protein crystals, st starch grains, va vacuole

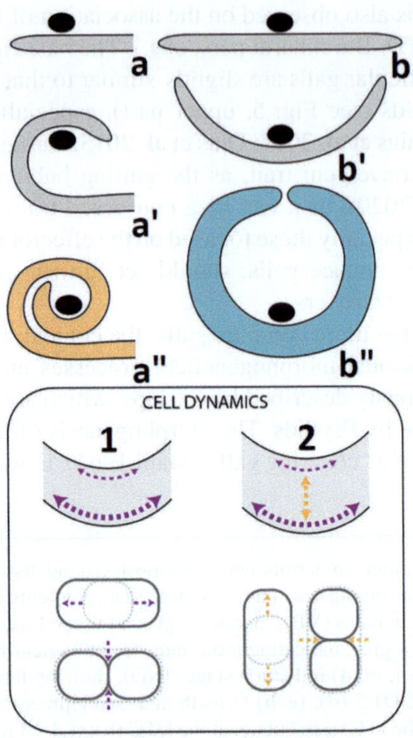

Fig. 4 Developmental patterns of galls induced by Neotropical psyllids with lower degrees of structural-functional complexities. (**a-a"**) Morphogenesis of leaf-roll galls. (**b-b"**) Morphogenesis of leaf-fold galls. (**a, b**) Gall-inducing immature placed on the adaxial surface of the leaf. (**a', b'**) Induction stage and initial morphogenesis of galls causing the leaf margin to roll upwards (**a'**) or the entire leaf to fold upwards (**b'**). (**a", b"**) Maturation stage and final morphogenesis of galls that caused the leaf margin to completely roll upwards (**a"**) or the entire leaf to fold upwards (**b"**). Note that mature galls are open, and the ostiole of the leaf-fold gall is not completely closed (see Fig. 1c for further detail). The cell dynamics involved in such morphogenetical processes (lower box) are: (**1**) periclinal cell elongation and/or anticlinal cell divisions (lower left cell diagrams) occurring at different intensities (purple arrows with different sizes in the upper left diagram) along the morphogenesis of leaf-roll galls. (**2**) Similar morphogenetical steps described in (**1**), with additional anticlinal cell elongation and/or periclinal cell divisions (lower right cell diagrams) for the thickening of the gall cortex (orange arrow in the upper right diagram) along the morphogenesis of leaf-fold galls. In both galls, the adaxial surface grows less than the abaxial surface, which lead them to grow upwards. Galls may also be induced on the abaxial leaf surface, which should lead them to grow downwards (not shown; see Fig. 5 for further details). Figure executed by Marcelo Kubo

around and upward over the developing gall inducer. The insect becomes enclosed in a chamber with a small obliterated ostiole on the tip. In the case of the bivalve-shaped and conical galls, the emergences that grow around the gall inducer, instead of simply covering it, continue to grow and form either a bivalve-shaped or cone-shaped structures. Cellular processes involved in such morphogenesis are mainly anticlinal cell elongation and periclinal cell divisions (Fig. 5c, d). This seems to be the case of the galls found on the leaves of *Schinus longifolia* [as *S. polygama*]

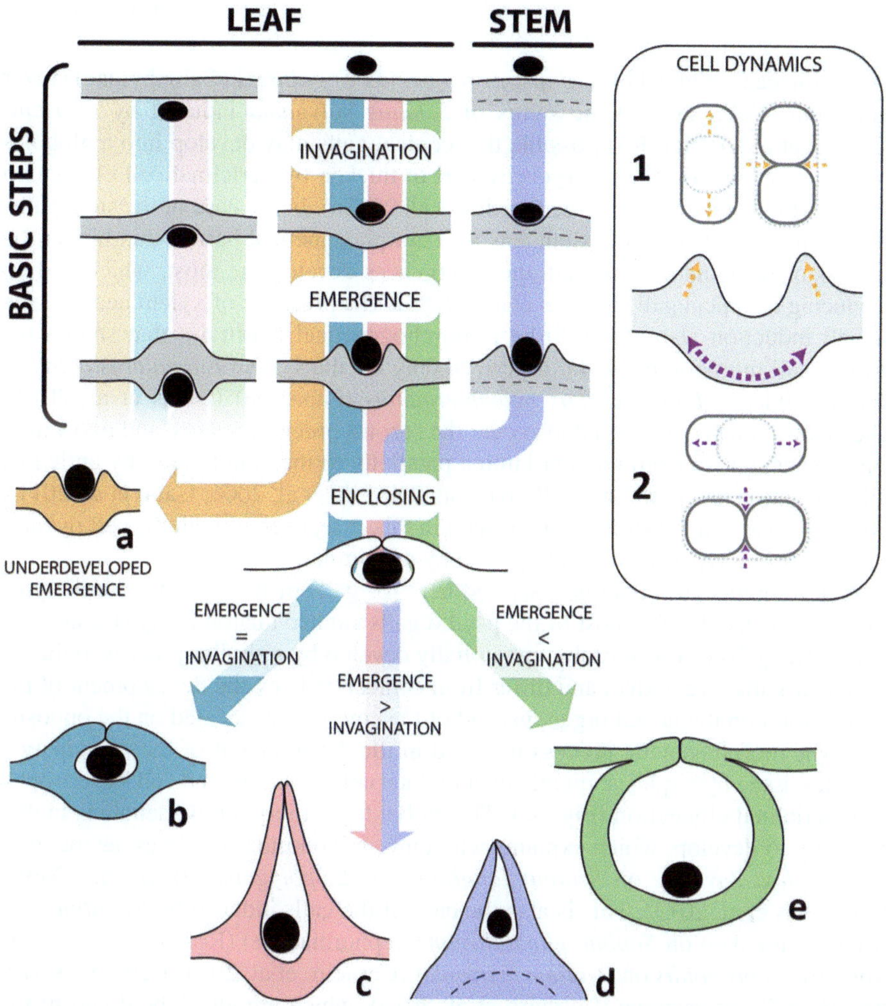

Fig. 5 Developmental patterns of galls induced by the Neotropical Psylloidea with higher degrees of structural-functional complexities. On the upper-left side of the figure, the basic morphogenetical steps of galls induced on leaves and stems are observed. Leaf galls may be induced either on the adaxial (shown at the center; bright coloured arrows show complete morphogenetical steps) or on the abaxial surface (shown at the left; faded coloured arrows show morphogenetical steps only partially). Gall induction and initial development involve the formation of invaginations below the inducer's body, followed by the formation of emergences around it, both on leaf galls and on stem galls. The cell dynamics involved in such morphogenetical processes (upper right box) are: (1) anticlinal cell elongation and/or periclinal cell divisions for the development of the emergences (orange arrows, and upper cell diagrams). (2) Periclinal cell elongation and/or anticlinal cell divisions for the development of the invaginations (purple arrows, and lower cell diagrams). These are general aspects; other developmental processes of emergence and invagination morphogenesis may involve further cell dynamics (see text). (**a**) Pocket-shaped gall, with well-developed invagination and underdeveloped emergence that does not cover the insect's body, for it is an open gall morphotype. (**b–e**) closed galls, which undergo the processes of closing by the growth of the emergences. (**b**) Lenticular gall, with evenly-developed invagination and emergence. (**c, d**) Bivalve-shaped gall and conical gall, in which the emergences develop more than the invaginations. (**e**) globoid gall, in which the invagination develops more than the emergence. Figure executed by Marcelo Kubo

(Anacardiaceae) induced by *Calophya* sp. nov. [as *C. gallifex*] (Calophyidae) (Lizer and Molle 1945) and on the stems of *Schinus polygama* induced by *C. rubra* (Guedes et al. 2018a). It is possible that conical galls may develop into a globoid-like structure in case the emergences start to thicken by periclinal cell elongation and anticlinal cell divisions (see Guedes et al. 2018a). In the case of the stem galls, the secondary xylem of the stem acts as a morphogenetical barrier that hinder the process of invagination; instead, the emergences develop the most, which end up producing a conical gall (Guedes et al. 2018b). The presence of xylem near the site of gall induction also seems to be a morphogenetical constraint that shapes the bivalve gall morphotype, so far described only for the system *Euphalerus ostreoides* (Psyllidae)—*Lonchocarpus muehlbergianus* (Fabaceae) (Isaias et al. 2013). The bivalve-shaped gall is induced over the second-order leaf veins, and two valves (emergences) arise from the leaf lamina parallelly to the vein mainly by anticlinal cell elongation and periclinal cell divisions (Oliveira et al. 2006; Isaias et al. 2011). Thus, no cone-shaped structure is formed in this case, even though the cell dynamics involved in hyperplasia and hypertrophy are similar.

A developmental model is described here for galls of the globoid morphotype, which accounts also for most of the psyllid galls studied anatomically (Isaias et al. 2013). The galls of this morphotype initially develop by a similar process to that of lenticular galls (see above) and differ from conical galls by the development of the invagination on the galled organ, instead of the emergences located on the opposite side. The morphogenetic process involved in the development of the invagination includes tissue hyperplasia, mainly by anticlinal cell divisions, and cell hypertrophy with periclinal elongation (Fig. 5e). The globoid gall morphotype demands higher plasticity to develop, which explains why they are common on leaves, as the galls of *Calophya duvauae* on *Schinus longifolia* [as *C. polygama*] (Lizer and Molle 1945; Dias et al. 2013). This is also the case of the galls induced by *Neolithus fasciatus* (Triozidae) on *Sapium glandulosum* (Euphorbiaceae) (Rosa et al. 2024), by *Nothotrioza myrtoidis* on *Psidium myrtoides* (Carneiro et al. 2014a), and by *N. cattleiani* on *P. cattleyanum* (Carneiro et al. 2015), which are all globoid and fit the developmental model described here. The studies on *Nothotrioza* spp. also illustrate how similar morphotypes may have slightly different developmental processes, by the differences on the dynamics of cell expansion in the cortex (Carneiro et al. 2015). To date, seven species of *Nothotrioza* are known all of which induce globoid galls on *Psidium* species (taxa from South America), except for *N. longipedis*, a Nearctic species developing on *Mosiera longipes* (Myrtaceae), which induces pocket-shaped galls (Burckhardt et al. 2022). Most of the *Nothotrioza* galls await to be studied anatomically.

The morphogenesis of a gall is tightly linked to the development of the inducer. This is the reason why gall development is traditionally divided into four stages: induction, growth and development, maturation, and senescence (Rohfritsch 1992; Isaias et al. 2014). While most of the information so far provided in this chapter comes from observations of the first three stages, the mechanisms by which galls undergo senescence are still poorly known. The few reports of the senescent stages of psyllid galls reveal interesting morphogenetic capabilities of galls, especially regarding mechanisms of dehiscence. As psyllids lack chewing mouthparts, they rely on efficient dehiscence mechanisms for the adult to emerge from the gall. In open galls like the pocket-shaped the adult emerges from the last instar still sitting in the gall. In leaf rolls, the fifth-instar usually leaves the gall and the adult emerges somewhere on the leaf blade or the gall may start to unroll and release the last instar or the adult (D. Burckhardt, pers. obs.). Similarly, leaf-fold galls start to unfold during senescence, possibly due to increased cell hypertrophy of the innermost cell layers and formation of necrotic sites near the gall ostiole (see Arduin et al. 2005). In the bivalve-shaped gall, similar events are observed, and are accompanied by lignification of cells lining the gall chamber and shrivelling (sings of cell death) of the outermost cell layers that drive the valves (emergences) apart from each other, thus opening the gall (Oliveira et al. 2006, Isaias et al. 2011).

In lenticular, conical and globoid gall morphotypes whose ostioles are just partially closed, the opening of the gall occurs by the enlargement of the pre-existing aperture, possibly involving increment of hypertrophy of the innermost cell layers and shrivelling of the outermost cell layers of the emergences. Further events involve loss of trichomes and suberization (Dias et al. 2013; Guedes et al. 2018a, b), cytological degradation, and overall tissue deterioration (Oliveira and Isaias 2010; Martini et al. 2020; Rosa et al. 2024) from the tip of the emergences toward the base. Cracks are often formed along the opening emergences and, eventually, all tissues die and dry out. In globoid galls that are permanently closed or have a very obliterated ostiole, similar events occur, but not on the emergences that formed the ostiole, as outlined above. In the latter cases, the gall cortex formed by the invagination of leaf lamina cracks open by different mechanisms: (1) by the development of necrotic sites and shrivelling of the outermost cell layers, accompanied by increased hypertrophy of innermost cell layers (Carneiro et al. 2014a, 2015); or (2) by the drying of the thin-walled parenchyma cells alternating with groups of sclereids on the gall wall, which cracks open by the tension caused by parenchyma death (Raman 1991). Much is yet to be discovered and described about the dehiscence mechanisms of galls, and psyllid galls seem particularly interesting models.

In summary, psyllid galls range from simple to complex structures. Factors that influence the degree of complexity are: (1) open or closed structure; (2) induction of the gall by a single or by multiple immatures; (3) cecidogenetic field restricted to the tissues surrounding the inducer or extending to the whole galled organ; and (4) if the gall has morphological parts enrolled in specialized functions (e.g. alterations near the ostiole, specialized feeding sites, dehiscence mechanisms) along gall development (Table 2). Further discussions concerned to the structural-functional complexity of psyllid galls are addressed next, based on ultrastructural and histochemical data.

4 Cyto-Histochemical Profiles Confirm the Complex Morpho-Functional Nature of Psyllid Galls

Galls of the sap-sucking psyllids are allegedly non-nutritive, unlike the metabolic profile of nutritive cells typically found in galls of Cecidomyiidae and Cynipidae (Sharma and Raman 2022). The features commonly observed in nutritive cells include abundant organelle-rich cytoplasm with poorly differentiated plastids and mitochondria, and hypertrophied nucleus, denoting high metabolism (Bronner 1992, and references therein). A cytological gradient is formed toward the larval chamber—the closer to the chamber, the more intense the cytological activity is. Histochemically, it has been demonstrated that the nutritive cells accumulate RNA, proteins and carbohydrates, as well as enzymes related to their metabolism. These features collectively ensure a high metabolic rate in the nutritive cells, facilitating intense proteosynthesis, proteolysis and active transport of carbohydrates toward the feeding sites of the gall inducer (Bronner 1992).

The first histochemical evidence of the accumulation of nutritious compound such as lipids and starch in cells near the chamber and vascular tissues comes from the study of the bivalve-shaped gall induced by *Euphalerus ostreoides* on *Lonchocarpus muehlbergianus* (Oliveira et al. 2006). These authors also report the detection of secondary metabolites such as phenolic and flavonoid compounds. The main conclusion therein is that primary and secondary metabolites are related to the metabolic maintenance of the gall structure and to plant defenses against herbivory, respectively. Oliveira et al. (2006) are the first to mention "nutritive tissue" in a psyllid-induced gall based on starch accumulation and suggest that secondary metabolites (phenolics) may have an important role in gall morphogenesis by the modulation of IAA levels, thus affecting patterns of cell hypertrophy. The same system was reviewed and studied from the ontogenetic perspective, approached anatomically and cytologically, which documented high metabolism of cells surrounding the gall chamber and the nearby vascular tissues (Isaias et al. 2011). Although the cytological characteristics found were similar to those reported by Bronner (1992) for other galls, the authors categorically state that they could not be named as "typical nutritive tissues" due to the presence of starch grains in these cells. They also detected the activity of invertases at the same sites; this enzyme catalyzes the breakdown of sucrose into hexoses, which may be used as food by the gall inducer but, by Isaias et al. (2011) was believed to be mobilized for the maintenance of the gall structure. The high metabolism of the cells in this gall was confirmed by the histochemical detection of ROS that, despite the strong detection, did not imply in cell degradation, suggesting that they play an important role in gall morphogenesis.

Table 2 Morphological characteristics of Neotropical psyllid gall morphotypes that correlate with different degrees of structural-functional complexity

Gall morphotypes	Criteria for complexity degrees in galls: Structure open/closed	Inducer solitary/colonial	Cecidogenetic field restricted/extended	Morphological and/or developmental specialization	Complexity degree
Pocket-shaped	Open	Solitary	Restricted	Establishment of feeding sites. Small-sized	Low
Leaf-roll	Open	Solitary/Colonial	Restricted/Extended	Establishment of feeding sites. Small- to big-sized	Low
Leaf-fold	Open	Solitary/Colonial	Extended	Establishment of feeding sites, thickening of gall wall, minor alterations near the ostiole. Big-sized	Medium
Lenticular	Closed	Solitary	Restricted	Establishment of feeding sites, thickening of gall wall, major alterations near the ostiole, dehiscence mechanisms via enlarged ostiole. Small-sized	High
Conical	Closed	Solitary	Restricted	Establishment of feeding sites, thickening of gall wall, major alterations near the ostiole, dehiscence mechanisms via enlarged ostiole Small-sized	High
Bivalve-shaped	Closed	Solitary	Restricted	Establishment of feeding sites, thickening of gall wall, major alterations near the ostiole, dehiscence mechanisms via enlarged ostiole. Big-sized	High
Globoid	Closed	Solitary	Restricted	Establishment of feeding sites, thickening of gall wall, major alterations near the ostiole, dehiscence mechanisms via enlarged ostiole or cracking of the gall wall. Big-sized	High

The second study on the histochemical and cytological basis of psyllid galls were the lenticular galls of *Pseudophacopteron aspidospermi* on *Aspidosperma australe* (Oliveira and Isaias 2010), who describe for the first time the establishment of cytological and histochemical gradients in psyllid galls. These galls exhibit the cytological features forming a gradient toward the gall chamber, matching the profiles described for nutritive tissues of Cecidomyiidae and Cynipidae (Bronner 1992). The histolocalization of DNA, RNA, proteins, starch, sugars and lipids reinforces the similarity of the innermost cell layers of these galls to nutritive tissues, which is mentioned by the authors, but they also argue that this metabolically active profile of cells is related to the maintenance of the gall, not to the nutrition of the insect (Oliveira and Isaias 2010). The enzymatic profile, in turn, reveals the activity of acid phosphatase, phosphorylase, glucose-6-phosphatase, invertase and sucrose synthase, that regulate source-sink relationships, thus balancing carbohydrate storage and remobilization processes along gall development. Reactive oxygen species were also intensively detected in these galls, specifically in metabolically-active cells with antioxidant organelles (see previous sections and Fig. 3 for details), reinforcing the developmental role played by ROS in galls.

In the study of globoid leaf galls induced by *Nothotrioza* on *Psidium*, further aspects on the complexity of psyllid galls were revealed by the cyto-histochemical evaluation of the systems *Nothotrioza myrtoidis—Psidium myrtoides* (Carneiro et al. 2014b; Carneiro and Isaias 2015a) and *N. cattleiani—P. cattleyanum* (Carneiro and Isaias 2015b). *Nothotrioza myrtoidis* galls have a simpler histochemical profile when compared to the previous galls, accumulating phenolics, proanthocyanidins, and alkaloids as secondary metabolites, but also lipids and reducing sugars as primary metabolites. Enzyme activity was low, with the detection of acid phosphatase, while ROS were detected forming a weak gradient together with other histochemical detections, mostly toward the gall chamber (Carneiro et al. 2014b). Due to low content of photosynthetic pigments and little accumulation of metabolites, these galls were considered to be photosynthesis-deficient, and thus, with low metabolism. Accordingly, cytological profiles reinforce such assumption, as the parenchyma cells of galls have few intercellular spaces, and poorly developed chloroplasts that lack plastoglobules (Carneiro et al. 2014b), differently to what had been observed by Oliveira and Isaias (2010). Vascular and perivascular parenchymatous cells, in turn, present very peculiar cytological features, even though no gradient is reported; they are metabolically-active, with hypertrophied nuclei, abundant mitochondria, and endoplasmic reticulum (Carneiro et al. 2014b). These tissues were named "nutritive-like tissues", as they were similar, to a certain extent, to the descriptions by Bronner (1992), Isaias et al. (2011), and Oliveira and Isaias (2010). In the *N. cattleiani—P. cattleyanum* system, cyto-histochemical profiles also show a lack of cytological gradients, even though histochemical gradients for reducing sugars, proteins, glucose-6-phosphatase, invertases, phenolics, proanthocyanidins, and ROS were found, intensifying toward the gall chamber (Carneiro and Isaias 2015b). Once again, the vascular and perivascular parenchyma, nearest to the gall chamber, were the most metabolically-active tissues, with well-developed ROS-scavenging apparatus, different from its co-generic counterpart (Carneiro et al. 2014b, Carneiro and Isaias 2015a) and similar to the galls studied by Oliveira and Isaias (2010), suggesting that these are convergent functional gall traits. The authors also argue that perivascular parenchyma cells in the galls of the *N. cattleiani—P. cattleyanum* system are "true nutritive cells"—not a nutritive tissue, *per se,* as no continuous tissue layer was observed, but collection of cells with high protein synthesis and enzymatic activity that ensure carbohydrate mobilization, from the outermost toward the innermost cell layers, as described by Bronner (1992).

Galls induced by *Calophya* species on *Schinus polygama* in Chile have been studied histochemically for the presence of both secondary and primary metabolites (Guedes et al. 2018a, b). In terms of primary metabolites, the presence of lipids has been reported to occur in *C. rubra* and *C. mammifex* galls (Guedes et al. 2018a, b), while reducing sugars, proteins, in addition to lipids, were found in the phloem and vascular parenchyma of *C. mammifex* galls, indicating high metabolic activity of a "nutritive-like tissue", as reported previously by Carneiro and Isaias (2015a), and revised by Ferreira et al. (2017). The presence of lipids in both galls is mentioned by Guedes et al. (2018a, b) as an indication that nutrition-related tissues in *Calophya* galls should have similar features to those described by Bronner (1992) and other psyllid galls studied until then (Oliveira et al. 2006; Oliveira and Isaias 2010; Carneiro et al. 2014b; Carneiro and Isaias 2015a, b). It is worth mentioning that the accumulation of phenolic substances is well studied in these galls, being discussed in terms of responses to biotic stress and defense against natural enemies (Guedes et al. 2018b), but also as important molecules enrolled in processes of gall development an in the complex plant-gall-environment interactions (Guedes et al. 2019). Agudelo et al. (2018) also provide some insights into these complex interactions by using an analytical chemical approach for the study of *Baccharopelma* spp. galls on *Baccharis spicata* from Argentina; similar conclusions as to the defensive/protective roles of phenolics in galls are provided.

Detailed histochemical profiles are also described for the pocket-shaped galls induced by *Ceropsylla pouteriae* on the leaves of *Pouteria ramiflora* (Oliveira et al. 2019), which accumulate both primary (starch, reducing sugars, proteins and lipids) and secondary (proanthocyanidins, alkaloids and phenolics) metabolites mostly forming gradients toward the gall chamber and near the vascular and perivascular tissues; auxins and ROS are also detected. The authors recognize the structural simplicity of the "pit galls" (=pocket-shaped galls), lacking nutritive tissues, but mention functional compartmentalization of tissues (*sensu* Bragança et al. 2017) for the first time in psyllid galls, as an indication of metabolic complexity. Oliveira et al. (2019) discuss the importance of vascular tissue neoformations in this gall for the feeding activity of the inducer. Secondary metabolites, ROS and auxins are also discussed as very important substances enrolled in gall morphogenesis and metabolism (Oliveira et al. 2019), as suggested by previous studies (Isaias et al. 2014).

Structural-functional aspects of psyllid galls also have been addressed by the study of *Neolithus fasciatus* galls on the leaves of *Sapium glandulosum*, investigated with biochemical, histochemical, cytological and histological technics (Rosa et al. 2024). The parenchymatous cortex of these galls has histological differences between the outermost and innermost cell layers, and is functionally compartmentalized, similar to the report by Oliveira et al. (2019). Galls accumulate different amounts of both primary (starch, proteins and lipids) and secondary (alkaloids and phenolic compounds) metabolites, mostly in the inner cortical parenchyma and vascular bundles. Cytological analyses show that cells of the innermost layers have thin and sinuous walls, dense cytoplasm with poorly-developed chloroplasts, mitochondria, Golgi apparatus and hypertrophied nuclei (see Fig. 3 for further details). The authors acknowledge the similarities between their findings and those described for Cecidomyiidae and Cynipidae (Bronner 1992) and for the neotropical psyllids studied so far (Isaias et al. 2011; Carneiro et al. 2014b; Carneiro and Isaias 2015a, b). Rosa et al. (2024) define the nutritious and metabolically-active inner tissue compartment of these galls as "nutritive-like tissues", which has been used by Ferreira et al. (2019) to unify the terminology and to distinguish them from "typical nutritive tissues". In this sense, typical nutritive tissues are the direct feeding sites of other gall-inducing taxa, while nutritive-like tissues, allegedly, are not.

As far as the feeding behavior of Neotropical psyllids is concerned, there are no studies systematically investigating the feeding habits and sites of the gall inducers. Studies from other biogeographic realms show that psyllids feed both directly from the vascular tissues of from the parenchyma sheathing the vascular bundles (Hodkinson 1984; Sharma and Raman 2022). Combining this information and the histochemical and cytological data on Neotropical psyllid galls, it seems likely that the metabolically-active cells observed nearest to the vascular tissues and chamber are the actual feeding sites of the psyllids. While feeding, psyllids inject saliva between gall tissues and cells, forming a "salivary sheath", which can be observed cytologically as cell wall enlargements and/or cell from cell detachment where an apoplastic movement of nutrients occurs (Sharma and Raman 2022). This cytological feature can be observed in the vascular and perivascular parenchyma cells of *Nothotrioza myrtoidis* galls (R.G.S. Carneiro, pers. obs.). Histologically, salivary sheaths have also been found around the vascular bundles and adjacent parenchyma in the galls induced by *Baccharopelma dracunculifoliae* (Arduin et al. 2005). A plausible assumption is that psyllid galls have nutritive tissues, which are mostly cyto-histochemically similar to that of other gall-inducing taxa (Bronner 1992; Ferreira et al. 2017, 2019). These tissues are already named as "nutritive tissues" in some papers on psyllid galls (Sharma and Raman 2022). The fact that they do not fit previous models perfectly does not undermine the evidence accumulated so far, especially the data on histochemical profiles. Peculiarities have been found in the nutritive tissues of various neotropical galling insects (Ferreira et al. 2019) and the gall-inducing psyllids are no exceptions.

5 Conclusions

Studies on Neotropical psyllid galls reveal that they represent simple and complex structures depending on functional and developmental specialization of their morphology and on the extension of the organ affected by the cecidogenetic field established by gall morphogenesis. Anatomical variations observed in psyllid galls extend to all tissue systems—epidermal cells are less modified, but are often functionally altered; the formation of an ostiole alters the epidermal cells the most. Ground and vascular systems are altered to a greater extent, with intense hyperplasia and/or hypertrophy to form a homogenous parenchymatous cortex with neoformed vascular bundles closest to the gall chamber. Cytological profiles confirm that these cells are metabolically active and functionally complex, and not only constitute the feeding sites of the gall inducers, but also play important developmental roles during the gall morphogenesis. An antioxidant apparatus is often observed in metabolically-active tissues that allows them to cope with high oxidative stress, thus avoiding premature cell death. During gall development, a coordinated series of dynamic cell processes is responsible for the determination of structural patterns in different gall morphotypes. Variations in the patterns of cell division (hyperplasia) and elongation (hypertrophy) explain how emergences and invaginations are the two main morphogenetical processes involved in the formation of psyllid galls. The least studied developmental stage of psyllid galls is the senescence, during which complex dynamics of cell hypertrophy and processes of cell death take part in the dehiscence mechanisms, which should be further explored anatomically. The cyto-histochemical processes of the psyllid galls show the accumulation of primary and secondary metabolites, display enzymatic activity related to the metabolism of proteins and carbohydrates, and the cytological profiles ensure a high metabolism. Though slightly different to the nutritive tissues from other galling insects, we proposed here that the terminology "nutritive-like tissue" should be replaced by the simpler and more popular term "nutritive tissue", which is widely accepted worldwide to refer to nutrition-related gall tissues.

References

Agudelo I, Cogoi L, Filip R, Kuzmanich N, Wagner ML, Ricco RA (2018) Anatomy, histochemistry, and comparative analysis of hydroxycinnamic derivatives in healthy leaves and galls induced by *Baccharopelma* spp. (Hemiptera: Psyllidae) in *Baccharis spicata* (Lam.) Baill (Asteraceae). Biochem Syst Ecol 77:22–30

Andreas P, Kisiala A, Emery RN, De Clerck-Floate R, Tooker JF, Price PW, Miller DG III, Shen MS, Connor EF (2020) Cytokinins are abundant and widespread among insect species. Plan Theory 9(2):208

Arduin M, Fernandes GW, Kraus JE (2005) Morphogenesis of galls induced by *Baccharopelma dracunculifoliae* (Hemiptera: Psyllidae) on *Baccharis dracunculifolia* (Asteraceae) leaves. Braz J Biol 65:559–571

Bragança GP, Oliveira DC, Isaias RMS (2017) Compartmentalization of metabolites and enzymatic mediation in nutritive cells of Cecidomyiidae galls on *Piper arboreum* Aubl. (Piperaceae). J Plant Stud 6(1):11–22

Brazil Flora Group et al (2022) Brazilian Flora 2020: Leveraging the power of a collaborative scientific network. Taxon 71:178–198

Bronner R (1992) The role of nutritive cells in the nutrition of cynipids and cecidomyiids. In: Shorthouse JD, Rohfritsch O (eds) Biology of insect-induced galls. Oxford University Press, New York, pp 192–211

Burckhardt D (2005) Biology, ecology, and evolution of gall-inducing psyllids (Hemiptera: Psylloidea). In: Raman A, Schaefer CW, Withers TM (eds) Biology, ecology, and evolution of gall-inducing arthropods. Science Publishers, Enfield, pp 143–157

Burckhardt D, Basset Y (2000) The jumping plant-lice (Hemiptera, Psylloidea) associated with *Schinus* (Anacardiaceae): Systematics, biogeography and host plant relationships. J Nat Hist 34:57–155

Burckhardt D, Queiroz DL (2012) Checklist and comments on the jumping plant-lice (Hemiptera: Psylloidea) from Brazil. Zootaxa 3571:26–48

Burckhardt D, Queiroz DL (2017) The jumping plant-lice of the Neotropical genus *Tainarys* (Hemiptera: Psylloidea) associated with Anacardiaceae. Zootaxa 4232:535–567

Burckhardt D, Ouvrard D, Queiroz DL, Percy DM (2014) Psyllid host-plants (Hemiptera: Psylloidea): resolving a semantic problem. Fla Entomol 97:242–246

Burckhardt D, Sharma A, Raman A (2018) Checklist and comments on the jumping plant-lice (Hemiptera: Psylloidea) from the Indian subcontinent. Zootaxa 4457:1–38

Burckhardt D, Ouvrard D, Percy DM (2021) An updated classification of the jumping plant-lice (Hemiptera: Psylloidea) integrating molecular and morphological evidence. Eur J Taxon 736:137–182

Burckhardt D, Carneiro RGS, Fernandes IS, Queiroz DL (2022) Gall inducing jumping plant lice of the genus *Nothotrioza* Burckhardt (Hemiptera: Psylloidea: Triozidae) associated with Myrtaceae. J Nat Hist 56:865–884

Carneiro RGS, Isaias RMS (2015a) Cytological cycles and fates in *Psidium myrtoides* are altered towards new cell metabolism and functionalities by the galling activity of *Nothotrioza myrtoidis*. Protoplasma 252:637–646

Carneiro RGS, Isaias RMS (2015b) Gradients of metabolite accumulation and redifferentiation of nutritive cells associated with vascular tissues in galls induced by sucking insects. AoB Plants 7:plv086

Carneiro RGS, Burckhardt D, Isaias RMS (2013) Biology and systematics of gall-inducing triozids (Hemiptera: Psylloidea) associated with *Psidium* spp. (Myrtaceae). Zootaxa 3620:129–146

Carneiro RGS, Oliveira DC, Isaias RMS (2014a) Developmental anatomy and immunocytochemistry reveal the neo-ontogenesis of the leaf tissues of *Psidium myrtoides* (Myrtaceae) towards the globoid galls of *Nothotrioza myrtoidis* (Triozidae). Plant Cell Rep 33:2093–2106

Carneiro RGS, Castro AC, Isaias RMS (2014b) Unique histochemical gradients in a photosynthesis-deficient plant gall. S Afr J Bot 92:97–104

Carneiro RGS, Pacheco P, Isaias RMS (2015) Could the extended phenotype extend to the cellular and subcellular levels in insect-induced galls? PLoS One 10(6):e0129331

Dias GG, Ferreira BG, Moreira GR, Isaias RMS (2013) Developmental pathway from leaves to galls induced by a sap-feeding insect on *Schinus polygama* (Cav.) Cabrera (Anacardiaceae). An Acad Bras Cienc 85:187–200

Ferreira BG, Álvarez R, Avritzer SC, Isaias RMS (2017) Revisiting the histological patterns of storage tissues: beyond the limits of gall-inducing taxa. Botany 95:173–184

Ferreira BG, Álvarez R, Bragança GP, Alvarenga DR, Pérez-Hidalgo N, Isaias RMS (2019) Feeding and other gall facets: patterns and determinants in gall structure. Bot Rev 85:78–106

Ferreira BG, Moreira GR, Carneiro RGS, Isaias RMS (2022) Complex meristematic activity induced by *Eucecidoses minutanus* on *Schinus engleri* turns shoots into galls. Am J Bot 109(2):209–225

Formiga AT, Silveira FAO, Fernandes GW, Isaias RMS (2015) Phenotypic plasticity and similarity among gall morphotypes on a superhost, *Baccharis reticularia* (Asteraceae). Plant Biol 17(2):512–521

Guedes LM, Aguilera N, Becerra J, Hernández V, Isaias RMS (2016) Leaf and stem galls of *Schinus polygama* (Cav.) Cabrera (Anacardiaceae): Anatomical and chemical implications. Biochem Syst Ecol 69:266–273

Guedes LM, Aguilera N, Ferreira BG, Becerra J, Hernández V, Isaias RMS (2018a) Anatomical and phenological implications of the relationship between *Schinus polygama* (Cav.) Cabrera and the galling insect *Calophya rubra* Blanchard. Plant Biol 20(3):507–515

Guedes LM, Aguilera N, Ferreira BG, Becerra J, Sáez K, Pérez C, Hernández V, Isaias RMS (2018b) Factors influencing the morphogenesis of galls induced by *Calophya mammifex* (Calophyidae) on *Schinus polygama* (Anacardiaceae) leaves. Botany 96(9):589–599

Guedes LM, Aguilera N, Ferreira BG, Riquelme S, Sáez-Carrillo K, Becerra J, Pérez C, Bustos E, Isaias RMS (2019) Spatiotemporal variation in phenolic levels in galls of calophyids on *Schinus polygama* (Anacardiaceae). J Plant Res 132:509–520

Guedes LM, Costa EC, Isaias RM, Sáez-Carillo K, Aguilera N (2023) Structural patterns of Lepidoptera galls and the case of *Andescecidium parrai* (Cecidosidae) galls on *Schinus polygama* (Anacardiaceae). J Plant Res 136(5):715–728

Harris MO, Pitzschke A (2020) Plants make galls to accommodate foreigners: some are friends, most are foes. New Phytol 225(5):1852–1872

Hodkinson ID (1984) The biology and ecology of the gall-forming Psylloidea (Homoptera). In: Ananthakrishnan TN (ed) Biology of gall insect. Oxford University Press, New Delhi, pp 59–77

Hodkinson ID (2009) Life cycle variation and adaptation in jumping plant lice (Insecta: Hemiptera: Psylloidea): a global synthesis. J Nat Hist 43:65–179

Houard C (1933) Les zoocécidies des plantes de L'Amérique du Sud et de L'Amérique Centrale. Hermann et Cie, Paris

Isaias RMS, Oliveira DC, Carneiro RGS (2011) Role of *Euphalerus ostreoides* (Hemiptera: Psylloidea) in manipulating leaflet ontogenesis of *Lonchocarpus muehlbergianus* (Fabaceae). Botany 89(9):581–592

Isaias RMS, Carneiro RGS, Oliveira DC, Santos JC (2013) Illustrated and annotated checklist of Brazilian gall morphotypes. Neotrop Entomol 42:230–239

Isaias RMS, Oliveira DC, Carneiro RGS, Kraus JE (2014) Developmental anatomy of galls in the neotropics: arthropods stimuli versus host plant constraints. In: Fernandes GW, Santos JC (eds) Neotropical Insect Galls. Springer Verlag, New York, pp 15–34

Isaias RMS, Oliveira DC, Moreira ASFP, Soares GLG, Carneiro RGS. (2015) The imbalance of redox homeostasis in arthropod-induced plant galls: mechanisms of stress generation and dissipation. Biochimica et Biophysica Acta, 1850(8):1509–1517.

Jorge NC, Souza-Silva ÉA, Alvarenga DR, Saboia G, Soares GLG, Zini CA, Cavalleri A, Isaias RMS (2018) Structural and chemical profiles of *Myrcia splendens* (Myrtaceae) leaves under the influence of the galling *Nexothrips* sp. (Thysanoptera). Front Plant Sci 9:402567

Korotyaev BA, Konstantinov AS, Lingafelter SW, Mandelshtam MY, Volkovitsh MG (2005) Gall-inducing coleoptera. In: Raman A, Schaefer CW, Withers TM (eds) Biology, ecology, and evolution of gall-inducing arthropods. Science Publishers, Enfield, pp 239–271

Kuster VC, Cardoso JCF, Moreira ASFP, Isaias RMDS, Oliveira DC (2022) Assessing oxidative stress and photosynthetic activity in leaf galls induced by different species of galling insect on *Aspidosperma* spp. Biologia 77(12):3441–3454

Lizer CA, Molle CC (1945) Estructura anatómica de filocecidias neotrópicas. Lilloa 16:153–188

Malenovský I, Burckhardt D, Queiroz DL, Isaias RMS, Oliveira DC (2015) Descriptions of two new *Pseudophacopteron* species (Hemiptera: Psylloidea: Phacopteronidae) inducing galls on *Aspidosperma* (Apocynaceae) in Brazil. Acta Entomologica Musei Nationalis Pragae 55(2):513–538

Mani MS (1964) Ecology of plant galls. W Junk, The Hague

Martini VC, Moreira ASFP, Kuster VC, Oliveira DC (2019) Galling insects as phenotype manipulators of cell wall composition during the development of galls induced on leaves of *Aspidosperma tomentosum* (Apocynaceae). S Afr J Bot 127:226–233

Martini VC, Moreira ASF, Fuzaro L, Gonçalves LA, Oliveira DC (2020) *Pseudophacopteron longicaudatum* (Hemiptera) induces intralaminar leaf galls on *Aspidosperma tomentosum* (Apocynaceae): a qualitative and quantitative structural overview. An Acad Bras Cienc 92:e20181002

Martins GS, Santos MG, Reis A, Ferreira BG (2023) Alterations induced by *Tortrimosaica polypodivora* on the stems of *Microgramma vacciniifolia*: simple or complex galls? Rodriguésia 74:e00532023

Oliveira DC, Isaias RMS (2010) Cytological and histochemical gradients induced by a sucking insect in galls of *Aspidosperma australe* Arg. Muell (Apocynaceae). Plant Sci 178(4):350–358

Oliveira DC, Christiano JDCS, Soares GLG, Isaias RMS (2006) Reações de defesas químicas e estruturais de *Lonchocarpus muehlbergianus* Hassl. (Fabaceae) à ação do galhador *Euphalerus ostreoides* Crawf. (Hemiptera: Psyllidae). Rev Bras Bot 29:657–667

Oliveira DC, Isaias RMS, Moreira ASFP, Magalhães TA, Lemos Filho JP (2011) Is the oxidative stress caused by *Aspidosperma* spp. galls capable of altering leaf photosynthesis? Plant Sci 180:489–495

Oliveira DC, Magalhães TA, Ferreira BG, Teixeira CT, Formiga AT, Fernandes GW, Isaias RMS (2014) Variation in the degree of pectin methylesterification during the development of *Baccharis dracunculifolia* kidney-shaped gall. PLoS One 9(4):e94588

Oliveira DC, Isaias RMS, Fernandes GW, Ferreira BG, Carneiro RGS, Fuzaro L (2016) Manipulation of host plant cells and tissues by gall-inducing insects and adaptive strategies used by different feeding guilds. J Insect Physiol 84:103–113

Oliveira DC, Burckhardt D, Calácio TDF, Kuster VC, Queiroz DL (2019) *Ceropsylla pouteriae* Burckhardt sp. nov. (Hemiptera: Psylloidea: Triozidae), a new species of jumping plant-louse inducing galls on the leaves of *Pouteria ramiflora* (Mart.) Radlk. (Sapotaceae): taxonomy, gall structure and histochemistry. J Nat Hist 53(31–32):1923–1950

Ouvrard D, Chalise P, Percy DM (2015) Host-plant leaps versus host-plant shuffle: a global survey reveals contrasting patterns in an oligophagous insect group (Hemiptera, Psylloidea). Syst Biodivers 13(5):434–454

Picanço MM, Silva RS, Azevedo AA, Lima LL, Ramos HJ, Souza OF, Picanço MC (2022) Assessment of *Triozoida limbata* (Hemiptera: Triozidae) attacks: morphological and biochemical changes on *Psidium guajava* plants. Austral Entomol 61(4):472–479

Raman A (1991) Cecidogenesis of leaf galls on *Syzygium cumini* (L.) Skeels (Myrtaceae) induced by *Trioza jambolanae* Crawford (Homoptera: Psylloidea). J Nat Hist 25:653–663

Rezende UC, Gonçalves PHP, Kuster VC, Carneiro RGS, Oliveira DC (2023) Every end is a new beginning: Histological features of galls induced on *Macairea radula* (Melastomataceae) allow a post-senescence colonization. Flora 306:152369

Rohfritsch O (1992) Patterns in gall development. In: Shorthouse JD, Rohfritsch O (eds) Biology of insect-induced galls. Oxford University Press, Oxford, pp 60–86

Rosa LMP, Silva MS, Carneiro RGS, Machado M, Kuster VC (2024) Hemiptera-induced galls of *Sapium glandulosum* have histological and cytological compartmentalization created with a large amount of carbohydrate. Protoplasma 261:593–606

Sharma A, Raman A (2022) Gall-inducing Psylloidea (Insecta: Hemiptera) – plant interactions. J Plant Interact 17(1):580–594

Sharma A, Raman A, Taylor GS, Fletcher M, Nicol H (2015) Feeding and oviposition behaviour of a gall-inducing species of *Glycaspis* (Synglycaspis) (Hemiptera: Psylloidea: Aphalaridae) and development of galls on the leaves of *Eucalyptus macrorhyncha* (Myrtaceae) in central western New South Wales, Australia. Eur J Entomol 112(1):75–90

Straka JR, Hayward AR, Emery RN (2010) Gall-inducing *Pachypsylla celtidis* (Psyllidae) infiltrate hackberry trees with high concentrations of phytohormones. J Plant Interact 5(3):197–203

A Complex But Few Explored Structure and Function of Lepidoptera Galls

Bruno Garcia Ferreira and Lubia María Guedes

1 Introduction

In the world of insects, Lepidoptera is one of the orders that includes representatives from several families capable of inducing galls. However, it's worth noting that their diversity in this regard is relatively smaller when compared to the gall-inducing species found in Diptera, Hymenoptera, and Hemiptera (Fernandes et al. 2012). Lepidopterans, like butterflies and moths, undergo complete metamorphosis, which encompasses distinct larval, pupal, and adult life stages (Savard et al. 2006). Their larvae, which induce and inhabit galls, are equipped with biting and chewing mouthparts. Typically, they possess eight pairs of legs, with the first three pairs classified as true legs, and the remaining as prolegs (Mani 1964). There is a documented case where an adult female of *Heliozela staneella* Fischer Von Röslerstamm (Heliozelidae) injects a liquid during oviposition to induce the gall (Miller 2005). Lepidoptera larvae are renowned for their voracious feeding habits, rapidly chewing the cells surrounding gall chambers. These galls are typically solitary—i.e., one single larva usually occupy each gall chamber (Cornell 1990). Another distinctive feature of lepidopteran galls is that the larval chamber has excessive feces due to their very active feeding activity (Fernandes et al. 2012).

Mani (1964) considered that only a hundred of galling lepidopteran species has been described. In a more recent analysis, this number has increased, with 352 known galls induced by lepidopterans (Miller 2005). The galling Lepidoptera may

B. G. Ferreira (✉)
Department of Botany, Universidade Federal do Rio de Janeiro (UFRJ), Rio de Janeiro, Brazil
e-mail: bgf@biologia.ufrj.br

L. M. Guedes
Facultad de Ciencias Forestales, Laboratorio de Semioquímica Aplicada, Universidad de Concepción (UdeC), Concepción, Chile

D. C. de Oliveira, R. M. dos Santos Isaias (eds.), *Plant Galls*,
https://doi.org/10.1007/978-3-031-80064-1_14

belong to 20 independently evolved distinct families of microlepidoptera. Among these, Gelechiidae and Tortricidae stand out, with 47 and 39 species, respectively, making them the microlepidopteran families with the greatest number of gall inducers (Fernandes et al. 2012; Luz et al. 2014). Most of these moths have hitherto been identified solely by their gall morphotype, as the taxonomic descriptions at the specific level for most of them are still pending (Miller 2005). About 180 galling Lepidoptera species are known (Fernandes et al. 2012). Although most species are yet unknown, it is supposed that tropical regions have the greatest diversity of microlepidoptera (Powell 1980). Some works on gall anatomy and ultrastructure reported the inducer as an unidentified lepidopteran larva (Vieira and Kraus 2007; Bedetti et al. 2013; Ferreira and Isaias 2013; Ferreira et al. 2015; Vecchi et al. 2013), which indicates the urgency of alpha-taxonomic work in moth lineages that induces galls on neotropics (Luz et al. 2014; Moreira et al. 2017).

Microlepidoptera is not a formal taxon, and it is not considered a monophyletic group. Such informal group includes the for most basal suborders of Lepidoptera and seven superfamilies of the derived superorder Ditrysia (Powell 1980). Microlepidoptera excludes the butterflies and specialized moths, which are considered macrolepidoptera. Microlepidopterans include several habits among lepidopterans, and they commonly developed an internal feed habit or shelter making against predators. Secondary adaptations are exclusive to macrolepidoptera, with caterpillars feeding exposed to the external environment, with cryptic resemblance and behavior, mimicry, urticating hairs, and glands, aposematic coloration, distastefulness, etc. The habits of microlepidopteran larvae include fungus and animal associates, scavengers, stem-and root-borers, leaf miners, and seed-and external foliage-feeders. Some of them independently developed the ability to induce galls (Powell 1980).

The formation of insect galls is very diverse and involve alterations associated with the mode of gall induction, inducer's feeding behavior, and evolutionary history of the gall inducers. In this context, we present hypotheses regarding the evolution of the galling habit on lepidopterans and explore how the life cycle and gall induction behaviors of these organisms are associated with anatomical, cytological, and physiological changes. Additionally, besides the action of gall inducers, other guilds are associated with Lepidoptera galls, prompting discussion on how they can influence gall development, from cellular levels to ecological controls. Our aim is to shed light on the complex interactions within gall systems and stimulate further research into their evolutionary and ecological significance.

2 Evolution of Galling Habit of Microlepidopterans

The evolution of galling habit among microlepidopterans is supposed to derive from leaf mining and stem and root boring habits (Mani 1964; Price et al. 1987; Guiguet et al. 2019). Within Gelechioidea, for instance, the family Gelechiidae typically consists of leaf-tiers, but also leaf and conifer needle-miners, gall-makers, and scavengers. Momphidae have leaf miners and gall inducers, while Walshiidae have miners, seed feeders, and gall inducers. Among Tortricoidea, the Tortricidae family are usually borers or gall makers in stems, roots, and seeds. Within the Sesioidea, the Sesiidae larvae are known to borers into stems or even into galls caused by other insects (Powell 1980; Menken et al. 2009). In the basal lepidopteran suborder Incurvarioidea, Cecidosidae, a small family comprising seven genera and 19 species, exhibits galling habits on Anacardiaceae (Table 1). The exception is *Xanadoses*, whose larvae are specialized as bark miners of Australian trees (Moreira et al. 2012, 2017; Silva et al. 2018a).

Therefore, it is natural to hypothesize that galling habit, i.e., the ability to induce changes in plant tissue growth and differentiation, evolved several times inside Lepidoptera due to the intricate relationship between some groups of leaf miners, stem, root, and seed borers and the plant tissues were their larvae feed and grow (Mani 1964; Price et al. 1987). There are reports of species in the Gracillariidae family of leaf miners and leaf miner-gallers which are capable of inducing cell redifferentiation, which could be an indication of an evolutionary step between leaf mining and gall-inducing habits (Guiguet et al. 2019; Isaias et al. 2021). The first and second instar of the Gracillariidae, *Caloptilia cecidophora* Kumata, 1966 construct mines in leaves of *Glochidion obovatum* von Siebold (Phyllanthaceae). The third instar is capable of inducing alterations such as cell hypertrophy, hyperplasia, and cell redifferentiation, indicating that the two first instars do not possess the stimuli necessary to promote gall formation, which is unusual among galling lepidopterans. *Caloptilia cecidophora* only complete its larval development if the third to six instar feed on gall nutritive tissue. In addition, this species is closely related to mining species uncapable to induce galls, but that construct a shelter by actively rolling the host plant leaf in the last stages (Guiguet et al. 2019). These results could indicate that gall-inducing habit could be evolved from mining habit in this lineage (Zhang et al. 2016; Guiguet et al. 2019; Isaias et al. 2021).

3 Inferring Life Cycles of Galling Micromoths

Reports on the life cycles of some gall inducers are scarce, especially on the Neotropics (Table 1). Some reported cases indicate that lepidopterans may oviposit upon plant surface or may inject the eggs more profoundly in the stem bark, as reported for Cecidosidae (Luz et al. 2014). After eclosion, first instar larva commonly enters in the host organ (usually the stem buds, but also young leaves, etc.) and begin to feed, inducing the gall growth. Some studies indicate that the galling larvae may survive several months inside the galls before the pupation and eclosion (Brown et al. 2003; San Blas and Davis 2013; Silva et al. 2018b). Luz et al. (2014) demonstrated significant variations in the pupal stage life history of *Palaeomystella* spp. (Momphidae), which last-instar larvae can either pupate within sessile or dehiscent galls (fallen galls) or they may choose to leave the galls and pupate in leaf litter, depending on the species.

Table 1 Some Neotropical galling lepidopterans, gall features, and life cycle details

Family	Species	Host plant—localities	Host organ	Gall morphotype and biological features	Larval instars	Life cycle	References
Cecidosidae	*Cecidonius pampeanus* (Moreira and Gonçalves 2017)	*Schinus weinmannifolius* Mart. ex Engl. (Anacardiaceae)—Brazil (South)	Stem	Initially cylindric (fusiform), occurring under the bark of the secondary stem; in the last larval instar, gall tissues disrupt bark and form a green smooth globoid outgrowth Not operculated—pupa opens an irregular orifice on the distal, weaker wall by frontal process and body contortions.	Five	Possibly univoltine: Larvae grown during summer; Galls with last larval instar occur mainly in autumn; Galls with last larval instar annealed at the base and fall in the soil; Larva diapauses for months inside the fallen gall Pupae appear in October (spring); Adults emerge and induce galls in late spring when plants produce several sprouts.	Moreira et al. (2017)
	Oliera argentinana (Brèthes 1916)	*S. polygama* (Cav.) Cabrera (Anacardiaceae)—Brazil (South)/*S. fasciculata* (Griseb.) I.M.Johnst. (Anacardiaceae)—Argentina (Mendoza)	Stem	Fusiform, initially covered by stem bark Not operculated—pupa opens an irregular orifice on the bark tissue of a weaker gall wall by frontal process and body contortions.	–	Univoltine: Adult emergence from late Spring to early Summer (November/ December)	Moreira et al. (2012)
	Oliera saizi Moreira & Vargas	*S. polygama* (Cav.) Cabrera (Anacardiaceae)	Stem	Fusiform, enclosed Larval chamber elliptical in shape	–	Univoltine Adult emergence occurs during late spring to early summer (November, December)	Silva et al. (2018b)

	Dicranoses capsulifex (Kieffer and Jörgensen 1910)	*S. fasciculata* (Griseb.) I.M.Johnst. and *S. johnstonii* F.A.Barkley (Anacardiaceae)—Argentina (Mendoza)	Stem	Cylindrical leaf-like galls, green Early gall globoid, reddish-brown Operculate	–	Female lays eggs from October to November, inside the stem (near bark); Few days after oviposition, first instar larvae hatch and enter the stems; Larvae remain in the first instar until May; Second instar larvae can be found in May, inside harder tissues, when gall begin to grow appearing through the bark; Gall elongates and the basal part constricts, resembling a rachis; In Semptember, gall shrink and pupation occurs; With pupation, a light brownish band is differentiated close to the gall apex; Pupa pushes apart the "hood" for adult emergence	San Blas and Davis (2013)

(continued)

Table 1 (continued)

Family	Species	Host plant—localities	Host organ	Gall morphotype and biological features	Larval instars	Life cycle	References
	Eucecidoses minutanus Curtis	*S. longifolia* (Lindl.) Speg. (Anacardiaceae)—Argentina (Buenos Aires)	Stem	Globoid; smooth surface and glabrous. Single spherical larval chamber Color green (young) and brown (later) With a circular operculum.	Four	No further information	Loetti et al. (2016)
	Cecidoses eremita Brèthes	*S. longifolia* (Lindl.) Speg. (Anacardiaceae)—Argentina (Buenos Aires)	Stem	Globoid; smooth surface and glabrous. Single spherical larval chamber Color green (young) and brown (later) With a circular operculum.	Four	No further information	Loetti et al. (2016)
	Andescecidium parrai Moreira & Vargas	*S. polygama* (Cav.) Cabrera (Anacardiaceae)	Stem buds	Elliptical to globoid The color varies between green to red and intermediate shades Pedunculate and operculate	–	Univoltine. Induction occurs in late spring (November) and reach senescence in May	Silva et al. (2018b)

Tortricidae	*Tortrimosaica polypodivora* (Brown and Baixeras 2004)	*Microgramma squamulosa* (Kaulf.) de la Sota (Polypodiaceae)—Brazil (Southeast)	Stem	Fusiform, with the same indumenta (scales) of the non-galled stem. Chamber is subdivided into two spaces, one of them with feces and exuviae, and another where the larva feeds. Only one larva per gall; larvae are strongly cannibalistic and may devour their exuviae.	Five	Possibly univoltine: Eggs deposited in late spring (October–December); Larvae hatch in ~45 days and bore into a terminal shoot, becoming internal feeders; An exit channel near the plant epidermis is occupied by a cocoon; Adults emerge in late spring (October–December)	Brown et al. (2003)
Momphidae	*Palaeomystella beckeri* (Moreira and Basilio 2015)	*Tibouchina trichopoda* (DC.) Baill. (Melastomataceae)—Brazil (South)	Stems	Globoid, woody, with external surface lacking conspicuous ornament. Brown Irregular, scaly surface; woody, usually unilocular and unilarval. Larval chamber as a tubular gallery circulating internally the stem cortex Without an operculum		Pupation occurs in the deepest portion of the gallery; In laboratory, galls bearing last-instar larvae that were collected during early winter (July); adult emergence started in early spring (September)	Basilio et al. (2015)

(continued)

Table 1 (continued)

Family	Species	Host plant—localities	Host organ	Gall morphotype and biological features	Larval instars	Life cycle	References
	Palaeomystella fernandesi (Moreira and Becker 2014)	*Tibouchina sellowiana* (Cham.) Cogn. (Melastomataceae)—Brazil (South)	Apical stem bud	Fusiform type, external surface without conspicuous ornament; reminiscent of stem ribs. Color green to violet as it ages. Fleshy, without uniformly defined internal chamber; unilocular, unilarval Circular operculum on upper half of gall wall constructed by the last instar larva prior to its pupation	–	Galls occur during spring (October) and summer (February) Galls free from kleptoparasite fall to the ground in late larval ontogeny and larva complete development on the ground Pupation occurs inside the gall Adult emerges presumably after the winter (September) through a circular operculum Kleptoparasitic gelechiid moth associated	Luz et al. (2014, 2015)
	Palaeomystella oligophaga (Becker and Adamski 2008)	*Macairea radula* (Bonpl.) DC./*M. thyrsiflora* DC. (Melastomataceae)—Brazil (Midwest and Southeast)	Stem, leaf blade (abaxial surface)	Two gall types: A large gall type—globoid to oblong, 10-15mm, pale green, pink to dark reddish brown on the side exposed to sun light, with emergences, sometimes with sticky secretion. The small gall type—oblong, about 6-7 mm in length, pale green to pinkish red Small chamber, unilocular Operculated	–	Bivoltine life cycle (induction at the end of rainy season – February/ March – and end of dry season – August/ September	Becker and Adamski (2008), Rezende et al. (2023)

	Palaeomystella rosaemariae (Moreira and Becker 2014)	*Tibouchina asperior* (Cham.) Cogn. (Melastomataceae) – Brazil (South)	Apical stem buds	Globoid with external surface covered with short spine-like projections. Green to reddish Unilocular, unilarval, pupates away from the gall	–	Mature last instar larva invariably made a lateral orifice by chewing the gall wall Not known life cycle duration in nature. In laboratory, larvae leave the gall and pupate at the bottom of plastic pots, and the adults emerged in May	Luz et al. (2014)
	Palaeomystella tavaresi (Becker and Moreira 2014)	*Tibouchina fissinervia* (Schrank & Mart. ex DC.) Cogn. (Melastomataceae)—Brazil (Northeast)	Stem apical/ terminal buds	Rosette and covered by clustered short leaves, and a cylindrical gall chamber. Green, progressively darkening during senescence, after emergence of the moth; unilocular, unilarval	–	Pupation occurs inside the gall Adults exit through the terminal leaves of the gall In laboratory, adults emerged in October (spring)	Luz et al. (2014)

(continued)

Table 1 (continued)

Family	Species	Host plant—localities	Host organ	Gall morphotype and biological features	Larval instars	Life cycle	References
Pterophoridae	*Adaina primulacea* (Meyrick 1929)	*Chromolaena odorata* (L.) R.M. King & H. Rob. (=*Eupatorium odoratum* L.) (Asteraceae)—USA (Florida), Central America	Young stems	Fusiform galls; operculated	Five	First instar larvae mine young stems, inducing the gall; The later instars are stem borers, primarily within the galls, and later beyond gall tissues Larvae were collected from plants in September-October (autumn—north hemisphere) Pupation occurs within a hollowed-out gall chamber formed by larval feeding on gall tissue Pupa is positioned headup toward a small operculum Gall phenology unknown	Mathews and Maharajh (2009)

Several gall inducers in Neotropics seem to have an univoltine life cycle, although the last larval instar in several cases remains in diapause inside the galls during the winter (Table 1). The life history of the gall inducers broadly varied among species of the same and distinct families. Galls induced by Cecidosidae are usually operculated, but galls of two cecidosids (*Cecidonius pampeanus* and *Oliera argentinana*) are not operculated (Moreira et al. 2017). In Momphidae, the species of *Palaeomystella* may have galls with or without an operculum, depending on the species (Luz et al. 2014, 2015; Basilio et al. 2015). In the case of Pterophoridae, *Adaina primulacea* galls have a very distinct circular operculum (Mathews and Maharajh 2009). Tortricidae and Momphidae described in Table 1 may dig an exit channel before pupation, but *Palaeomystella rosaemariae* last instar larvae leave the galls by chewing a lateral orifice and seem to pupate in the soil (Luz et al. 2014). The adults seem to emerge in late spring or early summer, although *Palaeomystella rosaemariae* emerged in May in laboratory conditions (late autumn in South hemisphere).

Some Lepidoptera galls containing pupa fall into the soil at the end of maturity as those induced by *Cecidonius pampeanus* and *Palaeomystella fernandesi*. In the case of Cecidosidae galls induced by *Scyrotis* spp. on leaves and young stems of *Rhus* spp. (Anacardiaceae) from South Africa, after the gall falls, the pupa moves actively, causing the gall jumping movement on the soil. This behavior was related to an adaptation for excessive heat dissipation. Such galls are called 'jumping beans' due to their movement caused by pupa occupying the fallen galls (Van Noort et al. 2007). It appears that the adults of cecidosids have a short lifespan and limited dispersal ability (Silva et al. 2018a). In some cases, adults have absented mouthparts, as in *Dicranoses capsulifex* Kieffer and Jörgensen, 1910 (San Blas and Davis 2013). Therefore, after adult emergence there is copulation and oviposition in buds or young stems.

As the galling Lepidoptera taxonomy lacks the description of most species, their life history is also little known. The brief demonstration of distinct life cycles and peculiarities of neotropical galling microlepidoptera (Table 1) indicate that variations occur even among species of same genus, with a broad range of distinctive features. Understanding the life cycles of galling microlepidoptera not only unveils novel adaptive traits but also provides researchers with essential information about different developmental stages, including larvae, pupae, and adults. This knowledge is crucial for taxonomists when describing and classifying these species.

4 Gall Induction and Development

The feeding habit and the life cycle of the gall inducers play a crucial role in determining the structural and histochemical patterns of galls induced by different taxa of arthropods (Bronner 1992; Rohfritsch 1992; Ferreira et al. 2019). Not only the gall-inducing organism, but also the host plant potentials, including the exact site on the host plant where the gall is induced, are important to determine the structure of gall tissues and metabolic activity of cells (Ferreira et al. 2017, 2019; Guedes et al.

2023). In the case of lepidopterans, the voracious chewing activity of larvae on host organs induce the formation of calli-like tissues, which meristematic cells capable of rapidly divide (Ferreira and Isaias 2013; Ferreira et al. 2015, 2022), which was already supposed to be simple regenerative responses of plant tissues to the chewing activity of larvae (Meyer and Maresquelle 1983; Dreger-Jauffret and Shorthouse 1992). Under these circumstances, the regenerative tissues were already considered poorly differentiated (Meyer and Maresquelle 1983). As histological studies of lepidopteran galls have advanced, typical nutritive cells had been described for some gall morphotypes. Examples include those with dense cytoplasm and fragmented vacuoles found in lepidopteran galls on *Senecio cacaliaster* Lamk (Meyer and Maresquelle 1983). In galls induced on *Limoniastrum guyonianum* Dur, nutritive cells near larval chamber are smaller and possess relatively large, sometimes duplicated lobbed nuclei and hypertrophic nucleoli (Meyer and Maresquelle 1983), as already observed in lepidopteran—*Marcetia taxifolia* (A.St.-Hil.) DC. galls (Ferreira and Isaias 2013; Ferreira et al. 2015) (Fig. 1) and *Palaeomystela oligophaga*—*Macairea radula* galls (Rezende et al. 2019). Nowadays, with the progress of structural studies of Neotropical galls, we know that several lepidopteran galls have typical nutritive cells (*sensu* Bronner 1992; Ferreira et al. 2017, 2019), commonly rich in lipids (Guedes et al. 2023).

4.1 Induction

The lepidopteran galls are induced by the larvae, usually the first instar, which enters in meristematic host plant tissues after egg eclosion (Mani 1964; Ferreira et al. 2022). Although Mani (1964) observed that galling lepidopteran eggs are always deposited on the host plant surface, there are some cases where the oviposition occurs inside the plant tissues. Cecidosidae bear a relatively long ovipositor of the piercing type capable of ovipositing within the plant cells (Moreira et al. 2012, 2017), such as inside lateral stem buds (pericycle cells) of *Schinus engleri* Barkley by *Eucecidoses minutanus* Brèthes (Ferreira et al. 2022) and in stem lenticels by *Cecidoses eremita* Curtis (Wille 1926). Ferreira and Isaias (2013) also suggested that oviposition by an unidentified gall-inducing Lepidoptera occurred in pericycle cells of *Marcetia taxifolia* developing stems. These authors observed the suberization of a cell layer around the egg, and the redifferentiation of pericycle cells into parenchymatic cells with tannins, which are hyperplasic and interrupt the

Fig. 1 (continued) (black arrows). (**g**) Transverse section of a mature gall, showing reminiscent pith cells, surrounding the replicative nutritive cells. The vascular system is on secondary growth. Abbreviations: Cx cortex, Fr fruit developed on the shoot above the gall, LC larval chamber, Lf leaf on the gall, NGS non-galled portion of stem, NT nutritive tissue, PV primary vascular system, RP reminiscent pith cells, SNT storage nutritive tissue, SP Secondary phloem, SX secondary xylem, Tr trichomes, TNT typical nutritive tissue. Bars: (**a**) 2 mm; (**b, c**) 5 mm; (**d**) 500 μm; (**e, g**) 200 μm; (**f**): 50 μm

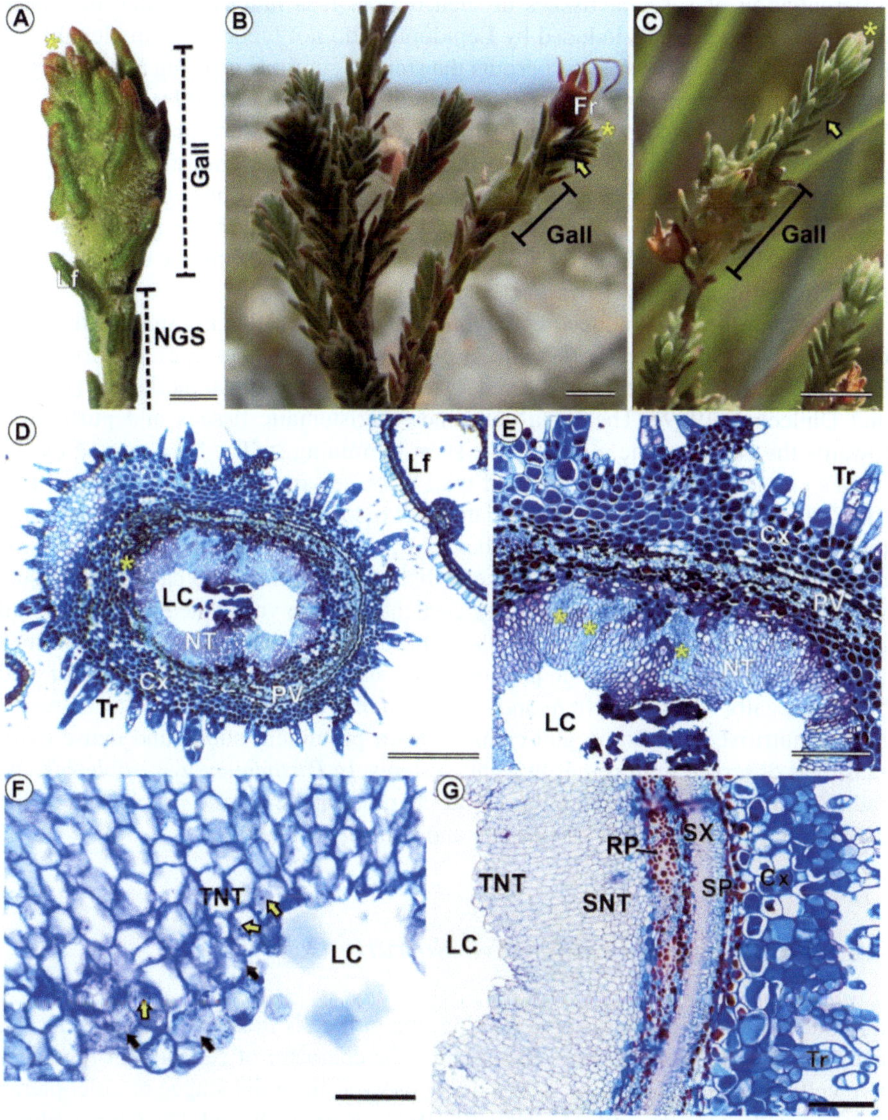

Fig. 1 Fusiform galls induced by an unidentified Lepidoptera species in the stems of *Marcetia taxifolia*. (**a**) The gall is induced on the stem pith, near the shoot apical meristem (asterisk), causing the stem swelling. (**b**, **c**) The activity of the shoot apical meristem (asterisk) is not impaired by the gall induction. The stem above the gall continues growing, forming new vegetative (arrow) and reproductive shoots. (**d–f**) Transverse section of the gall on the growth and developmental phase, on the median-to-apical portion. (**d**) The stem pith is redifferentiated into nutritive cells. The primary vascular cylinder of *M. taxifolia* is interrupted in the region where the larva entered and is replaced by parenchyma cells (asterisk). The trichomes are hypertrophied. (**e**) At this phase, neoformed phloem strands (asterisks) are observed immersed in the nutritive tissue. (**f**) The typical nutritive tissue cells lining gall chamber have evident nuclei (yellow arrows) and dense cytoplasm

differentiation of vascular tissues in such region. It is noteworthy that in certain instances, the stem galls induced by Lepidoptera do not hinder the functionality of the shoot apical meristem. This permits the stem to continue growing above the gall (Fig. 1a, b), even following gall senescence (Fig. 1c) (Ferreira and Isaias 2013). The hyperplasic pericycle cells influence local hyperplasia of surrounding cortex and pith stem cells (Fig. 1d, e). The interruption of vascular system differentiation was also noted in primary growth stems of *S. engleri* and lateral shoot galls induced by *Eucecidoses minutanus*. In this context, larva begins to feed in this region, moving towards the stem pith, increasing the larval chamber by chewing the differentiating parenchymatic pith cells (Ferreira et al. 2022).

Regarding the galls induced by *Epiblema strenuana* Walker, 1863 (Tortricidae) on the stems of the weed *Parthenium hysterophorus* L. (Asteraceae), the first instar larva begins to feed on leaf surface and moves towards the apical meristems (Raman and Dhileepan 1999). The larva consumes meristematic tissues and push itself towards the center of the meristematic stem by mining it. The hole created by the larva's entrance into the stem is filled with frass, serving as an external signal that the *Epiblema strenuana* larva has initiated gall induction. Regenerative parenchyma then occupies the region previously chewed by the larva. Once inside the stem axis, the larva feed on pith, derived from cells that redifferentiate into callus-like tissue, while cortical cells hypertrophy. These cortical cells are vacuolated and have small nuclei. The vascular traces appear disorganized due to the partial consumption of interfascicular parenchyma, a phenomenon also observed in *M. taxifolia* and *S. engleri* galls. *Epiblema strenuana—Parthenium hysterophorus* galls also present typical nutritive cells lining gall chamber with prominent nuclei and dense cytoplasm, as observed in the other two gall systems. In *Parthenium hysterophorus*, the radial walls of cells bordering the larval chamber are fractured, likely due to the caterpillar's chewing behavior (Raman and Dhileepan 1999).

4.2 Formation of Proliferative Nutritive Cells

Soon after the beginning of larva feeding, the pith cells in stems of both *Marcetia taxifolia* (Fig. 1) and *S. engleri* also assume the character of calli-like or regenerative cells (Ferreira and Isaias 2013), also observed in initial stages of coleopteran induced-galls (Meyer and Maresquelle 1983; Barnewall and De Clerck-Floate 2012). In the case of *S. engleri—E. minutanus* and *Cecidoses eremita* (Fig. 2) and *S. polygama* (Cav.) Cabrera—*Andescecidium parrai* Moreira and Vargas 2018 (Fig. 3), the galls are induced on young lateral shoots, impairing the further apical development of the lateral branch, by redirecting the fates of the cells to form the gall tissues. The nutritive tissues are voraciously consumed, and a meristematic activity is established, not only forming new nutritive cells, but also novel rows of vascular bundles, which are not present in the non-galled stems (Ferreira et al. 2022) (Figs. 2 and 3). The nutritive tissue in the young galls of an undescribed species of Lepidoptera induced on the stems of *M. taxifolia* also replaces the cells of the pith

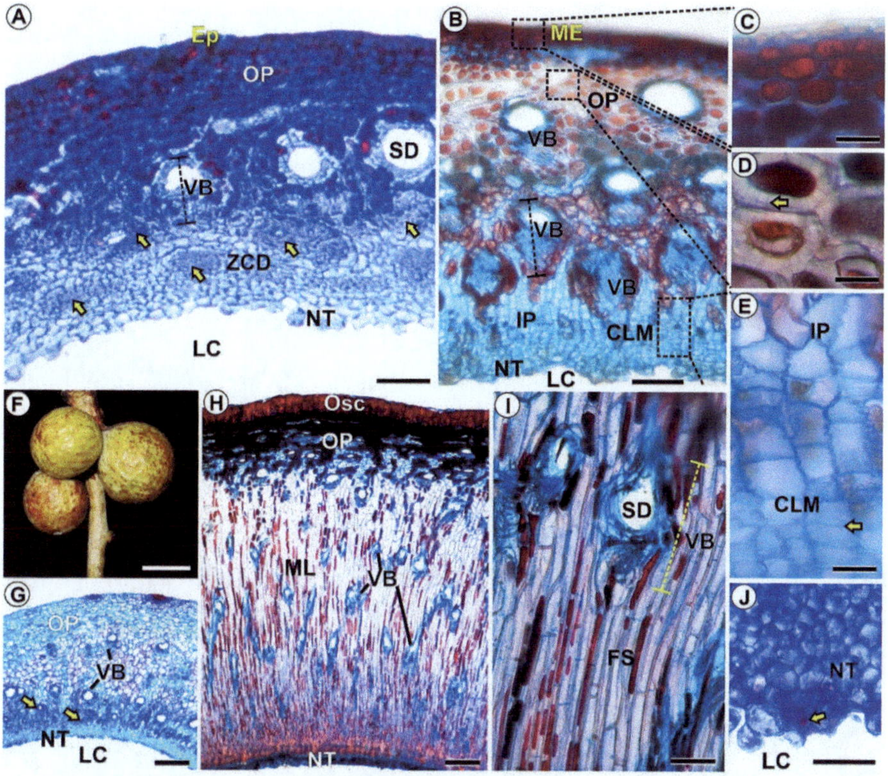

Fig. 2 Globoid galls induced by *Eucecidoses eremita* and *Cecidoses minutanus* (Cecidosidae) on the axillary buds stems of *Schinus engleri* (Anacardiaceae). (**a–e**) Transverse sections of *E. minutanus* galls. (**a**) Early growth and development phase. Arrows indicate new rows of procambial strands immersed in the zone of cell differentiation. (**b–e**) Maturation phase gall. Note the several rows of vascular bundles. Arrow indicates a pit in sclerified parenchyma cell wall. (**f–j**) *Cecidoses eremita* gall. (**f**) Macro view of a gall on maturation phase. (**g**) Transverse section of gall in early growth and development phase, with arrows indicating new rows of procambial strands differentiating in the zone of cell differentiation. (**h–j**) Transverse sections of gall on maturation phase. Note the several rows of vascular bundles. Arrow indicates the nutritive cell with prominent nucleus. Abbreviations: CLM cambium-like meristem, Ep epidermis, FS fibersclereids, IP inner parenchyma, LC larval chamber, ME multiple epidermis, ML median layers, NT nutritive tissue, OP outer parenchyma, Osc outer sclereids, SD secretory duct, VB vascular bundles, ZCD zone of cell differentiation. Bars: (**a, j**) 50 μm; (**b, i**) 100 μm; (**c–e**) 20 μm; (**f**) 1 cm; (**g–h**) 200 μm

lining the larval chamber (Fig. 1e, f), while the outer cells in the pith, called storage nutritive cells, within the pith exhibit characteristics of replicative cells (Ferreira and Isaias 2013). Strands of phloem differentiate towards the larval chamber (Fig. 1d, e).

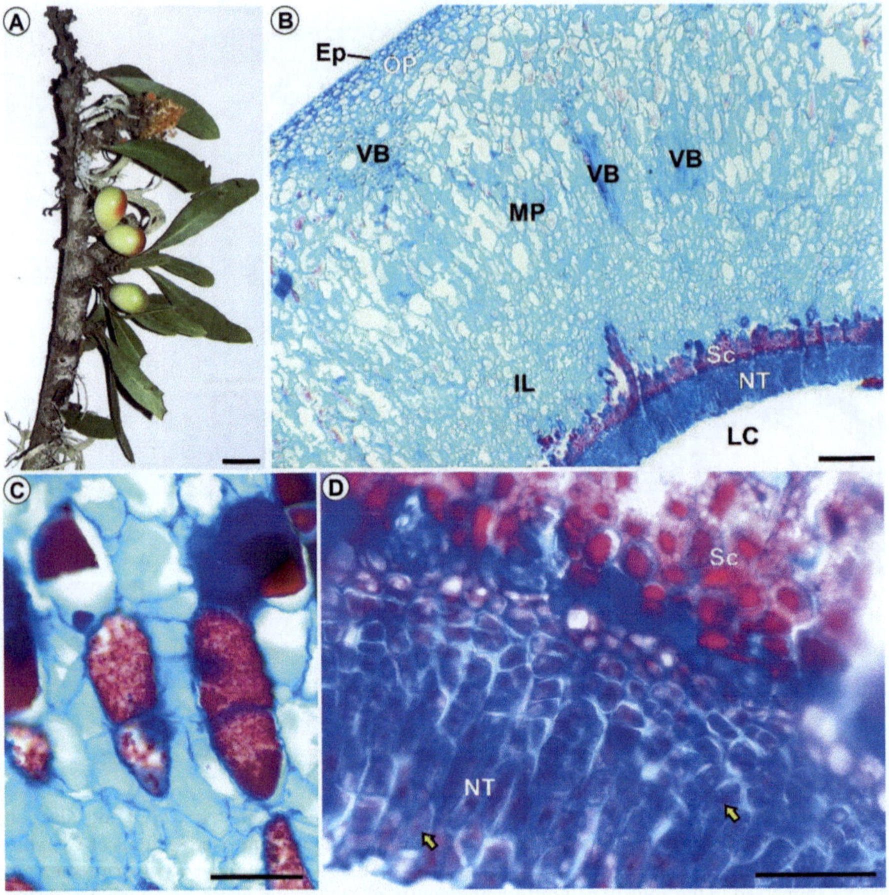

Fig. 3 Galls induced by *Anderscecidium parrai* (Cecidosidae) on the axillary buds stems of *Schinus polygama* (Anacardiaceae). (**a**) Macro view of bud galls. (**b–d**) Transverse sections. (**b**) Maturation phase gall. (**c**) Detail of the common storage tissue on the median layers. (**d**) Sclerenchyma band surrounding the nutritive tissue layers. The nutritive cells surrounding gall chamber are elongated, with dense cytoplasm and prominent nucleus (arrow). Abbreviations: Ep epidermis, IL inner layers of common storage tissue, LC larval chamber, MP median cells of common storage tissue, NT nutritive tissue, OP outer parenchyma, Sc sclerenchyma, VB vascular bundles. Bars: (**a**) 5 mm; (**b**) 200 µm; (**c, d**) 50 µm

4.3 Maturation Phase

During maturation stage, phloem strands are not present immersed in the nutritive tissue of *Marcetia taxifolia* galls, which indicates the consumption of a great part of nutritive cells by the larva (Fig. 1g). These nutritive cells accumulate lipids and terpenoids, while the replicative cells form a storage nutritive tissue which accumulate starch, reducing sugars, and phenolics (Ferreira and Isaias 2013). Therefore, the proliferative character of the storage nutritive tissue in *Marcetia taxifolia* galls is important to replace typical nutritive cells feed by the larva. The cells are smaller than the pith cells not only in growth and development stage, but also in maturation, due to their increased cell division rates. The storage nutritive cells also have prominent nuclei (Fig. 1f), as nutritive cells lining gall chamber, and these nuclei have sometimes two evident nucleoli (Ferreira et al. 2015). This feature was also observed in nutritive cells of other lepidopteran galls, which may have two nuclei or fused polyploid nuclei and duplicated nucleoli (Meyer and Maresquelle 1983), which probably occur in other lepidopteran galls.

The abovementioned stem galls on *S. engleri*, *Marcetia taxifolia*, and *Parthenium hysterophorus* are induced on the pith. However, the cellular changes occur not only in pith cells, with the formation of hypotrophied nutritive cells in *Marcetia taxifolia*, but also in the cortex, which presents hypertrophied cells in the three gall systems (Raman and Dhileepan 1999; Ferreira and Isaias 2013; Ferreira et al. 2022). The secondary vascular tissue of the stem galls induced on *Marcetia taxifolia* have hypertrophied xylem fibers, while in galls of *S. engleri*, the addition of rows of vascular bundles during the maturation phase increases the vascularization of the gall (Figs. 2 and 3). These developmental features indicate that *S. engleri* nutritive cells have a secondary (lateral) meristematic characteristic (Fig. 2), increasing gall thickness until the end of the gall maturation phase (Ferreira et al. 2022)—which is not commonly observed in other insect galls, since maturation phase is usually the main trophic phase of the gall inducers, but has already differentiated cells (Rohfritsch 1992). In *Parthenium hysterophorus* galls, there is also an intense division of the cells of the interfascicular region.

5 Developmental Patterns of Lepidoptera Galls

The comparative studies of lepidopteran galls with the non-galled host organs may indicate the main changes that lead to the organ swelling, characteristic of such galls. Changes in dermal, ground, and vascular systems are observed, as revised by Guedes et al. (2023). Some mature lepidopteran galls also have meristematic activity in nutritive tissues during the growth and development, and maturation phases (Ferreira et al. 2022) (Fig. 2). The meristematic activity of the inner cells, near the nutritive layer, differentiates not only into several new layers of nutritive cells and inner parenchyma cells but also new rows of procambial strands, as observed in *S. engleri*—*E. minutanus* young galls (Fig. 2a) and *S. engleri*—*Cecidoses eremita* galls (Fig. 2g). During the maturation phase, the cells between the nutritive tissue and the inner parenchyma cells act as a lateral meristem—called cambium-like meristem (Fig. 2e)—which deposit new layers of parenchyma and sclerenchyma tissues outwards and novel nutritive tissue layers inwards (Fig. 2d, h–j). In other insect galls, the maturation phase is characterized by the end of differentiation of gall cell types, with absence of tissue growth, but this phase is the main trophic phase of the galling larvae (Rohfritsch 1992). It's noteworthy that each system of these Cecidosidae galls has its own peculiarities. The galls of *E. minutanus* (Ferreira et al. 2022) have multiple non-lignified epidermis (Fig. 2b, c), isodiametric outer sclerenchyma cells (Fig. 2d), median layers of elongated sclerified cells in the last phase of maturation (not seen in the figure, see Ferreira et al. 2022), inner parenchyma cells, cambium-like meristem, and nutritive tissue (Fig. 2e).

The galls induced by *Cecidoses eremita* in maturation phase reach much thicker walls than *E. minutanus* and have sclereids coating the gall wall on the outside, a non-lignified outer parenchyma, median layers of elongated fibersclereids, a small replicative cell layer, and a nutritive tissue lining the gall chamber (Fig. 2h–j). The galls induced by *A. parrai* on *S. polygama* (Guedes et al. 2023) are other Cecidosidae galls induced on a host plant of the same genus. These galls have some similarities with the other two galls abovementioned, as the induction on axillary buds, the presence of more than one row of vascular bundles with the intraphloematic secretory duct typical of Anacardiaceae, and the presence of meristematic nutritive cells (Fig. 3). However, the meristematic nutritive cells in the *A. parrai* galls on maturation phase have a distinct feature of those induced by *E. minutanus*, since they only originate new nutritive cells (Fig. 3d). The peculiarities of Cecidosidae galls induced on *Schinus* spp. demonstrate that each galling species can induce peculiar anatomical features. The *A. parrai* galls have an outer storage tissue divided into outer parenchyma, median layers of spongy parenchyma with phenolic idioblasts (Fig. 3c), and inner layers of diminutive cells, surrounding the delimited sclerenchyma zone (Fig. 3d).

Lepidoptera galls may have increased trichome size when compared to non-galled organs, as observed on *Marcetia taxifolia* stem galls (Ferreira and Isaias 2013). Dense neoformed emergences, which are sometimes leafy-like projections, occur in some Lepidoptera galls on Melastomataceae, such as stem galls induced by *Palaeomystella oligophaga* Becker and Adamski 2008 on *Macairea radula* (Bonpl.) DC. (Rezende et al. 2023a), and those induced by an unidentified species on leaves of *Tibouchina pulchra* Cogn. (Vecchi et al. 2013).

As several lepidopteran galls are induced on buds or stem axis, they reveal distinct complexity levels, since stems have different tissue layers, typical of each plant species, and primary and secondary growth mediated by vascular cambium (Jayaraman 1989). Depending on where the galls are induced on stems, distinct tissue types participate in the gall formation. Some insect galls may be induced on the cortex, others on vascular cambium or secondary vascular tissues, or in the pith. Most stem galls are induced in the pith or in the transition among pith and primary vascular tissues, involving the participation of pericycle cells (Ferreira and Isaias 2013; Ferreira et al. 2022; Martins et al. 2023). Therefore, the pith cells are usually recruited to differentiate into nutritive cells.

6 Gall Functional Compartmentalization, Ultrastructure, and Cell Walls

The functional compartmentalization of complex galls has been histochemically demonstrated (Ferreira et al. 2019). Although there are few histochemical studies on lepidopteran galls, it has been corroborated that they accumulate lipids, proteins, starch, and terpenes (see Bedetti et al. 2013; Ferreira and Isaias 2013; Vecchi et al. 2013; Rezende et al. 2019; Ferreira et al. 2022; Guedes et al. 2023). In most histochemical studies, lipids and proteins were detected in the nutritive cells bordering the larval chamber (Bedetti et al. 2013; Vecchi et al. 2013; Ferreira et al. 2022; Guedes et al. 2023; Martins et al. 2023).

Previously, Rohfritsch (1992) described the non-differentiation of nutritive tissues in Lepidoptera galls. Probably, the non-observation of nutritive cells is related to the rapid consumption by the galling larvae. Later studies have shown that Lepidopteran galls are nutritive (see Guedes et al. 2023). These nutritive cells have a dense and granular cytoplasm, which are well represented in the galls induced on *Marcetia taxifolia* stems (Fig. 1) (Ferreira et al. 2014, 2015), *S. engleri* stems (Fig. 2) (Ferreira et al. 2022), and *S. polygama* buds (Fig. 3) (Guedes et al. 2023). The nutritive cells of these three lepidopteran galls also accumulated terpenes, as well as in *Bauhinia ungulata* L. (Fabaceae) leaf galls (Bedetti et al. 2013). However, the presence of terpenes in these galls has been related to a conservative feature of their host plants and not a peculiarity of lepidopteran galls (Guedes et al. 2023). The cyto-histological and histochemical features of the nutritive tissue are peculiar, and clearly distinguishable from other tissues described in lepidopteran galls.

The occurrence of storage tissues has also been described in lepidopteran galls, especially in the outermost layers (Fig. 3b, c), bordering the nutritive tissue (Ferreira and Isaias 2013; Rezende et al. 2019; Guedes et al. 2023). The hypertrophied cells of the storage tissues accumulate mostly starch, reducing sugars, phenols, and lipids (Ferreira et al. 2015, 2022; Naves et al. 2021; Guedes et al. 2023). The differential accumulation of metabolites, the topographical localization, and morphological and cytohistological features are used to differentiate both tissue types. Both nutritive and storage tissues have the same functions as in other insect galls. The nutritive cells are the food source of the larvae, which due to the voracity of the larvae, are systematically replaced (Ferreira et al. 2015, 2022), while the storage tissue guarantees the nutrient supply from the nutritive cells and guarantee the water status and gall maintenance (Oliveira et al. 2016).

Although the ultrastructure of Lepidoptera galls has been little studied, some features have been delights. For example, Ferreira et al. (2015) described that the nutritive cells of fusiform galls induced by a species of Lepidoptera on *Marcetia taxifolia*, have walls with heterogeneous thickenings, ribosomes, polysomes, mitochondria, Golgi apparatus, endoplasmic reticulum, and clearly evident nucleoli. In *Palaeomystella oligophaga* galls induced on *Macairea radula*, the presence of endoplasmic reticulum, Golgi apparatus, mitochondria, large and lobed nuclei with conspicuous nucleoli, and fragmented vacuoles have been described (Rezende et al.

2019). These two studies indicate that nutritive tissues of Lepidopteran galls are rich in organelles and metabolically active, which have also been described in other taxa of galling insects with chewing mouthparts such as Cecidomyiidae and Cynipidae, so they are not a peculiarity of Lepidopteran galls. The presence of oleosomes and ribosomes in the nutritive cells of these two galls confirm the histochemical results that reveal the presence of lipids and proteins. In the galls induced on *T. pulchra* leaves, Vecchi et al. (2013) demonstrated nutritive cells with dense cytoplasm, but anucleate, suggesting that metabolic maintenance of these cells occur by communication with the outer parenchyma cells.

Research on the cell wall components of galls induced by *Palaeomystella oligophaga* on *Macairea radula* has revealed dynamic changes in the composition of both outer (storage) and inner (nutritive) tissues from larval to pupal stages (Santos et al. 2024), providing further evidence of the impact of larval feeding activity on cell metabolism regulation. Within the nutritive cells of galls hosting larvae and pupae, the cell walls consist of both sterified and unesterified homogalacturonan pectins, crucial for maintaining the flexibility of cell walls during growth. Methyl-esterified homogalacturonans are exclusively present in the larval-stage galls' outer compartments, demonstrating juvenility maintenance, possibly due to the site of gall induction—the buds (Santos et al. 2024). Arabinan and galactan epitopes are found in both gall compartments across larval and pupal stages, serving not only to enhance extensibility but also to increase porosity, facilitating the transport of metabolites for insect feeding and cell replication.

In nutritive cells of *Palaeomystella oligophaga* galls, hemicelluloses such as heteromannans and xylans are detected in the cell walls, potentially serving as storage carbohydrates and supporting larval nutrition (Santos et al. 2024). Glycoproteins, like extensins and arabinogalactan-proteins, are also identified in nutritive cell walls, indicating their significant role in controlling oxidative stress. The process of cell wall remodeling is essential for the functional adaptation of insect gall cells, and these findings underscore the capability of galling lepidopterans to actively regulate cell wall composition, thereby promoting gall maintenance and sustaining their own feeding activity. Studies on other Lepidopteran gall systems may demonstrate variations in such patterns.

The cytological and metabolic peculiarities of the Lepidoptera galls need more studies, which allow comparison with the characteristics of other chewing insect galls. These studies will allow us to elucidate the steps of how the phenotype of the galls is reflected by the metabolism and cytology of their nutritive cells. Likewise, the metabolic characteristics of storage tissue deserve more in-depth studies.

7 Impact of Galling Lepidoptera on Host Plant Performance

Little is known about the effect of galling Lepidoptera on the physiological processes of their host organ. In general, gall-forming insects induce an increase in photosynthesis rates, although some studies show the opposite (Desnitskiy et al.

2023). For example, in the system *Parthenium hysterophorus*—*E. strenuana*, the establishment of the lepidopteran on stems induces a decrease in leaf-water potential, photosynthetic and transpiration rates, and stomatal conductance (Florentine et al. 2005). The galling Lepidoptera exert a suppressive effect on photosynthesis in the affected leaves as other groups of galling insects and mites (Desnitskiy et al. 2023). The negative effects on photosynthesis rates are determined by the down-regulation of genes related to photosynthesis, as demonstrated in leaves of two host plant species (*G. obovatum*, Phyllanthaceae and *Eurya japonica* Thunb, Pentaphylacaceae) affected by two species of Gracillariidae (*Caloptilia cecidophora* and *Borboryctis euryae* Kumata and Kuroko 1988) compared to non-galled leaves (Takeda et al. 2019). The galling lepidopterans did not affect the concentration of ions (K, Mg, Mn, N, P, S, and Zn), except the chloride concentration, which increased significantly (Florentine et al. 2005). In this same host plant-galling Lepidoptera system, Raman et al. (2006) determined a positive effect of the inducer on the carbon content, but the results were opposite in the nitrogen content and there was no effect on the total content of non-structural carbohydrates. Both Florentine et al. (2005) and Raman et al. (2006) results support the hypothesis of galls as nutrient sinks from the producing organs to the sites of gall formation. These photo-assimilates are stored in reserve tissues and can be used for gall maintenance or mobilized to nutritive tissues.

Due to their impact on plant yield, the galling lepidopterans were studied as biological control of weeds, such as *Parthenium hysterophorus* with galls induced by *E. strenuana* (Navie et al. 1998; Dhileepan 2003; Dhileepan et al. 2005). In greenhouse, the gallers applied precociously (35 days after germination) impacted the quantitative production of capitula and seeds, although no effects on seed quality were observed. The plant height was also affected when the gall inducers were introduced in the early phase of plant development (Navie et al. 1998). However, the reduction in plant performance is not evident in the field. The effects of parasitoids on *E. strenuana* indicate that these natural enemies may diminish the biological control potential of the gall inducer, thereby balancing the ecological relationships (Dhileepan et al. 2005). It appears that the impact of the gall inducer on *Parthenium hysterophorus* performance is limited and dependent on weather conditions each year (Dhileepan 2003).

The alterations in plant growth are related with accumulation of plant hormones in galls, although few studies in Lepidoptera galls are known on this matter (Tooker and Helms 2014). Indol-acetic acid (IAA) accumulation was associated with phenolic compounds in galls, as they act as inhibitors of IAA-oxidases (Hartley 1999). In galls induced by *Gnorimoschema gallaesolidaginis* Riley, 1869 (Gelechiidae) on *Solidago altissima* L. (Asteraceae) stems, auxin levels are higher in galls than non-galled stems. The galling larvae of *G. gallaesolidaginis* also accumulate higher levels of auxins when compared to non-galling caterpillars (Tooker and De Moraes 2011). Even with removal of stem apices and leaves, the galls continued to grow, revealing that galls are the source of IAA and this hormone is related to gall growth (Tooker and Helms 2014). Furthermore, research on the leaf miner *Phyllonorycter blancardella* Fabricius, 1781 (Gracillariidae) suggests that endosymbiotic bacteria can provide cytokinins, creating green islands within the mines. This phenomenon, observed in insect galls, is associated with an enhanced source of cytokinins, contributing to the prolonged maintenance of tissue vitality (Tooker and Helms 2014). Similar results are described in fusiform galls of *Parapodia sinaica* Frauenfeld, 1859, which had higher levels of auxins and cytokinins in both dry and wet seasons when compared to the non-galled stems of *Tamarix aphyllla* (L.) H. Karst (Fadl et al. 2023). Gibberellins also increased in the wet season in *Parapodia sinaica* galls, but in the dry season, such hormones decreased, which may be related to slower growth of gall tissues in this phase (Fadl et al. 2023), probably the mature phase of the gall. In the study by Tooker and De Moraes (2011), it was observed that the galls induced by *G. gallaesolidaginis* maintains the abscisic acid (ABA) content, whereas the non-galling caterpillar led to an accumulation of this hormone. This suggests that the formation of galls also relies on the non-accumulation of stress hormones in plant tissues, a factor that could hinder gall formation and is to some extent regulated by the gall inducer.

While phenolic compounds have demonstrated an impact on auxin accumulation in certain gall systems (Tooker and Helms 2014), a meta-analysis primarily suggests that there is no significant difference in secondary metabolite contents between non-galled organs and lepidopteran-galls, when compared to galls induced by other taxa of insects. However, the authors recommend further research to establish a clearer trend in this matter (Hall et al. 2017). Nevertheless, the accumulation of phenolic contents in the outer compartment of galls induced by *Palaeomystella olygophaga* on *Macaeira radula* (Bonpl.) DC. (Melastomataceae) was found to be different, with higher levels compared with nutritive tissues (Naves et al. 2021). This phenol accumulation is believed to play a role in dissipating oxidative stress in the outer gall tissues (Naves et al. 2021).

The effects of galling lepidopterans on the physiological parameters of their host plants, as observed in plant hormone balance, have been insufficiently investigated, hindering the establishment of clear trends. Consequently, additional research is warranted to elucidate the molecular mechanisms underlying the impacts of Lepidoptera on the photosynthetic performance of their host plants.

8 Lepidoptera Galls as a Multitrophic Ecosystem

Although poorly studied, parasitoids and cecidophages may inhabit gall tissues and provoke changes in cellular metabolism, mainly on nutritive cells. In the *Palaeomystella olygophaga—Macaeira radula* system, the dense-protoplasm nutritive cells are impacted by the presence of these natural enemies. The nutritive cells present packaging and degradation of organelles, because of the paralysis of the galling larva by endoparasitoids (Rezende et al. 2019). In this system, the interaction with abiotic events, as fire, revealed fluctuations in the proportion of parasitoids and cecidophages. The fire events, commonly observed in the Cerrado (Brazilian savannahs), result in a decrease in the abundance of natural enemies during the subsequent gall induction cycle, thereby favoring the proliferation of gall populations (Rezende et al. 2023b). The influence of abiotic events on parasitoidism was also reported by Dhileepan et al. (2005) on *E. strenuana* galls, as abovementioned.

Parasitoids impacting the anatomical profile of lepidopteran galls were documented in the fold galls induced on leaves of *Bauhinia ungulata* (Bedetti et al. 2013). In parasitized galls, the nutritive cells undergo hypertrophy, generating clusters of elongated nutritive cells within the larval chamber. These cells are abundant in lipids and terpenoids. However, further research is needed to confirm the role of these cells in the nutrition of the parasitoids.

Besides parasitoids and cecidophages, lepidopteran galls are also shelter to diverse successor organisms, such as spiders and ants. *Palaeomystella olygophaga* galls are induced on stem buds and exhibit a histological connection to non-galled parts through a complex vascular system. This intricate connection allows the gall to persist on the host plant for extended periods. Consequently, these galls serve as habitats for successor organisms, turning them into ecosystem engineers (Rezende et al. 2023a). Similarly, the protective shelter provided by senescent Lepidoptera galls was observed on the fern *Microgramma squamulosa* (Kaulf.) de la Sota (Polypodiaceae), attracting six recorded species of ants (Santos et al. 2019).

The inquiline *Rhoophilus loewi* Mayr, 1881 (Hymenoptera, Cynipidae) was documented within the galls induced by *Scyrotis* (Cecidosidae) on *Rhus* spp. (Anacardiaceae) (Van Noort et al. 2007). Inquilines, as discussed by Luz and Mendonça-Júnior (2019), can alter gall tissues. In instances where *R. loewi* attacks the galls, the multiple chambers produced by these inquilines induce tissue expansion, ultimately resulting in the demise of the gall inducer *Scyrotis*, as reported by Van Noort et al. (2007). Consequently, inquilines may also be referred to as endogallers, given their role in modifying gall tissues, which they invariably target. In other gall systems, inquiline wasps of the genus *Allorhogas* (Braconidae) attack galls induced by *Cecidonius pampeanus* on stems of *S. weinmannifolia* Engl. (Moreira et al. 2017).

While much remains unknown about the effects of other trophic levels in most lepidopteran galls, diverse parasitoid species were observed in the galls induced by *Cecidoses eremita* and *E. minutanus* on *Schinus* spp. (Loetti et al. 2016) and by *Cecidonius pampeanus* on *S. weinmannifolius* (Moreira et al. 2017). A kleptoparasite, *Locharcha opportuna* Moreira and Becker 2015 (Gelechiidae) was reported in galls induced by *Palaeomystella fernandesi* Moreira and Becker 2014 on *T. sellowiana* (Cham.) Cogn. (Melastomataceae) (Luz et al. 2015). The occurrence of inquilines, parasitoids, and cecidophages within lepidopteran galls appears to be widespread, although few reports of such interactions in literature. Further research into the dynamics among these organisms, the galling Lepidoptera, and their host plants is warranted. Additionally, exploring the impacts of these guilds on gall structure and function remains an open area of inquiry, promising the discovery of numerous novel insights.

9 Concluding Remarks

Lepidopteran galls warrant further investigation from both structural and functional perspectives. In the last years, several novelties in the structural complexity of these galls, with formation of typical nutritive tissues, neoformed meristems, new cell types, and dehiscence and fall mechanisms make them intriguing subjects for further histological, physiological, and cytological studies. Although physiological studies demonstrated the impacts of these gall inducers in photosynthesis and hormonal balance, they remain limited, especially when we consider the number of novel lepidopteran species described and other gall systems with undescribed inducers. The recent recognition of the role of these galls as ecosystem maintainers underscores the importance of understanding the effects of various guilds on their structure, which may unveil additional layers of influence on host development and metabolism. Parasitoids, inquilines, and successors seem to influence lepidopteran gall life cycle synchronism and morphogenesis, as well as the abiotic variables. Given the volume of new perspectives produced in recent years, novel investigations into the influence of galling lepidopterans interacting with biotic and abiotic variables hold promise for advancing structural, functional, and ecological studies.

References

Barnewall EC, De Clerck-Floate RA (2012) A preliminary histological investigation of gall induction in an unconventional galling system. Arthropod–Plant Interact 6:449–459. https://doi.org/10.1007/s11829-012-9193-4

Basilio DS, Casagrande MM, Bordignon SAL, Moreira GRP (2015) Description and life history of a new cecidogenous species of *Palaeomystella* Fletcher (Lepidoptera, Momphidae) from Brazil. Rev Bras Entomol 59:188–196. https://doi.org/10.1016/j.rbe.2015.06.005

Becker VO, Adamski D (2008) Three new cecidogenous *Palaeomystella* Fletcher (Lepidoptera, Coleophoridae, Momphinae) associated with Melastomataceae in Brazil. Rev Bras Entomol 52:647–657

Bedetti CS, Ferreira BG, Castro NM, Isaias RMS (2013) The influence of parasitoidism on the anatomical and histochemical profiles of the host leaves in a galling Lepidoptera—*Bauhinia ungulata* system. Braz J Biosci 11:242–249

Bronner R (1992) The role of nutritive cells in the nutrition of Cynipids and Cecidomyiids. In: Shorthouse JD, Rohfritsch O (eds) Biology of insect-induced galls. Oxford University Press, Oxford, pp 118–137

Brown JW, Baixeras J, Solórzano-Filho JA, Kraus JE (2003) Description and life history of an unusual fern-feeding tortricid moth (Lepidoptera: Tortricidae) from Brazil. Ann Entomol Soc Am 97:865–871

Cornell HV (1990) Survivorship, life history, and concealment: a comparison of leaf miners and gall formers. Am Nat 136:581–597

Desnitskiy AG, Chetverikov PE, Ivanova LA, Kuzmin IV, Ozman-Sullivan SK, Sukhareva SI (2023) Molecular aspects of gall formation induced by mites and insects. Life 13:1347

Dhileepan K (2003) Seasonal variation in the effectiveness of the leaf-feeding beetle *Zygogramma bicolorata* (Coleoptera: Chrysomelidae) and stem-galling moth *Epiblema strenuana* (Lepidoptera: Tortricidae) as biocontrol agents on the weed *Parthenium hysterophorus* (Asteraceae). Bull Entomol Res 93:393–401. https://doi.org/10.1079/BER2003255

Dhileepan K, Lockett CJ, McFadyen RE (2005) Larval parasitism by native insects on the introduced stem-galling moth *Epiblema strenuana* Walker (Lepidoptera: Tortricidae) and its implications for biological control of *Parthenium hysterophorus* (Asteraceae). Aust J Entomol 44:83–88. https://doi.org/10.1111/j.1440-6055.2005.00446.x

Dreger-Jauffret F, Shorthouse JD (1992) Diversity of gall-inducing insects and their galls. In: Shorthouse JD, Rohfritsch O (eds) Biology of insect-induced galls. Oxford University Press, Oxford, pp 60–86

Fadl A, Kamel M, Morsy A, Gamal E (2023) Assessment of phytohormone of *Tamarix aphylla* L. infected with *Parapodia sinaica* (Gelechiidea: Lepidoptera) in south of Sinai. J Environm Sci 52:1–11. https://doi.org/10.21608/jes.2023.204691.1508

Fernandes GW, Carneiro MAA, Isaias RMS (2012) Gall-inducing insects: from anatomy to biodiversity. In: Panizzi AR, Parra JRP (eds) Insect bioecology and nutrition for integrated Pest management. CRC Press, Boca Raton

Ferreira BG, Isaias RMS (2013) Developmental stem anatomy and tissue redifferentiation induced by a galling Lepidoptera on *Marcetia taxifolia* (Melastomataceae). Botany 91:752–760. https://doi.org/10.1139/cjb-2013-0125

Ferreira BG, Teixeira CT, Isaias RMS (2014) Efficiency of the polyethylene-glycol (PEG) embedding medium for plant histochemistry. J Histochem Cytochem 62:577–583. https://doi.org/10.1369/0022155414538265

Ferreira BG, Carneiro RGS, Isaias RMS (2015) Multivesicular bodies differentiate exclusively in nutritive fast-dividing cells in *Marcetia taxifolia* galls. Protoplasma 252:1275–1283. https://doi.org/10.1007/s00709-015-0759-8

Ferreira BG, Álvarez R, Avritzer SC, Isaias RMS (2017) Revisiting the histological patterns of storage tissues: beyond the limits of gall-inducing taxa. Botany 95:173–184

Ferreira BG, Álvarez R, Bragança GP, Alvarenga DR, Pérez-Hidalgo N, Isaias RMS (2019) Feeding and other gall facets: patterns and determinants in gall structure. Bot Rev 85:78–106. https://doi.org/10.1007/s12229-019-09207-w

Ferreira BG, Moreira GRP, Isaias RMS, Carneiro RGS (2022) Complex meristematic activity induced by *Eucecidoses minutanus* on *Schinus engleri* turns shoots into galls. Am J Bot. https://doi.org/10.1002/ajb2.1798

Florentine SK, Raman A, Dhileepan K (2005) Effects of gall induction by *Epiblema strenuana* on gas exchange, nutrients, and energetics in *Parthenium hysterophorus*. BioControl 50:787–801

Guedes LM, Costa EC, Isaias RMS, Sáez-Carillo K, Aguilera N (2023) Structural patterns of Lepidoptera galls and the case of *Andescecidium parrai* (Cecidosidae) galls on *Schinus polygama* (Anacardiaceae). J Plant Res 136:715–728. https://doi.org/10.1007/s10265-023-01472-6

Guiguet A, Ohshima I, Takeda S, Laurans F, Lopez-Vaamonde C, Giron D (2019) Origin of gall-inducing from leafmining in *Caloptilia* micromoths (Lepidoptera, Gracillariidae). Sci Rep 9:6794. https://doi.org/10.1038/s41598-019-43213-7

Hall CR, Carroll AR, Kitching RL (2017) A meta-analysis of the effects of galling insects on host plant secondary metabolites. Arthropod Plant Interact 11:463–473. https://doi.org/10.1007/s11829-016-9486-0

Hartley SE (1999) Are gall insects large Rhizobia? Oikos 84:333–342

Isaias RMS, Jorge NC, Ferreira BG, Fochezato J, Moreira GRP (2021) How cells and tissues of *Daphnopsis fasciculata* (Thymelaeaceae) react to the leaf-mining habit of *Phyllocnistis hemera* (Lepidoptera: Gracillariidae). J Plant Res 134(3):535–541. https://doi.org/10.1007/s10265-021-01268-6

Jayaraman P (1989) Histogenetic patterns and classification of the shoot-axis galls. Proc Indiana Acad Sci 99:391–403. https://doi.org/10.1007/BF03053583

Loetti V, Valverde A, Rubel DN (2016) Galls of *Cecidoses eremita* Curtis and *Eucecidoses minutanus* Brèthes (Lepidoptera: Cecidosidae) in Magdalena, Buenos Aires Province: preliminary study and associated fauna. Biota Neotrop 16:e20160161. https://doi.org/10.1590/1676-0611-BN-2016-0161

Luz FA, Mendonça-Júnior MS (2019) Guilds in insect galls: who is who. Florida Entomol 102:207–210

Luz FA, Gonçalves GL, Moreira GRP, Becker VO (2014) Three new cecidogenous species of *Palaeomystella* Fletcher (Lepidoptera, Momphidae) from the Brazilian Atlantic Rain Forest. ZooKeys 433:97–127. https://doi.org/10.3897/zookeys.433.7379

Luz FA, Gonçalves GL, Moreira GRP, Becker VO (2015) Description, molecular phylogeny, and natural history of a new kleptoparasitic species of gelechiid moth (Lepidoptera) associated with Melastomataceae galls in Brazil. J Nat Hist 49:1–27

Mani MS (1964) Ecology of plant galls. Dr. W. Junk Publishers, The Hague

Martins GS, Santos MG, Reis A, Ferreira BG (2023) Alterations induced by *Tortrimosaica polypodivora* on the stems of *Microgramma vacciniifolia*: simple or complex galls? Rodriguésia 74:e00532023

Mathews DL, Maharajh BV (2009) *Adaina primulacea* Meyrick, 1929: a gall-inducing plume moth of siam weed from South Florida and the Neotropics (Lepidoptera: Pterophoridae). Trop Lepid Res 19:64–70

Menken SBJ, Boomsma JJ, van Nieukerken EJ (2009) Large-scale evolutionary patterns of host plant associations in the Lepidoptera. Evolution 64:1098–1119. https://doi.org/10.1111/j.1558-5646.2009.00889.x

Meyer J, Maresquelle HJ (1983) Anatomie des Galles. Gebrüder Bornträger, Berlin

Miller WE (2005) Gall-inducing Lepidoptera. In: Raman A, Schaefer CA, Withers T (eds) Biology, ecology, and evolution of gall-inducing arthropods. Science Publishers, Enfield, pp 431–465

Moreira GRP, Gonçalves GL, Eltz RP, San Blas G, Davis DR (2012) Revalidation of *Oliera* Brèthes (Lepidoptera: Cecidosidae) based on a redescription of *O. argentinana* and DNA analysis of Neotropical cecidosids. Zootaxa 3557:1–19

Moreira GRP, Eltz RP, Pase RB, Silva GT, Bordignon SAL, Mey W, Gonçalves GL (2017) *Cecidonius pampeanus*, gen. et sp. n.: an overlooked and rare, new gall-inducing micromoth associated with *Schinus* in southern Brazil (Lepidoptera, Cecidosidae). ZooKeys 695:37–74. https://doi.org/10.3897/zookeys.695.13320

Naves GDFS, Kuster VC, Machado M, Santos PD, Martini VC, Oliveira DC (2021) Metabolite investments and stress levels among tissue compartments of *Palaeomystella olygophaga* (Lepidoptera) galls on *Macairea radula* (Melastomataceae). Aust J Bot 69:131–142

Navie SC, Priest TE, McFadyen RE, Adkins SW (1998) Efficacy of the stem-galling moth *Epiblema strenuana* walk. (Lepidoptera: Tortricidae) as a biological control agent for ragweed *Parthenium* (*Parthenium hysterophorus* L.). Biol Control 13:1–8. https://doi.org/10.1006/bcon.1998.0638

Oliveira DC, Isaias RMS, Fernandes GW, Ferreira BG, Carneiro RGS, Fuzaro L (2016) Manipulation of host plant cells and tissues by gall-inducing insects and adaptive strategies used by different feeding guilds. J Insect Physiol 84:103–113

Powell JA (1980) Evolution of larval food preferences in microlepidoptera. Ann Rev Entomol 25(1):133–159. https://doi.org/10.1146/annurev.en.25.010180.001025

Price PW, Fernandes GW, Waring GL (1987) Adaptive nature of insect galls. Environ Entomol 16:15–24

Raman A, Dhileepan K (1999) Qualitative evaluation of damage by *Epiblema strenuana* (Lepidoptera: Tortricidae) to the weed *Parthenium hysterophorus* (Asteraceae). Ann Entomol Soc Am 92(5):717–723

Raman A, Madhavan S, Florentine SK, Dhileepan K (2006) Metabolite mobilization in the stem galls of *Parthenium hysterophorus* induced by *Epiblema strenuana* inferred from the signatures of isotopic carbon and nitrogen and concentrations of total non-structural carbohydrates. Entomol Exp Appl 119:101–107

Rezende UC, Cardoso JCF, Kuster VC, Goncalves LA, Oliveira DC (2019) How the activity of natural enemies changes the structure and metabolism of the nutritive tissue in galls? Evidence from the *Palaeomystella oligophaga* (Lepidoptera) – *Macairea radula* (Metastomataceae) system. Protoplasma 256:679–679

Rezende UC, Gonçalves PHP, Kuster VC, Carneiro RGS, Oliveira DC (2023a) Every end is a new beginning: histological features of galls induced on *Macairea radula* (Melastomataceae) allow a post-senescence colonization. Flora 306:152369. https://doi.org/10.1016/j.flora.2023.152369

Rezende UC, Cardoso JCF, Gonçalves PHP, Oliveira DC (2023b) The multitrophic system configuration on galls of *Macairea radula* (Melastomataceae) induced by *Palaeomystella oligophaga* (Lepidoptera) depends on abiotic events. Int J Trop Insect Sci 43:2095–2104. https://doi.org/10.1007/s42690-023-01104-5

Rohfritsch O (1992) Patterns in gall development. In: Shorthouse JD, Rohfritsch O (eds) Biology of insect-induced galls. Oxford University Press, Oxford, pp 60–86

San Blas G, Davis DR (2013) Redescription of *Dicranoses capsulifex* Kieffer and Jörgensen (Lepidoptera: Cecidosidae) with description of the immature stages and biology. Zootaxa 3682:371–384. https://doi.org/10.11646/zootaxa.3682.2.9

Santos MG, Porto GF, Lancellotti IR, Feitosa RM (2019) Ant fauna associated with *Microgramma squamulosa* (Kaulf.) de la Sota (Polypodiaceae) fern galls. Rev Bras Entomol 63:101–103. https://doi.org/10.1016/j.rbe.2019.02.004

Santos PD, Bragança GPP, Kuster VC, Rocha DI, Oliveira DC (2024) Cell wall epitope distribution in the functional compartments of galls induced by *Palaeomystella oligophaga* (Lepidoptera) in *Macairea radula* (Melastomataceae). Aust J Bot 72:BT23064. https://doi.org/10.1071/BT23064

Savard J, Tautz D, Richards S, Weinstock GM, Gibbs RA, Werren JH, Tettelin H, Lercher MJ (2006) Phylogenomic analysis reveals bees and wasps (Hymenoptera) at the base of the radiation of Holometabolous insects. Genome Res 16:1334–1338

Silva GT, San Blas G, Peçanha W, Moreira GRP, Gonçalves GL (2018a) Phylogeography of the gall-inducing micromoth *Eucecidoses minutanus* Brèthes (Cecidosidae) reveals lineage diversification associated with the Neotropical Peripampasic Orogenic Arc. PLoS One 13:e0201251. https://doi.org/10.1371/journal.pone.0201251

Silva GT, Moreira GRP, Vargas HA, Gonçalves GL, Mainardi MD, San Blas G, Davis D (2018b) Overlooked gall-inducing moths revisited, with the description of *Andescecidium parrai* gen. et sp. n. and *Oliera saizi* sp. n. from Chile (Lepidoptera, Cecidosidae). ZooKeys 795:127–157. https://doi.org/10.3897/zookeys.795.27070

Tooker JF, De Moraes CM (2011) Feeding by a gall-inducing caterpillar species increases levels of indole-3-acetic and decreases abscisic acid in *Solidago altissima* stems. Arthropod-Plant Interact 5:115–124

Tooker JF, Helms AM (2014) Phytohormone dynamics associated with gall insects, and their potential role in the evolution of the gall-inducing habit. J Chem Ecol 40:742–753. https://doi.org/10.1007/s10886-014-0457-6

Van Noort S, Stone GN, Whitehead VB, Nieves-Aldrey JL (2007) Biology of *Rhoophilus loewi* (Hymenoptera: Cynipoidea: Cynipidae), with implications for the evolution of inquilinism in gall wasps. Biol J Linn Soc 90:153–172. https://doi.org/10.1111/j.1095-8312.2007.00719.x

Vecchi C, Menezes NL, Oliveira DC, Ferreira BG, Isaias RMS (2013) The redifferentiation of nutritive cells in galls induced by Lepidoptera on *Tibouchina pulchra* (Cham.) Cogn. reveals predefined patterns of plant development. Protoplasma 250:1363–1368. https://doi.org/10.1007/s00709-013-0519-6

Vieira AC, Kraus E (2007) Biologia e estrutura da galha do pedicelo de *Byrsonima sericea* DC. (Malpighiaceae). Rev Bras Biociências 5:402–404

Wille J (1926) *Cecidoses eremita* curt. und ihre galle an *Schinus dependens* Ortega. Zeitschrift für Morphologie und Ökologie der Tiere 7(1-2):1–101. https://doi.org/10.1007/BF00540718

Zhang H, Dugé de Bernonville T, Body M, Glevarec G, Reichelt M, Unsicker S, Bruneau M, Renou J-P, Huguet E, Dubreuil G (2016) Leaf-mining by *Phyllonorycter blancardella* reprograms the host-leaf transcriptome to modulate phytohormones associated with nutrient mobilization and plant defense. J Insect Physiol 84:114–127. https://doi.org/10.1016/j.jinsphys.2015.06.003

Structure and Function of Oak Gall Wasps

Juli Pujade-Villar, George Melika, and Víctor Cuesta-Porta

1 Introduction

The Cynipidae include a vast range of gall inducers and gall inquilines in Fagaceae, Rosaceae and herbaceous plants. Oak gall wasps (Hymenoptera: Cynipidae: Cynipini) are specialized gall-inducers on some genera of Fagaceae, mainly on *Quercus* L. (Pujade-Villar et al. 2001; Stone et al. 2002, 2009). As obligate parasites, oak cynipid wasps follow the distribution of their host plant. Fagaceae and, hence, Cynipini are mainly found in the temperate zone of the Northern Hemisphere (Holarctic) extending into tropical areas in the Americas (Neotropics) and Southeast Asia (Oriental Region). Oak gall wasps comprise more than 1000 species worldwide and are by far the most speciose tribe within gall wasps (Melika and Abrahamson 2002; Csóka et al. 2005; Ronquist et al. 2015; Pénzes et al. 2018; Buffington et al. 2020; Fang et al. 2020).

The biology of oak cynipids is remarkably complex and exhibits several attributes. First, the oak gall wasps are strongly host-specific. Each species induces galls in a reduced group of phylogenetically related species of plants (Csóka et al. 2005; Stone et al. 2009). The cynipid wasps that induce galls on *Quercus* are generally associated with a particular section within it (Sections *Cerris, Cyclobalanopsis, Ilex, Lobatae, Quercus, Ponticae, Protobalanus,* or *Virentes*). Some species can alternate their hosts between sections in parts of their life cycles (Stone et al. 2002; Abe et al. 2007). Furthermore, oak gall wasps also exhibit great specificity in the organ or tissue they attack. Galls can be found on any plant organ (acorns, branches, buds, flowers, leaves, roots, twigs) and some galls are specific for a certain region of

J. Pujade-Villar (✉) · V. Cuesta-Porta
Facultat de Biologia, Universitat de Barcelona, Barcelona, Catalonia
e-mail: jpujade@ub.edu

G. Melika
Plant Health Diagnostic National Reference Laboratory, Budapest, Hungary

D. C. de Oliveira, R. M. dos Santos Isaias (eds.), *Plant Galls*,
https://doi.org/10.1007/978-3-031-80064-1_15

the organ (e.g.: upper or under sides of leaves, male or female flowers, acorn's cupule or seed).

Second, most oak gall wasps undergo complex cyclically parthenogenetic life cycles (heterogony), an obligate succession of one sexual and one asexual generation (Pujade-Villar et al. 2001; Stone et al. 2002; Csóka et al. 2005). The sexual generation emerges out of the gall during the summer and the beginning of fall, and the asexual generation mainly overwinters inside the gall and emerges in spring. For each generation, most species exhibit notable differences in the morphology of both adults and galls (Pujade-Villar et al. 2001; Stone et al. 2002; Csóka et al. 2005). Each generation can induce galls in different organs of the host plant, and the host-alternating species switch sections of *Quercus* between generations. The evolutionary origin of this life cycle likely derives from an ancestral bivoltine cycle of bisexual adults, in which one of the cycles switched into parthenogenesis (Pujade-Villar et al. 2001; Stone et al. 2008). Heterogony has been a taxonomical burden for the Cynipini. The morphological disparity between asexual and sexual adults blurs out the connection between generations and linking both parts of the life cycle is still an unresolved issue for most species. Only through experimental or molecular evidence, those relationships can be unraveled (Folliot 1964a, b; Melika et al. 2021; Nicholls et al. 2022; Cuesta-Porta et al. 2023b).

Finally, Cynipini galls are the most complex among phytophagous-induced galls. The shape and structure of galls are determined by the wasp species and are so specific that galls are used as an extended phenotype of the wasp (Stone and Cook 1998; Stone et al. 2002; Csóka et al. 2005; Bailey et al. 2009). Certain gall morphologies can be used to identify a genus, a group of species, or even a certain species. The distinctiveness of galls varies among regions and is highly influenced by the oak diversity and the current knowledge of the cynipid in each region. The galls of the Nearctic and the Neotropics exhibit the greatest morphological diversity among regions, which is analogue to the great oak diversity in the Americas (Stone et al. 2002; Csóka et al. 2005). Some galls are characteristic of a species or a genus but many gall morphologies have independently appeared several times across gall wasps' phylogeny (Stone et al. 2002, Csóka et al. 2005). The Palearctic harbors lower gall diversity, but there are distinct differences between the west and the east. The Western Palearctic is the area with the most unequivocal gall-adult relationships, likely caused by a lower oak diversity and an extensively studied cynipid fauna. The Eastern Palearctic and Oriental Region exhibits a bigger oak diversity, and some gall morphologies are similar among genera which hinders the identification of gall wasp species (Stone et al. 2002, Csóka et al. 2005).

Galls are the basis of complex communities. There is a vast array of gall-associated fauna that organizes in several trophic levels inside galls and can be majorly divided into predators, parasitoids, inquilines and opportunists (Stone et al. 2002; Csóka et al. 2005; Nicholls et al. 2017; Ward et al. 2022). Among predators, galls attract both herbivores and insectivores. Galls can be nutritious for numerous herbivores ranging from other insects (mostly moths and beetles) to vertebrates. Herbivores can hinder the survival of the cynipid by collateral ingestion of the larvae, compromising the structural integrity of the gall or opening ways for the attack of other enemies. The Cynipini larvae inside galls are the target of a great number of specialized enemies, which produce a high mortality rate among cynipids. The most common of them are parasitoid wasps whose eggs are injected through the gall into the larval chamber and their larvae feed on the cynipid's body (Stone et al. 2002, Csóka et al. 2005). Despite the protection galls offer, some enemies are adapted to trespass those defenses. The parasitoid wasps specialized in galls generally attack different species of cynipids (Stone et al. 2002). Another part of the communities around galls are inquilines, species that take advantage of a developing gall to induce new larval chambers so their offspring can develop (Ronquist 1994; Stone et al. 2002; Csóka et al. 2005; Lobato-Vila et al. 2022). Most inquilines induce larval chambers in the outer surface of the galls and do not harm the gall inducer. Some species are lethal inquilines (Csóka et al. 2005). They develop nearby the inducer's larval chamber and the spatial competition can cause the collateral death of the gall inducer. Moving up in the trophic relationships, other parasitoid communities are specialized in attacking inquilines or other parasitoids inside galls.

The evolutionary drivers that have led to the diversification of galls are summarized in three main hypotheses (a) Protection from natural enemies; (b) Nutrition availability; (c) Microenvironment (Price et al. 1987; Stone et al. 2002; Stone and Schönrogge 2003). Protecting from predators and parasitoids is likely to be the strongest evolutionary driver for cynipids. Despite the existing defensive strategies, the gall wasp's survival rate can reach 20% (Bailey et al. 2009; Lobato-Vila et al. 2019). The great diversification of gall morphologies and their correlation with some parasitoids' morphological attributes (e.g.: ovipositor lengths) can be the result of the coevolution between gall wasps and their parasitoid community. This hypothesis is the most tested and some studies have found evidence for supporting this hypothesis, as similar parasitoid communities are associated with cynipids with a certain gall morphology (Price et al. 1987; Bailey et al. 2009; László and Tóthmérész 2013; Ide and Koyama 2023).

Gall induction produces a re-distribution of the host plant's secondary metabolites and resources into the gall tissues. The accumulation of nutrients in the gall increases the food availability and, thus, improves the fitness of the gall wasp larva (Price et al. 1987; Diamond et al. 2008). The galls offer protection to the cynipid larva against external environmental conditions. The great diversity of gall wasps in desertic areas may support the adaptation to more extreme environments by increasing the thickness of the gall walls (Stone et al. 2002; Miller et al. 2009). All these hypotheses are not exclusive to each other, but further research will evaluate to which extent each hypothesis is affecting every species or gall morphology.

2 Cycles and Models of Galls

It is difficult, if not impossible, to explain the exact reasons causing the heterogony in gall wasps. According to Kinsey (1920, 1922), the alternation of generations in Cynipini is only due to extreme stational dimorphism. He based this statement on the similarity of asexual and sexual generations in some cynipid species, for example, in the genus *Neuroterus* Hartig, 1840 (Kinsey 1920), and on the supposed absence of sexual forms in the group of *Andricus quercuscalifornicus* (Bassett 1881) in year-round warm regions. Thus, Kinsey states that in Southern California alternate forms are absent because of the short time given between the emergence of adults and the appearance of young galls, while in more northern forms this time interval is longer, so, he supposed a "normal" alternation of generations (Kinsey 1922). This hypothesis has not been proved yet, but some species, like *A. quadrilineatus* (Folliot 1964a, b), can bypass the alternation of generations, which complicates this issue.

Contrary to this hypothesis, Patterson (1928) found non-functional males in the asexual generation of two North American species, *Neuroterus quercusrileyi* (Bassett 1880) and *N. contortus* (Weld 1921), which he interpreted as the vestige of a primitive sexual condition. Assuming this is true, we could imagine a scenario of an ancestral bivoltine species with two sexual generations that eventually produced an asexual generation in one of the cycles after changing the host plant or target organ (Cotté 1926). Nevertheless, this hypothesis has been normally overlooked since the parthenogenetic generation is normally considered a basal condition within Cynipoidea. Even fossil forms are also considered parthenogenetic, for example, Gerocynipidae, known from the Upper Cretaceous (Kovalev 1995).

Another factor affecting life cycles in some Cynipoidea is the induced parthenogenesis by the endosymbiotic bacteria from the genus *Wolbachia*. These bacteria affect the insect's chromosome behavior (Stouthamer et al. 1999) and cause thelythokous (diploid females from non-fertilised eggs and absence or very few rare males) and arrhenotokous populations (males from nonfertilized eggs and females from fertilized ones) in Aylacini and Diplolepididae (Stille 1984; Plantard et al. 1998; Plantard and Solignac 1998; Field et al. 1999). The combination of the parthenogenesis-inducing microbes with a possibly bivoltine ancestor may be related to the origin of heterogony in cynipids. The *Wolbachia* is present in some species of Cynipini (Plantard, pers. com.) but the exact role of *Wolbachia* in the Cynipini remains unknown and may be unrelated to the parthenogenesis cycle (Stone et al. 2002). The *Wolbachia* infections in *Biorhiza pallida* do not interrupt cyclical partenogenesis (Rokas et al. 2001).

An interesting question is what benefit do cynipids get from such a complicated life cycle. One possibility is to avoid parasitoids by evolving an heteroecious life cycle. However, parasitoids and inquilines have diversified to adapt to gall models instead of specific gall wasp species. Thus, most gall-inducing species have different parasitoid and inquiline complexes for each generation. This hypothesis is hard to test and some authors have named it as the "ghost of past-parasitism" (Price and Pschorn-Walcher 1988; Berdegue et al. 1996; Stone and Cook 1998).

The alternation of generations in oak gall wasps mainly follows the model consisting of two sets of asexual females, androphores and gynephores. One set lays haploid meiotic eggs giving rise to haploid males (androphores), while the other lays diploid mitotic eggs that will become diploid females (gynephores). Sexual females lay diploid fertilized eggs that will only produce one set of asexual females, either androphores or gynephores. The unfertilized eggs of sexual females cannot parthenogenetically develop into haploid males (Folliot 1964a, b; Pujade-Villar et al. 2001; Stone et al. 2002). Some Cynipini can vary this life cycle model. The most studied case is *Biorhiza pallida* in which some asexual females are gynandrophores, they can produce both males and females from the same set of unfertilized eggs. A proportion of sexual females also can give rise to both androphores and gynephores, and other sexual females can reproduce without mating. In the last case, unfertilized eggs produce diploid asexual females and not haploid males (Pujade-Villar et al. 2001; Stone et al. 2002; Atkinson et al. 2003; Pujade-Villar 2013).

The only way of studying the life cycles is through experimental rearing and tracking the offspring of individualized females (Folliot 1964a; Askew 1984) or by molecular studies (Melika et al. 2021; Nicholls et al. 2022; Cuesta-Porta et al. 2023b). The mechanisms underlying the alternation of generations and the models of life cycles within oak gall wasps are still unknown. Most Cynipini species are only known for one generation, this raises the question if any of those species have suffered a secondary loss of sexuality (Hebert 1987; Pujade-Villar et al. 2001). There are few proven cases of sustained parthenogenetic asexuality in oak gall wasps, but only one with a complete loss of a sexual generation, *Andricus targionii* (Abe 1986; Pujade-Villar et al. 2001; Stone et al. 2002).

Most species of oak gall wasps develop their sexual galls (galls containing the sexual adults) between spring and early summer. The asexual galls develop during the late summer and fall. Asexual females can either overwinter inside the asexual gall and emerge before spring, or emerge during fall and lay eggs that will be in lethargy until the next spring. The time of emergence exhibits a great plasticity to environmental fluctuations. Thus, many species have both overwintering asexual females and lethargic sexual galls (Pujade-Villar et al. 2001; Stone et al. 2002).

The phenology of the alternate generations may create two different scenarios, where each generation is under differential selective pressures leading to the morphological divergence of those generations. The overall asexual galls tend to grow bigger and be more conspicuous, while sexual galls are usually smaller and more cryptic in the organ or tissue they are growing in. A similar disparity is found in adult cynipid wasps where the asexual females are generally bigger and stouter, usually accompanied by a thicker cuticle, rougher sculpture and denser pilosity. The sexual generation of most cynipids tends to be smaller and more delicate, with smoother and glabrous parts. Regardless of that trend, there are numerous lineages that diverge from that generalization. Some Nearctic *Neuroterus* species induce almost inconspicuous galls during both their cycles and the adult morphology is highly delicate (Figs. 1a and 2f) (Pujade-Villar et al. 2018), while the *Amphibolips* species tend to exhibit big-sized galls and adults on both generations (Figs. 1b and 3g) (Cuesta-Porta et al. 2020).

Fig. 1 Asexual galls. (**a**) *Neuroterus saltarius* (Source: Gallformers.org; Credit to Adam Kranz), (**b**) *Amphibolips hidalgoensis* (Credit to Edith Estrada Venegas), (**c**) *Andricus quercuslaurina* (Credit to David Cibrián-Tovar), (**d**) *Amphibolips cooki,* different gall structures marked with an arrow. (Source: iNaturalist; Credit to Bill Sheehan)

2.1 Structure of Cynipid Gall Wasps

Mature galls are structured in three main layers: larval chamber, parenchyma, and epidermis (Fig. 1c). The larval chamber contains the inducer's larva and it is delimited by a sclerenchyma cell layers whose internal walls are limited by the nutritive tissue. Galls can either contain a single chamber (monolocular) or many chambers in a single gall (multilocular). The monolocular galls are generally found isolated from other galls, yet some species can induce several galls in small areas of tissue creating aggregations of monolocular galls. Clusters of galls differ from multilocular galls by traceability of the limits between cells, aggregations are easily divided by the individual galls with their own gall structure, while the multilocular galls share the same parenchymatic and epidermal tissues (Stone et al. 2002). The interactions between tightly clustered galls can alter their morphology (Fig. 2a), up to the extent of creating secondary structures mimicking multilocular galls (Figs. 1c and 2b, c). Multi-chambered galls usually reach bigger sizes than monolocular galls, which directly correlates to the number of eggs laid by the female (Shorthouse and Brooks 1998; Stone et al. 2002; Ferreira et al. 2019). Despite most galls are induced by a single female, molecular evidence reported eggs from different mothers in galls of several *Andricus* species, which may indicate a collaborative effort to create bigger and safer galls against enemies (Atkinson et al. 2003; Stone and Schönrogge 2003). Independently of the number of wasp larvae inhabiting the galls, the morphology of the larval chamber is the most consistent among all gall structures (Figs. 1d, 3a, b, and 3f, g). In terms of shape and function, all larval chambers follow the same purpose, protecting the cynipid larvae (Shorthouse and Rohfritsch 1992; Stone and Cook 1998; Stone et al. 2002). In contrast, the parenchyma and epidermis are the tissues with the most morphological diversification among species (Cornell 1983; Stone and Schönrogge 2003).

Parenchyma is the tissue surrounding and holding the larval chambers. This tissue may be highly diverse among galls with remarkable variations in thickness, density and hardness. The size of the gall is mainly determined by the amount of parenchyma. Large galls are more conspicuous to predators but can offer better isolation from the environment and better protection from parasitoids. In contrast, small galls are more vulnerable to attacks but can be more cryptic (Stone et al. 2002). Most galls have uniformly distributed solid parenchyma, which can range from soft and spongy to completely solid and strongly lignified.

The growth of the parenchyma along gall development can also change. During the early stages of some galls, there is an accumulation of water in the parenchyma cells creating succulent galls. Once they mature, these cells become drier and usually harder, yet few galls maintain the succulence until senescence. The maturation of galls can result in many different morphologies, one of which has internal cavities caused by controlled apoptosis of parenchyma cells (Fig. 3f, g). Those cavities are usually disposed as either small air chambers mimicking larval chambers, or as a big hollow space around the larval chamber. Within the latter, the larval chamber can be held in place by radiating filaments anchoring the larval chamber to the outer shell of the gall, or the parenchyma is almost entirely cleared out releasing the larval chamber to roam free inside the gall's hollow space (Fig. 6b; Stone and Cook 1998; Stone et al. 2002).

Finally, the epidermis is the outer layer of the gall, and is by far the tissue amassing most of gall's variability (Stone et al. 2002). The epidermis can range from a thin brittle layer of cells to a thick lignified shell. The outer surface of the epidermis can be smooth to verrucose or spiny. Different developmental stages of the gall are accompanied by different external assets, such as coloration changes, varying degrees of pubescence (from short, to felt-like, to woolly long pubescence), or the

Fig. 2 Asexual galls. (**a**) *Disholcaspis spongiosa.* (Source: iNaturalist; Credit to Michelle [pufferchung]), (**b**) *Striatoandricus furnessae* in Cuesta-Porta et al. (2022). (Credit: George Melika), (**c**) *Disholcaspis fungiformis.* (Source: iNaturalist; Credit to threeagoutdoors), (**d**) *Antron douglasii.* (Source: iNaturalist; Credit to Gart Harwood), (**e**) *Disholcaspis pruniformis.* (Source: iNaturalist; Credit to Michelle [pufferchung]), (**f**) *Neuroterus fragilis*. (Source: iNaturalist; Credit to Cricket Raspet)

Fig. 3 Asexual galls. (**a–f**): (**a**) *Druon quercusflocci* in Cuesta-Porta et al. (2022) with detail of the larval chambers underlying the pubescence, (**b**) *Amphibolips nigra.* (Credit: Juli Pujade-Villar), (**c**) *Feron tetyanae* in Cuesta-Porta et al. 2023a, b. (Credit to George Melika), (**d**) *Callirhytis lanata.* (Source: gallformers. Org; Credit to Adam Kranz), (**e**) *Striatoandricus* sp. (Source: iNaturalist; Credit to Mark Richman), (**f**) *Atrusca* sp. (Source: iNaturalist.org; Credit to crothfels). Sexual gall: (**g**) *Amphibolips quercusostensackenii.* (Source: Gallformers.org; Credit to Adam Kranz)

production of sweet or sticky secretions (Fig. 5a, b). The aspect of galls can be influenced by the organ where the gall was induced, galls in leaves or buds can induce the growth of leaflets or bracts around the gall forming a multilayered epidermis (Fig. 5i), or galls in twigs, roots or branches can become hard woody masses (Figs. 1c and 6e, g, h).

The external shape of galls depends on differential cell division and growth rates on determined regions of the affected plant tissues (Stone et al. 2002; Isaias et al. 2014; Magalhães et al. 2014; Oliveira et al. 2016; Ferreira et al. 2017). The complexity of gall shapes can be outstandingly high (Figs. 2c, d, 5b–d, and 5i, j), but most galls follow the simplest growing patterns. Those simple shapes are either round, such as, *Amphibolips hidalgoensis,* and *Disholcaspis terrestris* (Figs. 1b, 2e, 3f, g, and 6a–c); or amorphous gall bodies, like the tumoral *Andricus* species-group (Nearctic) or several Nearctic *Callirhytis* species (Figs. 1c, and 6e, g, h).

The attachment to the host-plant organ can be tight or fragile, which confers galls the characteristic of being detachable or integral to the plant tissues (Russo 2007). Detachable galls usually grow without significantly affecting the attacked organ and are generally constricted at the base and connected to the host plant by a thin and small area (Figs. 1d, 2d, e, and 3f, g). Some species use the detachability to protect the gall from winter temperatures hiding under the foliage. Integral galls usually deform bigger parts of the attacked organ and can only be extracted by severely damaging the host plant organ's tissues, generally taking advantage of the organ's tissues to defend against enemies such as integral stem galls or galls on acorns (Figs. 1a, c, 2f, and 6d).

All the different attributes that galls exhibit can be found in different combinations among the cynipids, which is the result of multiple events of convergent evolution (Stone et al. 2002). For instance, densely pubescent galls are induced by several independently evolving lineages, are found on different plant organs, can be monolocular or multilocular, and exhibit different dispositions of internal tissue (Fig. 3a–e) (Cuesta-Porta et al. 2022, 2023a, b). Nevertheless, some of the combinations do not occur in oak gall wasps. The diversity of brittle round galls with internal radiating filaments sustaining the larval chamber does not include multilocular galls (Figs. 1d, 3f–g, and 6a).

3 Cecidogenesis and Gall Tissue Structure

Histological studies on gall wasps are very scarce (Lacaze-Duthiers 1853; Beyerinck 1882; Fockeu 1889; Houard 1903; Pujade-Villar 1987; Brooks and Shorthouse 1998; Garbin et al. 2005; Jara-Chiquito et al. 2021). There are different factors that can influence the Cynipidae cecidogenesis, such as the mechanical action on the plant (sting in oviposition), the introduction of substances (secretion of the glands of the oviposition apparatus) or even the intervention of DNA particles in the form of commensal bacteria and/or viruses (Folliot 1977; Cornell 1983; Hearn et al. 2019).

The mechanical action is supported because the artificial production of small wounds in plants can cause deformations of a tumoral nature that involve phenomena that appear in cecidogenesis such as hypertrophy (abnormal cell growth), hyperplasia (abnormal cell multiplication) (Csóka et al. 2005; Garbin et al. 2005; Oliveira and Isaias 2010; Dias et al. 2013; Carneiro et al. 2014, 2015; Suzuki et al. 2015; Giron et al. 2016; Ferreira et al. 2017; Hearn et al. 2019) and stopping of cell differentiation. But the mechanical action, by itself, does not explain the diversity of the galls so the chemical factor has been considered essential. The insertion of an egg by the gallicolous insect would be accompanied by the introduction of substances, whose action would result in the formation of gall. However, in cases of Cynipidae-induced galls, in which the laying of an egg is the origin of a gall, the external histological signs of cecinogenesis do not appear until the egg has hatched; for this reason, the importance of the substances injected with the laying tends to be minimized. Nevertheless, such process has been considered again (Guiguet et al. 2023).

The Cynipidae elaborate very complex galls and the importance of the behavior of the larva in the formation of the gall and in its maintenance has been verified, by means of the active secretion of substances as it feeds (Hough 1953; Beyerinck 1882); so, for example, if the larva dies in the process of the gall development, the cecidogenesis stops. The nature of the substances responsible for cecidogenesis has been studied by numerous authors (Hartley and Lawton 1992; Hartley 1998; Schönrogge et al. 1994, 1998, 2000; LeBlanc and Lacroix 2001; Harper et al. 2004; Choudhary and Sharma 2013; Giron et al. 2016; Gätjens-Boniche 2019), who have endeavored to artificially obtain galls using a wide variety of chemical substances. Deserves special consideration, among these, heteroauxin or indoleacetic acid, a growth hormone extracted from certain plants. It is known that the role of auxins is to allow cellular elongation (therefore hypertrophy), stimulate meristematic cell divisions and exert inhibitory actions on growth, when concentrations are high. The possibility that the heteroauxin may be involved in cecidogenesis is justified by the fact that its detection is accompanied by hypertrophy and hyperplasia phenomena (essential to consider a deforming structure as a gall) and also intervenes in processes of inhibition (which cause in summary a cellular reorganization). However, the purely chemical processes involved in cecidogenesis are not sufficient to explain all aspects of gall development which, apart from the already mentioned phenomena of hypertrophy and hyperplasia, involves biochemical modifications (at the level of RNA production), inhibitions of differentiation followed sometimes by complex re-differentiation processes, etc. (Hearn et al. 2019). The complex nature of the histology of these galls, their specificity of location and concrete form as well as their structural diversity, suggest a complex genetic control through some form of fundamental sequence of development, determinant of the structural fate of the cells.

Insects also carry endosymbiotic and commensal bacteria, and are infected by viruses (Gibson and Hunter 2010), and plant manipulation may involve signal production and/or delivery by these co-bionts. It is known that parasitoid wasps export proteins that suppress insect host immune responses (Cornell 1983; Bézier et al. 2009) by packaging them within virus-like-particles (Cornell 1983). These virus-like-particles components are derived from viral genes that have been incorporated into the parasitoid genome (Bézier et al. 2009). Then, galling wasps could use a similar delivery system. Hearn et al. (2019), after checking the genomic and transcriptomic data from *Biorhyza pallida* galls, found no evidence of the involvement of symbiotic viruses or microorganisms in gall induction, or any utilization of virus-like particles for delivering the gall-inducing stimuli. Neither they found evidence for gall wasp production of plant hormone homologues. Nevertheless, many of the differentially and highly expressed genes in young larvae were predicted to have secretory leader peptides, and may be exported to plant tissues as part of the induction system. Young larvae showed high expression of two chitinase genes, which may act on some host proteins to induce a somatic embryogenesis-like state. These gall wasp larvae expressed genes encoding for plant cell wall degrading enzymes. Then, according to Hearn et al. (2019), the host tissue adopts a cellular state with some somatic-embryo-like features, which is controlled by nuclear genes of the galling-wasp larva.

Markel et al. (2024) affirms that cynipid wasps can radically change the metabolite profile and restructure the cell walls to enable the formation of galls, providing insights into the mechanism of gall induction and the extent to which plants can be entirely reprogrammed to form unique structures and organs. Also, Chen (2024) compares the metabolic profile of two species of morphologically different oak leaf galls induced by cynipid wasps and concludes that their profiles are different, so each form of gall would have its own metabolic profile that defines each gall morphology.

Cynipid gall induction is initiated after a female oviposits into species-specific locations in meristematic host tissues. The female determines the host plant, the location of the oviposition on the plant organ, and the number of larvae that will develop. The eggs may stay some days, weeks or months inside the tissues without any external visible transformation in the plant organ although there is a lysis phenomenon (rupture of the cell wall) of the plant cells in contact with the egg introduced by the insect in the ovipositing process. When the larva hatches (1st stage), the surrounding plant cells de-differentiate to form a chamber lined with undifferentiated tissue (Meyer and Maresquelle 1983; Rohfritsch 1992) that will quickly become the nutritive tissue. The larva, after hatching, is involved by plant tissues and live in this chamber. Each larva develops in its own chamber, so these galls can be multilocular (with different chambers) or monolocular (with a single larva inside); in the laying of the asexual forms, the larva that will give rise to the sexual generation can remain in its 1st stage for several months, until the gall begins to form.

Gall development can be divided into three stages: early, growth, and mature (Meyer and Maresquelle 1983; Brooks and Shorthouse 1998). In early-stage galls, the gall wasp larvae are very small (<1 mm). Host tissues lining the larval chamber divide rapidly and differentiate into specialized nutritive cells (neoplasy) that provide all larval food; in practically all of the galls of cynipids, the larval chamber is totally lined by nutritive cells that the larva will consume during its development. Traditionally, these cells are considered to be the only food source for the cynipid larvae during the entire larval development. However, recent non-invasive studies of galls suggest that the larval chamber is filled with liquid on early stages and is likely to contain nutrients, which the first larval instar may be absorbing since they have underdeveloped mandibles and are unable to chew on the nutritive tissue (Markel et al. 2024). A thin wall of sclerenchyma separates the inner gall (larval chamber) from non-nutritive outer gall tissues (Hough 1953; Bronner 1992; Shorthouse and Rohfritsch 1992; Csóka et al. 2005). In growing galls, the larvae remain very small, but the development and differentiation of outer gall tissues are extensive (Fig. 4a–d) (Meyer and Maresquelle 1983; Bagatto et al. 1996). The nutritive cells multiply (hyperplasia) and enlarge (hypertrophy), may become multinucleate, undergo endocycling, and accumulate high concentrations of lipids, carbohydrates and nitrogenous compounds (Rey 1966, 1967, 1971; Folliot 1977; Bronner 1992; Schönrogge et al. 2000; Harper et al. 2004). The nutritive tissue gradually reduces its thickness, thus increasing the internal space of the larval chamber. Outside the nutritive core, gall growth proceeds by fast hyperplasia and hypertrophy of parenchyma layers (Csóka et al. 2005; Garbin et al. 2005; Jara-Chiquito et al. 2021). The cellular reorganization that will form the tissues of the gall starts to form. Extensive development of vascular tissues facilitates the translocation of nutrients and water from the host plant organ. In many galls of oak trees, the tissues surrounding the larval chamber are lignified, the external epidermis is multi-stratified, the vascular tissues (phloem and xylem) are disorderly grouped, and the medullar tissue lengthens (Fig. 4a–d), a consequence of the swelling produced by the gall and the subsequent rupture of the vascular bundles in groups during the cecidogenous process. In the transition to the gall mature stage, the larvae grow rapidly until they exhaust the nutritive tissue; the feeding is continuous until the sclerenchyma is reached. Once the nutritive tissue is completely consumed, the larvae start pupation and the gall enters full maturity, when the outer gall tissues often lignify. After weeks or years (depending on the species and the alternating generation) the adults bite the tissues that separate them from the outside world, which can sometimes be very hard, and emerge from the gall.

4 Defenses of the Galls Against Parasitoids, Inquilines and Predators

The Cynipini are the gall-forming group where the structural sophistication and diversity of their galls reach the highest degrees. For this reason, several authors have suggested that this diversity explains the enemy hypothesis, that is, that galls have evolved in response to selection to try to exclude or hinder attacks by natural enemies (Price et al. 1995; Price and Pschorn-Walcher, 1988; Stone and Cook 1998). Natural enemies can inflict high mortality on gall-inducing larvae of cynipids (Askew 1984; Stone et al. 1995, Plantard et al. 1996; Schönrogge et al. 1996a, b), but different chemical, morphological, and locational traits of galls confer or

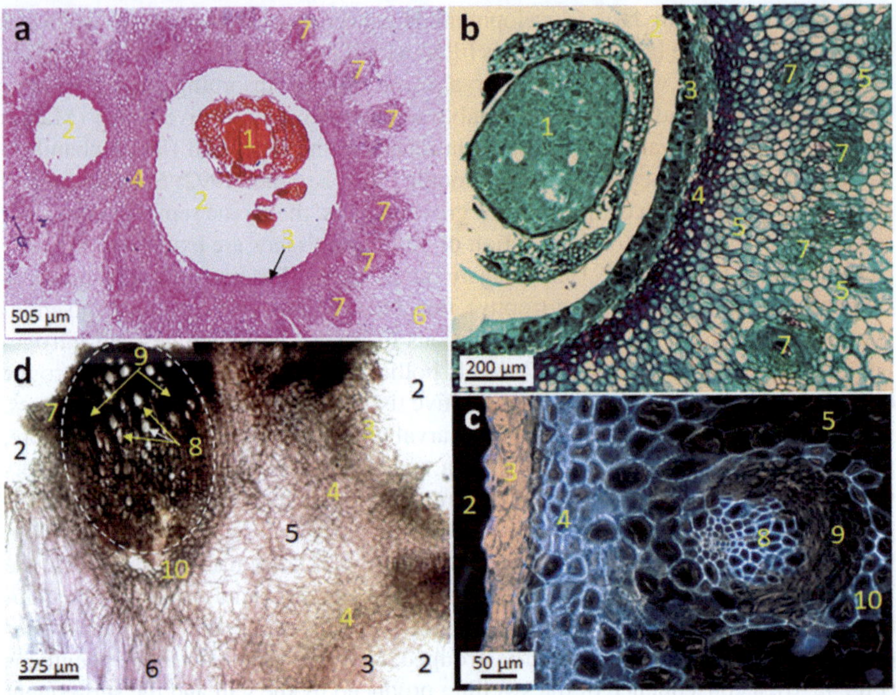

Fig. 4 Tissue reorganization of a gall at the level of the larval chambers of (**a–c**) *Dryocosmus kuriphilus* Yasumatzu (Cynipini on *Castanea sativa* - Fagaceae), (**d**) *Diastrophus rubi* (Diastrophini on *Rubus*). 1. larva, 2. larval chamber, 3. nutritive tissue, 4. Sclerenchyma layers surrounding the larval chambers, 5. storage parenchyma, 6. Medullar parenchyma (storage), 7. vascular bundles, 8. xylem, 9. phloem. 10. Sclerenchyma cells surrounding the vascular bundles. A similar organization can be observed even though the hosts are different and the cynipids belong to two different tribes

may confer protection against the attack by natural enemies, modifications that have not only propagated through natural selection but have repeatedly appeared in independently evolving lineages (Stone and Cook 1998). Structural examples include spines, sticky resin coatings, internal air spaces, and larval chambers that roll freely within a hollow gall, size, scales, etc. However, no gall is 100% immune to parasitoidism or lethal inquilines, so no gall structure offers absolute shelter from its enemies. In addition, many parasitoids and inquilines attack a wide range of gall morphologies, which also favors the presence of these organisms. On the other hand, the morphological variety of the parasitoids, for example in the length of the ovipositor, means that there are certain preferences or specificities in the attack of the galls. In addition, there are parasitoid species like *Torymus flavipes* (Walker 1833) (= *T. auratus* Geoffroy, 1785), which present different lengths of ovipositor in their two annual generations (Askew 1965), which expands the range of galls capable of attacking. For these reasons, although parasitoids may not be completely excluded, it remains possible that specific gall traits reduce mortality inflicted by specific members of the community.

Next, different chemical, physical, or gall morphological factors that could affect parasitoid communities will be commented and the effect these factors may have on the parasitism or lethal inquilinism of cynipid galls will be discussed.

4.1 Gall Secretions as a Mechanism of Defenses

Galls can secrete two types of chemical compounds that line their exterior: nectar and sticky substances. Some oak galls belonging to *Andricus*, *Disholcaspis* and *Dryocosmus* genera (for example) induce nectar-secreting galls that recruit ants (Fig. 5a), similar to the relationship established between communities of some free-living aphid and ant species; this act significantly reduces the parasitoid and inquiline attacks on these galls preventing the parasitoid or inquiline from ovipositing (Washburn 1984; Redfern and Askew 1992; Seibert 1993; Stone and Cook 1998; Fernandes et al. 1999; Nicholls et al. 2017). On the other hand, galls such as those induced by *Andricus dentimitratus* (Fig. 5b) or *A. pictus* (in the Western Palearctic) produce a sticky substance that stuck to parasitoids or inquilines on the surface of the gall when they land, making oviposition impossible (Stone and Cook 1998). Finally, there are species that may have secretions at the time of growth of the gall, but these secretions disappear when the gall has reached its almost definitive size. This fact supports the premise that these secretions probably represent the most obvious protection against the natural enemies of oak galls (Pujade-Villar 1994; Stone et al. 2002; Csóka et al. 2005), at least during the growth phase.

4.2 External Deterrent Characteristics

Although its effectiveness has not yet been proven, many galls change color during development, usually from green to red (Stone et al. 2002). The extent to which the cynipid larva controls pigment synthesis is unknown, but there is evidence that parasitoid females recognize color and use it to assess host plant quality (Askew 1961a; Czeczuga 1977), since the galls are green during early larval stages. Therefore, coloration could deter parasitism at some developmental stages. Alternatively, it could be a camouflage mechanism of the gall when it is green (Lev-Yadun and Inbar 2023) or could be related to the variation in light regimes that creates differential anthocyanin accumulation in red galls (Bomfim et al. 2019).

Coverings of hairs and spines of various lengths and densities have also evolved repeatedly and independently on the galls of oak cynipids. These structures are very common in North America, being found in many species of the genera *Amphibolips* (Fig. 5f, g), *Antron* (Fig. 5d), *Druon* (Fig. 5h), *Striatoandricus* (Fig. 5e) and *Feron* (Fig. 5c), for example; in the Palearctic, these coverings are much rarer and there are few represented in *Cerroneuroterus*, *Cycloneuroterus* and *Trichagalma* galls (in Asia), in *Andricus* (in Europe, Fig. 6f) and *Cynips* galls (in Asia Minor). Dense

Fig. 5 External defensive strategies to attack parasitoids. (**a**) *Disholcaspis quercussuccinipes* with defending ants; (**b**) *Andricus dentimitratus* with sticky surface (picture made by Josep Mª Lesma from Biodiversitat virtual); (**c**) *Feron crystallinum* with spines (picture made by David Cibrián), (**d**) *Antron acraspiformis* with spines; (**e–h**) pubescens galls: (**e**) *Striatoandricus furnessulus* (picture made by Alondra G. Pérez-García), (**f, g**) *Amphibolips ufo* (pictures made by Carlos Carrillo), (**h**) *Druon quercuslanigerum* var. *quercuslanigerum*; (**i, j**) *Andricus mukaigawae* with scates. (Pictures made by Csoka Gyuri)

coverings of hairs or spines can influence the attack of the parasitoids, making it difficult for them to oviposit. For example, *Amphibolips* galls of the '*niger*' group (Fig. 5f, g) can have a very dense layer of pilosity, which easily exceeds 10 mm in thickness, around the central gall, which is also extremely hard and protects a single larval chamber (Kinsey 1937; Cuesta-Porta et al. 2023a).

Other galls, cause hypertrophy of their host bracts so that they enclose the gall, which is usually unilocular as occurs in some *Andricus* galls in the Palaearctic (Fig. 5i, j) or *Erytres* galls in North America. To access the gall, the parasitoid must pierce each bract lamina, which can influence the final direction of oviposition. We observed that parasitoids of the genus *Torymus* often spend much more time laying the egg and often make more than one sting attempt. In some of these species, when the galls are mature, they are pushed from the base until they fall to the ground where the larvae metamorphose.

Finally, size can also be a limiting factor for parasitoidism. The galls of cynipids are very variable in size, ranging from galls that barely exceed 1 mm (many of them correspond to individual galls of sexual generations that are in the catkins, for example) to galls that can exceed 15 cm (such as *Andricus costaricensis* that attacks the branches of *Q. costaricensis* in Costa Rica) (Melika et al. 2009a). The length of the ovipositor and the ideal moment to attack the internal larva will mark the success or failure of the parasitoidism. Probably the species with the longest ovipositor corresponds to *Torymus lasallei*, specialized in attacking large spongy galls of the genus *Amphibolips* of the *nassa* group (Bubeníková et al. 2020).

4.3 Gall Hardness May Avoid Enemy's Attack

Cynipid galls do not appear as hardened galls from the moment of formation, but rather they harden in the growth process. In the first stages, the galls are not attacked by the parasitoid since the larval resource is insufficient (small cynipid larvae) for the parasitoid to develop, so the cynipid larvae do not begin to grow until the wall of the gall is fully developed. Harder galls are thought to be more difficult to attack than softer ones due to the physical difficulty of piercing the harder galls with an ovipositor and the risk of predation associated with prolonged oviposition (Rossi et al. 1992). By studying the chemical composition of the valves of the ovipositors, it has been found that there are species such as *Ormyrus nitidulus* and *Torymus auratus* (Mueller 1764) (= *T. nitens* Walker, 1833), capable of attacking hard galls; these species have particularly high concentrations of manganese in the valves of the ovipositor (Quicke et al. 1998). This is associated with increased chitin hardness and may be an adaptive response to gall hardness.

4.4 Internal Deterrent Characteristics

Many oak galls contain internal air spaces. These can be false chambers that are next to the larval chamber (deterrent chambers), where a possible spawning will not lead to the larval development of the parasitoid due to a lack of food source (the inducer larva is not in this chamber), or large tissue-free areas between the larval chamber and the outer tissues of the galls. In the latter case, for example, the larval chamber is held in the center by some radiating filaments derived from the initial spongy parenchyma as *Atrusca*, *Kinseyella* (Fig. 6a), some *Amphibolips* galls. In other galls, the larval chamber separates from this attachment and rolls free inside the gall, as in some *Disholcaspis* galls in America (Fig. 6b) or *Andricus* galls (Fig. 6c) in Europe; others have a large air chamber with the larval chamber fixed by a basal point, as occurs in *Trichagalma* galls, (Fig. 6d). Airspaces, of any kind, are a problem for the parasitoid since the ovipositor is unsupported for part of its length, which reduces the force a parasitoid can exert with the tip of its ovipositor without deforming (Quicke et al. 1998). After traversing this airspace, an ovipositor must penetrate the thin layer of sclerenchyma surrounding the larval chamber, often the

Fig. 6 Internal defensive structures and multilocular galls: (**a**) *Kinseyela quercusobtusata* with central larval chamber held by radial fibers, (**b**) *Disholcaspis crystalae* with the free larval chamber, (**c**) *Trichagalma serratae* asexual with larval chamber attached to the basal part of the gall, (**d**) *Andricus inflator* sexual with a false chamber located above the larval chamber; (**e**) *Andricus quercusradicis* asexual; (**f**, **g**) *Andricus lucidus*, (**h**) criptic multilocular gall of *Plagiotrochus amenti*. (Pictures (**c**–**g**) made by Csoka Gyuri)

hardest tissue in the gall. Therefore, only parasitoid species possessing a particularly hardened ovipositor can penetrate the sclerenchyma in these galls.

4.5 The Number of Larval Chambers per Gall

As already mentioned, the galls can be unilocular or multilocular. The advantage of being unilocular is that the cynipid disperses its eggs and therefore the parasitoid has to look for gall to gall to parasitize them, which entails a very large investment of time, while the free-living phase of a parasitoid is short (5–10 days). Multilocular galls have cynipid laying concentrated in a small space, which is an advantage for the parasitoid, but these galls have different models of organization that make parasitoidism difficult. Multilocular galls can have larval chambers placed at different depths (Fig. 6e, g), representing a defensive system compared to unilocular or multilocular galls with larval chambers located on the same plane where all larval chambers are equally accessible.

The purpose of this layered multilocularity is to protect part of the offspring from the attack by parasitoids. The larvae in the peripheral chambers are vulnerable to parasitoid attack, but the deeper ones are protected by a thicker shield of gall tissues and by having other larval chambers located above them (Jones 1983). We must also bear in mind that the more larval chambers the gall contains, the larger this structure can be, which can also increase the peripheral tissue, thus also making it difficult for the parasitoids to attack, which are limited by the length of the ovipositor. Multilocular galls with surface cameras placed in the same plane tend to be cryptic (Fig. 6h), thus making it difficult for the parasitoid to locate them. It would be the case of some species of *Loxaulus* galls (in America) or of *Callirhytis* and *Plagiotrochus* galls in Europe, which are induced on the stems of different species of *Quercus* without forming an obvious dilatation of the branch.

4.6 Location of the Galls

The location of the galls on the host plant, the time required for their development and their persistence or not on the host plant affect the richness of parasitoid species (Stille 1984; Schönrogge, et al. 1994, 1995, 1996a; Plantard et al. 1996; Plantard and Hochberg 1998). So, it is not the same whether the gall is in the stems or the catkins, in terms of hardness, for example. The sexual galls have a very fast growth (a few weeks) compared to the asexual forms that often emerge the following year and when, therefore, the time of exposure to the attack is much longer. And finally, some galls finish their development on the ground where it is more difficult for the parasitoids to search for. All these factors affect parasitoidism.

On the other hand, galls of the same species that are capable of attacking different organs have different phenology (galls on catkins are shed or dry earlier than those on leaves, for example). This double location of the galls is not a common occurrence in cynipids but occurs in some species of the genera *Neuroterus*, *Plagiotrochus* and *Andricus* in the Palearctic (for example). This is the case of the galls of the sexual generation of *Neuroterus quercusbaccarum*, for example, which can develop on leaves or catkins. The parasitoid attack rates and species richness are lower in galls developed on catkins than in those developed on leaves (Plantard and Hochberg 1998). Because of all this, the effects of the location of the galls and the time it takes to develop cannot be separated since this also affects the percentage of parasitoidism.

4.7 Defenses Against Predation and Pathogens

Galls accumulate part of the host plant resources, which makes them potential targets for some herbivores, mostly insects and vertebrates. The presence of robust spines is probably one of the most effective physical and visual deterrents for vertebrate predators (Stone et al. 2002). Galls also accumulate chemical compounds to defend against herbivores and pathogens (mostly fungi and bacteria). The most abundant and largely studied defense-related secondary metabolites are phenolic compounds (Hartley 1998; Barbehenn and Constabel 2011; Kot et al. 2018; Tataroğlu et al. 2023; Chen 2024). Some studies show that plants with high incidence of galls are not or less consumed by herbivores (Margalef 1977). This low consumption is due to phenolic compounds (e.g.: phenylpropanoids, tannins, and lignins), which can be found in large quantities in galls mostly during growth and maturing phases (Kariñho-Betancourt et al. 2019). The location of those compounds within the gall is regulated by phenolic-related genes, which have been reported to be highly expressed in outer layers of the gall and suppressed around the larval chambers (Hearn et al. 2019; Martinson et al. 2022; Schultz and Stone 2022). The differential expression of those genes are thought to generate an expression gradient in the gall with higher expression of defense-related genes on the outer layers of the gall and higher expression of genes related to the production of nutritive substances towards the larval chamber (Giron et al. 2016; Martinson et al. 2022; Schultz and Stone 2022). The only defense-related genes expressing close to the larval chamber are the ones encoding for anti-bacterial or anti-fungal proteins (Martinson et al. 2022; Schultz and Stone 2022).

The production of secondary metabolites is dependent on the plant species and the environmental conditions. Hence, the patterns of expression of galls induced by the same cynipid species can significantly vary if induced on different host-plant species or different locations (Hartley 1998; Hall et al. 2017). Given the great variety of cynipid diversity, each gall wasp species will induce different responses on a certain oak species (Kariñho-Betancourt et al. 2019). Despite numerous advances in the understanding of the induced metabolic response of host plants by gall induction, the study of the metabolic signal of galls is a multi-variant issue and difficult to predict.

Tannins are generally toxins that significantly reduce the growth and survival of many herbivores when added to their diet as they can inactivate some digestive enzymes (Robbins et al. 1987). These effects are often combined with aposematic colorations, which help to the rejection of galls by greater herbivores. .

As far as insects are concerned, the presence of tannins is also a disadvantage for phytophages since phenols make digestion difficult, and they are thought to reducing the nutritional value of plant tissues, so that only a small group of species have the ability to feed from the galls (Berenbaum 1983). Often small-sized xylophagous organisms can use this resource and always occasionally. No species is known to exclusively feed on galls.

5 Oak Galls in the Ecosystem

The faunal community associated with the galls can be divided into three large groups: primary, secondary, and tertiary fauna (Fig. 7). The primary fauna is made up of that community of insects directly linked to the galls or to their inhabitants through a trophic relationship. This fauna settles in the galls during its growth and maturation process. This is the case of inquilines (also Cynipidae) and parasitoids (different families of the Chalcidoidea). The inquilines may or may not provoke the death of the gall inducer's larva, but the parasitoids always cause it. The secondary and tertiary fauna is mentioned by several authors as successors (Mani 1964; Luz

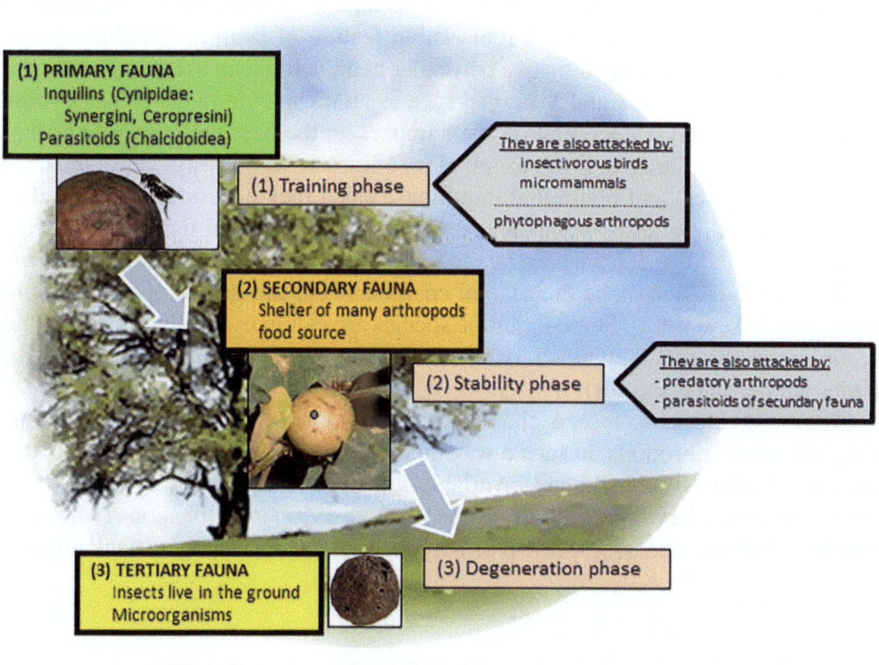

Fig. 7 Summary of the groups of organisms associated with galls (according to Massana-Canals et al. 2013)

and Mendonça 2019); it corresponds to all the fauna that can inhabit the galls after the emergence of the primary fauna and include all the other fauna that degrades the gall by using it as a food resource including its parasitoids.

5.1 Primary Fauna

The inquiline fauna is a set of cynipid species that have lost the ability to induce their own galls 'de novo', they are nevertheless able to induce the development of larval chambers lined with nutritive tissue inside cynipid galls. They are exclusively phytophagous. Inquilinism was described as a special form of kleptoparasitism between close relatives named agastoparasitism (Ronquist 1994), a hypothesis that has been strongly supported by Blaimer et al. (2020), and it has traditionally been considered to represent an obligatory advantageous relationship that benefits the inquilines only (Askew 1984; Ronquist 1994, 1999; Csóka et al. 2005). In oak gall wasps, the cynipid inquilines are included in the tribe Synergini (*Agastoroxenia* only from Panama, *Lithosaphonecrus* distributed in Far Eastern Palaearctic and the Oriental regions, *Saphonecrus* in Palaearctic, *Synophrus* in Western Palaearctic, *Synergus* in Holarctic and the Neotropics and *Ufo* in Eastern Palaearctic) and Ceroptresini (*Ceroptres* distributed in Holarctic region and *Buffingtonella* only know from USA) (Lobato-Vila et al. 2019, 2022). Although they cannot initiate gall formation, some inquiline cynipids substantially modify the size and/or shape of the host oak galls (*Synergus clandestinus* in the galls of *Andricus legitimus* and *Ceroptres clavicornis* in galls of *Andricus hispanicus*, for example (Wiebes-Rijks 1979 and Pujade-Villar 1991). A remarkable feature of cynipid inquilines is that they are generally more specific to a particular plant taxon than to a certain host gall inducer. Many *Ceroptres* and *Synergus* species attack a wide range of oak cynipid galls (Melika 2006). Depending on their impact on the host galls, inquilines have been classified by Duffett (1969) as lethal inquilines (if they always cause the death of the gall-inducer) and non-lethal inquilines (when they never cause the death of the host cynipid, because the larvae are found in the peripheral tissues). The lethal inquilines are those species whose larvae live inside the inducer chamber competing for food, or those that make their larval chambers near the inducer's chamber, collapsing it (Askew 1984). Some inquiline species are facultative (lethal or non-lethal) according to the gall they attack or the final position of the chambers (if the gall wasps are the same species).

Parasitoids are those insects characterized by laying their eggs in the immature stages of other arthropods, in such a way that after the emergence of the larvae, they consume the host organism, and invariably kill it; to complete the cycle they only need to feed on a larva. Parasitoids can be (Quicke 1997) monophagous (when they attack exclusively one host species) or polyphagous (when they can attack different hosts). They will be primary (when they directly attack the inducer or an inquiline of the galls), secondary, tertiary, etc. (depending on whether they attack another parasitoid or the parasitoid of another parasitoid, etc., in this case called hyperparasitoids). Nevertheless, there are cases where a parasitoid can be primary or secondary at the same time, so they can attack two different larval models (Askew 1961b). All cynipid gall parasitoids are koinobionts (they allow the complete development of their hosts after the attack), and the majority are ectoparasites, although there are also endoparasitic species/genera (Askew 1961b). All the parasitoids of oak galls correspond to different families of chalcidoids (Eupelmidae, Ormyridae, Pteromalidae, Eulophidae, Ormyridae, Torymidae and Megastigmidae, the latter absent until now in the Nearctic attacking the galls of Cynipidae). Complex trophic networks are established that relate inducers, inquilines, parasitoids and hyperparasitoids (Askew 1961b, 1984), and to cite an example, the gall of *Andricus hispanicus* (asexual generation) presents a trophic network in which 27 species of parasitoids and inquilines are found implicated (Ros-Farré and Pujade-Villar 1998).

5.2 Secondary and Tertiary Fauna

The secondary fauna is made up of organisms indirectly or collaterally related to the galls, generally when they have been abandoned (both in the tree and on the ground). This usually happens in large, woody galls (Pujade-Villar 2013). These galls are a permanent or temporary refuge and can also be a food source for other phytophagous/xylophagous arthropods capable of feeding on gall tissues. The channels, chambers, scales, setosity, etc., provide a wide variety of refuges and accommodation for a multitude of arthropods. Not only juvenile forms take advantage of these shelters but they are also used by small adults and some wintering arthropods (Torossian 1971a; Maia 2013; Massana-Canals et al. 2013; Pujade-Villar 2013): such as spiders, pseudoscorpions, mites (regarding chelicerates, Fig. 8a), Thysanoptera, Psocoptera, aphids, myrmecophilous coccids, Lepidoptera, Coleoptera, Hymenoptera, Collembola, etc. (referring to insects, Fig. 8b) and sometimes also some myriapods. Among all these fauna, ants nesting in abandoned galls stand out for its importance (Torossian 1971b, 1972, Espadaler and Nieves 1983). Some species of ants occupy the gall alone or with a few workers (1–10), which represent the initial phases of the colony (foundations). Other times, the female is surrounded by dozens of workers, so it constitutes a miniature colony. The presence of colonial foundations in the galls facilitates the dispersion of the species.

Finally, tertiary fauna, which is very little studied, is in the degenerative phase of the gall, that is to say that it falls to the ground and gradually degrades until it disappears. The arthropod fauna is made up of hypogeal or epigeal organisms linked to the soil that will transform organic into inorganic matter.

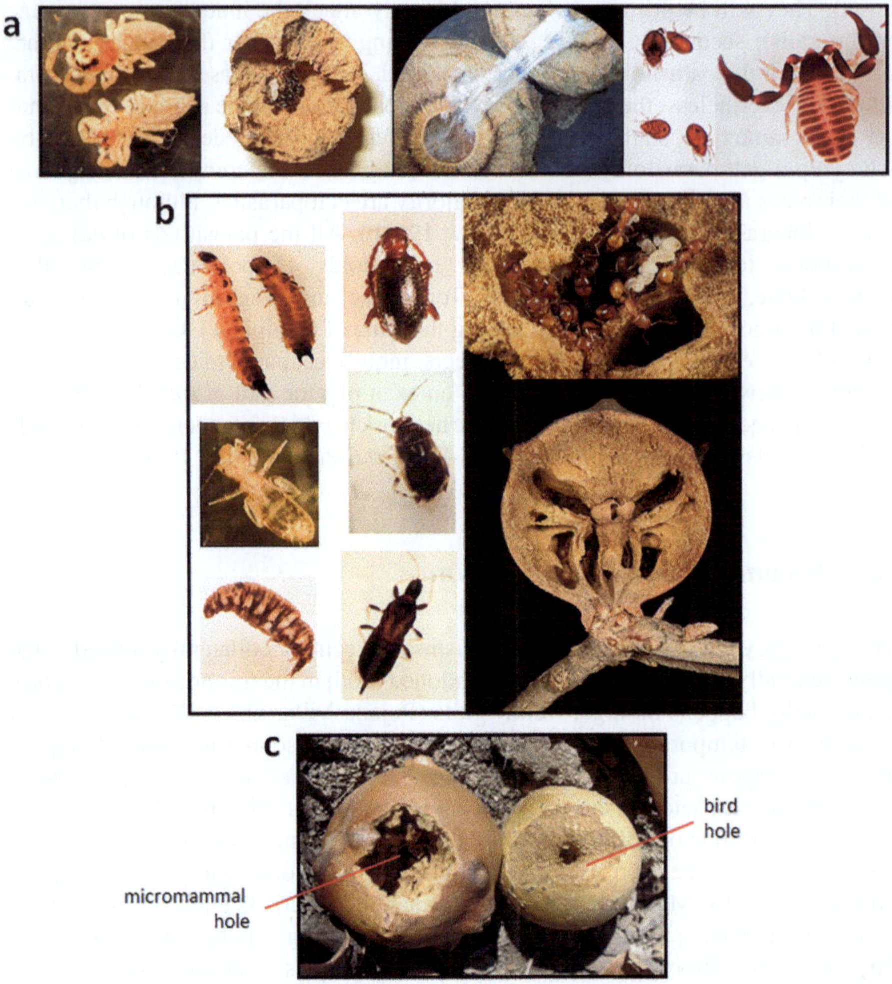

Fig. 8 Some examples of secondary fauna of oak galls (**a**) chelicerates and (**b**) insects. (**c**) Galls depredated, the holes made by birds are delicate while those made by small mammals are rough. (Pictures of galls with ants made by Csoka Gyuri)

5.3 Gall's Importance in the Ecosystem, Function

Oak galls are a miniature ecosystem (Pujade-Villar 2013) and their presence plays an important role in the forest ecosystem. We must keep in mind that galls are a very complex way of isolating the phytophagous within a plant structure so that its phytophagic action affects the plant as little as possible (Margalef 1977). There is therefore an adaptive value and a mutual benefit between the plant and the gallicolous insect (Price et al. 1986, 1987).

The primary fauna is responsible for controlling the populations of gall-inducers, so that population explosions do not occur; when this associated fauna declines or disappears (usually due to human intervention), the cynipid can become a pest. Some species of oak gall wasps could be harmful to the oak trees they develop on (Brussino et al. 2002; Battisti et al. 2014; Pujade-Villar 2014), especially those that attack the branches (preventing the flow of sap), those that attack the acorns (preventing the regeneration of the oaks) or those that cause a loss of tree vigor (due to the decrease in photosynthetic capacity). So, for example, *Plagiotrochus amenti* can be a dangerous pest of *Q. suber* in the Mediterranean area (Benia et al. 2009; Pujade-Villar et al. 2010), particularly in northeastern Spain (Garbin et al. 2005), and in North and South America, where *Q. suber* has been introduced to provide cork for the wine industry (Bailey and Stange 1966; Díaz 1972; Zuparko 1996). *Disholcaspis cinerosa* induces stem galls in Texas, USA, sometimes strongly affecting urban ornamental live of oak trees (Frankie et al. 1992). *Andricus quercuslaurinus* produced a drastic change in the floristic composition of the ecosystems of Hidalgo, Mexico, (Melika et al. 2009a, b; Pujade-Villar et al. 2012) until its managed control by the introduction of parasitoids and inquilines for about 12 years. The traditional control of the pests produced by cynipids is very complicated for many reasons, but some of the most important are that: (i) the larvae are not accessible as they are protected by the plant tissues, (ii) the adults live very little and the agamic forms are usually the most dangerous (Jara-Chiquito et al. 2022), and (iii) we cannot use pheromones, once pheromones could be used in sexual generations, but in many species, it is unknown what this generation is.

The secondary fauna has many roles in the ecosystem and the galls are places of protection with an important role to conserve the diversity. Once the primary fauna has emerged, the secondary fauna can use galls on five different purposes (Massana-Canals et al. 2013): (i) pseudoscorpions, spiders, hemipterans, thysanopterans, among other groups, use the gall as an occasional refuge (for example, to hide from different predators, during molting processes when they are most vulnerable or to take advantage of adverse weather conditions), (ii) the ants, as a temporary shelter (to spend the winter there) building an incipient colony; (iii) mites and psocopterans can use it to feed on the hyphae that may be inside or on other organic matter; (iv) coleopteran larvae that are predators look for possible prey inside the galls, and (v) other larvae that are xylophagous feed on the tissues that make up the galls.

Galls with fauna inside, whether primary or secondary, have another function in the ecosystem, that is to provide protein to the insectivorous vertebrates (micromammals or birds, Fig. 8c) so that these vertebrates look for and extract the larvae from their interior to feed especially in winter periods when resources are scarce (Pujade-Villar 2013). The tertiary fauna participates in closing the cycle of the matter, since as decomposers, crumble the galls ingesting small fragments that they digest, favoring in this way decomposition and mineralization; also, increasing the surface areas of available resources (feces) that stimulate the colonization of microbes and fungi, that end up converting the organic matter into inorganic, thus mineralizing the soil 'de novo' (Coleman and Wall 2015). These events are very common in Cynipini galls but not exclusive, it can also occur in some lepidopteran galls (Rezende et al. 2023).

6 Final Considerations

Oak galls are highly complex and variable morphological structures with one or more chambers, where the larvae are locked and cannot leave. A priori, the galls are defenseless against the attack of parasitoids, and have developed a vast array of adaptations to prevent parasitoidism and, parallelly, parasitoids have adapted to favor their attack. These adaptations have independently appeared in different gall wasp genera. Oak galls provide an excellent framework for investigating the coevolutionary dynamics between cecidogenous organisms as they contend against the parasitic abilities of their parasitoids. Once the galls are abandoned by inducers and their associated fauna, they represent a temporary refuge for many small arthropods thus contributing to maintaining the biodiversity of the ecosystem.

References

Abe Y (1986) Taxonomic status of the *Andricus mukaigawae* complex and its speciation with geographic parthenogenesis (Hymenoptera: Cynipidae). Appl Entomol Zool 21(3):436–447. https://doi.org/10.1303/aez.21.436

Abe Y (1992) The advantage of attending ants and gall aggregation for the gall wasp Andricus symbioticus (Hymenoptera: Cynipidae). Oecologia 89(2):166–167. https://doi.org/10.1007/BF00317214

Abe Y, Melika G, Stone GN (2007) The diversity and phylogeography of cynipid gallwasps (Hymenoptera: Cynipidae) of the Oriental and eastern Palearctic regions, and their associated communities. Orient Insects 41(1):169–212. https://doi.org/10.1080/00305316.2007.10417504

Askew RR (1961a) A study of the biology of species of the genus Mesopolobus Westwood (hymenoptera: Pteromalidae) associated with Cynipid galls on Oak. Trans R Entomol Soc London 113(8):155–173. https://doi.org/10.1111/j.1365-2311.1961.tb00806.x

Askew RR (1961b) On the biology of the inhabitants of oack galls of Cynipidae (Hymenoptera) in Britain. Trans Soc Br Entomol 14(11):237–268

Askew RR (1965) The biology of the British species of the genus Torymus Dalman (Hymenoptera: Torymidae) associated with galls of Cynipidae (Hymenoptera) on oak, with special reference to alternating forms. Trans Soc Br Entomol 16(9):217–232

Askew RR (1984) The biology of gall wasps. In: Ananthakrischan TN (ed) Biology oof gall insects. Oxford and IBH Publishing Co., New Delhi/Bombay/Calcutta, pp 223–271

Atkinson RJ, Brown GS, Stone GN (2003) Skewed sex ratios and multiple founding in galls of the oak apple gall wasp Biorhiza pallida. Ecol Entomol 28(1):14–24. https://doi.org/10.1046/j.1365-2311.2003.00483.x

Bagatto G, Paquette LC, Shorthouse JD (1996) Influence of galls of Phanacis taraxaci on carbon partitioning within common dandelion, Taraxacum officinale. Entomol Exp Appl 79(1):111–117. https://doi.org/10.1111/j.1570-7458.1996.tb00815.x

Bailey SF, Stange LA (1966) The Twig Wasp of Cork Oak—its biology and control. J Econ Entomol 59(3):663–668. https://doi.org/10.1093/jee/59.3.663

Bailey R, Schönrogge K, Cook JM, Melika G, Csóka G, Thuróczy C, Stone GN (2009) Host Niches and defensive extended phenotypes structure Parasitoid Wasp Communities. PLoS Biol 7(8):e1000179. https://doi.org/10.1371/journal.pbio.1000179

Barbehenn RV, Constabel CP (2011) Tannins in plant–herbivore interactions. Phytochemistry 72(13):1551–1565. https://doi.org/10.1016/j.phytochem.2011.01.040

Bssett HF (1880) The structure and development of certain Hymenopterous galls. Am Entomol 3:284
Bassett HF (1881) New species of Cynipidae. Can Entomol 13(3):51–57. https://doi.org/10.4039/Ent1351-3
Battisti A, Benvegnù I, Colombari F, Haack RA (2014) Invasion by the chestnut gall wasp in Italy causes significant yield loss in Castanea sativa nut production. Agric For Entomol 16(1):75–79. https://doi.org/10.1111/afe.12036
Benia F, Khelil MA, Pujade-Villar J (2009) *Plagiotrochus amenti*, une espèce gallicole potentiellement dangereuse pour le chêne liège (*Quercus suber* L.) trouvée pour la première fois en Algérie (Hymenoptera, Cynipidae). Nouvelle Revue d'Entomologie 25:291–296
Berenbaum MR (1983) Effects of tannins on growth and digestion in two species of papilionids. Entomologia Experimentalis et Applicata 34(3):245–250. https://doi.org/10.1111/j.1570-7458.1983.tb03329.x
Berdegue M, Trumble JT, Hare JD, Redak RA (1996) Is it enemy-free space? The evidence for terrestrial insects and freshwater arthropods. Ecol Entomol 21(3):203–217. https://doi.org/10.1111/j.1365-2311.1996.tb01237.x
Beyerinck MW (1882) In: Müller J (ed) Beobachtungen ueber die ersten Entwicklungsphasen einiger Cynipidengallen, Amsterdam, p 198
Bézier A, Annaheim M, Herbinière J, Wetterwald C, Gyapay G, Bernard-Samain S, Wincker P, Roditi I, Heller M, Belghazi M, Pfister-Wilhem R, Periquet G, Dupuy C, Huguet E, Volkoff A-N, Lanzrein B, Drezen J-M (2009) Polydnaviruses of Braconid Wasps Derive from an Ancestral Nudivirus. Science 323(5916):926–930. https://doi.org/10.1126/science.1166788
Blaimer BB, Gotzek D, Brady SG, Buffington ML (2020) Comprehensive phylogenomic analyses re-write the evolution of parasitism within cynipoid wasps. BMC Evol Biol 20:1–22. https://doi.org/10.1186/s12862-020-01716-2
Bomfim PMS, Cardoso JCF, Rezende UC, Martini VC, Oliveira DC (2019) Red galls: the different stories of two gall types on the same host. Plant Biol 21(2):284–291. https://doi.org/10.1111/plb.12915
Bronner N (1992) The role of nutritive cells in the nutrition of cynipids and cecidomyiids. In: Shorthouse JD, Rohfritsch O (eds) Biology of insect-induced galls. Oxford University Press, New York, pp 118–140
Brooks SE, Shorthouse JD (1998) Developmental morphology of stem galls of Diplolepis nodulosa (Hymenoptera: Cynipidae) and those modified by the inquiline Periclistus pirata (Hymenoptera: Cynipidae) on Rosa blanda (Rosaceae). Can J Bot 76(3):365–381. https://doi.org/10.1139/b98-001
Brussino G, Bosio G, Baudino M, Giordano R, Ramello F, Melika G (2002) Pericoloso insetto esotico per il castagno europeo. Informatore Agrario 58(37):59–62
Bubeníková K, Pujade-Villar J, Janšta P (2020) Description of Torymus lasallei, sp. nov. (Hymenoptera: Torymidae), a species with an unusual ovipositor. J Nat Hist 54(9–12):791–800. https://doi.org/10.1080/00222933.2020.1769763
Buffington ML, Forshage M, Liljeblad J, Tang C-T, van Noort S (2020) World Cynipoidea (Hymenoptera): a key to higher-level groups. Insect Syst Div 4(4):1–69. https://doi.org/10.1093/isd/ixaa003
Carneiro RGS, Oliveira DC, Isaias RMS (2014) Developmental anatomy and immunocytochemistry reveal the neo-ontogenesis of the leaf tissues of Psidium myrtoides (Myrtaceae) towards the globoid galls of Nothotrioza myrtoidis (Triozidae). Plant Cell Rep 33(12):2093–2106. https://doi.org/10.1007/s00299-014-1683-7
Chen J (2024) It's gall relative: metabolic profiling of two morphologically distinct oak leaf galls induced by cynipid wasps. Plant Physiol 195(1):248–250. https://doi.org/10.1093/plphys/kiae032
Choudhary S, Sharma IR (2013) Quantitative estimation of some metabolites and enzymes in *Trioza jambolanae* induced leaf galls of *Syzygium cumini* Linn. Int J Pure Appl Biosci 1(3):94–98

Coleman DC, Wall DH (2015) Chapter 5 - Soil Fauna: occurrence, biodiversity, and roles in ecosystem function. In: Paul EA (ed) Soil microbiology, ecology and biochemistry, 4th edn. Academic, Boston, pp 111–149

Cornell HV (1983) The secondary chemistry and complex morphology of galls formed by the Cynipinae (Hymenoptera): why and how? Am Midl Nat 110(2):225–234. https://doi.org/10.2307/2425263

Cotté J (1926) Considérations sur l'hétérogonie chez les Cynipides Cécidogènes (alternation of generations among the gall-making Cynipids). Marcellia 22(1–6):89–119

Csóka G, Stone G, Melika G (2005) Biology, ecology and evolution of gall-inducing Cynipidae. In: Raman A, Schaefer CW, Withers TW (eds) Biology, ecology and evolution of gall-inducing arthropods. Science Publishers, Inc, Enfield, pp 573–642

Cuesta-Porta V, Equihua-Martínez A, Estrada-Venegas EG, Cibrián-Tovar D, Barrera-Ruíz UM, Ordaz Silva S, López Sánchez IV, Melika G, Pujade-Villar J (2020) Revision of the *Amphibolips* species of the "*nassa*" complex from Mexico and Central America (Hymenoptera: Cynipidae). Zootaxa 4877(1):1–50. https://doi.org/10.11646/zootaxa.4877.1.1

Cuesta-Porta V, Melika G, Nicholls JA, Stone GN, Pujade-Villar J (2022) Re-establishment of the Nearctic oak cynipid gall wasp genus *Druon* Kinsey, 1937 (Hymenoptera: Cynipidae: Cynipini), with description of five new species. Zootaxa 5132(1):1–92. https://doi.org/10.11646/zootaxa.5132.1.1

Cuesta-Porta V, Melika G, Nicholls JA, Stone GN, Equihua-Martinez A, Estrada-Venegas E, Pujade-Villar J (2023a) Revision of the *Amphibolips* "*Niger*" group with the description of a new species (Hymenoptera: Cynipini). Zootaxa 5360(4):451–486. https://doi.org/10.11646/zootaxa.5360.4.1

Cuesta-Porta V, Melika G, Nicholls JA, Stone GN, Pujade-Villar J (2023b) Re-establishment of the Nearctic oak cynipid gall wasp genus *Feron* Kinsey, 1937 (Hymenoptera: Cynipidae: Cynipini), including the description of six new species. Zootaxa 5366(1):1–174. https://doi.org/10.11646/zootaxa.5366.1

Czeczuga B (1977) Investigations of carotenoids in some animals of the Adriatic Sea V. Echinodermata. Hydrobiologia 54(2):177–180. https://doi.org/10.1007/BF00034992

da Carneiro RGS, Pacheco P, dos Isaias RMS (2015) Could the extended phenotype extend to the cellular and subcellular levels in insect-induced galls? PLoS One 10(6):e0129331. https://doi.org/10.1371/journal.pone.0129331

Diamond SE, Blair CP, Abrahamson WG (2008) Testing the nutrition hypothesis for the adaptive nature of insect galls: does a non-adapted herbivore perform better in galls? Ecol Entomol 33(3):385–393. https://doi.org/10.1111/j.1365-2311.2007.00979.x

Dias GG, Moreira GRP, Ferreira BG, Isaias RMS (2013) Why do the galls induced by Calophya duvauae Scott on Schinus polygamus (Cav.) Cabrera (Anacardiaceae) change colors? Biochem Syst Ecol 48:111–122. https://doi.org/10.1016/j.bse.2012.12.013

Díaz NB (1972) Una nueva plaga del alcornoque en la República Argentina. Revista de la Sociedad Entomológica Argentina 34(1–2):85–88

Duffett GH (1969) Some new inter-relationships of Hymenoptera over-wintering within the galls of *Andricus kollari* (Hartig). Entomol Month Magaz 105:1259–1261

Espadaler X, Nieves-Aldrey JL (1983) Hormigas (Hymenoptera, formicidae) pobladoras de agallas abandonadas de cinipidos (Hymenoptera, cynipidae) sobre Quercus SP en la Peninsula Iberica. Boletín de la Estación Central de Ecología 12:89–93

Fang Z, Tang C-T, Nicholls JA, Zhu Y, Xiong T, Hearn J, Sinclair F, Melika G, Nieves-Aldrey JL, Csóka G, Mikolajczak KM, Stone GN, Fang S (2020) A new genus of Oak Gallwasp, *Heocynips* Fang, Nieves-Aldrey, and Melika (Hymenoptera: Cynipidae: Cynipini), from China. Proc Entomol Soc Wash 122(4):787–804. https://doi.org/10.4289/0013-8797.122.4.787

Fernandes GW, Fagundes M, Woodman RL, Price PW (1999) Ant effects on three-trophic level interactions: plant, galls, and parasitoids. Ecol Entomol 24(4):411–415. https://doi.org/10.1046/j.1365-2311.1999.00218.x

Ferreira BG, Álvarez R, Avritzer SC, Isaias RMS (2017) Revisiting the histological patterns of storage tissues: beyond the limits of gall-inducing taxa. Botany 95(2):173–184. https://doi.org/10.1139/cjb-2016-0189

Ferreira BG, Álvarez R, Bragança GP, Alvarenga DR, Pérez-Hidalgo N, Isaias RMS (2019) Feeding and other gall facets: patterns and determinants in gall structure. Bot Rev 85(1):78–106. https://doi.org/10.1007/s12229-019-09207-w

Field LM, James AA, Plantard O, Rasplus J-Y, Mondor G, Le Clainche I, Solignac M (1999) Distribution and phylogeny of *Wolbachia* inducing thelytoky in Rhoditini and "Aylacini"(Hymenoptera: Cynipidae). Insect Mol Biol 8(2):185–191. https://doi.org/10.1046/j.1365-2583.1999.820185.x

Fockeu HLJB (1889) Contribution a l'histoire des galles: étude anatomique de quelques espèces. C. Robbe Lille, Paris, France, p 110

Folliot R (1964a) Contribution à l'étude de la biologie des Cynipides gallicoles (Hyménoptères, Cynipoidea). Ann Sci Nat Zool 12 ser. (6):407–564

Folliot R (1964b) Les insectes cécidogènes et la cécidogenèse. Traité de zoologie : anatomie, systématique, biologie 8:389–429

Folliot R (1977) Les insectes cécidogènes et la cécidogenèse. In: Grassé PP (ed) Traité de zoologie anatomie, systématique, biologie Vol. 8 (5-B): Embryologie, cécidogenise, insectes venimeux. Masson Editeurs, Paris, pp 389–429

Frankie GW, Morgan DL, Grissell EE (1992) Effects of urbanization on the distribution and abundance of the cynipid gall wasp, *Disholcaspis cinerosa*, on ornamental live oak in Texas, USA. In: Shorthouse JD, Rohfritsch O (eds) Biology of insect-induced galls. Oxford Univesity Press, New York/Oxford, pp 258–279

Garbin L, Durfort M, Diaz NB, Pujade-Villar J (2005) Histological modifications on Quercus suber Twigs (Fagaceae) caused by the Gallwasp Plagiotrochus suberi (Hymenoptera: Cynipidae). Entomologia Generalis 28(2):91–102. https://doi.org/10.1127/entom.gen/28/2005/91

Gätjens-Boniche O (2019) The mechanism of plant gall induction by insects: revealing clues, facts, and consequences in a cross-kingdom complex interaction. Rev Biol Trop 67(6):1359–1382. https://doi.org/10.15517/rbt.v67i6.33984

Gibson CM, Hunter MS (2010) Extraordinarily widespread and fantastically complex: comparative biology of endosymbiotic bacterial and fungal mutualists of insects. Ecol Lett 13(2):223–234. https://doi.org/10.1111/j.1461-0248.2009.01416.x

Giron D, Huguet E, Stone GN, Body M (2016) Insect-induced effects on plants and possible effectors used by galling and leaf-mining insects to manipulate their host-plant. J Insect Physiol 84:70–89. https://doi.org/10.1016/j.jinsphys.2015.12.009

Guiguet A, Tooker JF, Deans AR, Mikó I, Ning G, Schwéger S, Hines HM (2023) Comparative anatomy of venom glands suggests a role of maternal secretions in gall induction by cynipid wasps (Hymenoptera: Cynipidae). Insect Syst Diver 7(5):1–22. https://doi.org/10.1093/isd/ixad022

Hall CR, Carroll AR, Kitching RL (2017) A meta-analysis of the effects of galling insects on host plant secondary metabolites. Arthropod Plant Interact 11(4):463–473. https://doi.org/10.1007/s11829-016-9486-0

Harper LJ, Schönrogge K, Lim KY, Francis P, Lichtenstein CP (2004) Cynipid galls: insect-induced modifications of plant development create novel plant organs. Plant Cell Environ 27(3):327–335. https://doi.org/10.1046/j.1365-3040.2004.01145.x

Hartley SE (1998) The chemical composition of plant galls: are levels of nutrients and secondary compounds controlled by the gall-former? Oecologia 113(4):492–501. https://doi.org/10.1007/s004420050401

Hartley SE, Lawton JH (1992) Host-plant manipulation by gall-insects: a test of the nutrition hypothesis. J Anim Ecol 61(1):113–119. https://doi.org/10.2307/5514

Hearn J, Blaxter M, Schönrogge K, Nieves-Aldrey J-L, Pujade-Villar J, Huguet E, Drezen J-M, Shorthouse JD, Stone GN (2019) Genomic dissection of an extended phenotype: Oak gall-

ing by a cynipid gall wasp. PLoS Genet 15(11):e1008398. https://doi.org/10.1371/journal.pgen.1008398
Hebert PDN (1987) Genotypic characteristics of cyclic parthenogens and their obligately asexual derivatives. In: Stearns SC (ed) The evolution of sex and its consequences. Experientia Supplementum, Birkhäuser, Basel, pp 175–195
Houard CAV (1903) Recherches anatomiques sur les galles de tiges: Pleurocécidies. Bulletin scientifique de la France et de la Belgique 38:140–419
Hough JS (1953) Studies on the common Spangle Gall of Oak: Ii. a general consideration of past work on gall induction. New Phytol 52(3):218–228. https://doi.org/10.1111/j.1469-8137.1953.tb07512.x
Ide T, Koyama A (2023) The formation of a rolling larval chamber as the unique structural gall of a new species of cynipid gall wasps. Sci Rep 13(1):18149. https://doi.org/10.1038/s41598-023-43641-6
Isaias RMS, Oliveira DC, Carneiro RGS, Kraus JE (2014) Developmental anatomy of galls in the neotropics: arthropods stimuli versus host plant constraints. In: Fernandes GW, Santos JC (eds) Neotropical insect galls. Springer Netherlands, Dordrecht, pp 15–34
Jara-Chiquito JL, Dolader JH, i Villar JP (2022) La poda d'escarida primaveral com a tractament per a la reducció dels danys de la vespeta del castanyer i comentaris sobre altres mètodes de control utilitzats. Butlletí de la Institució Catalana d'Història Natural 86(1):13–24. https://doi.org/10.2436/20.1502.01.100
Jara-Chiquito J-L, Pujade-Villar J, Ferreira BG, Álvarez R (2021) Histological changes induced by the cynipid wasp Dryocosmus kuriphilus (Hymenoptera: Cynipidae) in leaves of the chestnut Castanea sativa (Fagaceae): mechanisms of galling impact on host vigor. Arthropod Plant Interact 15(2):223–233. https://doi.org/10.1007/s11829-021-09810-y
Jones D (1983) The influence of host density and gall shape on the survivorship of Diastrophus kincaidii Gill. (Hymenoptera: Cynipidae). Can J Zool 61(9):2138–2142. https://doi.org/10.1139/z83-280
Kariñho-Betancourt E, Hernández-Soto P, Rendón-Anaya M, Calderón-Cortés N, Oyama K (2019) Differential expression of genes associated with phenolic compounds in galls of Quercus castanea induced by Amphibolips michoacaensis. J Plant Interact 14(1):177–186. https://doi.org/10.1080/17429145.2019.1603404
Kinsey AC (1920) Life histories of American Cynipidae. Bull Am Mus Nat Hist 42:319–357, pls. 28–31. https://doi.org/10.5962/bhl.title.122602
Kinsey AC (1922) Studies of some new and described Cynipidae (Hymenoptera). Indiana Univ Stud 9:3–171
Kinsey AC (1937) New Mexican Gall Wasps (Hymenoptera, Cynipidae) II. Revista de Entomologia 74:428–471
Kot I, Jakubczyk A, Karaś M, Złotek U (2018) Biochemical responses induced in galls of three Cynipidae species in oak trees. Bull Entomol Res 108(4):494–500. https://doi.org/10.1017/S0007485317001055
Kovalev OV (1995) New taxa of fossil cynipoids (Hymenoptera, Cynipoidea) from the Cretaceous and Palaeogene. Amber Fossils 1(1):9–16
Lacaze-Duthiers H (1853) Recherches pour servir à l'historie des galles. Annales des Sciences Naturelles, Botanique 19:273–354
László Z, Tóthmérész B (2013) The enemy hypothesis: correlates of gall morphology with parasitoid attack rates in two closely related rose cynipid galls. Bull Entomol Res 103(3):326–335. https://doi.org/10.1017/S0007485312000764
LeBlanc DA, Lacroix CR (2001) Developmental potential of galls induced by *Diplolepis rosaefolii* (Hymenoptera: Cynipidae) on the leaves of *Rosa virginiana* and the influence of *Periclistus* species on the *Diplolepis rosaefolii* galls. Int J Plant Sci 162(1):29–46. https://doi.org/10.1086/317900
Lev-Yadun S, Inbar M (2023) The potential of visual and olfactory signals in gall defence. Indian J Entomol 86(1):287–294. https://doi.org/10.55446/IJE.2023.871

Lobato-Vila I, Pujade-Villar J (2019) Revision of world Ceroptresini (Hymenoptera: Cynipidae) with the description of a new genus and five new species. Zootaxa 4685(1):zootaxa-4685. https://doi.org/10.11646/zootaxa.4685.1.1

Lobato-Vila I, Cibrián-Tovar D, Barrera-Ruíz UM, Equihua-Martínez A, Estrada-Venegas EG, Buffington ML, Pujade-Villar J (2019) Review of the Synergus Hartig species (Hymenoptera: Cynipidae: Synergini) associated with tuberous and other tumor-like galls on oaks from the New World with the description of three new species from Mexico. Proc Entomol Soc Wash 121(2):193–255. https://doi.org/10.4289/0013-8797.121.2.193

Lobato-Vila I, Bae J, Roca-Cusachs M, Kang M, Jung S, Melika G, Pénzes Z, Pujade-Villar J (2022) Global phylogeny of the inquilinous gall wasp tribe Synergini (Hymenoptera: Cynipoidea: Cynipidae): first insights and establishment of a new cynipid tribe. Zool J Linnean Soc 195(4):1338–1354. https://doi.org/10.1093/zoolinnean/zlab085

Luz FA, de Mendonça MS Jr (2019) Guilds in insect galls: who is who. Fla Entomol 102(1):207–210. https://doi.org/10.1653/024.102.0133

Magalhães TA, Oliveira DC, Suzuki AYM, Isaias RMS (2014) Patterns of cell elongation in the determination of the final shape in galls of Baccharopelma dracunculifoliae (Psyllidae) on Baccharis dracunculifolia DC (Asteraceae). Protoplasma 251(4):747–753. https://doi.org/10.1007/s00709-013-0574-z

Maia VC (2013) Insect galls of São Tomé das Letras (MG, Brazil). Biota Neotropica 13:164–189. https://doi.org/10.1590/S1676-06032013000400017

Mani MS (1964) Ecology of plant galls. Springer, Netherlands, Dordrecht, p 434

Margalef R (1977) Ecología, 2nd edn. Omega, Barcelona, p 951

Markel K, Novak V, Bowen BP, Tian Y, Chen Y-C, Sirirungruang S, Zhou A, Louie KB, Northen TR, Eudes A, Scheller HV, Shih PM (2024) Cynipid wasps systematically reprogram host metabolism and restructure cell walls in developing galls. Plant Physiol 195(1):698–712. https://doi.org/10.1093/plphys/kiae001

Martinson EO, Werren JH, Egan SP (2022) Tissue-specific gene expression shows a cynipid wasp repurposes oak host gene networks to create a complex and novel parasite-specific organ. Mol Ecol 31(11):3228–3240. https://doi.org/10.1111/mec.16159

Massana-Canals N, Arnal J, Pujade-Villar J (2013) Dades preliminars de la fauna secundària associada a gales de la forma asexual d'Andricus hispanicus (Hartig, 1856)(Hymenoptera: Cynipidae). Butlletí de la Institució Catalana d'Història Natural 77:95–104

Melika G (2006) Gall wasps of Ukraine. Vols. 1 & 2. Vestnik Zoologii, Suppl 21(1–2):1–644

Melika G, Abrahamson W (2002) Review of the world genera of oak cynipid wasps (Hymenoptera: Cynipidae, Cynipini). In: Melika G, Thuróczy C (eds) Parasitic wasps: evolution, systematics, biodiversity and biological control. Agroinform, Budapest, pp 150–190

Melika G, Cibrián-Tovar D, Cibrián-Llanderal VD, Tormos J, Pujade-Villar J (2009a) New species of oak gallwasp from Mexico (Hymenoptera: Cynipidae: Cynipini)– a serious pest of *Quercus laurina* (Fagaceae). Dugesiana 16(2):67–73

Melika G, Pérez-Hidalgo N, Hanson P, Pujade-Villar J (2009b) New species of oak gallwasp from Costa Rica (Hymenoptera: Cynipidae: Cynipini). Dugesiana 16(1):35–39

Melika G, Nicholls JA, Abrahamson WG, Buss EA, Stone GN (2021) New species of Nearctic oak gall wasps (Hymenoptera: Cynipidae, Cynipini). Zootaxa 5084(1):1–131. https://doi.org/10.11646/zootaxa.5084.1.1

Meyer J, Maresquelle HJ (1983) *Anatomie des galles*. Encyclopedia of Plant Anatomy, Band XIII Teil 1, Berlin, Germany, p 662

Miller DG, Ivey CT, Shedd JD (2009) Support for the microenvironment hypothesis for adaptive value of gall induction in the California gall wasp, Andricus quercuscalifornicus. Entomol Exp Appl 132(2):126–133. https://doi.org/10.1111/j.1570-7458.2009.00880.x

Nicholls JA, Melika G, Stone GN (2017) Sweet tetra-trophic interactions: multiple evolution of nectar secretion, a defensive extended phenotype in Cynipid gall wasps. Am Nat 189(1):67–77. https://doi.org/10.1086/689399

Nicholls JA, Melika G, Digweed SC, Stone GN (2022) Pairing of sexual and asexual generations of Nearctic oak gallwasps, with new synonyms and new species names (Hymenoptera: Cynipidae, Cynipini). Zootaxa 5145(1):1–79. https://doi.org/10.11646/zootaxa.5145.1.1

Oliveira DC, Isaias RMS (2010) Redifferentiation of leaflet tissues during midrib gall development in *Copaifera langsdorffii* (Fabaceae). S Afr J Bot 76(2):239–248. https://doi.org/10.1016/j.sajb.2009.10.011

Oliveira DC, Isaias RMS, Fernandes GW, Ferreira BG, Carneiro RGS, Fuzaro L (2016) Manipulation of host plant cells and tissues by gall-inducing insects and adaptive strategies used by different feeding guilds. J Insect Physiol 84:103–113. https://doi.org/10.1016/j.jinsphys.2015.11.012

Patterson JT (1928) Sexes in the Cynipidæ and male-producing and female-producing lines. Biol Bull 54(3):201–211. https://doi.org/10.2307/1536853

Pénzes Z, Tang C-T, Stone GN, Nicholls JA, Schwéger S, Bozsó M, Melika G (2018) Current status of the oak gallwasp (Hymenoptera: Cynipidae: Cynipini) fauna of the Eastern Palaearctic and oriental regions. Zootaxa 4433(2):245–289. https://doi.org/10.11646/zootaxa.4433.2.2

Plantard O, Hochberg ME (1998) Factors affecting parasitism in the Oak-Galler Neuroterus quercusbaccarum (Hymenoptera: Cynipidae). Oikos 81(2):289–298. https://doi.org/10.2307/3547049

Plantard O, Solignac M (1998) *Wolbachia*-induced thelytoky in cynipids. In: Csóka G, Mattson WJ, Stone GN, Price PW (eds) The biology of Gall-inducing arthropods. U.S. Department of Agriculture, Forest Service, North Central Research Station, Minnesota, pp 111–121

Plantard O, Rasplus JY, Hochberg ME (1996) Resource partitioning in the parasitoid assemblage of the oak galler Neuroterus quercusbaccarum L.(Hymenoptera: Cynipidae). Acta Oecol 17(1):1–15

Plantard O, Rasplus J-Y, Mondor G, Clainche IL, Solignac M (1998) *Wolbachia*–induced thelytoky in the rose gallwasp *Diplolepis spinosissimae* (Giraud)(Hymenoptera: Cynipidae), and its consequences on the genetic structure of its host. Proc R Soc Lond Ser B Biol Sci 265(1401):1075–1080. https://doi.org/10.1098/rspb.1998.0401

Price PW, Pschorn-Walcher H (1988) Are galling insects better protected against parasitoids than exposed feeders?: a test using tenthredinid sawflies. Ecol Entomol 13(2):195–205. https://doi.org/10.1111/j.1365-2311.1988.tb00347.x

Price PW, Waring GL, Fernández GW (1986) Hypotheses on the adaptive nature of galls. Proc Entomol Soc Wash 88(22):361–363

Price PW, Fernandes GW, Waring GL (1987) Adaptive nature of insect galls. Environ Entomol 16(1):15–24. https://doi.org/10.1093/ee/16.1.15

Price PW, Craig TP, Roininen H (1995) Working toward theory on galling sawfly population dynamics. In: Cappuccion N, Price PW (eds) Population dynamics: new approaches and synthesis. Academic Press, San Diego, pp 321–338

Pujade-Villar J (1987) Organografia i malalties del cecidi de *Diastrophus rubi* (Bouché, 1834) Hartig, 1840 (Hymenoptera: Cynipoidea: Cynipinae) a la tija de l'esbarzer, *Rubus caesius* L. (Rosaceae). Butlletí de la Institució Catalana d'Història Natural 54:65–71

Pujade-Villar J (1991) Consideracions sobre la morfologia de la gal· la d'Andricus kollari (Htg) (Insecta: Hymenoptera: Cynipidae)(1a part). La Sitja del Llop, Revista del Montseny 2:8

Pujade-Villar J (1994) Estratègia defensiva de la fase larvària d'*Andricus viscosus* f. a. (Hym., Cynipidae). Aigüerola 1:10

Pujade-Villar J (2013) Las agallas de los encinos: un ecosistema en miniatura que hace posible estudios multidisciplinares. Entomologia Mexicana 12(1):1–20

Pujade-Villar J (2014) Espèces de Cynipini (Hymenoptera, Cynipidae) nuisibles ou potentiellement nuisibles aux forêts de *Quercus*. Integrated Protection in Oak Forests IOBC-WPRS Bulletin 101:195–203

Pujade-Villar J, Bellido D, López GS, Melika G (2001) Current state of knowledge of heterogony in Cynipidae (Hymenoptera, Cynipoidea). Sessió Conjunta d'Entomologia:87–107

Pujade-Villar J, Grami M, Ben Jamâa ML (2010) The gall wasp *Plagiotrochus amenti* potentially dangerous for the cork oak found for the first time in Tunisia. Tunis J Plant Prot 5:225–230

Pujade-Villar J, Equihua-Martínez A, Estrada-Venegas EG, Lomelí-Flores JR, Serrano-Muñoz M, Cabral O, Treto R, Landa L, Carrillo C, Cibrián-Tovar D, Cibrián-Llanderal VD (2012) Aportaciones de 2010-2011 en el conocimiento de los Cynipidae mexicanos (Hym., Cynipidae, Cynipini) y perspectivas futuras. Entomología Mexicana 11(2):1057–1062

Pujade-Villar J, Cibrián-Tovar D, Barrera-Ruíz UM, Cuesta-Porta V (2018) Las especies de *Neuroterus* (Hartig, 1840) de México, con la descripción de una especie nueva (HYM., CYNIPIDAE). Entomología Mexicana 5:453–461

Quicke DL (1997) Parasitic wasps. Chapman & Hall Ltd, London, p 470

Quicke DLJ, Fitton MG, Tunstead JR, Ingram SN, Gaitens PV (1994) Ovipositor structure and relationships within the Hymenoptera, with special reference to the Ichneumonoidea. J Nat Hist 28(3):635–682. https://doi.org/10.1080/00222939400770301

Quicke DL, Wyeth P, Fawke JD, Basibuyuk HH, Vincent JFV (1998) Manganese and zinc in the ovipositors and mandibles of hymenopterous insects. Zool J Linnean Soc 124(4):387–396. https://doi.org/10.1111/j.1096-3642.1998.tb00583.x

Redfern M, Askew RR (1992) Plant galls. Richmond Publishing, Slough, UK, 99 pp.

Rey L (1966) La galle de *Biorhiza pallida* Ol. Étude préliminaire du bourgeon de *Quercus pedunculata* Ehrh. au moment de l'attaque du parasite. Bulletin de la Société Botanique de France 113(9):503–514. https://doi.org/10.1080/00378941.1966.10835558

Rey L (1967) La galle de *Biorhiza pallida* Bosc. Sur la racine de *Quercus pedunculata* Ehrh.: observation de la ponte et de les premières réactions de l'hôte. Bulletin de la Société Botanique de France 113(9):503–514. https://doi.org/10.1080/00378941.1966.10835558

Rey L (1971) La galle de *Biorhiza pallida* Ol., stades ultérieurs de développement. Marcellia 37:193–218

Rezende UC, Gonçalves PHP, Kuster VC, Carneiro RGS, Oliveira DC (2023) Every end is a new beginning: histological features of galls induced on Macairea radula (Melastomataceae) allow a post-senescence colonization. Flora 306:1–10. https://doi.org/10.1016/j.flora.2023.152369

Robbins CT, Mole S, Hagerman AE, Hanley TA (1987) Role of Tannins in defending plants against ruminants: reduction in dry matter digestion? Ecology 68(6):1606–1615. https://doi.org/10.2307/1939852

Rohfritsch O (1992) Patterns in gall development. In: Shorthouse JD, Rohfritsch O (eds) Biology of insect-induced galls. Oxford University Press, New York, pp 60–86

Rokas A, Atkinson RJ, Brown GS, West SA, Stone GN (2001) Understanding patterns of genetic diversity in the oak gallwasp Biorhiza pallida: demographic history or a Wolbachia selective sweep? Heredity 87(3):294–304. https://doi.org/10.1046/j.1365-2540.2001.00872.x

Ronquist F (1994) Evolution of parasitism among closely related species: phylogenetic relationships and the origin of inquilinism in gall wasps (Hymenoptera, Cynipidae). Evolution 48(2):241–266. https://doi.org/10.1111/j.1558-5646.1994.tb01310.x

Ronquist F (1999) Phylogeny, classification and evolution of the Cynipoidea. Zool Scr 28(1–2):139–164. https://doi.org/10.1046/j.1463-6409.1999.00022.x

Ronquist F, Nieves-Aldrey J-L, Buffington ML, Liu Z, Liljeblad J, Nylander JAA (2015) Phylogeny, evolution and classification of gall wasps: the plot thickens. PLoS One 10(5):e0123301. https://doi.org/10.1371/journal.pone.0123301

Ros-Farré P, Pujade-Villar J (1998) Relacions tròfiques en gal· les de la forma agàmica d'*Andricus kollari* (Hartig, 1843). Butlletí de la Institució Catalana d'Història Natural 66:95–108

Rossi AM, Stiling PD, Strong DR, Johnson DM (1992) Does gall diameter affect the parasitism rate of Asphondylia borrichiae (Diptera: Cecidomyiidae)? Ecol Entomol 17(2):149–154. https://doi.org/10.1111/j.1365-2311.1992.tb01172.x

Russo R (2007) Field guide to plant galls of California and other Western states. University of California Press, Berkerley, p 415

Schönrogge K, Stone GN, Cockrellt B, Crawley MJ (1994) The communities associated with the galls of Andricus quercuscalicis (Hymenoptera: Cynipidae) an invading species in Britain: a geographical view. In: Williams MAJ (ed) Plant galls: organisms, interactions, populations. Oxford University Press, pp 369–390

Schönrogge K, Stone GN, Crawley MJ (1995) Spatial and temporal variation in guild structure: Parasitoids and inquilines of Andricus quercuscalicis (Hymenoptera: Cynipidae) in its native and alien ranges. Oikos 72(1):51–60. https://doi.org/10.2307/3546037

Schönrogge K, Stone GN, Crawley MJ (1996a) Alien herbivores and native parasitoids: rapid developments and structure of the parasitoid and inquiline complex in an invading gall wasp Andricus quercuscalicis (Hymenoptera: Cynipidae). Ecol Entomol 21(1):71–80. https://doi.org/10.1111/j.1365-2311.1996.tb00268.x

Schönrogge K, Stone GN, Crawley MJ (1996b) Abundance patterns and species richness of the Parasitoids and inquilines of the Alien Gall-Former Andricus quercuscalicis (Hymenoptera: Cynipidae). Oikos 77(3):507–518. https://doi.org/10.2307/3545940

Schönrogge K, Harper LJ, Brooks SE, Shorthouse JD, Lichtenstein CP (1998) Reprogramming plant development: two approaches to study the molecular mechanism of gall formation. In: Csóka G, Mattson WJ, Stone GN, Price PW (eds) The biology of gall-inducing arthropods. USDA, St. Paul, MN, pp 153–160

Schönrogge K, Harper LJ, Lichtenstein CP (2000) The protein content of tissues in cynipid galls (Hymenoptera: Cynipidae): similarities between cynipid galls and seeds. Plant Cell Environ 23(2):215–222. https://doi.org/10.1046/j.1365-3040.2000.00543.x

Schultz JC, Stone GN (2022) A tale of two tissues: probing gene expression in a complex insect-induced gall. Mol Ecol 31(11):3031–3034. https://doi.org/10.1111/mec.16482

Seibert TF (1993) A nectar-secreting gall wasp and ant mutualism: selection and counter-selection shaping gall wasp phenology, fecundity and persistence. Ecol Entomol 18(3):247–253. https://doi.org/10.1111/j.1365-2311.1993.tb01097.x

Shorthouse JD, Brooks SE (1998) Biology of the galler *Diplolepis rosaefolii* (Hymenoptera: Cynipidae), its associated component community, and host shift to the shrub rose thérèse bugnet. Canad Entomol 130(3):357–366. https://doi.org/10.4039/Ent130357-3

Shorthouse JD, Rohfritsch O (eds) (1992) Biology of insect-induced galls. Oxford University Press, New York, Oxford, 285 pp.

Stille B (1984) The effect of hosptlant and parasitoids on the reproductive success of the parthenogenetic gall wasp Diplolepis rosae (Hymenoptera, Cynipidae). Oecologia 63(3):364–369. https://doi.org/10.1007/BF00390666

Stone GN, Cook JM (1998) The structure of cynipid oak galls: patterns in the evolution of an extended phenotype. Proc R Soc Lond B Biol Sci 265(1400):979–988. https://doi.org/10.1098/rspb.1998.0387

Stone GN, Schönrogge K (2003) The adaptive significance of insect gall morphology. Trends Ecol Evol 18(10):512–522. https://doi.org/10.1016/S0169-5347(03)00247-7

Stone GN, Schönrogge K, Crawley MJ, Fraser S (1995) Geographic and between-generation variation in the parasitoid communities associated with an invading gallwasp, Andricus quercuscalicis (Hymenoptera: Cynipidae). Oecologia 104(2):207–217. https://doi.org/10.1007/BF00328585

Stone GN, Schönrogge K, Atkinson RJ, Bellido D, Pujade-Villar J (2002) The population biology of oak gall wasps (Hymenoptera: Cynipidae). Annu Rev Entomol 47(1):633–668. https://doi.org/10.1146/annurev.ento.47.091201.145247

Stone GN, Atkinson RJ, Rokas A, Nieves-Aldrey JL, Melika G, Acs Z, Csoka G, Hayward A, Bailey R, Buckee C (2008) Evidence for widespread cryptic sexual generations in apparently purely asexual Andricus gallwasps. Mol Ecol 17(2):652–665. https://doi.org/10.1111/j.1365-294X.2007.03573.x

Stone GN, Hernandez-Lopez A, Nicholls JA, Di Pierro E, Pujade-Villar J, Melika G, Cook JM (2009) Extreme host plant conservatism during at least 20 million years of host plant pursuit by oak gallwasps. Evol Int J Organ Evol 63(4):854–869. https://doi.org/10.1111/j.1558-5646.2008.00604.x

Stouthamer R, Breeuwer JAJ, Hurst GDD (1999) *Wolbachia pipientis:* microbial manipulator of arthropod reproduction. Ann Rev Microbiol 53(1):71–102. https://doi.org/10.1146/annurev.micro.53.1.71

Suzuki AYM, Bedetti CS, dos Isaias RMS (2015) Detection and distribution of cell growth regulators and cellulose microfibrils during the development of *Lopesia* sp. galls on *Lonchocarpus cultratus* (Fabaceae). Botany 93(7):435–444. https://doi.org/10.1139/cjb-2015-0012
Tataroğlu M, Aksoy ÖK, Katılmış Y, Mammadov R (2023) Determination of secondary metabolite in galls of some cynipid wasps (Hymenoptera: Cynipidae) and characterization of the phenolic compound. Int J Sec Metabol 10(4):555–569. https://doi.org/10.21448/ijsm.1300763
Torossian C (1971a) Faune secondaire des galles de cynipidae: I. — Étude systématique des fourmis et des principaux arthropodes récoltés dans les galles. Insect Soc 18(3):135–154. https://doi.org/10.1007/BF02223806
Torossian C (1971b) Étude biologique des fourmis forestières peuplant les galles de cynipidae des chênes. Insect Soc 18(4):193–201. https://doi.org/10.1007/BF02223174
Torossian C (1972) Étude biologique des fourmis forestières peuplant les galles de Cynipidae des chênes: III. — Rôle et importance numérique des femelles fondatrices. Insect Soc 19(1):25–38. https://doi.org/10.1007/BF02223213
Ward AK, Bagley RK, Egan SP, Hood GR, Ott JR, Prior KM, Sheikh SI, Weinersmith KL, Zhang L, Zhang YM (2022) Speciation in Nearctic oak gall wasps is frequently correlated with changes in host plant, host organ, or both. Evolution 76(8):1849–1867. https://doi.org/10.1111/evo.14562
Washburn JO (1984) Mutualism between a Cynipid gall wasp and ants. Ecology 65(2):654–656. https://doi.org/10.2307/1941429
Wiebes-Rijks AA (1979) The identity of the gall-wasp causing stunted acorns (Hymenoptera, Cynipidae). Netherlands J Zool 30(2):243–253. https://doi.org/10.1163/002829679X00403
Zuparko RL (1996) Hymenoptera reared from *Plagiotrochus suberi* (Hymenoptera: Cynipidae) galls in California. Pan Pac Entomol 72(1):27–30

The Structural Development and Functional Traits of Diptera-Induced Galls

Elaine Cotrim Costa

1 Introduction

Many organisms have acquired the ability to redirect plant development to induce galls (Schultz et al. 2019; Harris and Pitzschke 2020). The emergence of gall development habitats in plants represents a fascinating evolutionary phenomenon, having independently evolved multiple times across diverse taxa (Miller III and Raman 2019; Labandeira 2021). Among these, Diptera galls stand out as a notable example, with their origins dating back to the Upper Cretaceous period, coinciding with the major radiation of angiosperms (Dorchin et al. 2019). Within the order Diptera, the Cecidomyiidae emerges as one of the largest families, with approximately 6600 described species, and thousands of undescribed species worldwide (Gagné and Jaschhof 2021). These insects manipulate the development patterns of host plants and form diverse and complex galls, with the structures often exhibiting repetitive and symmetrical patterns in nature (Raman 2007).

Anatomical, histochemical, chemical, and immunocytochemical investigations tracking gall development in both host organs and galls have provided valuable insights into the influence of the feeding habitat of the gall-inducing insects. The morphogenetic potentialities and constraints of host plants in the new structural-functional profiles in the Diptera galls (Isaias et al. 2014, 2024; Ferreira et al. 2019) provide food, shelter, and protection against environmental conditions (Stone and Schönrogge 2003) to the gall-inducing insect. In the last century, the Diptera gall development has been documented in classical literatures (Meyer 1957; Mani 1964; Meyer and Maresquelle 1983; Meyer 1987; Bronner 1992; Rohfritsch 1992). In this chapter, we revisit these classical studies, alongside integrating contemporary

E. C. Costa (✉)
Instituto de Ciências Biológicas, Universidade Federal do Rio Grande (FURG), Rio Grande, Rio Grande do Sul, Brazil
e-mail: elainecosta@furg.br

D. C. de Oliveira, R. M. dos Santos Isaias (eds.), *Plant Galls*,
https://doi.org/10.1007/978-3-031-80064-1_16

findings and insights, to provide a current view on the state of knowledge regarding the structural and functional traits during Diptera gall development. Specifically, the chapter encompasses an array of topics, including the signaling mechanisms implicated in gall developmental stages, the structural and functional features of gall tissues, and the intriguing phenomenon of ambrosia galls. Moreover, the chapter addresses the influence of abiotic factors on the structural and functional traits of Diptera galls, shedding light on their role as dynamic microenvironments within broader ecological contexts.

2 Signaling Mechanisms for Gall Development

Cecidologists have been seeking to understand the initial triggers in the formation of insect galls since the last century. Gall-inducing insects are believed to have the ability to secrete certain effectors, including phytohormones, into plant tissues using their mouthparts or ovipositors to induce gall formation in the host plants (Desnitskiy et al. 2023). Additionally, the induction process can be triggered by the insertion of exogenous genetic elements into the genome of plant cells through an endosymbiotic bacterium originating from the gall-inducing Diptera (Gätjens-Boniche 2019; Gätjens-Boniche et al. 2023). To induce complex gall structures in plant tissues, gall-inducing Diptera must manipulate the cellular machinery of their host plants, including development, metabolism, chemistry, and physiology to hijack host gene expression programs in favor of those of the insect (Takeda et al. 2021). The mechanical and/or chemical stimuli of the egg or larva when it hatches and feeds on plant tissues appears to be the initial trigger for the chemical and molecular responses that promote hyperplasia and cell hypertrophy, forming the gall (Rohfritsch 1992). Studies with Diptera galls have demonstrated that the phytohormones, such as indole-3-acetic acid (IAA) (Bedetti et al. 2014, 2017; Suzuki et al. 2015), cytokinins (CKs) (Bedetti et al. 2018), abscisic acid, jasmonic acid, and salicylic acid (Tooker and Helms 2014) are involved in the gall induction, as well as in gall developmental processes (De Bruyn et al. 1998; Tanaka et al. 2013; Gätjens-Boniche 2019). In addition to phytohormones, the phenolic compounds and reactive oxygen species (ROS) seem to participate in the gall induction mechanism (Raman 2011; Formiga et al. 2011; Bartlett and Connor 2014; Tooker and Helms 2014). Other indications of the participation of phytohormones in gall development rely on the high levels of IAA and trans-zeatin family CKs in the bodies and salivary glands of the cecidomyiid and tephritid larvae (Tanaka et al. 2013; Andreas et al. 2020; Ponce et al. 2021). General hypothesis suggests that gall formation is triggered by the action of chemical substances secreted by the gall inducer, including plant growth regulators such as IAA, CKs and other types of compounds. During gall induction, there is an increase in ROS and alterations in protein content, as a hypersensitive defense response of the host plant i.e. localized cell death and necrosis around the induction site to isolate the herbivore (Fernandes 1990; Detoni et al. 2011). However, the ROS, which have the role of blocking auxin activity (Suzuki et al. 2015; Isaias et al.

2015), are captured by antioxidant molecules such as flavonoids, allowing auxin activity and cell hypertrophy (Bedetti et al. 2014, 2017), a key process in gall development (Rohfritsch 1992).

Molecular studies have indicated that a considerable number of phytohormone-regulatory genes are upregulated in developing Diptera galls (Shih et al. 2018; Takeda et al. 2021; Desnitskiy et al. 2023). Commonly, genes related to developmental processes, cell cycle, cell wall compound synthesis, ROS generation and phytohormones are upregulated along gall development (Takeda et al. 2019). In addition, Darwin proposed that the shapes of complex insect galls resemble those of flowers or fruits (Darwin 1868). Indeed, many remarkable flower- and fruit-like structures have been observed in insect galls in various host plant species, suggesting that the formation of gall tissues is similar to the development of flowers or fruits (Schultz et al. 2019). This hypothesis has been partly supported by comparative transcriptome analysis using four different insect galls induced on leaves, among them those induced by *Rhopalomyia yomogicola* Matsumura, 1931 (Cecidomyiidae) on *Artemisia montana* (Nakai) Pampanini (Asteraceae) leaves (Takeda et al. 2019). A comparison of four transcriptomes of gall tissues showed that among the several thousand genes analyzed, there were about 38 "common genes" active in all four cases of gall development (Takeda et al. 2019). In particular, several key peptide regulators in plant growth and development (*CLE44*, *BAM3*, *WOXs*), and maintenance floral organ development (*SEPALLATA*, *AGAMOUS*, and *APETALA1*) were commonly upregulated in *A. montana* galls (Takeda et al. 2019, 2021). In contrast, the photosynthetic genes are downregulated in galls (Shih et al. 2018; Desnitskiy et al. 2023), supporting the hypothesis that galls are converted to sink organs rather than source organs on leaves (Fig. 1).

3 Anatomical Development of the Galls

Gall development involves the increased cell proliferation (hyperplasia), cell hypertrophy, tissue neoformation, and cell redifferentiation (Mani 1964; Rohfritsch 1992; Isaias et al. 2024). Dipteran gall ontogenesis is influenced by the feeding habits (Rohfritsch 1992), life cycles (Carneiro et al. 2017; Costa et al. 2021a) and sex (Gonçalves et al. 2022), but also by the induction site, the origin of cell lineages that forms the gall tissues (Mani 1964; Bragança et al. 2020b), and by the morphogenetical potentialities of the host plants (Isaias et al. 2014; Formiga et al. 2015; Amorim et al. 2017). In addition, the environment (biotic and abiotic interactions) also influences the gall development (Nobrega et al. 2021; Costa et al. 2022b; Rezende et al. 2022; Isaias et al. 2024). Classically, gall development comprises the induction, growth and development, maturation, dehiscence (Rohfritsch 1992) and/or senescent stages (Arduin and Kraus 1995; Kraus et al. 1996). The structural and functional arrangement of galls influence the survivorship of the inducers, providing a more nutritious source, and a shelter against environmental conditions (Stone and Schönrogge 2003).

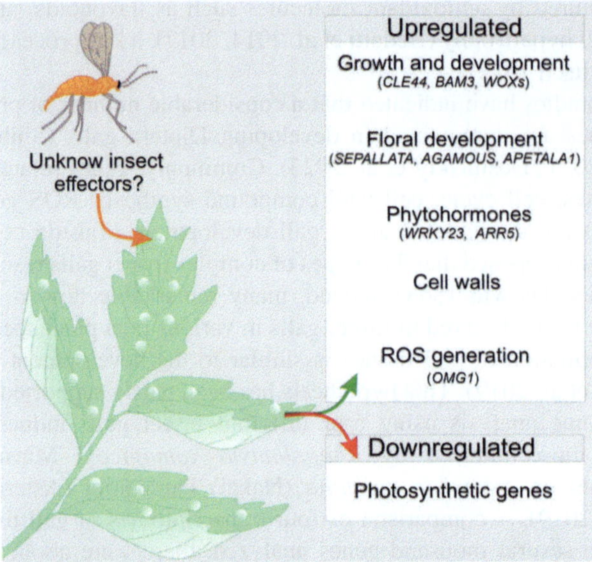

Fig. 1 Schematic representation of the "common genes" upregulated and downregulated during the gall development induced by *Rhopalomyia yomogicola* Matsumura, 1931 (Cecidomyiidae) on *Artemisia montana* (Nakai) Pampanini (Asteraceae) leaves

3.1 Induction Stage

Gall induction depends on synchronization strategies of the life cycle of gall-inducing Diptera with available induction sites and food resources during the phenology of their host plants (Yukawa 2000; Tokuda et al. 2014; Oliveira et al. 2016; Yukawa and Uechi 2021). Gall-inducing Diptera can have univoltine (Nogueira et al. 2022), bivoltine or multivoltine life cycles (Mendonça Jr. and Romanowski 2012; Oliveira et al. 2013; Costa et al. 2021a) and can induce galls on Bryophyta, Monilophyta, Lycophyta, Gymnosperms and angiosperms (Mani 1964). Among angiosperms, Eudicotyledones, especially Asteraceae, Fabaceae, and Myrtaceae, are the main hosts of Cecidomyiidae-Diptera galls in the world (Maia 2018; Santos-Silva and Araújo 2020; Gagné and Jaschhof 2021; Flor et al. 2022). This pattern may be associated with the richness and wide distribution of these botanical families in different biomes around the world, as well as the evolution of the secondary metabolites of these families. For example, the cecidomyiid adult females in some cases use sensitive olfactory systems to detect relevant host volatiles and avoid unsuitable hosts, and choose their host plant (Hall et al. 2012; Molnár et al. 2018). Typically, the most species of gall-inducing Diptera exhibit monophagy, inducing galls on one plant single species, thus maintaining a species-specific relationship (Yukawa et al. 2021). In this case, each gall has a peculiar and conservative ontogenesis, resulting in a specific morphotype induced in one single plant species (Suzuki et al. 2015), or in distinct gall morphotypes induced by different Diptera species on the super-host plant species (Oliveira et al. 2013; Amorim et al. 2017; Costa et al. 2022a; Teixeira et al. 2022). However, some gall-inducing Diptera are oligophagous, inducing galls on one plant genus or one plant family, and a few species are polyphagous and induce galls on different plant families (Yukawa et al. 2021).

Diptera galls are induced on all plant groups, affecting both vegetative and reproductive organs (Mani 1964; Haarder et al. 2016). Young organs are primarily targeted for gall induction by Diptera due to their undifferentiated and meristematic cells, which possess a high capacity for cell division and cellular redifferentiation (Meyer 1987). While less common, Diptera can also induce galls in mature organs, facilitated by the dedifferentiation capacity of mature parenchyma cells (Oliveira and Isaias 2009, 2010; Oliveira et al. 2013). Upon locating the induction site on the host plant organ, Diptera females can oviposit either on the epidermal cells or internally among parenchyma cells that originate from the ground or vascular meristems (procambium and vascular cambium) (Rohfritsch 1992; Formiga et al. 2011; Raman 2003). In the first scenario, the larvae hatch and initially feed on the epidermal cells, then become covered by the gall tissues, forming what are known as "cover galls" (Rohfritsch 1992). In the second scenario, the larvae penetrate ground tissues of the host plant and subsequently induce gall formation, resulting in "mark galls" (Rohfritsch 1992). In both cases, the manipulation of larval feeding behavior causes the tissues of the host plant organ to cease their typical developmental pattern of differentiation and assume new fates, leading to gall formation (Arduin and Kraus 1995; Oliveira and Isaias 2010; Costa et al. 2018). The cells of the host organs undergo cytological rejuvenation and become metabolically active around the larva, in which have a cytoplasm with a higher abundance of organelles such as dictyosomes, ribosomes, and endoplasmic reticulum, as well as enlargement of the nucleus and nucleolus (Rohfritsch and Shorthouse 1982). This cell modification is referred to as "metaplasia" (Meyer 1957). Thus, the typical growth polarity of the host organ is changed, leading to slight cell hypertrophy around the induction site (Rohfritsch 1992). The removal of the larva during the induction stage results in the disappearance of cytological modifications, restoring the tissues to resemble those of the non-galled organ (Rohfritsch 1992).

Growth and Development Stage During the growth and development stage, a combination of cell division and cell growth occurs throughout the gall tissues (Rohfritsch 1992). Once the larva is enclosed by the growing tissues, further differentiation occurs within the cell layers, leading to the formation of the typical nutritive tissue, which marks the beginning of the growth and development stage of galls (Rohfritsch and Shorthouse 1982; Rohfritsch 1992). Gall development is attributed to the establishment of a cecidogenetic field from the typical nutritive tissue outward, maintained by the manipulation of the larval instars (Mani 1964; Raman 2003). The continuous feeding behavior of the Diptera larval instars in the nutritive cells is the stimulus that maintains all metabolic and functional dynamics necessary for gall growth. The formation of the ground system of the galls involves anticlinal and periclinal divisions of the host organ's ground system, with the accumulation of calcium oxalate crystals observed in the parenchyma and nutritive cells (Mani 1964; Kraus et al. 1996; Costa et al. 2018), potentially linked to chemical signaling (Isaias et al. 2015) and cell wall dynamics (Nogueira et al. 2022). Furthermore, it is at this stage that the parenchyma and/or procambial cells from the host organs originate the procambial strands of the galls, establishing the continuity of the gall's vascularization with that of the organ induction site (Meyer 1969; Bragança et al. 2020b). Cell division and expansion, and the differentiation of new vascular cells are mediated by phytohormones and ROS (Bragança et al. 2020a), as well as the structural and functional dynamics of cell wall components, such as pectins (Formiga et al. 2013; Teixeira et al. 2018), cellulose microfibrils (Suzuki et al. 2015), hemicelluloses, and glycoproteins (Nogueira et al. 2022), which provide the cell walls with greater flexibility during gall morphogenesis (Souza et al. 2024). Trichome overexpression is common in the gall epidermis, particularly in the ostiole region (Meyer 1987).

Maturation The mature stage is characterized by cell differentiation and the establishment of histo-functional compartments in the Diptera galls (Meyer 1987; Rohfritsch 1992; Bragança et al. 2017). The outer tissues of the galls make up the covering tissue, which may be the epidermis or periderms, storage parenchyma (reserve), primary or secondary vascular tissues and a mechanical tissue formed by cells with lignified secondary walls that limit the outer and inner tissue zones (Meyer 1987; Rohfritsch 1992). The outer tissues are associated with storage and transport of substances, as well as chemical and mechanical defense functions against biotic and abiotic factors (Stone and Schönrogge 2003). In some cases, the outer parenchyma cells perform efficient photosynthetic activity (Oliveira et al. 2011b; Kuster et al. 2022). The inner tissue layers located around the larval chamber constitutes the typical nutritive tissue that assumes the role of accumulation of primary metabolites, which are helpful for the diet of the Diptera larvae (Bronner 1992; Qi et al. 2020).

The tissue compartmentalization and metabolite accumulation are influenced by the complex mechanism of cell division and expansion, as well as the processes of cell hypertrophy and tissue hyperplasia, which are associated with the dynamic properties of the cell walls (Teixeira et al. 2018; Costa et al. 2021b; Nogueira et al. 2022; Souza et al. 2024). The dynamics of cell wall components also determine the final shape of the gall morphotype during its maturation. The gall final shape is associated with the anisotropic and isotropic dynamics of cell expansion (Oliveira and Isaias 2010) promoted by the cellulose microfibrils, which direct anticlinal, periclinal, and isodiametric cell elongation patterns (Suzuki et al. 2015; Bragança et al. 2020a). Diptera galls are mainly globoid (Fig. 2a) (Formiga et al. 2013;

Fig. 2 (continued) (Sapindaceae) leaf. (**g**) Rosette gall on *Declieuxia* sp. (Rubiaceae). (**h**) Leaf fold gall induced by undetermined species of *Clinodiplosis* on *Mikania hoehnei* B.L.Rob. (Asteraceae). (**i**). Marginal roll gall induced by *Lopesia marginalis* Maia, 2001 on *Couepia ovalifolia* (Schott.) Bentham (Chrysobalanaceae). (**j**) Clavate gall induced by undetermined species of *Lopesia* on *M. gemmulata*. (**k**) Cup-shaped galls on *C. oblongifolia* Mart. (Fabaceae). (**l**) Horn-shaped galls on *C. langsdorffii* Desf. (Fabaceae)

Fig. 2 Gall morphotypes induced by species of Cecidomiiydae (Diptera). (**a**) Globoid gall on *Copaifera luetzelburgii* Harms (Fabaceae) leaf. (**b**) Fusiform gall on *Lantana camara* L. (Verbenaceae) stem. (**c**) Lenticular galls on *Anacardium humile* A.St.-Hil. (Anacardiaceae) leaves. (**d**) Bivalve-shaped gall induced by undetermined species of *Lopesia* on *Mimosa gemmulata* Barneby (Fabaceae) leaves. (**e**) Conical gall on *Byrsonima* sp. (Malpighiaceae) leaf. (**f**) Cylindrical gall induced by *Paulliniamyia ampla* Maia, 2001 on *Paullinia weinmanniifolia* Mart.

Teixeira et al. 2022), fusiform (Fig. 2b) (Bragança et al. 2022), and lenticular (Fig. 2c) (Bedetti et al. 2014; Bragança et al. 2017), but they can also be bivalve-shaped (Fig. 2d) (Costa et al. 2022a; Nogueira et al. 2022), conical (Fig. 2e) (Amorim et al. 2017), cylindrical (Fig. 2f) (Souza et al. 2024), in a rosette (Fig. 2g) (Fleury et al. 2015), leaf fold (Fig. 2h) (Jayaraman 1989), marginal roll (Fig. 2i) (Staszak et al. 2023), clavate (Fig. 2j) (Costa et al. 2021a) and may also assume bizarre shapes, such as the cup-shaped gall (Fig. 2k) (Huang et al. 2015; Shih et al. 2018) and horn-shaped gall (Fig. 2l) (Carneiro et al. 2017).

In a general way, galling insects can induce only one gall morphotype on a host plant species. However, in some cases the gall shape changes along its development, as observed in the cecidomyiid galls associated with *Copaifera langsdorffii* Desf. (Fabaceae) leaves that are globoid in the growth and development stage. The activity of gall meristems promotes a change toward the horn-shape of the gall at the maturation stage (Carneiro et al. 2017). Nevertheless, the evidence of dimorphic and polymorphic galls (Ganaha et al. 2007; Mishima and Yukawa 2007; Lin et al. 2019) induced by a single species of Cecidomyiidae on a single host plant species is rare. The case of dimorphic galls may be related to sexual dimorphism (Mishima et al. 2014), in which the female and male manipulate the gall developmental pathways of different ways, resulting in two gall morphotypes (Fig. 3) (Gonçalves et al. 2022).

Senescence and Dehiscence After the larval instars, the Diptera enter the pupal stage, which is a period of diapause until the adult emerges from the gall. In some instances, Diptera development initiates within the galls, but the larvae eventually drop and pupate in the soil (Rohfritsch 1992), where the adults emerge. In other cases of host plant-Diptera galler interactions, leaves with developing galls undergo a process of local abscission (Fernandes et al. 2008), leading to the death of the gall-inducer. However, in some systems, a mechanism named the 'green island effect,' mediated by cytokinin signaling, occurs (Mapes and Davies 2001; Naseem et al. 2014; Tooker and Helms 2014). In the green island effect, the galls and adjacent regions remain photosynthetically active while the rest of the leaf senesces and decomposes, as seen in the two lenticular galls induced by an unidentified species of *Bruggmannia* (Cecidomyiidae) on *Neea madeirana* Standl. (Nyctaginaceae) (Fernandes et al. 2008). In this case, the development of the galls and galling insects also ends in the soil when the galls are not attacked by natural enemies, such as fungi (Fernandes et al. 2008).

Regardless of the gall development process, during the pupal stages the inducer stops feeding, causing the activities of the gall to cease, leading to dehiscence (Rohfritsch 1992) and/or senescence (Arduin and Kraus 1995). In the senescent process, starch grains are typically found in gall tissues after the gall inducer halts or reduces its feeding activity, which ends up in a reduction in gall cell metabolism (Bronner 1992; Kraus et al. 1996), resulting sometimes in cell death programmed to occur by the time of the end of the gall cycle (Oliveira et al. 2010). Consequently, this stage is often characterized by the collapse of nutritive cells and outer

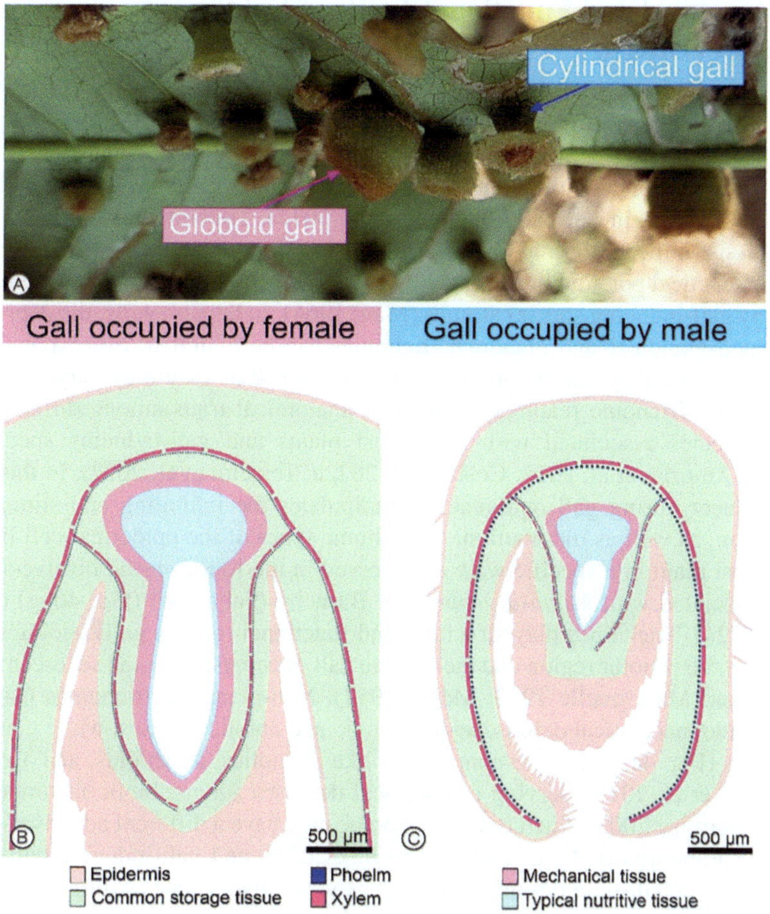

Fig. 3 Dimorphic galls of *Lopesia mataybae* Garcia and Urso-Guimaraes, 2018 (Cecidomyiidae) on *Matayba guianensis* Aubl. (Sapindaceae) leaflets. (**a**) Globoid gall occupied by female (pink arrow) and cylindrical gall occupied by male (blue arrow). (**b**) Schematic diagram of the gall occupied by female in cross section evidencing anatomical features. (**c**) Schematic diagram of the gall occupied by male in cross section evidencing anatomical features

parenchyma cells adjacent to the mechanical tissue. The collapse leads to loss of water, which seems to be linked to the gall opening mechanisms allowing the inducers to exit (Meyer and Maresquelle 1983). In other scenarios, the insects excavate an exit channel visible in the gall anatomy as a small hole, forming the ostiole in the gall wall (Mani 1964). The emergence of adults marks the culmination of the gall-inducing process and the beginning of a new generation of gall-inducing Diptera.

4 Structural and Functional Traits of Epidermis and Periderm

Gall outer epidermis and their appendages originate from the host organ's protoderm and serves as the primary defensive barrier (Fig. 4) (Meyer and Maresquelle 1983). It often contains pigments such as anthocyanins (Fig. 4a, b) (Mani 1964), which may result in red galls (Connor et al. 2012). Trichome development from trichoblasts in the epidermis of Diptera galls is a common trait, influenced by the morphogenetical potentialities of the host plants and environmental conditions, triggered by gall-inducing species stimuli (Mani 1964). In Diptera galls, the expression of the non-glandular and glandular trichomes can be inhibited, overexpressed, or neoformed, and they can be of various types, as well as they may vary in density (Fig. 4a–g). Trichome features are distinct anatomical traits among different gall morphospecies associated with super-host plants and gall-inducing species of Diptera (Amorim et al. 2017; Costa et al. 2022a; Teixeira et al. 2022). In this case, each inducer during gall ontogenesis manipulates the inhibiting and stimulating expression, as well as distribution, of trichome types of the epidermal cell lineage of the host plant in a specific way, as observed in the four *Lopesia* bivalve-shaped galls associated with *Mimosa gemmulata* Barneby (Fabaceae) (Fig. 4d–g) (Costa et al. 2022a). Trichomes play structural and functional roles in galls, aiding in gall closure in the ostiolar region and facilitating gall opening during the senescent stage (Meyer and Maresquelle 1983; Meyer 1987). Non-glandular trichomes (Fig. 4b) can provide mechanical defense against biotic and abiotic stresses, while glandular trichomes (Fig. 4c, h, i) store compounds like phenolics, terpenoids, and alkaloids (Nogueira et al. 2018), serving as chemical defenses against biotic factors (Stone and Schönrogge 2003). Each type of trichome may have a different adaptive significance (Karabourniotis et al. 2020; Muravnik 2021), and gall-inducing herbivores may control cell differentiation mechanisms toward diverse fates within a single cell lineage.

In the outer epidermis of Diptera galls, stomata differentiation may be repressed (Costa et al. 2022a) or result in malfunctioning (Fig. 4j) and lower density compared to non-galled organs (Sá et al. 2009; Castro et al. 2012; Nobrega et al. 2021; Teixeira et al. 2022). The epidermis also can contain stomata that are different in shape and size compared to stomata in non-galled leaf tissues (Guimarães et al. 2013). These stomata changes reduce gas exchange ability, configuring the new structural-functional patterns in the galls (Huang et al. 2015). Stomata in some Diptera galls facilitate tissue aeration to prevent hypercarbia (high CO_2 levels), while their absence may lead to compensatory mechanisms such as slower photosynthetic rates (Haiden et al. 2012; Huang et al. 2014, 2015; Pincebourde and Casas 2016). In galls associated with host plants of environments with water stress and high light (Arriola et al. 2018; Castro et al. 2023), the stomata may enhance gas exchange rates and tissue aeration, contributing to thermal balance through transpiration (Lawson and Blatt 2014).

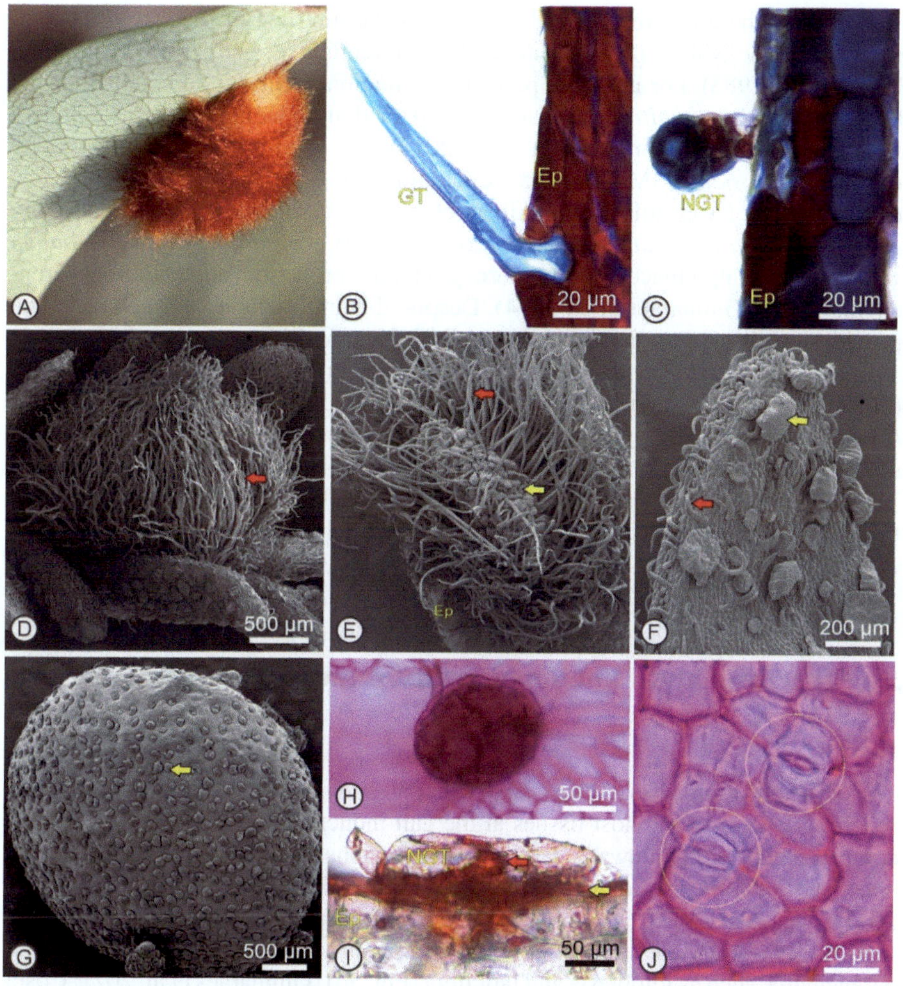

Fig. 4 Structural traits of the epidermis of Diptera galls. (**a**) Macroscopic aspect of the globoid galls induced by *Myrciariamyia admirabilis* Maia, 2007 on *Erythroxylum suberosum* A.St.-Hil. (Erythroxylaceae) showing the density of trichomes. (**b-c**) Transverse sections of cylindrical galls induced by *Paulliniamyia ampla* Maia, 2001 on *Paullinia weinmanniifolia* Mart. (Sapindaceae) leaves showing epidermal features. (**b**) Uniseriate epidermis showing the common epidermal cells accumulating phenolic compounds in the vacuole and the glandular trichomes. (**c**) Non-glandular trichomes. (**d-g**) Scanning electron micrographs of four bivalve-shaped galls induced by specific undetermined species of *Lopesia* on *Mimosa gemmulata* Barneby (Fabaceae). (**d**) Brown lanceolate bivalve-shaped gall with hypertrophied glandular multicellular filiform trichomes (red arrow). (**e**) Green lanceolate bivalve-shaped gall with glandular multicellular peltate (yellow arrow) and hypertrophied non-glandular unicellular filiform trichomes (red arrow). (**f**) Lenticular bivalve-shaped gall with non-glandular unicellular filiform (red arrow) and glandular multicellular peltate trichomes (yellow arrow). (**g**) Globoid bivalve-shaped gall with glandular multicellular peltate trichomes (yellow arrow). (**h**) Epidermis dissociation showing a glandular multicellular peltate trichome of globoid bivalve-shaped gall. (**i**) Detail of epidermis of globoid bivalve-shaped gall showing thin cuticle (yellow arrow) and lipid droplets in common epidermal cells and glandular trichomes (red arrow). (**j**) Detail of stomata by dissociation of the epidermis (yellow dotted circle). *Ep* epidermis, *GT* glandular trichomes, *NGT* non-glandular trichomes

Although the epidermis typically serves as the first line of defense in Diptera galls, in some galls' parenchyma cells differentiate a periderm (Meyer and Maresquelle 1983). For instance, periderm formation occurs in stem galls induced by *Tetradiplosis panghitruz* Martínez, 2013 (Cecidomyiidae) on *Prosopis caldenia* Burkart (Fabaceae) (Corró-Molas et al. 2014) and in midvein galls induced by *Odinadiplosis odinae* Mani, 1935 (Cecidomyiidae) on *Lannea coromandelica* (Houtt.) Merr. (Anacardiaceae) (Albert et al. 2013). In floral galls induced by *Bruggmanniella byrsonimae* Maia & Couri, 1993 (Cecidomyiidae) on *Byrsonima sericea* DC. (Malpighiaceae), a periderm gradually forms from subepidermal parenchyma layers (Guimarães et al. 2014). Despite the origin, the neoformed periderm provides mechanical protection to the galls (Meyer and Maresquelle 1983; Meyer 1987). Additionally, in the senescent stem-galls on *Eremanthus erythropappus* (DC.) MacLeish (Asteraceae) induced by stimuli of an undescribed species of *Neolasioptera* (Cecidomyiidae) occupied by ants and fungi, the new interaction seems to stimulate the differentiation of an ectopic phellogen (Jorge et al. 2022b). The ants apparently use gall tissues for food and shelter, but senescent gall tissues may be attacked by fungi that also serve as a food resource (Mani 1964; Jorge et al. 2022b).

5 Structural and Functional Traits of the Common Storage Tissue

The common storage tissue originates from parenchyma cells of the ground and/or vascular systems of the host tissues at the gall induction site (Rohfritsch 1992). It comprises cells with thin walls, large vacuole, peripheral cytoplasm with few organelles, and reduced intercellular spaces (Oliveira et al. 2010, 2011a; Ferreira et al. 2015; Costa et al. 2022c; Jorge et al. 2022a), forming a homogeneous tissue in most galls (Fig. 5a–c) (Mani 1964). Internal secretory structures differentiate within the parenchyma, such as idioblasts and laticifers (Fig. 5d) (Guimarães et al. 2021; Costa et al. 2022b). The structural traits of the common storage tissue are associated with its role as a sink and water storage (Meyer and Maresquelle 1983), as well as the accumulation of primary (Bronner 1992) and secondary metabolites (Hall et al. 2017; Kuster et al. 2020). Carbohydrates, proteins, and lipids accumulate in the protoplast and together with the activity of several enzymes maintain the gall's metabolism (Bragança et al. 2017; Guimarães et al. 2021). Water and energetic molecules accumulating increase gall succulence (tissue thickness), which may support the long lifespan of some species of Cecidomyiidae (Costa et al. 2021a, b, 2022a).

Depending on the taxonomic group of the host plant, the common storage tissue may also accumulate (poly)phenols, proanthocyanidins, alkaloids, and terpenoids, which contribute to the defense of the gall inducer against natural enemies (Hartley 1998), abiotic factors (Formiga et al. 2011; Hall et al. 2017; Kuster et al. 2020; Guimarães et al. 2021), antioxidant mechanisms (Isaias et al. 2015), and to

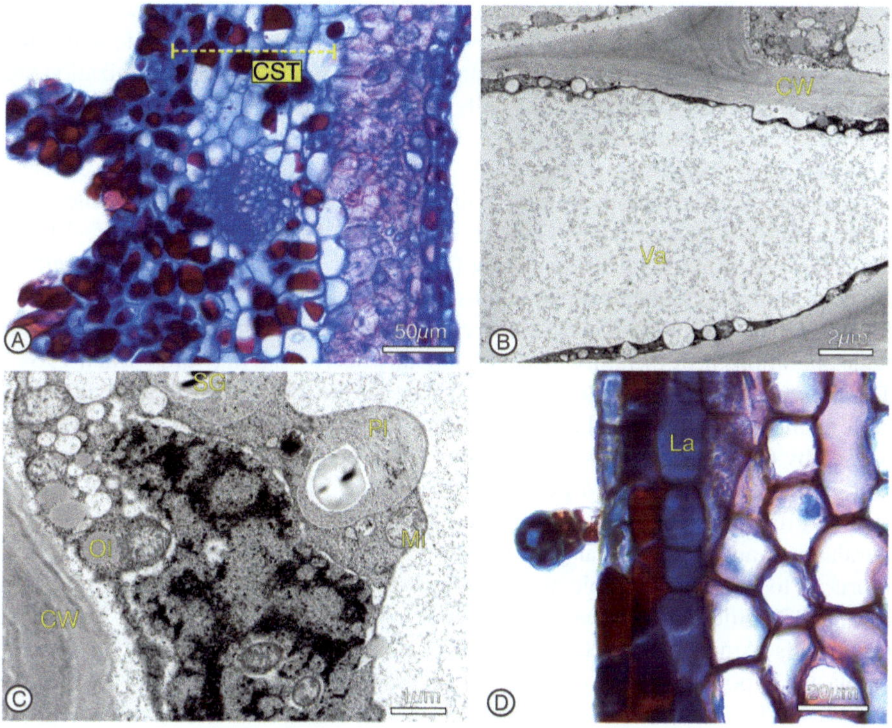

Fig. 5 Structural traits of the common storage tissue of Diptera galls. (**a-c**) Mature bivalve-shaped galls induced by an undetermined species of *Lopesia* on *Mimosa gemmulata* (Fabaceae). (**a**) General view of a transversely sectioned gall, highlighting the common storage tissue. (**b-c**) Cytological features evidencing thin cell walls, large vacuole, and peripheral cytoplasm with few organelles, such as mitochondria, plastids with starch grains, oleosomes, and free ribosomes. (**d**) Detail in transverse sections of cylindrical galls induced by *Paulliniamyia ampla* Maia, 2001 on *Paullinia weinmanniifolia* Mart. (Sapindaceae) leaves showing laticifers immersed in common storage tissue. *CST* common storage tissue, *CW* cell walls, *La* laticifers, *Mi* mitochondria, *Ol* oleosomes, *Pl* plastids, *SG* starch grains, *Va* vacuole

hormonal regulation during gall growth (Bedetti et al. 2014, 2017, 2018; Suzuki et al. 2015; Qi et al. 2020). For instance, (poly)phenols are typically detected during the early stages of gall development (Carneiro et al. 2017) and may inhibit the IAA-oxidases, thereby increasing auxin levels involved in cell expansion (Bedetti et al. 2014, 2017), via acidification of the cell wall (Bragança et al. 2020a). In addition, in Diptera galls, the herbivore-induced plant volatiles have been evidenced (Caselli et al. 2022), and the galling insects can change their host organs' chemical composition, either by inhibiting or eliciting chemical compound synthesis (Guimarães et al. 2013; Guedes et al. 2023a), which may have allelopathic potentials (Jorge et al. 2022c), antioxidant activity (Guedes et al. 2024), and attract natural enemies, indirectly influencing plant defenses (Izzo et al. 2006; Borges 2018).

The structural features of the common storage tissue are associated with the loss or reacquisition of the photosynthetic capacity in gall structure. In the leaf galls induced by Diptera, the loss of dorsiventrality and parenchyma homogenization are common responses, particularly in galls with extralaminar development (Oliveira and Isaias 2010; Amorim et al. 2017; Carneiro et al. 2017; Arriola et al. 2018; Bragança et al. 2020b; Costa et al. 2022a; Nogueira et al. 2022). However, galls with intralaminar development may maintain a continuity with the non-galled leaf lamina, palisade parenchyma (Oliveira et al. 2010; Bragança et al. 2017, 2020b; Teixeira et al. 2022), and collenchyma (Nobrega et al. 2021, 2022). This maintenance is attributed to the host plant constraints on the gall induction sites and to the feeding activity of gall inducers, which create a cecidogenic field with greater metabolic alterations near the gall inducers, gradually decreasing outwards (Mani 1964). The presence of chlorophyllian tissues whose plastids have well-developed thylakoid membranes indicates the preservation of the photosynthetic capacity, as evidenced by the similar photosynthetic rates observed in the Diptera galls and in the non-galled leaves of *Aspidosperma australe* Müll.Arg. (Apocynaceae) (Oliveira et al. 2011b). However, a decrease in photosynthetic rates typically occurs in Diptera intralaminar and extralaminar galls (Castro et al. 2012; Rezende et al. 2018), which is crucial to avoid hypercarbia and hypoxia (low O_2 levels) in gall tissues (Carneiro et al. 2017). Additionally, the rare condition of developing aerenchyma in cecidomyiid galls (Amorim et al. 2017; Teixeira et al. 2022) may also help to avoid hypercarbia and hypoxia since the intercellular spaces increase gas diffusion (Pincebourde and Casas 2016).

6 Structural and Functional Traits of the Mechanical Tissue

Although the formation of the mechanical tissue is a common structural trait of Diptera galls, in some cases, such as in flower galls (Ferreira and Isaias 2014; Guimarães et al. 2014) or in galls on aquatic host plants (Jäger-Zürn et al. 2013), it does not occur. The mechanical tissue is composed of thick-walled and lignified cells redifferentiated from the parenchyma or vascular cells (Mani 1964; Meyer and Maresquelle 1983; Rohfritsch 1992; Dorchin et al. 2002). Its formation starts with the deposition of the secondary cell walls, and its components, such as xylans, during gall growth and development stage, and it is completely established during the maturation stage (Fig. 6) (De Bruyn et al. 1998; Bragança et al. 2020a, b; Costa et al. 2021b, 2022a ; Nogueira et al. 2022), providing mechanical support to the gall structure. In some stem galls, the mechanical support is guaranteed by xylem cells (Corró-Molas et al. 2014; Jorge et al. 2022a).

The mechanical tissue is usually formed in the outer compartment of the gall in the subepidermal layers (Teixeira et al. 2022) or near the layer of vascular bundles, surrounding the typical nutritive tissue (Lalonde and Shorthouse 1984; Kraus et al. 1996; De Bruyn et al. 1998; Mishra and Patni 2008). Positioned near the feeding site of the cecidozoan, the lignified cells may remain alive, and maintain several

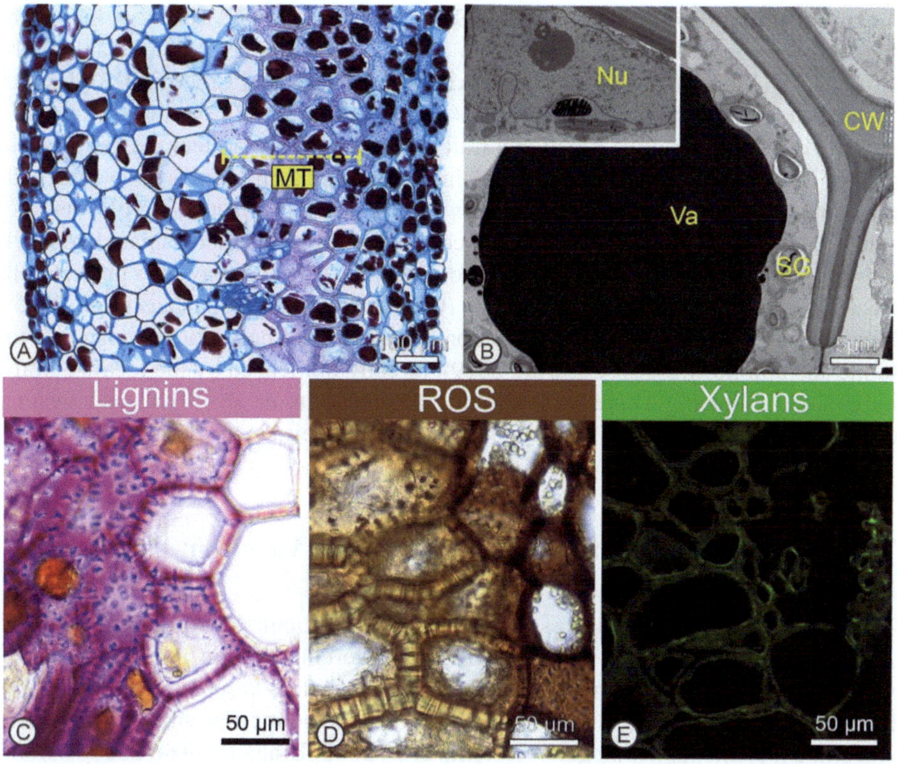

Fig. 6 The structural traits of the mechanical tissue of the bivalve-shaped galls induced by an undetermined species of *Lopesia* on *Mimosa gemmulata* (Fabaceae). (**a**) General view of a transversely sectioned gall, highlighting the mechanical tissue. (**b**) Cytological features evidencing thick cell walls, large vacuole with phenolics, and peripheral nucleus and cytoplasm with mitochondria, plastids with starch grains, and free ribosomes. (**c**) Detail of the secondary cell wall showing histochemical detection of lignins. (**d**) Detail of the secondary cell wall showing histochemical detection of hydrogen peroxide. (**e**) Xylans labeled by LM10 in the secondary cell walls of the mechanical cells. *CW* cell walls, *MT* mechanical tissue, *Nu* nucleus, *Pl* plastids, *SG* starch grains *Va* vacuoles

mitochondria, endoplasmic reticulum, free ribosomes, Golgi apparatus, multivesicular bodies (Costa et al. 2022c), and heavily pitted walls (Fig. 6b, c) (Lalonde and Shorthouse 1984) allowing the translocation of solutes by the symplast pathway (Rohfritsch and Shorthouse 1982). In this case, the mechanical tissue enables the influx of molecules of primary and secondary metabolites from the common storage tissue toward the nutritive tissue (Oliveira et al. 2010; Nogueira et al. 2018; Costa et al. 2021b). When the gall reaches the dehiscence stage, the mechanical tissue may be involved in the opening process (Meakin and Roberts 1991) in the ostiolar area, allowing the adult inducer to escape (Meyer and Maresquelle 1983; Corró-Molas et al. 2017). The mechanical cell layers perform other functions. Traditionally interpreted as conferring protection to the gall inducer against parasitoids or opportunistic predators or herbivores (Stone and Schönrogge 2003), it has been shown that some gall inducers may suffer over mortality due to parasitoid and predator attacks (Mendonça and Romanowski 2002; Roubos and Liburd 2013) despite the presence of the hard cell wall layers in the gall outer compartment (Rezende et al. 2018). An alternative functional trait of this layer is to protect the gall tissues against ROS originated from the gall-inducing organism's respiration and cell metabolism (Isaias et al. 2015). In the formation of the mechanical cell layers, the synthesis of lignin and its deposition in cell walls consume and dissipate hydroxyl radicals (Fig. 6c, d) (Liu et al. 2018; Zhong et al. 2019) and may also protect gall tissues against the high oxidative stress (Oliveira et al. 2011b; Isaias et al. 2015; Rezende et al. 2018). The lignification process may also be a reactive response to biotic and abiotic stresses (Cesarino 2019), directly influenced by the feeding stimulus of gall-inducing Diptera on a biotic basis (Isaias et al. 2015), and by the decrease in water resources on the abiotic basis (Nogueira et al. 2022).

7 Structural and Functional Traits of the Vascular Tissues

The neoformation of vascular tissues is a common structural trait of Diptera galls (Meyer 1969). Neoformed vascular cells can differentiate from the procambium, vascular cambium, or parenchyma cells of the ground or vascular systems of the host organs. This process alters cell fates, leading to the redifferentiation of xylem and phloem elements within galls (Dorchin et al. 2002; Bragança et al. 2020b; Aloni 2023). In the xylem, the tracheary elements exhibit annular or helical patterns of wall thickening (Fig. 7a), allowing cell elongation during the final growth of the cecidomyiid galls (Raman and Devadas 1977; Rohfritsch and Shorthouse 1982). The vascular bundles in Diptera galls are distributed within the common storage tissue or associated with the mechanical layers, enveloping the typical nutritive tissue. They establish a connection with the vascular system of the host organ, often influencing the host plant vascular system some distance from the larval chamber (Meyer 1969). Additionally, a continuous vascularization has been associated with the gall organogenesis as an adventitious new organ (Corró-Molas et al. 2017; Nogueira et al. 2022).

The collateral bundle organization is commonly observed in Diptera galls (Fig. 7b), indicating a conservative vascular pattern of the host plant organs. This pattern is associated with gall induction sites such as leaves (Oliveira and Isaias 2010; Suzuki et al. 2015; Amorim et al. 2017), stems (Dorchin et al. 2002), buds (Fleury et al. 2015) and flowers (Guimarães et al. 2014). Some Diptera galls on angiosperms exhibit bicollateral bundles (Mani 1964; Teixeira et al. 2022), while those on ferns demonstrate concentric vascular bundles (Bragança et al. 2023), reflecting the host plant's vascular traits within the galls. The induction site influences the secondary growth potential of the vascular system (Fig. 7c), as observed in stem (Corró-Molas et al. 2014; Jorge et al. 2022a; Guedes et al. 2023b) and leaflet vein galls (Bragança et al. 2020b). For instance, in *Tetradiplosis panghitruz*

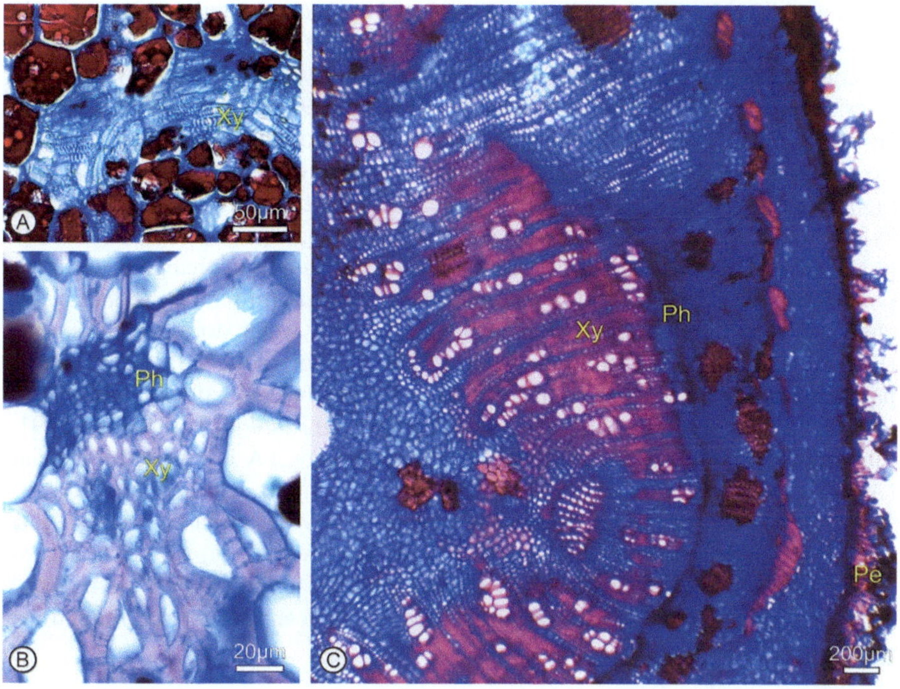

Fig. 7 The structural traits of the vascular tissues. (**a**) Cross-section detail of the bivalve-shaped galls induced by an undetermined species of *Lopesia* on *Mimosa gemmulata* Barneby (Fabaceae) leaves showing tracheary elements exhibiting annular patterns of wall thickening, and (**b**) collateral bundle organization of the vascular bundles. (**c**) Cross-sectional detail of the vascular system of stem galls induced by an undetermined species of *Neolasioptera* on *Eremanthus erythropappus* (DC.) MacLeish (Asteraceae) showing secondary growth. *Ph* phloem, *Xy* xylem

Martínez, 2013 (Cecidomyiidae) stem galls on *Prosopis caldenia* Burkart (Fabaceae), the larva stimulates cell hypertrophy and hyperplasia on the cambial zone toward gall formation, resulting in the production of anomalous secondary xylem, with scarce vessel elements, and abundant parenchyma cells and thick-walled fibers (Corró-Molas et al. 2014). In another aspect, the larvae of the cecidomyiid species associated with the stem galls on *Suaeda monoica* Forssk. ex J.F. Gmel (Chenopodiaceae) inhibit the secondary thickening of vascular tissues within the gall, retaining the primary state of the vascular bundles (Dorchin et al. 2002). However, the portions of the gall above and below the stem gall undergo normal secondary differentiation (Dorchin et al. 2002).

The vascular connection and reconfiguration of the xylem cells play a crucial role in the development and maintenance of gall structures by facilitating the water and nutrient uptake (Aloni 2021). In addition to photoassimilates, the gall phloem transports proteins, enzymes, and plant hormones, such as cytokinins and auxins, which are involved in the neovascularization process, as well as in the gall development and metabolism (Oliveira et al. 2011b; Bragança et al. 2017; Bedetti et al.

2018; Aloni 2023). Some studies have explored the reconfiguration of xylem cells in cecidomyiid galls (Bragança et al. 2020b; Nobrega et al. 2022), testing the vascular constriction hypothesis, which argues that larger vessel elements within galls prioritize water supply for gall tissues (Aloni et al. 1995; Ullrich and Aloni 2000). However, contrary evidence exists, such as in cecidomyiid galls on leaves of *Avicennia schaueriana* Stapf & Leechm. ex Moldenke (Acanthaceae) and *Inga ingoides* (Rich.) Willd. (Fabaceae), which has reduced hydraulic conductivity in the xylem due to fewer and smaller vessel elements (Bragança et al. 2020b; Nobrega et al. 2022). These alterations in vascular tissue configuration change water transport routes inside the galls, allowing the expansion and proliferation of gall parenchyma cells, acting as a compensatory mechanism for water storage and transport (Bragança et al. 2020b; Nobrega et al. 2022). The investment in water storage parenchyma is also a structural-functional mechanism in stem gall vascular systems (Guedes et al. 2023b). For instance, in the *Tomoplagia rudolphi* Lutz & Lima, 1918 (Tephritidae) stem galls on *Vernonia polyanthes* (Spreng.) A.J. Vega & M. Dematt. (Asteraceae), the hypertrophy of the medullary parenchyma cells and the disorganization of the vascular tissues occur (Andrade et al. 1995). Similarly, the stem galls on *Eremanthus erythropappus* (DC.) MacLeish (Asteraceae) (Fig. 7b) have a lower vessel element density but similar-sized elements to non-galled stems, with increased vessel numbers in the proximal regions, suggesting improved water transport efficiency (Jorge et al. 2022a).

8 Structural and Functional Traits of the Typical Nutritive Tissue

Typical nutritive tissues (Fig. 8a), which develop in dipteran galls, may originate from the redifferentiation (metaplasia) of epidermal, parenchyma (Meyer 1957; Mani 1964; Larew 1981; Rohfritsch and Shorthouse 1982), procambial (Lalonde and Shorthouse 1984), and cambial cells (Corró-Molas et al. 2014) of the host organ induction sites. The mixed origin of the typical nutritive tissue from the epidermal and parenchyma cells has been demonstrated in leaf galls (Oliveira and Isaias 2010; Nogueira et al. 2022) and appears to be a consequence of the intense feeding stimuli and high nutritional demand of the larvae during their lifespan (Carneiro et al. 2017; Costa et al. 2022a). Based on the cell origin of the globoid stem galls of *Urophora cardui* Linnaeus, 1758 (Tephritidae) on *Cirsium arvense* (L.) Scop. (Asteraceae), the formation of primary and secondary nutritive tissues occurs (Lalonde and Shorthouse 1984). The primary nutritive tissue originates from redifferentiation of parenchyma cells, while the secondary nutritive tissue originates from procambial cells (Lalonde and Shorthouse 1984).

The cytological profiles of the nutritive cells have been associated with cell divisions, remodeling of cell wall components, synthesis of energetic molecules directed to the nutrition of the gall-inducing larvae, and gall structural metabolism (Bronner

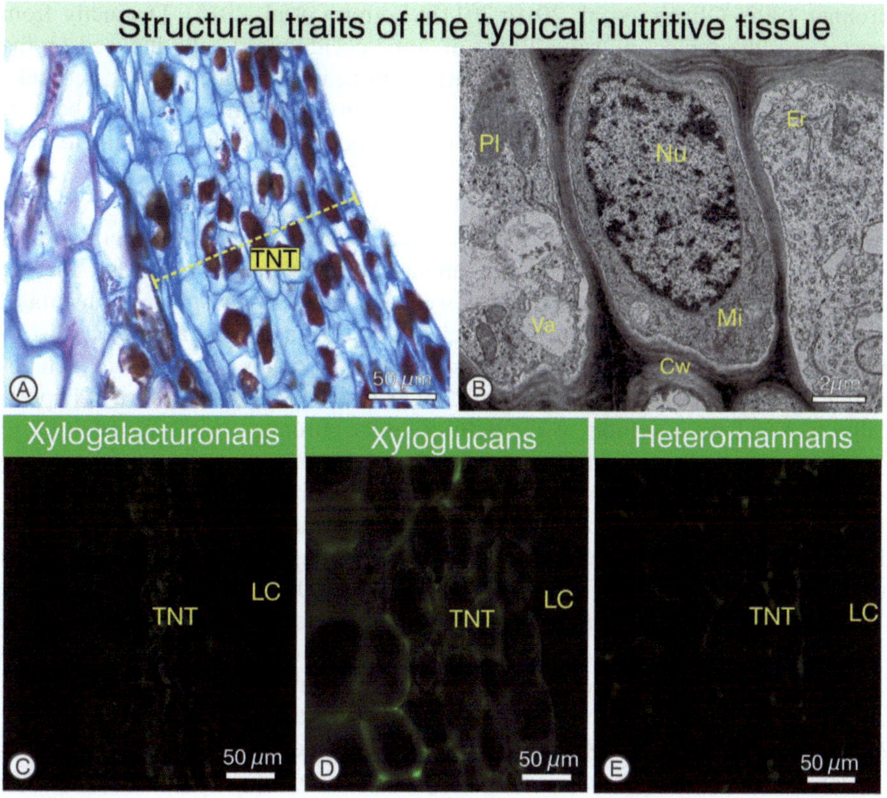

Fig. 8 The structural traits of the typical nutritive tissue. (**a**) Cross-sectional detail of nutritive gall cells induced by an undetermined species of Cecidomyiidae on *Byrsonima coccolobifolia* A.Juss. (Malpighiaceae) leaves. (**b**) Cytological features of the lenticular galls induced by an undetermined species of Cecidomyiidae on *Inga ingoides* (Rich.) Willd. (Fabaceae) leaves evidencing thin cell walls, a large nucleus, and dense cytoplasm with several mitochondria, endoplasmic reticulum, plastids, fragmented vacuoles, and free ribosomes. (**c-e**) Immunocytochemical profiles of the nutritive cells. (**c**) Xylogalacturonans labeled by LM8 in the nutritive cell walls of the bivalve-shaped galls induced by *Lopesia mimosae* Maia, 2009 (Cecidomyiidae) on *Mimosa tenuiflora* (Willd.) Poir. (Fabaceae) leaves. (**d**) Xyloglucans labeled by LM15 in the nutritive cell walls of the bivalve-shaped galls induced by an undetermined species of *Lopesia* on *M. gemmulata* Barneby (Fabaceae) leaves. (**e**) Heteromannans labeled by LM21 in the nutritive cell walls of the bivalve-shaped galls induced by an undetermined species of *Lopesia* on *M. gemmulata* leaves. *Cw* cell walls, *Er* endoplasmic reticulum, *LC* larval chamber, *Mi* mitochondria, *Nu* nucleus, *Pl* plastids, *TNT* typical nutritive tissue, *Va* vacuoles

1992; Harris et al. 2006; Oliveira et al. 2010, 2011a, b). Based on the ultrastructural aspects, the nutritive cells usually exhibit a large nucleus, conspicuous nucleolus, fragmented vacuoles, and dense cytoplasm rich in organelles, such as free ribosomes, oleosomes, rough and smooth endoplasmic reticulum, Golgi apparatus, several mitochondria, plastids with well-defined thylakoid membranes, as well as thin cell walls, and the absence of a cuticle (Fig. 8b) (Rohfritsch and Shorthouse 1982;

Bronner 1992; Oliveira et al. 2010, 2011b; Ferreira et al. 2015). Distinctly from most cases, the typical nutritive tissue of some cecidomyiid galls induced on the leaves of the *Copaifera langsdorffii* Desf. (Fabaceae) contained enucleated cells (Oliveira et al. 2011a). In this case, the metabolism of nutritive cells may be controlled by adjacent cells, similarly to the role of companion cells over phloem sieve elements (Oliveira et al. 2011a).

The typical nutritive tissue is enzymatically active, and numerous hydrolase activities have been localized by histochemical techniques. Nutritive cells nearest the larval chamber contain the highest amylase, protease, aminopeptidase, sucrose synthase and phosphatase activities, as well as glucose-6-phosphatase, invertase, and phosphorylase (Bronner 1992; Oliveira et al. 2011a, b; Bragança et al. 2017). The activity of the sucrose synthase and invertases mediates carbohydrate metabolism in the nutritive cells (Bragança et al. 2017), which is a classic pattern of Diptera galls, favoring the nutrition of the galling insect (Bronner 1992). In addition to carbohydrates, such as reducing sugars and starch grains, proteins and lipids are energetic molecules that may accumulate in the nutritive cells and favor the nutrition of the gall-inducing Diptera. The lipids in the nutritive cells of Diptera galls on different host plants seem to be linked to the intrinsic metabolism of the host plants and are food resources for the larva (Moura et al. 2008; Guimarães et al. 2013; Ferreira and Isaias 2014; Amorim et al. 2017; Nogueira et al. 2018; Bragança et al. 2022; Costa et al. 2022c). Lipids may become available by the activity of a lipase-like protein expressed in the salivary glands of Diptera larvae during the feeding activity and involved in extra-oral digestion (Hatchett et al. 1990; Shukle et al. 2009; Al-Jbory et al. 2018). On the other hand, the accumulation of proteins (Rezende et al. 2018), as well as secondary metabolites, such as phenolics, proanthocyanidins, and terpenoids in the protoplasm of the nutritive cells of Diptera galls may be involved in the maintenance of redox-potential homeostasis (Oliveira et al. 2011b; Bragança et al. 2022; Costa et al. 2022c).

Another important chemical aspect of Diptera galls relies on cell wall compounds, which have been studied by immunocytochemical techniques. Such studies have demonstrated the dynamics of pectins, and hemicelluloses associated with the organization of cellulose microfibrils, conferring flexibility and stability to the nutritive cell walls, guaranteeing cell elongation and expansion, which are important processes involved in gall growth (Suzuki et al. 2015; Teixeira et al. 2018; Bragança et al. 2020a; Nogueira et al. 2022; Souza et al. 2024). In addition to the structural roles, the xylogalacturonans (pectins) (Nogueira et al. 2022) and the xyloglucans and heteromannans (hemicelluloses) have been associated with additional nutritive resources that favor the feeding activity of the larvae (Fig. 8c–e) (Bragança et al. 2020a; Costa et al. 2022c; Nogueira et al. 2022; Souza et al. 2024) once their long molecules may be broken by enzymatic activity. The monosaccharides of the pectins and hemicelluloses can become available as a feeding resource through the activity of pectinase, α-D-glucosidase and α-D-and β-D-galactosidase, which may be found in the salivary glands and/or midgut of Diptera larvae (Grover et al. 1988).

9 Ambrosia Galls

Some gall-inducing Cecidomyiidae establish a symbiotic association with specific fungi that proliferate their mycelium within the larval chambers and on which the larvae feed, rather than on the nutritive cells (Fig. 9) (Yukawa and Rohfritsch 2005; Rohfritsch 2008; Lebel et al. 2012; Raman and Suryanarayanan 2017) or in addition to them (Chao and Liao 2013), forming ambrosia galls (Meyer 1987; Bissett and Borkent 1988). Due to their mixed diet (phytomycetophagy), the inducers of ambrosia galls perform a unique step in the evolution of gall-inducing Cecidomyiidae, whose habit is considered plesiomorphic (Dorchin et al. 2019). Currently, it has been reported in the three taxonomically separate tribes: Asphondyliini, Alycaulini, and Lasiopterini, which belong to the subfamily Cecidomyiinae (Dorchin et al. 2019).

It is believed that the main role of the fungus in ambrosia galls is to serve as a food source, but it may also have other roles during gall initiation and morphogenesis (Rohfritsch 1997; Arduin and Kraus 2001; Dorchin et al. 2002; Sá et al. 2009; Teixeira et al. 2022). In this symbiotic interaction, the adult cecidomyiid females lay eggs and inoculate fungal spores into plant tissues, afterward hatched larvae initiate gall development and at the same time germinated spores spread their hyphae adjacent to the larval chamber, and the larvae feed on these fungal hyphae (Yukawa and Rohfritsch 2005). The fungus may be feeding intra- and intercellularly without killing host plant cells (Rohfritsch 2008). When the adult cecidomyiid females come out of the galls, they carry fungal spores by the mycangia (Heath and

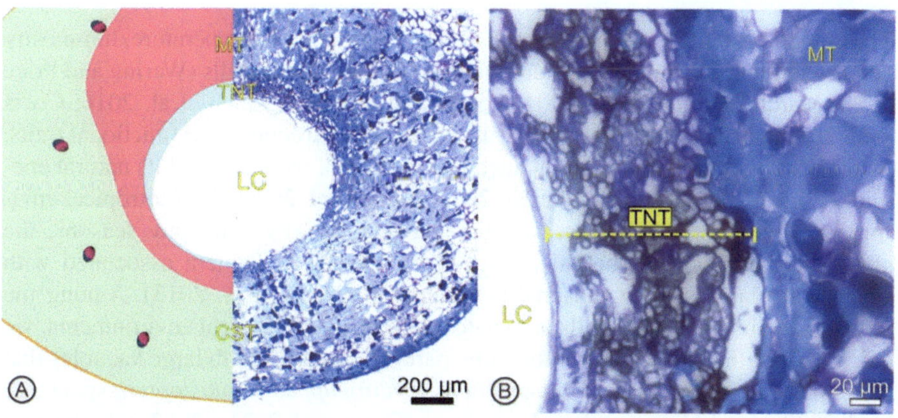

Fig. 9 The structural traits of the ambrosia galls induced by an undetermined species of Cecidomyiidae on *Inga ingoides* (Rich.) Willd. (Fabaceae) leaves. (**a**) Schematic and anatomical overview of the gall in cross section, showing epidermis (orange), common storage tissue (green), vascular bundles (blue and pink), mechanical tissue (pink), and typical nutritive tissue (blue) surrounding the larval chamber. (**b**) Detail of nutritive cells with fungi. *CST* common storage tissue, *LC* larval chamber, *MT* mechanical tissue, *TNT* typical nutritive tissue

Stireman 2010) and induce new ambrosia galls synchronizing their life cycles with the phenology of the host plant (Tokuda 2012; Yukawa et al. 2013).

The structural and functional traits of ambrosia galls may vary according to their associated inducing species and induction sites, and they may be histologically complex structures consisting of outer neoformed tissues, a mechanical layer, and (Chao and Liao 2013; Pan et al. 2015) vascular tissues (Dorchin et al. 2002; Bragança et al. 2020a). In the ambrosia galls without a nutritive tissue, the nutritive resources are provided by the fungal mycelia, whose cells have dense cytoplasm and accumulate lipids and proteins in a so-called 'pseudoparenchyma' (Rohfritsch 2008). The fungal mycelia are intimately associated with the surrounding walls of gall parenchyma cells with dense cytoplasm, which provide water and nutrients to the fungi (Meyer 1987; Rohfritsch 2008). Thus, the nutrients reach the larva through a structure composed of plant cells associated with protein or lipid-rich fungal cells (Rohfritsch 2008). In some ambrosia galls, not only fungal hyphae but also the differentiation of nutritive cells have been demonstrated (Chao and Liao 2013; Pan et al. 2015; Bragança et al. 2022). In addition to the nutrition of the larvae, the fungal hyphae may contribute to the degradation of nutritive cell walls, obtaining substrate for their growth, and facilitating the access of the Cecidomyiidae larvae to the carbohydrates, lipids, and proteins in the protoplast of nutritive cells (Bragança et al. 2022). The structure and function of the tissues in the ambrosia galls may change due to trophic interactions with natural enemies (Sugiura et al. 2004; Tokuda 2012).

10 Abiotic Factors and Diptera Galls

The climatic and local conditions such as water availability, temperature, luminosity and fire influence the abundance and richness of the Diptera galls (Waring and Price 1990; Marchosky and Craig 2004; Formiga et al. 2009; Kelch et al. 2016; Costa et al. 2021b; Kuzmanich et al. 2023), as well as their structural and functional traits, which confer a microenvironment adequate to the gall inducers, and its natural enemies (Price et al. 1987; Boukili et al. 2007; Baine et al. 2024). In Neotropical environments with well-marked seasonality between the dry and rainy seasons, the abundance and richness of galls induced by species of Diptera associated with super-host plants present seasonal syndromes (Oliveira et al. 2013). Among the structural traits of Diptera galls associated with plants of tropical environments, we can highlight the gall wall constituted by parenchyma cells with large vacuoles that store water, as observed, for instance, in *E. erythropappus-Neolasioptera* sp. (Jorge et al. 2022a), and *M. gemmulata-Lopesia* sp. (Costa et al. 2021a, 2022c). In the case of the *Mimosa tenuiflora* (Willd.) Poir. (Fabaceae) - *Lopesia mimosae* Maia, 2009 (Cecidomyiidae) system, cell wall porosity mediated by methyl-esterified homogalacturonans (pectins) seems to favor water flux toward the storage parenchyma, favoring gall development in the Caatinga environment (Nogueira et al. 2022).

In aquatic plants or those adapted to experience flooding events, the development of aerenchyma facilitates gas diffusion in submerged organs (Evans 2004). The formation of aerenchyma may help to avoid hypoxia and hypercarbia in gall tissues, as the intercellular spaces increase gas diffusion (Pincebourde and Casas 2016). Such structural trait developed in the outer compartment of cecidomyiid galls is, nevertheless, a rare condition (Amorim et al. 2017; Teixeira et al. 2022). Despite the functions of aeration attributed to aerenchyma, in cecidomyiid galls (Amorim et al. 2017; Teixeira et al. 2022), it may be considered part of the defensive tissue selected by evolutionary pressures such as the attack of specialist parasitoids (Bailey et al. 2009). In another aspect, the globoid leaf galls of *Meunieriella* sp. (Cecidomyiidae) associated with *Avicennia schaueriana* Stapf & Leechm. ex Moldenke (Acanthaceae), a plant extremely adapted to mangroves, the galls have a high investment in water-storage parenchyma, a site of ion accumulation (Nobrega et al. 2021). In these galls, excess salt is removed by epidermal salt glands, favoring internal ion balance (Nobrega et al. 2021), a specific environmental adaptation to the variable salinity and flooding of the mangrove environment. The light intensity also influences size and density of epidermal cells and stomata, as well as the thickness of common storage tissue of the galls, as observed in the sun and shade galls of *Clinodiplosis profusa* Maia, 2001 (Cecidomyiidae) on *Eugenia uniflora* L. (Myrtaceae) leaves (Castro et al. 2023). Furthermore, the lignification process requires ROS, and culminates in the formation of the mechanical cell layers of Diptera galls, which are commonly associated with host plants growing in environments with high luminosity where lignification may contribute to the protection of gall tissues against the oxidative stress (Arriola et al. 2018; Costa et al. 2022b; Nogueira et al. 2022).

11 Conclusions and Remarks

The structural development and functional traits of Diptera-induced galls are influenced by the morphogenetic potential of the host plant (e.g., induction site and age of the host organ), the galling insect (e.g., life cycle and sex), and environmental conditions. Although the molecular triggers of gall induction are not fully understood, the roles of reactive oxygen species (ROS) and growth regulatory hormones are well established. Despite the ontogenetic specificities of each gall, the review confirms the structural and functional compartmentalization in Diptera galls. Trichomes and the outer epidermal cells of the galls, as well as the periderm, provide protection against environmental conditions. The stomata and common storage tissue are involved in photosynthetic activity, preventing hypercarbia and hypoxia. Additionally, the common storage tissue is involved in the synthesis, storage, and/or transport of primary and secondary metabolites, enzymes, and phytohormones, which are essential for gall development and metabolism. The storage parenchyma and vascular tissues act in the storage and transport of water and nutrients, promoting gall succulence and creating a favorable microenvironment for the galling insect. Besides providing structural support, mechanical tissue is crucial in dissipating oxidative stress and mediating nutrient transport to the typical nutritive tissue. In addition to carbohydrates and lipids in the protoplasm of nutritive cells, the xylogalacturonans, xyloglucans, and heteromannans in the cell walls can also provide nutrition to the larva. Future research on how the structure-function relationships of different Diptera galls respond to environmental conditions may shed light on the adaptation of plant tissues to climate change. Finally, detailed analysis of anatomical, ultrastructural, and genetic changes in cell lineages may help us better understand how galls are induced and grow, as well as further clarify the metabolic functions of each tissue compartment.

References

Albert S, Switi R, Gandhi D (2013) Anatomy and ontogenesis of foliar galls induced by *Odinadiplosis odinae* (Diptera: Cecidomyiidae) on *Lannea coramandelica* (Anacardiaceae). Acta Entomol Serbica 18(1/2):161–175

Al-jbory Z, Anderson KM, Harris MO, Mittapalli O, Whitworth RJ, Chen MS (2018) Transcriptomic analyses of secreted proteins from the salivary glands of wheat midge larvae. J Insect Sci 18(1):1–5

Aloni R (2021) Vascular differentiation and plant hormones. Springer, Chan

Aloni R (2023) How moving hormonal signals regulate plant vascular differentiation, adaptation, and evolution: hypotheses and evidence. In: Progress in botany. Springer, Berlin/Heidelberg

Aloni R, Pradel KS, Ullrich CI (1995) The three-dimensional structure of vascular tissues in *Agrobacterium tumefaciens*-induced crown galls and in the host stem of *Ricinus communis* L. Planta 196:597–605

Amorim DO, Ferreira BG, Fleury G (2017) Plant potentialities determine anatomical and histochemical diversity in *Mikania glomerata* Spreng. galls. Braz J Bot 40:517–527

Andrade GI, Silva IM, Fernandes GW, Scatena VL (1995) Aspectos biológicos das galhas de *Tomoplagia rudolphi* (Diptera: Tephritidae) em *Vernonia polyanthes* (Asteraceae). Rev Bras Biol 55(4):819–829

Andreas P, Kisiala A, Emery RJN, Clerck-Floate R, Tooker JF, Price PW, Miller DG III, Chen MS, Connor EF (2020) Cytokinins are abundant and widespread among insect species. Plan Theory 9(2):1–23

Arduin M, Kraus JE (1995) Anatomia e ontogenia de galhas foliares de *Piptadenia gonoacantha* (Fabales, Mimosaceae). Bol Bot Univ São Paulo 14:109–130

Arduin M, Kraus JE (2001) Anatomia de galhas de ambrosia em folhas de *Baccharis concinna* e *Baccharis dracunculifolia* (Asteraceae). Braz J Bot 24(1):63–72

Arriola IA, Melo-Júnior JC, Ferreira BG, Isaias RMS (2018) Galls on *Smilax campestris* Griseb. (Smilacaceae) protect the insects against restinga constraints, but do not provide enriched nutrition. Braz J Bot 41:145–153

Bailey R, Schönrogge K, Cook JM, Melika G, Csóka G, Thuróczy C, Stone GN (2009) Host niches and defensive extended phenotypes structure parasitoid wasp communities. PLoS Biol 7(8):e1000179

Baine Q, Casares EE, Hughes DW, Martinson VG, Martinson EO (2024) Arthropod communities associated with gall-inducing *Aciurina bigeloviae* and *Aciurina trixa* (Diptera: Tephritidae) in New Mexico. Ann Entomol Soc Am 117(2):107–117

Bartlett L, Connor EF (2014) Exogenous phytohormones and the induction of plant galls by insects. Arthropod Plant Interact 8:339–348

Bedetti CS, Modolo LV, Isaias RMS (2014) The role of phenolics in the control of auxin in galls of *Piptadenia gonoacantha* (Mart.) MacBr (Fabaceae: Mimosoideae). Biochem Syst Ecol 55:53–59

Bedetti CS, Bragança GP, Isaias RMS (2017) Influence of auxin and phenolic accumulation on the patterns of cell differentiation in distinct gall morphotypes on *Piptadenia gonoacantha* (Fabaceae). Aust J Bot 65(5):411–420

Bedetti CS, Jorge NC, Trigueiro FCG, Bragança GP, Modolo LV, Isaias RMS (2018) Detection of cytokinins and auxin in plant tissues using histochemistry and immunocytochemistry. Biotech Histochem 93(2):149–154

Bissett J, Borkent A (1988) Ambrosia galls: the significance of fungal nutrition in the evolution of the Cecidomyiidae (Diptera). In: Pirozynski KA, Hawksworth DL (eds) Coevolution of fungi with plants and animals. Academic, London, pp 203–225

Borges RM (2018) The galling truth: limited knowledge of gall-associated volatiles in multitrophic interactions. Front Plant Sci 9:370973

Boukili VKS, Hoopes MF, Briggs CJ (2007) Effect of microenvironment on development of a gall midge. Environ Entomol 36(2):441–450

Bragança GP, Oliveira DC, Isaias RMS (2017) Compartmentalization of metabolites and enzymatic mediation in nutritive cells of Cecidomyiidae galls on *Piper arboreum* Aubl. (Piperaceae). J Plant Stud 6(1):11–22

Bragança GPP, Alencar CF, Freitas MSC, Isaias RMS (2020a) Hemicelluloses and associated compounds determine gall functional traits. Plant Biol 22(2):981–991

Bragança GPP, Freitas MSC, Isaias RMS (2020b) The influence of gall position over xylem features in leaflets of *Inga ingoides* (Rich.) Willd. (Fabaceae: Caesalpinioideae). Trees 35:199–209

Bragança GPP, Ferreira BG, Isaias RMS (2022) Distinct cytological mechanisms for food availability in three *Inga ingoides* (Fabaceae)—Cecidomyiidae gall systems. Protoplasma 259(1):155–162

Bragança GPP, Costa EC, Arriola IA, Sanín D, Isaias RMS (2023) As soft as silk: structural and chemical traits can help with the identification of *Niphidium crassifolium* (Polypodiaceae) gall inducers. Rodriguésia 74:e00572023

Bronner N (1992) The role of nutritive cells in the nutrition of cynipids and cecidomyiids. In: Shorthouse JD, Rohfritsch O (eds) Biology of insect-induced galls. Oxford University Press, New York, pp 118–140

Carneiro RG, Isaias RMS, Moreira AS, Oliveira DC (2017) Reacquisition of new meristematic sites determines the development of a new organ, the Cecidomyiidae gall on *Copaifera langsdorffii* Desf. (Fabaceae). Front Plant Sci 8:1–10

Caselli A, Favaro R, Petacchi R, Angeli S (2022) Infestation of the gall midge *Dasineura oleae* provides first evidence of induced plant volatiles in olive leaves. Bull Entomol Res 112(4):481–493

Castro AC, Oliveira DC, Moreira ASFP, Lemos-Filho JP, Isaias RMS (2012) Source–sink relationship and photosynthesis in the horn-shaped gall and its host plant *Copaifera langsdorffii* Desf. (Fabaceae). S Afr J Bot 83:121–126

Castro RRM, Barbosa PEF, Sant'Anna LG, Pereira CMS, Ferreira BG (2023) Sun and shade galls of *Clinodiplosis profusa* (Cecidomyiidae) on *Eugenia uniflora* (Myrtaceae): are there differences in their establishment and growth? Flora 303:152281

Cesarino I (2019) Structural features and regulation of lignin deposited upon biotic and abiotic stresses. Curr Opin Biotechnol 56:209–214

Chao JF, Liao GI (2013) Histocytological aspects of four types of ambrosia galls on *Machilus zuihoensis* Hayata (Lauraceae). Flora 208(3):157–164

Connor EF, Bartlett L, O'Toole S, Byrd S, Biskar K, Orozco J (2012) The mechanism of gall induction makes galls red. Arthropod Plant Interact 6(4):489–495

Corró-Molas BM, Martínez JJ, Carbonell-Silletta L, Gallia MC, Alfonso GL (2014) Modificaciones morfológicas inducidas por *Tetradiplosis panghitruz* Martínez (Diptera: Cecidomyiidae) en tallos de *Prosopis caldenia* Burkart (Fabaceae). Rev Mus Argent Cienc Nat 16(1):1–12

Corró-Molas BM, Martínez JJ, Nacaratti S, Arrese JJ, Testa N, Álvarez MN (2017) Exomorfología y anatomía de la agalla bivalva foliar inducida por *Rhopalomyia* sp. (Insecta, Diptera, Cecidomyiidae) en *Prosopis caldenia* (Fabaceae). Bol Soc Argent Bot 52(3):535–547

Costa EC, Carneiro RGS, Silva JS, Isaias RMS (2018) Biology and development of galls induced by *Lopesia* sp. (Diptera: Cecidomyiidae) on leaves of *Mimosa gemmulata* (Leguminosae: Caesalpinioideae). Aust J Bot 66(2):161–172

Costa EC, Martini VC, Souza-Silva A, Lemos-Filho JP, Oliveira DC, Isaias RMS (2021a) How galling herbivores share a single super-host plant during their phenological cycle: the case of *Mimosa gemmulata* Barneby (Fabaceae). Trop Ecol:1–14

Costa EC, Oliveira DC, Ferreira DK, Isaias RMS (2021b) Structural and nutritional peculiarities related to lifespan differences on four *Lopesia* induced bivalve-shaped galls on the single super-host *Mimosa gemmulata*. Front Plant Sci 12:660557

Costa EC, Freitas MS, Carneiro RG, Oliveira DC, Isaias RMS (2022a) The ontogenesis of four *Lopesia* Rübsaamen (Cecidomyiidae) galls on the super-host *Mimosa gemmulata* Barneby (Fabaceae) reveals peculiar anatomical traits. Flora 287:151996

Costa EC, Bragança GPP, Arriola IA, Freitas MS, Isaias RMS (2022b) Host plant traits define the strategies of success to the Cecidomyiidae in a restinga environment. S Afr J Bot 146:751–759

Costa EC, Oliveira DC, Isaias RMS (2022c) Parasitoid impairment on the galling *Lopesia* sp. activity reflects on the cytological and histochemical profiles of the globoid bivalve-shaped gall on *Mimosa gemmulata*. Protoplasma 259(6):1585–1597

Darwin C (1868) The variation of animals and plants under domestication. John murray

De Bruyn L, Vandevyvere I, Jaminé D, Prinsen E (1998) The effects of gall formation by *Lipara lucens* (Diptera: Chloropidae) on its host *Phragmites australis* (Poaceae). In: Csóka G, Mattson WJ, Stone GN, Price PW (eds) The biology of gall-inducing arthropods. General technical report NC-199. United States Department of Agriculture Forest Service General Technical Report NC, pp 173–187

Desnitskiy AG, Chetverikov PE, Ivanova LA, Kuzmin IV, Ozman-Sullivan SK, Sukhareva SI (2023) Molecular aspects of gall formation induced by mites and insects. Life 13(6):1347

Detoni ML, Vasconcelos EG, Maia ACRG, Gusmão MAN, Isaias RMS, Soares GLG, Santos JC, Fernandes GW (2011) Protein content and electrophoretic profile of insect galls on susceptible and resistant host plants of *Bauhinia brevipes* Vogel (Fabaceae). Aust J Bot 59(6):509–514

Dorchin N, Freidberg A, Aloni R (2002) Morphogenesis of stem gall tissues induced by larvae of two cecidomyiid species (Diptera: Cecidomyiidae) on *Suaeda monoica* (Chenopodiaceae). Can J Bot 80(11):1141–1150

Dorchin N, Harris KM, Stireman JO III (2019) Phylogeny of the gall midges (Diptera, Cecidomyiidae, Cecidomyiinae): systematics, evolution of feeding modes and diversification rates. Mol Phylogenet Evol 140:106602

Evans DE (2004) Aerenchyma formation. New Phytol 161(1):35–49

Fernandes GW (1990) Hypersensitive: a neglected plant resistance mechanism against insect herbivores. Environ Entomol 19(5):1173–1182

Fernandes GW, Marco-Júnior P, Schönrogge K (2008) Plant organ abscission and the green Island effect caused by gallmidges (Cecidomyiidae) on tropical trees. Arthropod Plant Interact 2(2):93–99

Ferreira BG, Isaias RMS (2014) Floral-like destiny induced by a galling Cecidomyiidae on the axillary buds of *Marcetia taxifolia* (Melastomataceae). Flora 209(8):391–400

Ferreira BG, Carneiro RGS, Isaias RMS (2015) Multivesicular bodies differentiate exclusively in nutritive fast-dividing cells in Marcetia taxifolia galls. Protoplasma 252(5):1275–1283

Ferreira BG, Álvarez R, Bragança GP, Alvarenga DR, Pérez-Hidalgo N, Isaias RMS (2019) Feeding and other gall facets: patterns and determinants in gall structure. Bot Rev 85:78–106

Fleury G, Ferreira BG, Soares GLG, Oliveira DC, Isaias RMS (2015) Elucidating the determination of the rosette galls induced by *Pisphondylia brasiliensis* Couri and Maia 1992 (Cecidomyiidae) on *Guapira opposita* (Nyctaginaceae). Aust J Bot 63(7):608–617

Flor IC, Rodrigues AR, Silva SA, Proença B, Maia VC (2022) Insect galls on Asteraceae in Brazil: richness, geographic distribution, associated fauna, endemism and economic importance. Biota Neotropica 22(1):e20211250

Formiga AT, Gonçalves SJMR, Soares GLG, Isaias RMS (2009) Relações entre o teor de fenóis totais e o ciclo das galhas de Cecidomyiidae em *Aspidosperm aspruceanum* Müll. Arg. (Apocynaceae). Acta Bot Bras 23(1):93–99

Formiga AT, Soares GLG, Isaias RMS (2011) Responses of the host plant tissues to gall induction in *Aspidosperma spruceanum* Müell. Arg. (Apocynaceae). Am J Plant Sci 2(6):823–834

Formiga AT, Oliveira DC, Ferreira BG, Magalhães TA, Castro AC, Fernandes GW, Isaias RMS (2013) The role of pectic composition of cell walls in the determination of the new shape-functional design in galls of *Baccharis reticularia* (Asteraceae). Protoplasma 250(4):899–908

Formiga AT, Silveira FAO, Fernandes GW, Isaias RMS (2015) Phenotypic plasticity and similarity among gall morphotypes on a superhost, *Baccharis reticularia* (Asteraceae). Plant Biol 17(2):512–521

Gagné RJ, Jaschhof M (2021) A Catalog of the Cecidomyiidae (Diptera) of the world, 5th edn. [Digital]. — Systematic Entomology Laboratory, Agricultural Research Service, U.S. Department of Agriculture, U.S. National Museum, Washington, D.C.

Ganaha T, Nohara M, Sato S, Uechi N, Yamagishi K, Yamauchi S, Yukawa J (2007) Polymorphism of axillary bud galls induced by *Rhopalomyia longitubifex* (Diptera: Cecidomyiidae) on *Artemisia princeps* and *A. montana* (Asteraceae) in Japan and Korea, with designation of new synonyms. Entomol Sci. 10(2):157–169

Gätjens-Boniche O (2019) The mechanism of plant gall induction by insects: revealing clues, facts, and consequences in a cross-kingdom complex interaction. Rev Bio Trop 67(6):1359–1382

Gätjens-Boniche O, Jiménez-Madrigal JP, Whetten RW, Valenzuela-Diaz S, Alemán-Gutiérrez A, Hanson PE, Pinto-Tomás AA (2023) Microbiome and plant cell transformation trigger insect gall induction in cassava. Front Plant Sci 14:1237966

Gonçalves PHP, Rezende UC, Bomfim PMS, Maia VC, Oliveira DC (2022) Sexually dimorphic galls induced on leaflets of *Matayba guianensis* (Sapindaceae): a rare phenomenon occurring in Diptera (Cecidomyiidae). Sci Nat 109(10):1–11

Grover PB Jr, Ross DR, Shukle RH (1988) Identification and partial characterization of digestive carbohydrases in larvae of the Hessian fly, *Mayetiola destructor* (Say) (Diptera: Cecidomyiidae). Archiv Insect Biochem Physiol 8(1):59–72

Guedes LM, Gavilán E, Pérez C, Becerra J, Aguilera N (2023a) Impact of the galling *Dasineura* sp. on the structural and chemical profile of *Peumus boldus* stems. Trees 37(2):545–553

Guedes LM, Aguilera N, Gavilán E, Péndola JA, Villagrán NE (2023b) Vascular implications of *Dasineura* sp. galls' establishment on *Peumus boldus* stems. Plant Biol 25(6):965–972

Guedes LM, Aguilera N, Torres S, Gavilán E, Rosales N (2024) Gall inducer *Dasineura* sp. alters the polyphenol profile and antioxidant activity of *Peumus boldus* stems. Tree Physiol 44(3):tpae024

Guimarães ALA, Bizarri CHB, Barbosa LS, Nakamura MJ, Ramos MFS, Vieira ACM (2013) Characterisation of the effects of leaf galls of *Clusiamyia nitida* (Cecidomyiidae) on *Clusia lanceolata* Cambess. (Clusiaceae): anatomical aspects and chemical analysis of essential oil. Flora 208(3):165–173

Guimarães ALA, Cruz SMS, Vieira ACM (2014) Structure of floral galls of *Byrsonima sericea* (Malpighiaceae) induced by *Bruggmanniella byrsonimae* (Cecidomyiidae, Diptera) and their effects on host plants. Plant Biol 16(2):467–475

Guimarães A, Vieira R, Vieira A (2021) Structure of leaf galls in *Clusia fluminensis* Planch and Triana (Clusiaceae): sex-biased development in a dioecious host plant. Plan Theory 10(1):1–18

Haarder S, Bruun HH, Harris KM, Skuhravá M (2016) Gall midges (Diptera: Cecidomyiidae) new to the Danish fauna. Ent Tidskr 137(3):79–98

Haiden SA, Hoffmann JH, Cramer MD (2012) Benefits of photosynthesis for insects in galls. Oecologia 170(4):987–997

Hall DR, Amarawardana L, Cross JV, Francke W, Boddum T, Hillbur Y (2012) The chemical ecology of cecidomyiid midges (Diptera: Cecidomyiidae). J Chem Ecol 38(1):2–22
Hall CR, Carroll AR, Kitching RL (2017) A meta-analysis of the effects of galling insects on host plant secondary metabolites. Arthropod Plant Interact 11(4):463–473
Harris MO, Pitzschke A (2020) Plants make galls to accommodate foreigners: some are friends, most are foes. New Phytol 225(5):1852–1872
Harris MO, Freeman TP, Rohfritsch O, Anderson KG, Payne SA, Moore JA (2006) Virulent Hessian fly (Diptera: Cecidomyiidae) larvae induce a nutritive tissue during compatible interactions with wheat. Ann Entomol Soc Am 99(2):305–316
Hartley SE (1998) The chemical composition of plant galls: are levels of nutrients and secondary compounds controlled by the gall-former? Oecologia 113(4):492–501
Hatchett JH, Kreitner GL, Elzinga RJ (1990) Larval mouthparts and feeding mechanism of the Hessian fly (Diptera: Cecidomyiidae). Ann Entomol Soc Am 83(6):1137–1147
Heath JJ, Stireman JO III (2010) Dissecting the association between a gall midge, *Asteromyia carbonifera*, and its symbiotic fungus, *Botryosphaeria dothidea*. Entomol Exp Appl 137(1):36–49
Huang MY, Chou HM, Chang YT, Yang CM (2014) The number of cecidomyiid insect galls affects the photosynthesis of *Machilus thunbergii* host leaves. J Asia Pac Entomol 17(2):151–154
Huang MY, Huang WD, Chou HM, Chen CC, Chen PJ, Chang YT, Yang CM (2015) Structural, biochemical, and physiological characterization of photosynthesis in leaf-derived cup-shaped galls on *Litsea acuminata*. BMC Plant Biol 15(61):1–12
Isaias RMS, Oliveira DC, Carneiro RGS, Kraus JE (2014) Developmental anatomy of galls in the Neotropics: arthropods stimuli versus host plant constraints. In: Fernandes GW, Santos JC (eds) Neotropical insect galls. Springer, New York, pp 15–34
Isaias RMS, Oliveira DC, Moreira ASFP, Soares GLG, Carneiro RGS (2015) The imbalance of redox homeostasis in arthropod-induced plant galls: mechanisms of stress generation and dissipation. Biochim Biophys Acta Gen Subj 1850(8):1509–1517
Isaias RMS, Kraus JE, Costa EC, Carneiro RGS (2024) The anatomy of neotropical galls and the untold lessons about the morphogenetical potentialities of plants. Rodriguésia 75:e01542023
Izzo TJ, Julião GR, Almada ED, Fernandes GW (2006) Hiding from defenders: localized chemical modification on the leaves of an Amazonian ant-plant induced by a gall-making insect (Diptera: Cecidomyiidae). Sociobiology 48(2):417–426
Jäger-Zürn I, Spies M, Philbrick CT, Bove CP, Mora-Olivo A (2013) Plant galls (cecidia) in the neotropical water plant family Podostemaceae induced by larvae of Chironomidae. Spixiana 36(1):97–112
Jayaraman P (1989) Ontogenetic studies on some leaf-fold galls. Proc Indian Natn Sci Acad 55(1):57–62
Jorge NC, Freitas MSC, Caffaro RM, Vale FHA, Lemos-Filho JP, Isaias RMS (2022a) Vascular traits of stem galls: cell increment versus morphogenetic constraints in wood anatomy. Plant Biol 24(3):450–457
Jorge NC, Vasconcelos HL, Freitas MSC, Vale FHA, Isaias RMS (2022b) The peculiar post-senescence in globoid stem galls triggered by fungi and arboreal ants on *Eremanthus erythropappus* (DC.) McLeisch (Asteraceae). S Afr J Bot 150:1026–1030
Jorge NC, Guedes LM, Aguilera N, Becerra J, Isaias RMS (2022c) Allelopathic potential of the extracts of non-galled stems and globoid stem galls of *Eremanthus erythropappus* (DC) McLeish (Asteraceae). Biochem Syst Ecol 100:104379
Karabourniotis G, Liakopoulos G, Nikolopoulos D, Bresta P (2020) Protective and defensive roles of non-glandular trichomes against multiple stresses: structure–function coordination. J For Res 31(1):1–12
Kelch NS, Neves FS, Fernandes GW, Wirth R (2016) Mechanisms driving galling success in a fragmented landscape: synergy of habitat and top-down factors along temperate forest edges. PLoS One 11(6):e0157448
Kraus JE, Sugiura HC, Cutrupi S (1996) Morfologia e ontogenia em galhas entomógenas de *Guarea macrophylla* subsp. *tuberculata* (Meliaceae). Fitopatol Bras 21(3):349–356

Kuster VC, Rezende UC, Cardoso JCF, Isaias RMS, Oliveira DC (2020) How galling organisms manipulate the secondary metabolites in the host plant tissues? A histochemical overview in neotropical gall systems. In: Mérillon JM, Ramawat K (eds) Co-evolution of secondary metabolites. Springer, Cham, pp 823–842

Kuster VC, Cardoso JCF, Moreira ASFP, Isaias RMS, Oliveira DC (2022) Assessing oxidative stress and photosynthetic activity in leaf galls induced by different species of galling insect on *Aspidosperma* spp. Biologia 77(12):3441–3454

Kuzmanich N, Giorgis MA, Bernaschini L, Tavella J, Salvo A (2023) Galling insect communities mediate the effects of fire on their associated parasitoid communities. Insect Conserv Divers 16(6):817–827

Labandeira CC (2021) Ecology and evolution of gall-inducing arthropods: the pattern from the terrestrial fossil record. Front Ecol Evol 9:632449

Lalonde RG, Shorthouse JD (1984) Developmental morphology of the gall of *Urophora cardui* (Diptera, Tephritidae) in the stems of Canada thistle (*Cirsium arvense*). Can J Bot 62(7):1372–1384

Larew H (1981) A comparative anatomical study of galls caused by the major cecidogenetic groups, with special emphasis on the nutritive tissue [dissertation]. Oregon State University

Lawson T, Blatt MR (2014) Stomatal size, speed, and responsiveness impact on photosynthesis and water use efficiency. Plant Physiol 164(4):1556–1570

Lebel T, Peele C, Veenstra A (2012) Fungi associated with Asphondylia (Diptera: Cecidomyiidae) galls on *Sarcocornia quinqueflora* and *Tecticornia arbuscula* (Chenopodiaceae). Fungal Divers 55(1):143–154

Lin SF, Tokuda M, Yang MM (2019) Leaf gall polymorphism and molecular phylogeny of a new *Bruggmanniella* species (Diptera: Cecidomyiidae: Asphondyliini) associated with *Litsea acuminata* (Lauraceae) in Taiwan, with ecological comparisons and a species description. Entomol Sci 23(1):10–22

Liu Q, Luo L, Zheng L (2018) Lignins: biosynthesis and biological functions in plants. Int J Mol Sci 19(2):1–16

Maia VC (2018) Insect galls on Myrtaceae: richness and distribution in brazilian restingas. Biota Neotropica. 19(1):e20180526

Mani MS (1964) Ecology of plant galls. Dr Junk Publishers, The Hague

Mapes CC, Davies PJ (2001) Cytokinins in the ball gall of Solidago altissima and in the gall forming larvae of Eurosta solidaginis. New Phytol 151(1):203–212

Marchosky RJ Jr, Craig TP (2004) Gall size-dependent survival for *Asphondylia atriplicis* (Diptera: Cecidomyiidae) on *Atriplex canescens*. Environ Entomol 33(3):709–719

Meakin PJ, Roberts JA (1991) Anatomical and biochemical changes associated with the induction of oilseed rape (*Brassica napus*) pod dehiscence by *Dasineura brassicae* (Winn.). Ann Bot 67(3):193–197

Mendonça MS Jr, Romanowski HP (2002) Natural enemies of the gall-maker *Eugeniamyia dispar* (Diptera, Cecidomyiidae): predatory ants and parasitoids. Braz J Biol 62(2):269–275

Mendonça MS Jr, Romanowski HP (2012) Population ecology of the multivoltine Neotropical gall midge *Eugeniamyia dispar* (Diptera, Cecidomyiidae). Iheringia Ser Zoo 102(2):170–176

Meyer J (1957) Cécidogenèse comparée de quelques galles d'Arthropodes et évolution cytologique des tissus nourriciers [Thèse Doctorat d'Etat]. Strasbourg

Meyer J (1987) Plant galls and gall inducers. Gebrüder Borntraeger, Berlin

Meyer J, Maresquelle HJ (1983) Anatomie des Galles. Gebrüder Borntraeger, Berlin

Meyer. (1969) Irrigation vasculaire dans les galles. Bull Soc Bot Fr 116(1):75–97

Miller DG III, Raman A (2019) Host–plant relations of gall-inducing insects. Ann Entomol Soc Am 112(1):1–19

Mishima M, Yukawa J (2007) Dimorphism of leaf galls induced by *Pseudasphondylia neolitseae* (Diptera: Cecidomyiidae) on *Neolitsea sericea* (Lauraceae) and their distributional patterns in Kyushu, Japan. Bulletin of the Kyushu University Museum. pp 57–64

Mishima M, Sato S, Tsuda K, Yukawa J (2014) Sexual isolation between two known intraspecific populations of *Hartigiola* (Diptera: Cecidomyiidae) that induce leaf galls on upper and lower surfaces of *Fagus crenata* (Fagales: Fagaceae), indicating possible diversification into sibling species. Ann Entomol Soc Am 107(4):789–798

Mishra P, Patni V (2008) Anatomical changes and development of Cecidomyiid leaf gall of *Quercus leucotrichophora* Himalayan white oak. J Mycol Plant Pathol 38(2):203–207

Molnár BP, Boddum T, Hill SR, Hansson BS, Hillbur Y, Birgersson G (2018) Ecological and phylogenetic relationships shape the peripheral olfactory systems of highly specialized gall midges (Cecidomiiydae). Front Physiol 9:308834

Moura MZD, Soares GLG, Isaias RMS (2008) Species-specific changes in tissue morphogenesis induced by two arthropod leaf gallers in *Lantana camara* L. (Verbenaceae). Aust J Bot 56(2):153–160

Muravnik LE (2021) The structural peculiarities of the leaf glandular trichomes: a review. In: Ramawat KG, Ekiert HM, Goyal S (eds) Plant cell and tissue differentiation and secondary metabolites: fundamentals and applications. Springer, Cham, pp 1–35

Naseem M, Wölfling M, Dandekar T (2014) Cytokinins for immunity beyond growth, galls and green islands. Trends Plant Sci 19(8):481–484

Nobrega LP, Silva JB, Luna BN, Ferreira BG (2021) Modulation of anatomical adaptations of leaves of *Avicennia schaueriana* (Acanthaceae) by a galling *Meunieriella* (Cecidomyiidae). Flora 274:151750

Nobrega LP, Sá-Haiad B, Ferreira BG (2022) Mechanisms of hydraulic conductivity in the leaf galls of *Meunieriella* sp. (Cecidomyiidae) in *Avicennia schaueriana* (Acanthaceae): does vascularization explain the preferred sites of induction? Plant Biol 25(1):198–207

Nogueira RM, Costa EC, Silva JS, Isaias RMS (2018) Structural and histochemical profile of *Lopesia* sp. Rübsaamen 1908 pinnula galls on *Mimosa tenuiflora* (Willd.) Poir. in a Caatinga environment. Hoehnea 45(2):314–322

Nogueira RM, Costa EC, Silva JS, Isaias RMS (2022) A phenological trick and cell wall bricks toward adaptive strategies of *Mimosa tenuiflora-Lopesia mimosae* interaction in Caatinga environment. Flora 294:152121

Oliveira DC, Isaias RMS (2009) Influence of leaflet age in anatomy and possible adaptive values of the midrib gall of *Copaifera langsdorffii* (Fabaceae: Caesalpinioideae). Rev Biol Trop 57(1-2):293–302

Oliveira DC, Isaias RMS (2010) Redifferentiation of leaflet tissues during midrib gall development in *Copaifera langsdorffii* (Fabaceae). S Afr J Bot 76(2):239–248

Oliveira DC, Magalhães TA, Carneiro RGS, Alvim MN, Isaias RMS (2010) Do Cecidomyiidae galls of *Aspidosperma spruceanum* (Apocynaceae) fit the pre-established cytological and histochemical patterns? Protoplasma 242(1-4):81–93

Oliveira DC, Carneiro RGS, Magalhães TA, Isaias RMS (2011a) Cytological and histochemical gradients on two *Copaifera langsdorffii* Desf. (Fabaceae)—Cecidomyiidae gall systems. Protoplasma 248(4):829–839

Oliveira DC, Isaias RMS, Moreira ASFP, Magalhães TA, Lemos-Filho JP (2011b) Is the oxidative stress caused by *Aspidosperma* spp. galls capable of altering leaf photosynthesis? Plant Sci 180(3):489–495

Oliveira DC, Mendonça MS Jr, Moreira ASFP, Lemos-Filho JP, Isaias RMS (2013) Water stress and phenological synchronism between *Copaifera langsdorffii* (Fabaceae) and multiple galling insects: formation of seasonal patterns. J Plant Interact 8(3):225–233

Oliveira DC, Isaias RMS, Fernandes GW, Ferreira BG, Carneiro RGS, Fuzaro L (2016) Manipulation of host plant cells and tissues by gall-inducing insects and adaptive strategies used by different feeding guilds. J Insect Physiol 84:103–113

Pan LY, Chen WN, Chiu ST, Raman A, Chiang TC, Yang MM (2015) Is a gall an extended phenotype of the inducing insect? A comparative study of selected morphological and physiological traits of leaf and stem galls on *Machilus thunbergii* (Lauraceae) induced by five species of *Daphnephila* (Diptera: Cecidomyiidae) in Northeastern Taiwan. Zool Sci 32(3):314–321

Pincebourde S, Casas J (2016) Hypoxia and hypercarbia in endophagous insects: larval position in the plant gas exchange network is key. J Insect Physiol 84:137–153
Ponce GE, Fuse M, Chan A, Connor EF (2021) The localization of phytohormones within the gall-inducing insect Eurosta solidaginis (Diptera: Tephritidae). Arthropod Plant Interact 15(3):375–385
Price PW, Fernandes GW, Waring GL (1987) Adaptive nature of insect galls. Environ Entomol 16(1):15–24
Qi Y, Duan C, Ren L, Wu H (2020) Growth dynamics of galls and chemical defence response of *Pinus thunbergii* Parl. to the pine needle gall midge, *Thecodiplosis japonensis* Uchida & Inouye (Diptera: Cecidomyiidae). Sci Rep 10(1):12289
Raman A (2003) Cecidogenetic behavior of some gall-inducing thrips, psyllids, coccids, and gall midges, and morphogenesis of their galls. Orient Insects 37(1):359–413
Raman A (2007) Insect-induced plant galls of India: unresolved questions. Curr Sci 92(6):748–757
Raman A (2011) Morphogenesis of insect-induced plant galls: facts and questions. Flora 206(6):517–533
Raman A, Devadas C (1977) Morphology, anatomy and development of the midrib galls on the leaflets of *Lannea coramandelica* (Hoult.) Merrill (Anacardiaceae) caused by *Odinadiplosis odinae* Mani (Diptera). Proc Indian Acad Sci 86(3):159–165
Raman A, Suryanarayanan TS (2017) Fungus–plant interaction influences plant-feeding insects. Fungal Ecol 29:123–132
Rezende UC, Moreira ASFP, Kuster VC, Oliveira DC (2018) Structural, histochemical and photosynthetic profiles of galls induced by *Eugeniamyia dispar* (Diptera: Cecidomyiidae) on the leaves of *Eugenia uniflora* (Myrtaceae). Rev Biol Trop 66(4):1469–1480
Rezende UC, Jorge NC, Dolabela B, Isaias RMS, Hanson P, Oliveira DC (2022) Endoparasitoidism reflections on the structure of a Cecidomyiid-Gall on *Piptocarpha macropoda* (DC.) Baker (Asteraceae). Biol Bull 49(1):98–202
Rohfritsch O (1992) Patterns in gall development. In: Shorthouse JD, Rohfritsch O (eds) Biology of insect-induced galls. Oxford University Press, New York, pp 60–86
Rohfritsch O (1997) Morphological and behavioural adaptations of the gall midge *Lasioptera arundinis* (Schiner) (Diptera Cecidomyiidae) to collect and transport conidia of its fungal symbiont. Tijdschr Entomol 140(1):59–66
Rohfritsch O (2008) Plants, gall midges, and fungi: a three-component system. Entomol Exp Appl 128(1):208–216
Rohfritsch O, Shorthouse JD (1982) Insect galls. In: Kahl G, Schell JS (eds) Molecular biology of plant tumors. Academic, New York, pp 131–152
Roubos CR, Liburd OE (2013) Parasitism of *Dasineura oxycoccana* (Diptera: Cecidomyiidae) in North Central Florida. Environ Entomol 42(3):424–429
Sá CEMD, Silveira FA, Santos JC, Isaias RMS, Fernandes GW (2009) Anatomical and developmental aspects of leaf galls induced by *Schizomyia macrocapillata* Maia (Diptera: Cecidomyiidae) on *Bauhinia brevipes* Vogel (Fabaceae). Braz J Bot 32(2):319–327
Santos-Silva J, Araujo TJ (2020) Are Fabaceae the principal super-hosts of galls in Brazil? An Acad Bras Ciênc 92(2):e20181115
Schultz JC, Edger PP, Body MJ, Appel HM (2019) A galling insect activates plant reproductive programs during gall development. Sci Rep 9(1):1–17
Shih TH, Lin SH, Huang MY, Sun CW, Yang CM (2018) Transcriptome profile of cup-shaped galls in *Litsea acuminata* leaves. PLoS One 13(10):e0205265
Shukle RH, Mittapalli O, Morton PK, Chen MS (2009) Characterization and expression analysis of a gene encoding a secreted lipase-like protein expressed in the salivary glands of the larval Hessian fly, *Mayetiola destructor* (Say). J Insect Physiol 55(2):105–112
Souza AP, Oliveira DC, Dalvi VC, Kuster VC (2024) Nutritive tissue rich in reserves in the cell wall and protoplast: the case of *Manihot esculenta* (Euphorbiaceae) galls induced by *Iatrophobia brasiliensis* (Diptera, Cecidomyiidae). Protoplasma 261(3):513–525

Staszak AM, Ratajczak E, Leśniewska J, Piotrowska-Niczyporuk A, Kostro-Ambroziak A (2023) A broad spectrum of host plant responses to the actions of the gall midge: case study of *Robinia pseudoacacia* L. and *Obolodiplosis robiniae* (Haldeman). BMC Plant Biol 23(1):1–19

Stone GN, Schönrogge K (2003) The adaptive significance of insect gall morphology. Trends Ecol Evol 18(10):512–522

Sugiura S, Yamazaki K, Fukasawa Y (2004) Weevil parasitism of ambrosia galls. Ann Entomol Soc Am 97(1):184–193

Suzuki AYM, Bedetti CS, Isaias RMS (2015) Detection and distribution of cell growth regulators and cellulose microfibrils during the development of *Lopesia* sp. galls on *Lonchocarpus cultratus* (Fabaceae). Botany 93(7):435–444

Takeda S, Yoza M, Amano T, Ohshima I, Hirano T, Sato MH, Sakamoto T, Kimura S (2019) Comparative transcriptome analysis of galls from four different host plants suggests the molecular mechanism of gall development. PLoS One 14(10):e0223686

Takeda S, Hirano T, Ohshima I, Sato MH (2021) Recent progress regarding the molecular aspects of insect gall formation. Int J Mol Sci 22(17):9424

Tanaka Y, Okada K, Asami T, Suzuki Y (2013) Phytohormones in Japanese mugwort gall induction by a gall-inducing gall midge. Biosci Biotech Bioch 77(9):1942–1948

Teixeira CT, Oliveira DCD, Kuster VC, Isaias RMS (2018) Immunocytochemical demonstration of cell wall components related to tissue compartments in the globoid galls induced by *Clinodiplosis* sp. (Cecidomyiidae) on *Croton floribundus* Spreng. (Euphorbiaceae). Botany 96(1):9–18

Teixeira CT, Kuster VC, Carneiro RGS, Cardoso JCF, Isaias RMS (2022) Anatomical profiles validate gall morphospecies under similar morphotypes. J Plant Res 135(4):593–608

Tokuda M (2012) Biology of *Asphondyliini* (Diptera: Cecidomyiidae). Entomol Sci 15(4):361–383

Tokuda M, Yukawa J, Gôukon K (2014) Life history traits of *Pseudasphondylia rokuharensis* (Diptera: Cecidomyiidae) affecting emergence of adults and synchronization with host plant phenology. Environ Entomol 36(3):518–523

Tooker JF, Helms AM (2014) Phytohormone dynamics associated with gall insects, and their potential role in the evolution of the gall-inducing habit. J Chem Ecol 40(7):742–753

Ullrich CI, Aloni R (2000) Vascularization is a general requirement for growth of plant and animal tumours. J Exp Bot 51(353):1951–1960

Waring GL, Price PW (1990) Plant water stress and gall formation (Cecidomyiidae: *Asphondylia* spp.) on creosote bush. Ecol Entomol 15(1):87–95

Yukawa J (2000) Synchronization of gallers with host plant phenology. Popul Ecol 42(2):105–113

Yukawa J, Rohfritsch O (2005) Biology and ecology of gall inducing Cecidomyiidae (Diptera). In: Raman A, Schaefer CW, Withers TW (eds) Biology, ecology, and evolution of gall-inducing arthropods. Science Publishers, Enfield, pp 273–304

Yukawa J, Uechi N (2021) Life history traits. In: Yukawa J, Tokuda M (eds) Biology of gall midges: evolution, ecology, and biological interactions. Springer, Berlin, pp 119–149

Yukawa J, Nakagawa K, Saigou T, Awa T, Fukuda T, Higashi M (2013) Adult behavior of an ambrosia gall midge *Illiciomyia yukawai* (Diptera: Cecidomyiidae) and synchronization between its emergence and host plant phenology. Entomol Sci 16(4):400–412

Yukawa J, Tokuda M, Sato S, Ganaha-Kikumura T, Uechi N (2021) In: Yukawa J, Tokuda M (eds) Biology of gall midges: evolution, ecology, and biological interactions. Springer, Berlin, pp 11–52

Zhong R, Cui D, Ye ZH (2019) Secondary cell wall biosynthesis. New Phytol 221(4):1703–1723

Exploring the Intriguing Characteristics of Eriococcid-Induced Galls

Ana Flávia de Melo Silva, Mariana de Souza Costa Freitas, and Thiago Alves Magalhães

1 General Aspects

The Eriococcidae, commonly known as felt scales for their distinct felt-like wax covering, ranks as the fourth-largest family in the Coccoidea superfamily of the Hemiptera order (Miller and Gimpel 2000; Ben-Dov et al. 2023). Renowned for their diversity, Eriococcidae-induced galls are particularly abundant in countries of southern continents, notably in Australia and Brazil. While most Eriococcidae species reproduce sexually and exhibit sexual dimorphism in the adult stage (Gullan and Kosztarab 1997), some employ parthenogenesis due to the brief lifespan of adult males (Miller and Kosztarab 1979).

Galls induced by Eriococcidae may exhibit variations in morphology related to the sex of the gall inducer (Gullan and Kosztarab 1997; Gonçalves et al. 2005). Male and female galls can differ significantly in shape, structure, and may even be induced and develop in different host plant organs (Gullan and Kosztarab 1997). The eriococcids may exhibit characteristics such as sexual dimorphism, asexual reproductive strategies and variations in lifespan (Gullan and Kosztarab 1997). As consequence, eriococcid-induced galls display differences in their anatomical and chemical profiles, particularly in those where sexual dimorphisms occur. In the *Pseudobombax grandiflorum* (Malvaceae)-*Eriogallococcus isaias* system, the galling-eriococcid dimorphism does not reflect in the gall morphotype, but implies in subtle features at the anatomical level (Magalhães 2014). On the other hand, the sexual dimorphism in the *Annona dolabripetala* - *Pseudotectococcus rolliniae*

A. F. de Melo Silva (✉) · M. de Souza Costa Freitas
Department of Botany, Universidade Federal de Minas Gerais (UFMG), Belo Horizonte, Minas Gerais, Brazil

T. A. Magalhães
Department of Biology, Universidade Federal de Lavras (UFLA), Lavras, Minas Gerais, Brazil

D. C. de Oliveira, R. M. dos Santos Isaias (eds.), *Plant Galls*,
https://doi.org/10.1007/978-3-031-80064-1_17

(Annonaceae) system reflects in the gall morphotype, with female-induced galls being larger and globoid, while the males are smaller and conical (Gonçalves et al. 2005). In the *Matayba guianensis* (Sapindaceae) - *Bystracoccus mataybae* system, the female insects are parthenogenetic and induce globoid galls (Pfeffer et al. 2018).

The female-induced galls of the three host species share anatomical features, such as the lignified cell layers around the chamber, which results from the more intense and longer feeding activity of the females, mainly of *E. isaias* and *P. rolliniae* (Silva et al. 2023). Galls induced by insects with sexual dimorphism also exhibit differences in their chemical profile (Silva et al. 2023). Only galls induced by females of *P. rolliniae*, *B. mataybae*, and *E. isaias* accumulate starch grains in the cells of the inner cortical parenchyma (Silva et al. 2023). These metabolites may aid in the dissipation of ROS and maintenance of cell and tissue homeostasis, which are necessary to complete their life cycles, which are longer in female-induced galls than in male-induced galls (Silva et al. 2023). In an overall analysis, peculiar anatomical and morphological variation of the galls induced by Eriococcidae are associated with differences not only in their life cycle, but probably also to the molecular basis of the distinct and diverse eriococcid-induced galls.

2 Eriococcid-Inducing Gall Distribution and Their Hosts

The Eriococcidae family exhibits its greatest diversity in the southern continents, being well represented in Australia, New Zealand, and South America (Miller and Gimpel 2000). Up to 2023, the Eriococcidae has comprised 109 genera and 681 species (scalenet.info), within which 23 genera and 89 species are known as gall inducers (Table 1). Regarding the gall-inducing Eriococcidae, the high species diversity is associated with the Myrtaceae, which is one of the most important host plant families of eriococcids in Australia (Cook and Gullan 2004), and Brazil (Table 1). Accordingly, where the Myrtaceae occur, great is the chance for eriococcid occurrence. The significant portion of the Australian gall-inducing scale insect fauna, for instance, is associated with the diverse genus *Eucalyptus* (Gullan et al. 2005). In South America, eriococcid-inducing galls are also associated with Myrtaceae, but their taxonomic identity is unknown, even in genus level (Gonzalez et al. 2020; Table 1).

Peculiarly, the scale insects have limited mobility and gene exchange, often staying on a single host plant throughout their lives (Hanks and Denno 1993). This sedentary behavior, along with specific host preferences of the crawlers and the weak-flying capacity of the adult males, lead to genetically distinct populations adapted to their host plants (Gullan and Kosztarab 1997), which may explain the large occurrence of the *Eucalyptus*-Eriococcidae associations in Australia. These characteristics influence host finding and preference in nature, and impact the development of the associated gall morphotypes (Gullan and Kosztarab 1997).

Table 1 List of Eriococcid-inducing galls

Genera	Specie	Host plant	Morphotye	Host organ	Distribution/ Localization	References
Aculeococcus	*A. morrisoni*	Unkown	Conical (♀)	(♀) Leaves	Brazil	Hodgson and Miller (2010)
Apiomorpha	*A. minor*	*Eucalyptus*/ Myrtaceae	Enclosing	(♀) Stems / (♂) Leaves	Australia	Cook and Gullan (2001, 2004)
	A. munita tereticornuta	*Eucalyptus*/ Myrtaceae	Enclosing	Stems	Australia	Cook and Gullan (2001, 2004)
	A. spinifer	*Eucalyptus*/ Myrtaceae	Enclosing	(♀) Leaves / (♂) clustered attached to the female induced galls	Australia	Cook and Gullan (2001, 2004), Gullan (1984a)
	A. strombylosa	*Eucalyptus*/ Myrtaceae	Enclosing	(♀) Stems / (♂) Steams - Leaves	Australia	Cook and Gullan (2001, 2004)
	A. gullanae	*Eucalyptus*/ Myrtaceae	Enclosing	Leaves	Australia	Cook (2003)
	A. malleeacola	*Eucalyptus*/ Myrtaceae	Globoid (♀)	(♀) Stems, buds, fruits	Australia	Gullan (1984b), Mills et al. (2017)
	A. jucundacrispi	*Eucalyptus jucunda*/ Myrtaceae	Globoid	(♀) Stems/ (♂) Leaves	Australia	Mills et al. (2017)
	A. gongylocarpae	*Eucalyptus*/ Myrtaceae	(♀) Urn-shaped galls	(♀) Buds, stems	Australia	Mills et al. (2017)
	A. pomaphora	*Eucalyptus*/ Myrtaceae	(♀) Urn-shaped galls	(♀) Buds, stems	Australia	Mills et al. (2017)
	A. pileata	*Eucalyptus*/ Myrtaceae	(♀) Globoid /(♂) Cylindrical	(♀) Stems / (♂) Leaves	Australia	Gullan (1984b)
	A. duplex	*Eucalyptus*/ Myrtaceae	(♀) Fusiform / (♂) Cylindrical	(♀) Stems/ (♂) Leaves	Australia	Gullan (1984b)
	A. excupula	*Eucalyptus*/ Myrtaceae	(♀) Globoid / (♂) Cylindrical	(♀) Stems / (♂) Stem	Australia	Gullan (1984b)
	A. dipsaciformis	*Eucalyptus*/ Myrtaceae	(♀) Globoid / (♂) Cylindrical	(♀) Stems / (♂) Stem	Australia	Gullan (1984b), Cook et al. (2000)
	A. bauerleni	*Eucalyptus*/ Myrtaceae	(♀) Globoid	(♀) Stems / (♂) Unkown	Australia	Gullan (1984b)
	A. rosaeformis	*Eucalyptus*/ Myrtaceae	(♀) Cylindrical / (♂) Cylindrical or clustered attached to FG	(♀♂) Leaves	Australia	Gullan (1984b)
	A. pharetrata	*Eucalyptus*/Myrtaceae	(♀) Fusiform / (♂) Tubular or clustered attached to female-induced galls	(♀♂) Leaves	Australia	Gullan (1984b), Cook and Gullan (2008)
	A. macqueeni	*Eucalyptus*/Myrtaceae	(♀) Fusiform / (♂) Cylindrical	(♀) Stems / (♂) Leaves	Australia	Gullan (1984b), Cook et al. (2000)
	A. tepperi	*Eucalyptus*/Myrtaceae	(♀) Globoid	(♀) Stems / (♂) Leaves	Australia	Gullan (1984b)

(continued)

Table 1 (continued)

Genera	Specie	Host plant	Morphotye	Host organ	Distribution/ Localization	References
Apiomorpha	*A. maliformis*	*Eucalyptus*/Myrtaceae	(♀) Globoid / (♂) Tubular	(♀) Stems / (♂) Leaves	Australia	Gullan (1984b), Cook and Gullan (2001)
	A. annulata	*Eucalyptus acmenioides*/Myrtaceae	(♀) Cylindrical/ (♂) Cylindrical	(♀) Stems / (♂) Leaves	Australia	Gullan (1984b), Mills and Cook (2014)
	A. sessilis	*Eucalyptus*/ Myrtaceae	(♀) Cylindrical	(♀) Stems	Australia	Gullan (1984b)
	A. karschi	*Eucalyptus*/ Myrtaceae	(♀) Globoid/ (♂) Cylindrical	(♀) Stems / (♂) Leaves	Australia	Gullan (1984b)
	A. variabilis	*Eucalyptus*/ Myrtaceae	(♀) Pyriform	(♀) Stems	Australia	Gullan (1984b)
	A. sloanei	*Eucalyptus drepanophylla*/ Myrtaceae	(♀) Fusiform / (♂) Cylindrical	(♀) Stems / (♂) Leaves	Australia	Gullan (1984b)
	A. hilli	*Eucalyptus miniata*/ Myrtaceae	(♀) Globoid	(♀) Stems	Australia	Gullan (1984b)
	A. calycina	*Eucalyptus*/ Myrtaceae	(♀) Obconical to cylindrical / (♂) Conical	(♀) Stems / (♂) Stems, leaves or on the FG	Australia	Gullan (1984b)
	A. urnalis	*Eucalyptus*/ Myrtaceae	(♀) Urn-shaped / (♂) conical	(♀) Stems / (♂) Stems	Australia	Gullan (1984b)
	A. longiloba	*Eucalyptus*/ Myrtaceae	(♀) Amorphous	(♀) Stems	Australia	Gullan (1984b)
	A. densispinosa	*Eucalyptus*/ Myrtaceae	(♀) Fusiform / (♂) Cylindrical	(♀) Stems / (♂) Leaves	Australia	Gullan (1984b)
	A. helmsii	*Eucalyptus*/ Myrtaceae	(♀) Obconical to fusiform / (♂) Cylindrical	(♀) Stems / (♂) Leaves	Australia	Gullan (1984b)
	A. frenchi	*Eucalyptus*/ Myrtaceae	Globoid or cylindrical (♀♂)	(♀) Stems / (♂) Stems, leaves	Australia	Gullan (1984b)
	A. withersi	*Eucalyptus*/ Myrtaceae	(♀) Globoid / (♂) Cylindrical	(♀) Stems / (♂) Leaves	Australia	Gullan (1984b)
	A. ovicola	*Eucalyptus*/ Myrtaceae	(♀) Globoid to fusiform / (♂) Cylindrical	(♀) Stems / (♂) Leaves	Australia	Gullan (1984b)
	A. conica	*Eucalyptus*/ Myrtaceae	(♀) Amorphous / (♂) Cylindrical	(♀) Stems / (♂) Stems, leaves	Australia	Gullan (1984b)
	A. intermedia	*Eucalyptus*/ Myrtaceae	(♀) Cylindrical / (♂) Cylindrical or clustered attached to female-induced galls	(♀) Stems / (♂) Stems, fruits, leaves	Australia	Gullan (1984b)
	A. attenuata	*Eucalyptus*/ Myrtaceae	(♀) Fusiform / (♂) Cylindrical	(♀) Stems / (♂) Stems, leaves	Australia	Gullan (1984b)

Apiomorpha	*A. amarooensis*	*Eucalyptus*/ Myrtaceae	(♀) Fusiform / (♂) Cylindrical	(♀) Stems / (♂) Leaves	Australia	Gullan (1984b)
	A. pedunculata	*Eucalyptus*/ Myrtaceae	(♀) Fusiform pedunculate / (♂) Cylindrical	(♀) Stems / (♂) Stems, petiole, leaves	Australia	Gullan (1984b)
	A. regularis	*Eucalyptus*/ Myrtaceae	(♀) Fusiform pedunculate / (♂) Cylindrical	(♀) Stems / (♂) Leaves	Australia	Gullan (1984b)
	A. subconica	*Eucalyptus*/ Myrtaceae	(♀) Fusiform / (♂) Cylindrical	(♀) Stems / (♂) Leaves	Australia	Gullan (1984b)
	A. ovicoloides	*Eucalyptus*/ Myrtaceae	(♀) Globoid / (♂) Cylindrical	(♀) Stems / (♂) Leaves	Australia	Gullan (1984b)
Apiococcus	*A. asperatus*	Myrtaceae	Globoid (♀)	(♀) Stems	Brazil	González and Claps (2013)
	A. globosus	Myrtaceae	Globoid (♀)	(♀) Stems	Brazil	González and Claps (2013)
	A. gregarius	Myrtaceae	Globoid (♀)	(♀) Stems	Brazil	González and Claps (2013)
	A. singularis	Myrtaceae	Globoid (♀)	(♀) Stems	Brazil	González and Claps (2013)
Ascelis	*A. praemollis*	*Corymbia*/Myrtaceae	Enclosing	(♀) Leaves	Australia	Cook and Gullan (2004)
Bystracoccus	*B. mataybae*	*Mataybae guianensis*/ Sapindaceae	Globoid (♀)	Stems, (♀ Parthenogenetic) leaves	Brazil	Hodgson et al. (2013), Pfeffer et al. (2018)
Callococcus	*C. leptospermi*	*Leptospermum*/ Myrtaceae	Stem swelling (♀)	(♀) Stems	Australia	Cook and Gullan (2004)
Calycicoccus	*C. merwei*	*Apodytes*/ Icacinaceae	Enclosing	(♀) Leaves	South Africa	Cook and Gullan (2004), Gullan et al. (2006)
Capulinia	*C. crateraformis*	*Eugenia jaboticaba*/ Myrtaceae	Crater-shaped(♀)	(♀)Stems/ (♂) not described	Brazil	Kondo et al. (2016)
	C. orbiculata	*Metrosideros*/ Myrtaceae	Stem swelling (♀♂)	(♀) Stems/ (♂) Stems	New Zealand	Kondo et al. (2016)
Cylindrococcus	*C. casuarinae*	*Allocasuarina*/ Casuarinaceae	Rosette (♀)	(♀) Buds, stems	Australia	Cook and Gullan (2004)
	C. spiniferus	*Allocasuarina*/ Casuarinaceae	Rosette (♀)	(♀) Buds, stems	Australia	Cook and Gullan (2004)
Cystococcus	*C. echiniformis*	*Corymbia*/Myrtaceae	Enclosing	Stem	Australia	Cook and Gullan (2004)

(continued)

Table 1 (continued)

Genera	Specie	Host plant	Morphotye	Host organ	Distribution/ Localization	References
Dromedaricoccus	*D. hansoni*	*Astronium graveolens*/ Anacardiaceae	Enclosing	Stem/ Petioles/Leaves	Costa Rica	Hodgson and Miller (2010)
Eriococcus	*E. aconeae*	*Pittosporum eugenioides*/ *Pittosporaceae*	Pit/enclosing	Leaves	New Zealand	Henderson (2006)
	E. parsonsiae	*Parsonsia*/ Apocynaceae	Pit/enclosing	(♀) Leaves, stems	New Zealand	Henderson (2006)
	E. elytranthae	*Peraxilla tetrapetala*/ Loranthaceae	Enclosing (♀)	(♀) Leaves	New Zealand	Henderson (2006)
	E. CA1	*Olneya*/ Fabaceae	Leaf fold	Leaves	USA	Cook and Gullan (2004)
Hoheriococcus	*H. fionae*	*Hoheria*/ Malvaceae	Pit/enclosing (♀♂)	(♀♂) Leaves and stems	New Zealand	Henderson (2006)
Eriogallococcus	*Eriogallococcus isaias*	*Pseudobombax grandiflorum*/ Malvaceae	Conical (♀♂)	(♀♂) Leaves and stems	Brazil	Hodgson et al. (2011), Magalhães et al. (2014)
Lachnodius	*Lachnoodius sp.*	*Eucalyptus*/ Myrtaceae	Pit/enclosing	Leaves	Australia	Cook and Gullan (2004)
Opisthoscelis	*O. beardsleyi*	*Eucalyptus*/ Myrtaceae	(♀) Globoid/ (♂) Conical	(♀♂) Leaves	Australia	Hardy and Gullan (2010)
	O. mammularis	*Eucalyptus*/ Myrtaceae	Enclosing	(♀) Stems	Australia	Cook and Gullan (2004), Hodgson and Miller (2010)
	O. serrata	*Eucalyptus*/ Myrtaceae	(♀) Globoid/ (♂) Opening oblong	(♀♂) Leaves	Australia	Hardy and Gullan (2010)
	O. subrotunda	*Eucalyptus*/ Myrtaceae	(♀) Globoid/ (♂) Conical	(♀♂) Leaves	Australia	Hardy and Gullan (2010)
	O. thurgoona	*Eucalyptus*/ Myrtaceae	(♀) Lenticular	(♀) Leaves	Australia	Hardy and Gullan (2010)
	O. tuberculata	*Eucalyptus*/ Myrtaceae	(♀) Pit/enclosing	(♀) Leaves	Australia	Hardy and Gullan (2010)
	O. ungulifinis	*Eucalyptus*/ Myrtaceae	(♀) Rather nipple-like excrescences	(♀) Leaves	Australia	Hardy and Gullan (2010)
Pseudotectococcus	*P. rolliniae*	*Annona dolabripetala*/ Annonaceae	(♀) Globoid/ (♂) Conical	(♀♂) Leaves, stems	Brazil	Hodgson et al. (2004), Gonçalves et al. (2005)

Sphaerococcus	*S. ferrugineus*	*Melaleuca*/ Myrtaceae	(♀) Rosette	(♀) Stems	Australia	Cook and Gullan (2004)
	S. pustulans	*Eucalyptus*/ Myrtaceae	(♀) Goboid	(♀) Stems	Australia	Cook and Gullan (2004), Hardy and Gullan (2007)
	S. sociais	*Melaleuca*/ Myrtaceae	(♀) Rosette	(♀) Stems	Australia	Cook and Gullan (2004)
Stibococcus	*S. cerinus*	*Myrceugenia briggesii*/ Myrtaceae	Enclosing	Leaves	Chile	Miller and González (1975), Hodgson (2020)
Tanyscelis	*T. conica*	*Eucalyptus*/ Myrtaceae	(♀) Globoid	(♀) Leaves	Australia	Hardy and Gullan (2010)
	T. convexa	*Eucalyptus*/ Myrtaceae	(♀) Globoid -Cylindrical/ (♂) Conical	(♀♂) Stems and petioles	Australia	Hardy and Gullan (2010)
	T. grallator	*Eucalyptus*/ Myrtaceae	(♀) Thorn-like/ Conical	(♀) Leaves	Australia	Hardy and Gullan (2010)
	T. maculata	*Eucalyptus*/ Myrtaceae	(♀) Conical	(♀)Stems, petioles, midrib and fruits	Australia	Hardy and Gullan (2010)
	T. maskelli	*Eucalyptus*/ Myrtaceae	(♀) Conical/ (♂) Cylindrical	(♀♂) Stems	Australia	Hardy and Gullan (2010)
	T. megagibba	*Eucalyptus*/ Myrtaceae	Stem swelling (♀♂)	(♀♂) Stems	Australia	Hardy and Gullan (2010)
	T. mollicornuta	*Eucalyptus*/ Myrtaceae	(♀) Globoid	Leaves	Australia	Hardy and Gullan (2010)
	T. pisiformis	*Eucalyptus*/ Myrtaceae	(♀) Oblong/ (♂) Conical	(♀) Leaves/ (♂) Leaves and stems	Australia	Hardy and Gullan (2010)
	T. spinosa	*Eucalyptus*/ Myrtaceae	(♀) Thorn-like- Conical /(♂) Conical	(♀♂) Leaves	Australia	Hardy and Gullan (2010)
	T. tripocula	*Eucalyptus*/ Myrtaceae	Globoid (♀♂)	(♀♂) Leaves	Australia	Hardy and Gullan (2010)
	T. verrucula	*Eucalyptus*/ Myrtaceae	(♀) Globoid/ (♂) Cylindrical-Conical	(♀♂) Leaves, stems	Australia	Hardy and Gullan (2010)
	T. villosigibba	*Eucalyptus*/ Myrtaceae	(♀) Globoid	(♀) Stems	Australia	Hardy and Gullan (2010)
Tectococcus	*T. ovatus*	*Psidium cattleianum*/ Myrtaceae	(♀) Conical	(♀ Parthenogenetic) Leaves	Brazil	Vitorino et al. (2000), Gonzalez et al. (2020)
Neotectococcus	*N. lenticularis*	Unkown	Lenticular	Leaves	Brazil	González and Claps (2013)

3 Diversity of Morphotypes Associated with Eriococcid-Induced Galls

In most of the gall-inducing Eriococcidae, both males and females of the same species induce similar gall morphotypes (Stone and Cook 1998; Stone and Schonrogge 2003). However, some of the galls induced by the Eriococcidae may exhibit sex-related morphological variations (Gullan and Kosztarab 1997; Gonçalves et al. 2005, 2009). The morphology of eriococcid galls can range from simple structures like pits or leaf folds to more complex ones such as enclosed or rosette-like galls (Cook and Gullan 2004), but can also vary from irregular deformities, pockets, or sometimes mimetize epiphyllous buds or fruits (Beardsley 1984; Larew 1990; Meyer 1987). Among the various galling-eriococcid-host plant systems documented in literature, 15 distinct shapes have been reported, with the most frequently observed morphotypes being the cylindrical and the globoid (Table 1). In the most diverse genera *Apiomorpha*, the females induce woody, cylindrical, or globoid galls, often with a textured surface, arm-like extensions, or an apical cap (Gullan 1984b). Certain galls induced by these species appear to mimic the woody capsules of their host eucalypts (Gullan and Kosztarab 1997). Peculiarly, some species of *Apiomorpha* may have male-induced galls on existing female-induced galls (Rung et al. 2014).

Stems are the primary host organs in which the galling-eriococcids induce their galls, resulting in 11 distinct shapes. In contrast, leaves, while hosting a substantial number of galls (68 in total), exhibit a comparatively limited diversity of forms, with 8 different shapes (Table 1). The shape, complexity, and woodiness of these galls represent a selective advantage in reducing the predation or parasitism of scale insects (Gullan and Kosztarab 1997). It has been hypothesized that complex coccoid galls, especially thick-walled, woody ones induced on Myrtaceae, may have evolved as adaptations to protect the occupants from the fire events characteristic of eucalypt forests (Koteja 1986). In South America, prevalent gall morphotypes, including the globoid and conical shapes, are linked both to stems and leaves (Table 1). Species of the *Apiomorpha* and other gall-inducing Eriococcidae genus exhibit a species-specific diversity in gall shapes (Gullan 1984a, b, c), which can be influenced by the duration of the life cycle of the inducer (Gonçalves et al. 2005).

4 Biology of Eriococcid-Induced Gall

The metamorphic males of the eriococcid usually have five instars, four nymphal stages and one adult, whose body is small, and develop in a short time. These short lifespans favored the evolution of dormant male stages, which allow the synchronization of sexual maturation between males and females (Koteja 1985; Gullan and Kosztarab 1997). In contrast, females exhibit neoteny, retaining a nymph-like appearance, and have only two nymphal stages and one adult. They remain sessile within their galls, whose developmental time is longer than that of the males (Gullan et al. 2005). In general, the longest the insect life cycle is, the highest is the impact on the morphogenesis of the host organ (Gonçalves et al. 2005). Such impacts are observed in the anatomical and chemical profiles of sexually dimorphic galls and also in parthenogenetic galls studied in the Brazilian flora (Gonçalves et al. 2005; Magalhães et al. 2014; Pfeffer et al. 2018; Silva et al. 2023).

In the Neotropical region, among the 33 genera and 92 species of the Eriococcidae (Caballero et al. 2019), 16 genera and 33 species have been reported in Brazil (García Morales et al. 2016; Gonzalez et al. 2020). Among the Brazilian species of eriococcid, 7 genera and 12 species are gall-inducers (Table 1). Despite the considerable number of galling-eriococcid, there remains a scarcity of knowledge concerning their gall structure, anatomy and metabolism. The main biological, structural and functional studies on Eriococcidae-induced gall focus on three distinct host plant-galling insect systems occurring in Minas Gerais state, Brazil. These three systems involve distinct host plant species: *Annona dolabripetala* Raddi (= *Rollinia laurifolia* Schltdl.) (Annonaceae)—*Pseudotectococccus rolliniae* Hodgson and Gonçalves (Hodgson et al. 2004), *Pseudobombax grandiflorum* Cav.) A. Robyns (Malvaceae)—*Eriogallococcus isaias* Hodgson and Magalhães (Hodgson et al. 2011), and *Matayba guianensis* Aubl. (Sapindaceae)—*Bystracoccus mataybae* (Hodgson et al. 2013).

Pseudotectococcus rolliniae induces galls both on stems and leaves of the semi-deciduous host plant *A. dolabripetala* (Fig. 1). *Eriogallococcus isaias* induces exclusively leaflet galls on the deciduous host plant *P. grandiflorum* (Fig. 2), while *B. mataybae* induces galls both on stems and leaflets of the semi-deciduous host plant *Matayba guianensis* (Sapindaceae) (Fig. 3). In the case of the *A. dolabripetala—P. rolliniae* system, the gall morphology reflects the sexual dimorphism of the gall-inducing insect (Gonçalves et al. 2005). Conversely, in the *P. grandiflorum—E. isaias* system, the sexual dimorphism is not manifested in gall morphology, but just in the anatomical level (Magalhães 2014). In the *M. guianensis—B. mataybae* system, males are not found, the females are parthenogenetic and remain inside the gall throughout their life cycles (Pfeffer et al. 2018) (Fig. 1).

The host plants and their associated galling-Eriococcidae species exhibit synchronized life cycles. (Gonçalves et al. 2005; Magalhães et al. 2014; Pfeffer et al. 2018). The males of Brazilian Ericoccidae have varied stages, the males of *P. rolliniae* have five instars, whereas the males of *E. isaias* have four instars. Regarding the females, both of these two eriococcid species follow the neotenic pattern with only three instars (Hodgson et al. 2004, 2011, 2013). In the dry season, when the leaves of the three host plants fall down, the first instar nymphs of *P. rolliniae* and *B. mataybae* move to the stems of *A. dolabripetala* and *M. guianensis* host plants, respectively, inducing stem galls and enter dormancy (Gonçalves et al. 2009; Pfeffer et al. 2018). However, the first instar nymphs of *E. isaias* remain affixed to the stems of *P. grandiflorum* without inducing any cell or tissue transformation, i.e., not inducing stem galls (Magalhães et al. 2014). In the spring, the second instar nymphs of all the three Eriococcidae species trigger the development of galls on the immature

Fig. 1 *Annona dolabripetala—Pseudotectococcus rolliniae* system. (**a**) *Annona dolabripetala* tree, with a stem gall in detail (arrow). (**b**) Galls induced by females (globoid) and males (conical) on a leaf. (**c**) Schematic representation of the structure of female and male induced galls. (**d**) Anatomical transverse section of a female-induced gall, with the limit of the IC and the two chambers. (**e**) Anatomical transverse section of a male-induced gall with vascular bundles in the top of the chamber. *Abbrevivations:* CH Nymphal chamber, FI Female insect, IC Inner cortex, OC Outer cortex, SC Sclerenchyma layer, VB Vascular bundle, TC Trichomes

Fig. 2 *Pseudobombax grandiflorum—Eriogallococcus isaias* system. (**a**) *Pseudobombax grandiflorum* tree. (**b**) Galls induced by females (conical) and males (conical) on a leaflet. (**c**) Schematic representation of the structure of female-and-male induced galls. (**d**) Anatomical transverse section of a female-induced gall, with idioblast (mucilaginous cells) at the top of the gall. (**e**) Anatomical transverse section of a female-induced gall with the limit of the IC (sclerenchymatic layer) and MC cortices. (**f**) Trichomes in the ostiole of female-induced gall. (**g**) Anatomical transverse section of a male-induce gall with parenchymatic IC and a mucilaginous idioblast. *Abbrevivations:* CH Nymphal chamber, ID Idioblasts, IC Inner cortex, MC Median cortex, SC Sclerenchymatic layer, VB Vascular bundle, TC Trichomes

Fig. 3 *Matayba guianensis—Bystracoccus mataybae* system. (**a**) Stem galls (arrow). (**b**) Globoid gall induced on a leaflet. (**c**) Schematic representation of the structure of a parthenogenetic gall. (**d**) Anatomical transverse section showing the two chambers and the inducer in the inner chamber

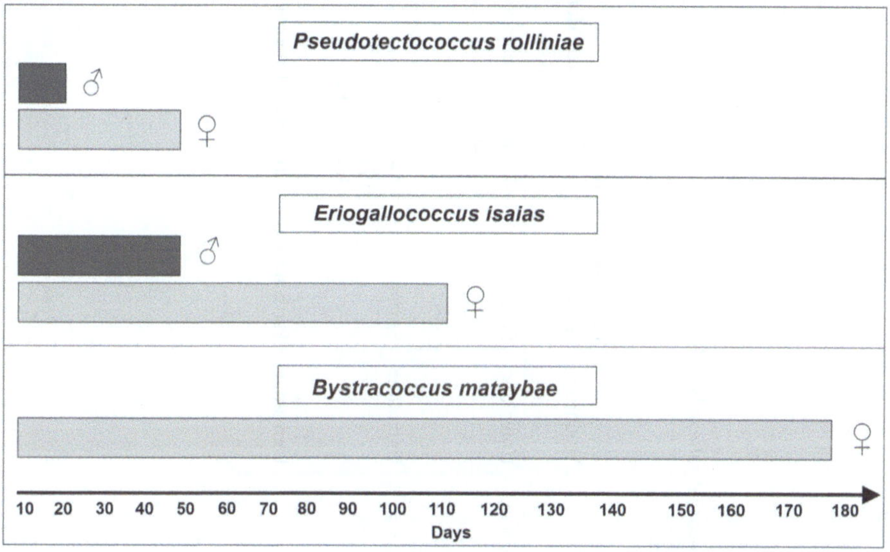

Fig. 4 Diagram of the life cycles duration of the eriococcid-inducing galls in Brazil

leaves of their host plants. On the leaves, the second instar nymphs of the three Eriococcidae species develop up to the adult stage, and the sexual dimorphism of *P. rolliniae* and *E. isaias* becomes evident. Males of *P. rolliniae* and *E. isaias* leave their galls to mate with females, that remain inside their galls throughout their life cycles. The males of *P. rolliniae* and *E. isaias* exhibit shorter life cycle durations than the females (Table 2 and Fig. 4) (Gonçalves et al. 2005; Magalhães et al. 2014).

Table 2 General features of eriococcid galls in Brazil

Host plants	General aspects						Anatomic features			
	Gall inducer	Voltinism	Reproductive strategie	Life cycle duration (days)	Gall Morphotype		Epidermis	Cortex	Number of chambers	Vascular bundles
					Stem	Leaves				
Annona dolabripetala	*Pseudotectococcus rolliniae*	Univoltine	Sexual	44–50 (♀) 14–20 (♂)	Concave	Globoid (♀) Conical (♂)	Unisseriate	Inner and outer cortex	Two (♀) One(♂)	Collateral bundles
Pseudobombax grandiflorum	*Eriogallococcus isaias*	Univoltine	Sexual	75–100 (♀) 36–48 (♂)	–	Conical (♀♂)	Unisseriate	Inner, median and outer cortex	One (♀♂)	Collateral bundles
Matayba guianensis	*Bystracoccus mataybae*	Univoltine	Parthenogenic	110–180	Concave	Globoid	Unisseriate	Inner and outer cortex	Two	Collateral bundles

Meanwhile, the parthenogenetic *B. mataybae* insect lasts a longer period inside their leaflet galls (Pfeffer et al. 2018) than the other two eriococcid species (Table 2 and Fig. 4).

5 Morphological Diversity of Eriococcid Galls in Brazil

Galls induced by individuals of the same species exhibit remarkable similarity in shape, while galls induced by different species on the same plant organ differ significantly (Mani 1964). This observation aligns with the hypothesis that insects, rather than the host plant, regulate gall shape (Cook and Gullan 2008), and that the shape of galls are result of the galling insects behavior conferring an adaptive significance in relation to the free-living ancestors (Stone and Schönrogge 2003). In the three Brazilian host plant-galling-eriococcid systems, three distinct morphotypes are observed: the globoid and the conical for leaf galls and the concave for the stem galls. The stem galls induced by the first instars of *P. rolliniae* and *B. mataybe* in their respective host plants are concave (Gonçalves et al. 2009; Silva et al. 2019). On the other hand, the galls induced on leaves by the parthenogenetic females of *B. mataybae* and the females of *P. rolliniae* are globoid, while those induced by the males are conical. Regarding *E. isaias*, the galls of both sexes are conical. The observed differences in gall morphotypes on *P. rolliniae* - *A. dolabripetala* and *E. isaias* - *P. grandiflorum* systems suggest that these differences are particularly linked to the temporal scale (life cycle duration) and the mechanical pressure exerted by the inducer body, making the gall assume its shape. The elongated-bodied males of *P. rolliniae* relates to the conical-shaped galls, mirroring the shape of their abdomen, while the females with their globoid shape, especially when inseminated, relate to the globoid leaf galls (Gonçalves et al. 2005).

Hypothetically, the absence of morphological differences in *E. isaias*-induced galls may be attributed to a behavioral trait regulated by haplodiploidy, a common characteristic of the Coccoidae, in which the males carry only their mother's genes (Yokogawa and Yahara 2009), which is yet to be corroborated. Moreover, the molecular basis of gall induction has been little explored in scale insects (Cook and Gullan 2008). In the case of *B. mataybae*, a parthenogenetic insect species triggers the development of two morphotypes within the same host-plant species. The configuration of the galls is determined by the early developmental stages of the galling insect on the host plant leaflets, undergoing notable alterations throughout gall maturation (Silva et al. 2019). Generally, it is the insect that controls the morphology of the gall in which it lives (Stone and Schönrogge 2003; Gullan et al. 2005; Cook and Gullan 2008), nevertheless, constraints imposed by the host plant tissues and the organ, where the gall is induced are noteworthy, as seen in the leaf and stem galls on *M. guianensis* and *P. rolliniae* (Gonçalves et al. 2009; Silva et al. 2019).

6 Anatomical Features of Eriococcid Galls in Brazil

Despite the gall-inducing organism control over gall development, the host plant tissues and organs, where the galls are induced can impose important constraints that can be detected using anatomical tools (Isaias et al. 2014). As an example, stem galls induced by the first instar nymphs of *P. rolliniae* on *A. dolabripetala* and *B. mataybae* on *M. guianensis* are less complex than the leaf galls (Gonçalves et al. 2009; Silva et al. 2019). In both systems, the stem concave galls (Figs. 1 and 3) have an increase of phellogen activity and the triggering of cell hypertrophy and phelloderm hyperplasia, but non-discernible changes occur in the cortical and vascular tissues (Gonçalves et al. 2009; Silva et al. 2019).

Leaf galls are induced by the second instar nymphs in the three-host plant species in Brazil. These galls are green and glabrous, with a homogeneous cortex in their initial stages (Gonçalves et al. 2005; Magalhães et al. 2014; Pfeffer et al. 2018; Silva et al. 2019). At the mature stage, the sexual dimorphism becomes evident and is pronounced in the galls induced by *P. rolliniae* (Gonçalves et al. 2005), but can be observed just in the anatomical level in galls induced by *E. isaias* (Magalhães et al. 2014). Despite their different modes of reproduction, female-induced galls of *P. rolliniae* have similarities with the parthenogenetic galls induced by *B. mataybae* (Table 2). The globoid galls induced by the females of *P. rolliniae* and *B. mataybae*, possess two chambers, and cortical parenchyma organized in outer (OC) and inner cortices (IC) (Figs. 1 and 3). The OC is limited by the epidermis and is composed just of parenchyma cells, while the IC is formed by parenchyma and sclerenchyma cells, surrounding the two chambers. Collateral vascular bundles are located on the lateral portions of the inner nymphal chamber and phloem bundles differentiate on top of this chamber. The conical male-induced galls of *P. rolliniae* have a single nymphal chamber, and cortical parenchyma organized in OC and IC, with the OC limited by the epidermis (Fig. 1c). Sclerenchyma cells and collateral vascular bundles occur in the IC on top of the nymphal chamber (Silva et al. 2023).

For galls induced by *E. isaias*, both female- and male-induced galls are conical, with the dimorphism of the insect being reflected in the shape of the unique chamber (Magalhães et al. 2014). The cortex of female- and male-induced galls have similar tissue organization, comprising outer (OC), median (MC), and inner cortices (IC). The OC and MC are parenchymatic, with mucilaginous idioblasts occurring in the OC, and collateral bundles occuring in the MC. In the galls induced by females, sclerenchyma cells differentiate at the top and lateral portions of the chamber, whereas in the male-induced galls, the sclerenchyma cells occur only on top of the chamber in the IC (Silva et al. 2023) (Fig. 2).

The primary sex-related distinctions among the eriococcid-induced galls rely in the presence and position of trichomes in the female-induced galls of *P. rolliniae* and *E. isaias*, as well as in the number of chambers among the three eriococcid species. The differentiation of trichomes that obliterate the opening on the abaxial surface of the galls induced by the females on *P. grandiflorum* and in the inner ostiole on *A. dolabripetala* can result from the stimuli of the female insects that spend more time inside the galls (Magalhães 2014). The position of the trichomes is different, but their differentiation may confer a stable microenvironmetal condition inside the nymphal chamber in a similar way (Stone and Schönrogge 2003). The trichomes may also facilitate the exit of the crawlers, as they do not have mouthpieces suitable for digging an escape channel (Gonçalves 2008; Magalhães et al. 2014). In addition, the presence of trichomes in galls induced by the females of *E. isaias* and *P. rolliniae* may allow the insertion of the male copulatory organ (Gonçalves 2008; Magalhães et al. 2014) and prevent the eggs from falling down from the gall (Gonçalves 2008), since the ostiole is in the abaxial leaf surface. Peculiarly, the galls on *M. guianensis* have the ostiole on the adaxial leaf face and the trichomes do not persist during gall development.

The development of one or two nymphal in the galls induced by the females of *P. rolliniae* and *B. mataybae* is a differential character, whose presence had not been identified as a pattern of Eriococcidae galls (Mani 1964; Meyer 1987). The two chambers observed in the female-induced galls of *P. rolliniae* and *E. isaias* provides an ideal environment for oviposition and hatching of the eggs, also conferring protection (Gonçalves et al. 2005; Pfeffer et al. 2018). The presence of mucilage cells in the cortical parenchyma of the galls induced on *P. grandiflorum* and of calcium oxalate crystals in galls on *A. dolabripetala* seems to be an intrinsic feature of the Malvaceae (Metcalfe and Clalk 1950) and Annonaceae (Sá et al. 2019), respectively, which has been conserved in the galls.

A common anatomical trait shared by the three Eriococcidae galls in Brazil is the formation of a lignified zone next to the larval chamber. As the degree of lignification depends on the presence of active forms of oxygen and the formation of phenoxide radicals, processes necessary for lignin biosynthesis (Boerjan et al. 2003), the sclerenchyma formation around the chambers in *P. rolliniae* (female- and male-induced galls), *B. mataybae*, and *E. isaias* (female-induced galls) has been related to the oxidative stress imposed by the feeding and respiratory activity of the insects (Silva et al. 2019, 2023). Differences in sclerified cells among the galls induced by the parthenogenetic females, and the female- and male-induced galls denote distinct metabolism in terms of free radical formation and oxy-reduction reactions, which can also imply in the local blockage of synthesis or deposition in cell walls.

The differences related to the distinct morphotypes in different organs of their hosts may indicate that the host-plant tissues can impose constraints to gall development, which can be considered together with the lifespan of gall-inducing insects (Stone and Schönrogge 2003). Among the gall-inducing eriococcid species, differences in their lifespans are evident, which correlates with the extent of their impact on gall structure (Gonçalves et al. 2005). In fact, the temporal scale has an impact on gall structure in all three systems, but especially in the two species with sexual reproduction. Due to their longer life cycle, galls induced by the females exhibit specialized structural characteristics, as the differentiation of trichomes and greater investment in lignification than the male-induced galls. All these characteristics represent an adaptive value of the gall structure for the insect survival and their offspring protection.

7 Chemical Profiles of Eriococcid Galls in Brazil

Sexually dimorphic galls also exhibit a metabolism linked to the sex of the inducers. The longer life cycles of females result in a higher demand for metabolites, as evidenced by the histochemical and anatomical profiles of the galls (Silva et al. 2023). Primary metabolites, such as starch and lipids, accumulate in the IC of the female-induced galls and may, together with the lignins, play a role in dissipating reactive oxygen species (ROS) (Silva et al. 2023). Secondary metabolites, specifically terpenoids, accumulate in the IC and MC of the male-induced galls of *P. rolliniae* and *E. isaias*, respectively. In female-induced galls, secondary metabolites accumulate in the OC, which can be potentially related to the sexual attraction of males to copulate with the sessile females inside the gall chamber. In fact, the female-induced galls on *A. dolabripetala* and *P. grandiflorum*, with their longer life cycles, exhibit a wider histochemical profile in terms of types of metabolites (Silva et al. 2023). In the *M. guianensis*—*B. mataybae* system, the histochemical profile are established during the early stages of development of the leaflet galls. Despite the structural conspicuous differences between the stem and leaflet galls, the histochemical profiles are similar (Silva et al. 2019).

Different patterns on the accumulation of primary and secondary metabolites in Eriococcid-induced galls are indicative of variations in the stress level imposed by the males and females. The distinct levels of stress interfere in the dissipation of the ROS that accumulate in the symplast and apoplast, and may involve distinct distributions of the chemical components of cell walls, as well. The investigations on the chemistry of cell walls in galls have focused mainly on the distribution of pectins, proteins, and hemicelluloses, which has functional implications. The principal distinction in the pectin composition of eriococcid galls relies on the presence of xylogalacturonans (XGAs) only in the cell walls of female-induced galls. The XGAs confer resistance to cell wall degradation by pectinases (Zandleven et al. 2006). Furthermore, the XGAs in the female-induced galls may indicate a greater metabolic demand imposed by the females along gall establishment as observed by Silva et al. (2023), since this pectin is generally found in storage tissues (Albersheim et al. 1996; Zandleven et al. 2006, 2007). The stress imposed by the female eriococcids on their gall tissues may also affect the activity of pectin methylesterases (PMEs), which can be increased by the ROS accumulation (Xiong et al. 2015). The ROS acts as a downstream signaling molecule of auxins, that induce PMEs activity, and decrease the degree of pectin methylesterification, resulting in the thickening of the cell walls and improving defense against stresses (Xiong et al. 2015). Together with the detection of ROS throughout the tissues of the mature female-induced galls of *P. rolliniae* and *E. isaias* (Silva et al. 2023), the increase of the PMEs activity may increase the hardness of the cell walls, which is yet to be proved.

8 Conclusion and Remarks

The remarkable diversity within the Eriococcidae, especially among gall-inducing species, highlights their adaptability to some environments and host plant families, with a notable affinity for Myrtaceae in both Australian and Brazilian ecosystems. The development of specific gall morphotypes within the Eriococcidae is intricately influenced by the sex of the inducer and the life cycle duration, leading to significant variations in gall structural and chemical profiles between male-and female-induced galls. The variations in morphology, anatomy, and chemistry underscore the dynamic nature of these host plant-galling eriococcid systems, providing a foundation for future research into the molecular mechanisms governing gall development orchestrated by the scale insects.

References

Albersheim P, Darvill AG, O'Neill MA, Schols HA, Voragen AGJ (1996) An hypothesis: the same six polysaccharides are components of the primary cell walls of all higher plants. In: Visser J, Voragen AGJ (eds) Pectins and Pectinases. Elsevier Science B.V, Amsterdam, pp 47–55

Beardsley JW (1984) Gall-forming Coccoidea. In: Ananthakrishnan TN (ed) Biology of gall insects. Edward Arnol, London, pp 79–106

Ben-Dov Y, Miller DR, Gibson GAP ScaleNet. Available from: http://www.sel.barc.usda.gov/scalenet/scalenet.htm. Last accessed July 2023

Boerjan W, Ralph J, Baucher M (2003) Lignin biosynthesis. Annu Rev Plant Biol 54:519–546. https://doi.org/10.1146/annurev.arplant.54.031902.134938

Caballero A, González P, Ramos Portilla AA (2019) A new species of *Acanthococcus* Signoret and new records for *Uhleria araucariae* (Maskell) (Hemiptera: Eriococcidae). J Insect Biodivers 12(1):1–8

Cook LG (2003) *Apiomorpha gullanae* sp. n., an unusual new species of gall-inducing scale insect (Hemiptera: Eriococcidae). Aust J Entomol 42:327–333

Cook LG, Gullan PJ (2001) Longevity and reproduction in *Apiomorpha* Rubsaamen (Hemiptera: Sternorrhyncha: Coccoidea). Bollettino di Zoologia Agraria e di Bachicoltura 33(3):259–265

Cook LG, Gullan PJ (2004) The gall-inducing habit has evolved multiple times among the eriococcid scale insects (Sternorrhyncha: Coccoidea: Eriococcidae). Biol J Linn Soc 83(4):441–452. https://doi.org/10.1111/j.1095-8312.2004.00396.x

Cook LG, Gullan PJ (2008) Insect, not plant, determines gall morphology in the *Apiomorpha pharetrata* species-group (Hemiptera: Coccoidea). Aust J Entomol 47:51–57. https://doi.org/10.1111/j.1440-6055.2007.00605.x

Cook LG, Gullan PJ, Stewart AC (2000) First-instar morphology and sexual dimorphism in the gall-inducing scale insect Apiomorpha Rübsaamen (Hemiptera: Coccoidae: Eriococcidae). J Nat Hist 34:879–894. https://doi.org/10.1080/002229300299291

García Morales M, Denno BD, Miller DR, Miller GL, Ben-Dov Y, Hardy NB ScaleNet. A literature-based model of scale insect biology and systematics 2016. Available in: http://scalenet.info. Last accessed, July 2023

Gonçalves SJMR (2008) Sincronia fenológica, dimorfismo sexual e ontogênese em galhas de *Pseudotectococcus rolliniae* Hodgson & Gonçalves (Hemiptera: Coccoidea: Eriococcidae) em *Rollinia laurifolia* Schltdl (Annonaceae). Tese de Doutorado. Minas Gerais, Brasil: Universidade Federal de Minas Gerais

Gonçalves SJMR, Isaias RMS, Vale FHA, Fernandes GW (2005) Sexual dimorphism of *Pseudotectococcus rolliniae* Hodgson and Gonçalves (Hemiptera: Coccoidea: Eriococcidae) influences gall morphology on Rollinia laurifolia Schltdl. (Annonaceae). Tropical. Zoology 18:161–169

Gonçalves SJMR, Moreira GRP, Isaias RMS (2009) A unique seasonal cycle in a leaf gall- inducing insect: the formation of stem galls for dormancy. J Nat Hist 43:843–854

González P, Claps LE (2013) El género *Apiococcus* Hempel (Coccoidea: Eriococcidae): redescripción de dos especies. Rev Bras Entomol 57(1):19–23

González P, Claps L, Juarez A (2020) Review of Brazilian Eriococcidae (Hemiptera: Coccomorpha) with redescription of a *Neotectococcus* and two *Acanthococcus* species and description of a new species. Revista Brasileira De Entomologia 64(1):1–9. https://doi.org/10.1590/1806-9665-RBENT-2018-131

Gullan PJ (1984a) A revision of the gall-forming coccoid genus *Cylindrococcus* Maskell (Homoptera: Eriococcidae). Aust J Zool 32:677–690

Gullan PJ (1984b) A revision of the gall-forming coccoid genus *Apiomorpha* Rübsaamen (Homoptera: Eriococcidae: Apiomorphinae). Aust J Zool 97:203

Gullan PJ (1984c) Taxonomy and biology of Australian gall-forming Coccoidea. 381–83

Gullan PJ, Kosztarab M (1997) Adaptations in scale insects. Annu Rev Entomol 42:23–50

Gullan PJ, Miller DR, Cook LG (2005) Gall-inducing scale insects (Hemiptera: Sternorrhyncha: Coccoidea). In: Raman A, Schaefer CW, Withers TM (eds) Biology, ecology, and evolution of gall-inducing arthropods. Science Publishers, New Hampshire, pp 159–229

Gullan PJ, Giliomee JH, Hodgson CJ, Cook LG (2006) The systematics and biology of the South African gall-inducing scale insect, *Calycicoccus merwei* Brain (Hemiptera: Coccoidea: Eriococcidae). African Entomology, 14(1):13–33.

Hanks LM, Denno RF (1993) The role of demic adaptation in colonization and spread of scale insect populations. In: Evolution of insect pests/patterns of variation. Wiley, New York, pp 393–411

Hardy NB, Gullan PJ (2007) A new genus and four new species of felt scales on *Eucalyptus* (Hemiptera: Coccoidea: Eriococcidae) in south-eastern Australia. Aust J Entomol 46(2):106–120. https://doi.org/10.1111/aen.2007.46.issue-2, https://doi.org/10.1111/j.1440-6055.2007.00576.x

Hardy NB, Gullan PJ (2010) Australian gall-inducing scale insects on *Eucalyptus*: revision of *Opisthoscelis* Schrader (Coccoidea, Eriococcidae) and descriptions of a new genus and nine new species. ZooKeys 58:1–74

Henderson (2006) Four new species and a new monotypic genus Hoheriococcus (Hemiptera: Coccoidea: Eriococcidae) associated with plant galls in New Zealand. N Z Entomol 29:37–57

Hodgson C (2020) A review of neococcid scale insects (Hemiptera: Sternorrhyncha: Coccomorpha) based on the morphology of the adult males. Zootaxa 4765:1–264

Hodgson CJ, Miller DR (2010) A review of the Eriococcid genera (Hemiptera: Sternorrhyncha: Coccoidea) of South America. Zootaxa 2459:1–101

Hodgson C, Gonçalves SJMR, Miller D, Isaias RMS (2004) A key to genera of Eriococcidae (Hemiptera: Coccoidea) from the Neotropical region and a revision of *Pseudotectococcus* Hempel (Eriococcidae), a gall inducing scale insect genus from Brazil, with a description of a new species. Lundiana 5(1):51–72

Hodgson C, Magalhães TA, Miller D (2011) Two new gall-inducing genera and species of Eriococcidae (Hemiptera: Sternorrhyncha: Coccoidea) of Malvaceae from the neotropics. Lundiana 10:53–72

Hodgson CJ, Isaias RMS, Oliveira DC (2013) A new gall-inducing genus and species of Eriococcidae (Hemiptera: Sternorrhyncha: Coccoidea) on Sapindaceae from Brazil. Zootaxa 3734(3):317–330

Isaias RMS, Oliveira DC, Carneiro RGS, Kraus JE (2014) Developmental anatomy of galls in the neotropics: arthopods stimuli versus host plant constraints. In: Fernandes GW, Santos JC (eds) Neotropical insect galls. Springer, Alemanha, pp 15–34

Kondo T, Gullan PJ, Cook LG (2016) A review of the genus *Capulinia* Signoret (Hemiptera: Coccoidea: Eriococcidae) with description of two new species. Zootaxa 4111(4):471–491

Koteja J (1985) Essay on the prehistory of the scale insects (Homoptera, Coccinea). Ann Zool 38:461–504

Koteja J (1986) Fire has given birth to complex plant galls. Boll Lab Entomol 43:35–39

Larew HG (1990) Gall formation. [nome do diário] 293–300

Magalhães TA (2014) Respostas subcelulares a indução de galhas por *Baccharopelma dracunculifoliae* (Psyllidae) em *Baccharia dracunculifolia* (Asteraceae) e por *Eriogallococcus isaias* (Eriococcidae) em *Pseudobombax grandiflorum* (Malvaceae). Tese de Doutorado. Minas Gerais, Brasil: Universidade Federal de Minas Gerais

Magalhães TA, Oliveira DC, Isaias RMS (2014) Population dynamics of the gall inducer *Eriogallococcus isaias* (Hemiptera: Coccoidea: Eriococcidae) on *Pseudobombax grandiflorum* (Malvaceae). J Nat Hist 64(1):1–9. https://doi.org/10.1080/00222933.2014.951083

Mani MS (1964) Ecology of plant galls. Junk, The Hague [diário, numeração, página]

Metcalfe CR, Chalk L (1950) Anatomy of the Dicotyledons. "Malvaceae". In: Anatomy of Dicotyledons, vol I. Clarendon Press, Oxford, UK

Meyer J (1987) Plant galls and gall inducers. Gebrüder Borntraeger 1:291

Miller DR, Gimpel ME (2000) A systematic catalogue of the Eriococcidae (felt scales) (Hemiptera: Coccoidea) of the world, vol 1. Intercept Ltd, p 589

Miller DR, Kosztarab M (1979) Recent advances in the study of scale insects. Rev Entomol 24:1–27

Mills PJ, Gullan PJ, Cook LG (2017) Nomenclatural changes in the Australasian gall-inducing genus *Apiomorpha* Rübsaamen (Hemiptera: Coccomorpha: Eriococcidae). Zootaxa 4250(5):484–488. https://doi.org/10.11646/zootaxa.4250.5.6

Mills PJ, Cook LG (2014) Rapid chromosomal evolution in a morphologically cryptic radiation. Mol Phylogenet Evol 77:126–135. https://doi.org/10.1016/j.ympev.2014.03.015

Miller DR, González RH (1975) A taxonomic analysis of the Eriococcidae of Chile. Rev Chil Entomol 9:131–163

Pfeffer L, Rezende UC, Barônio GJ, Oliveira DC (2018) Building two houses on a single host plant: galling insect synchronizes its life cycle with plant phenology. Oecol Aust 22:438–448. https://doi.org/10.4257/oeco.2018.2204.07

Rung A, Miller D, Parikh G (2014) Scale insects, edition 2, a tool for the identification of potential pest scales at U.S.A. ports-of-entry (Hemiptera, Sternorrhyncha, Coccoidea). ZooKeys 431:61–78. https://doi.org/10.3897/zookeys.431.7474

Silva AFM, Kuster VC, Rezende UC, Oliveira DC (2019) The early developmental stages of gall-inducing insects define final gall structural and histochemical profiles: the case of *Bystracoccus mataybae* galls on *Matayba guianensis*. Botany 97:427–438

Silva AFM, Oliveira DC, Isaias RMS (2023) Influence of the sex of galling insects on gall metabolism: the case of eriococcid galls. Biochem Syst Ecol 108:1–9. https://doi.org/10.1016/j.bse.2023.104643

Stone GN, Cook JM (1998) The structure of cynipid oak galls: patterns in the evolution of an extended phenotype. Biol Sci 265:979–988

Stone GN, Schonrogge K (2003) The adaptive significance of insect gall morphology. Trends Ecol Evol 18:512–522

Sá RD, Santana ASCO, Padilha RJR, et al. (2019) Oxalic acid content and pharmacobotany study of the leaf blades of two species of *Annona* (Annonaceae), Flora, 253, 10-16. https://doi.org/10.1016/j.flora.2019.02.011

Vitorino MC, Pedrosa-Macedo J, Smith CW (2000) The biology of *Tectococcus ovatus* Hempel (Heteroptera: Eriococcidae) and its potential as a biocontrol agent of *Psidium cattleianum* (Myrtaceae). Proceedings of the X Interantional Symposium on Biological Control of Weeds, 4–14th July, 1999, Montana State University, Bozenam, Montata, USA, 651–657

Xiong J, Yang YJ, Fu GF, Tao LX (2015) Novel roles of hydrogen peroxide (H_2O_2) in regulating pectin synthesis and demethylesterification in the cell wall of rice (*Oryza sativa*) root tips. New Phytol 206:118–126

Yokogawa T, Yahara T (2009) Mitochondrial phylogeny certified PGL (Paternal Genome Loss) is of single origin and haplodiploidy *sensu stricto* (arrhenotoky) did not evolve from PGL in the scale insects (Hemiptera: Coccoidea). Genes Genet Syst 84:57–66

Zandleven J, Beldman G, Bosveld M, Schols HA, Voragen A (2006) Enzymatic degradation studies of xylogalacturonans from apple and potato, using XGH. Carbohydr 65:495–503

Zandleven J, Sørensen SO, Harholt J, Beldman G, Schols HA, Scheller HV, Voragen AJ (2007) Xylogalacturonan exists in cell walls from various tissues of *Arabidopsis thaliana*. Phytochemistry 68:1219–1226

Plant Volatile Organic Compounds' Cues in Gall Induction: In the Search for a Semiochemical Framework Considering Developmental Emergent Properties and Ecological Context Dependency

Cristian Villagra

1 Introduction

Galls or "Cecidia" are derived plant structures that develop due to the interaction of plant tissue and cecidogenic organisms, *i.e.*, gall inducers (Mani 1965). The latter comprehend a varied group of microorganisms (*e.g.*, viruses, bacteria, and fungi) and invertebrates (*e.g.*, nematodes, acari, and insects) that depend on and complete their development within the gall and often correspond to endophagous parasites and herbivores (Fernandes and Santos 2014; Hardy and Cook 2010; Shorthouse et al. 2005). Due to synchronized coaction between plants and gall-inducer stimulation (chemical and physical), plant morphogenesis and metabolism are modified, producing lasting structural and physiological effects on the plant-modified tissues, which end sheltering and nourishing a foreign entity or entities at the expense of plant resources (Carneiro et al. 2017; Isaias et al. 2014). Galls are formed preferably in meristem-rich areas of plant structure by increasing the number (hyperplasia) and/or size (hypertropia) of cells, resulting in the accumulation of specialized tissue with nourishing and protective characteristics (Bedetti et al. 2018; Isaias and De Oliveira 2012). This results in novel, derived and complex organ-looking structures in affected plants (Raman 2005a). Along with these changes, the plant's secondary metabolites are also modified (Motta et al. 2005; Qi et al. 2020; Ye et al. 2019). For instance, these changes may alter plants' tolerance towards cecidogenic organisms and provide an adequate microenvironment for gall inducers (Carneiro et al. 2014; de Silva et al. 2019; Tooker and De Moraes 2008; Tooker et al. 2008). For example, tannins and phenolic compounds, frequently found in external layers of cecidia,

C. Villagra (✉)
Instituto de Entomología, Universidad Metropolitana de Ciencias de la Educación, Santiago, Chile
e-mail: cristian.villagra@umce.cl

D. C. de Oliveira, R. M. dos Santos Isaias (eds.), *Plant Galls*,
https://doi.org/10.1007/978-3-031-80064-1_18

have been found in increased amounts after gall induction in several systems. These metabolites have been proposed to increase mechanical support and provide an aposematic defense (Hall et al. 2017; Kuster et al. 2019). Moreover, gall-derived structures may accumulate carotenoids and polyphenols in elevated concentration compared to plant tissues unaffected by gall induction (Ferreira et al. 2018; Guedes et al. 2024; Qi et al. 2020). It has been proposed that these compounds are associated with increased protection against abiotic stresses like damaging UV-range wavelengths and sunlight energy dissipation as heat (Ferreira et al. 2014, 2018). In this way, the changes in plant secondary metabolite expression related to gall induction may benefit directly (or indirectly) the organisms hosted inside gall tissues by providing biotic and abiotic protection (Mescher 2012).

1.1 Plant Volatile Organic Compounds in Galls

Volatile organic compounds (VOCs) are a highly variable part of the battery of plant's secondary metabolites that can be produced and released by different plant organs and structures (Bouwmeester et al. 2019). VOCs operate not only within plant tissues, promoting or triggering biochemical and physiological responses (Dudareva et al. 2006), but also in several fundamental ecological interactions, including (but not only) plant defense (Ferreira et al. 2014; Hahn and Maron 2016; Neilson et al. 2013). It has been widely demonstrated that these small metabolites participate in the semiochemical interaction between plants and other organisms, including other plants, animals, and microorganisms (Bouwmeester et al. 2019; Rosenkranz et al. 2021; Wang et al. 2024).

Several arthropods, including Insects, can detect plant volatiles and react to these as a stimulus that may entice or repel their attraction (Qian et al. 2024); this includes endophagous herbivores, such as arthropod gall inducers (Etl et al. 2022; Kanjana et al. 2024). The fact that cecidogenic arthropods can react to plant regular VOCs is suggested as the base of plant-gall inducer semiochemical interaction (Zhu et al. 2020). Over this battery of plant-derived volatile stimuli, changes in emissions of varied nature, including hereditary differences, plastic responses, or physiological conditions, such as the effect of herbivores and gall inducers, are incorporated as additional layers of complexity of semiochemical cues (with varied ecological outcomes) to the interaction between plants and its associated community, including their specialized endophages such as cecidogenic arthropods (Barônio and Oliveira 2019).

Plant volatile blends may experience changes as a result of host plants and cecidogenic organisms' developmental interactions (Fig. 1). Qualitative and quantitative changes in VOC emissions often accompany cecidogenesis (Besten et al. 2014; Borges 2018; Jiang et al. 2018; Rostás et al. 2013). This response may correspond to a specialized subset of interactions found between plants and free-living herbivores.

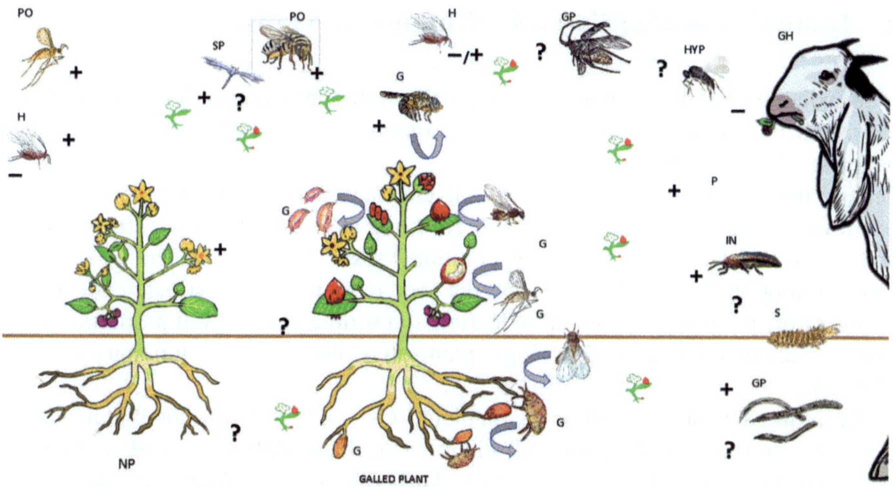

Fig. 1 Summary of plant-animal and plant-plant interactions where VOCs could be mediating its ecological outcome. The galled plant is represented by a generic plant with different galls on its organs (bud, leaf, stem, and root galls). These are accompanied by their gall-former organisms (G) coupled with a curved gray arrow representing their need for gall induction to complete their life cycle. The spectrum of association considered: Pollinators (PO), herbivores (H), seed predators (SP), gall predators (GP), parasitoids (PA), hyperparasitoids (HYP), inquilines (IN), successors (S), gall herbivores (GH), and nearby plants (NP). Green arrows and the green plant icon represent non-galled plant VOCs. Gall-modified VOCs are illustrated with red, purple, and brown arrows, each representing a different type of gall: bud gall, leaf, stem, and root galls, respectively. Positive effects of receivers, such as attraction, are shown with a (+) sign, while harmful consequences (such as deterrence) are represented by a (−). Depending on the case, variable results are shown with (−/+). Cases where the effects of galled plant VOCs on associated organisms are unknown or barely explored are represented with a question mark (?). Illustration by Cristian Villagra

Under a chemoecological context, these changes in the host's VOCs are suggested to contribute to plant indirect defenses (Cardé and Bell 1995; Schoonhoven et al. 2013; War et al. 2012), as in many cases, the alteration of volatile blend composition or quantity of emission can be detected by gall-inducer natural enemies (*e.g.*, predators and parasitoids) (Borges 2018; di Castro et al. 2018; Huang et al. 2022; Jiang et al. 2018; Villagra et al. 2021). Changes such as the increase in VOC emissions have also been found to deter herbivory on galls by browsing vertebrates (Rostás et al. 2013). Conversely, volatile changes in galls may contribute to the well-being of gall-inducers. For instance, in cases where cecidogenesis leads to a reduction (or suppression) of emissions, this helps gall inhabitants avoid detection by their natural enemies (Tooker et al. 2008; Zhao et al. 2015).

2 Induction of Plant's VOCs Changes in Galls

Plant volatile phenotypes may change based on the interactions between plant tissues, gall inducers, and their associated organisms. Nonetheless, several authors have established that there is still a need for explanatory mechanisms supporting a general theoretical background for the diverse semiochemical functions and their resulting ecological patterns (Borges 2018; Jiang et al. 2020; Raman 2011; Tooker and Helms 2014). This deficit is the combined result of a small scientific community working on this specific topic and the need to identify the distinct novel contribution of the phenomenon of gall induction as a source of emergent properties to be assessed under the chemical ecology, plant physiology, and evolutionary context (Table 1).

Regarding cecidogenic influence on plant VOCs, it has been found that the coaction between the host plant and gall-inducer (mechanical, biochemical, and developmental) influences the emission of standard volatile compounds related to direct and indirect defense strategies, like terpenoids and VOCs from the lipoxygenase pathway (Barônio and Oliveira 2019; Jiang et al. 2018; Uesugi et al. 2016).

From the side of the host plant, it has been found that gall induction involves a chain of variable patterns of plant gene expression associated with the main histological and morphological modification of plant tissues affected by cecidogenesis (Desnitskiy et al. 2023). This can be a part of the consequences of the general gall formation induction process triggered by chemical cues provided by the gall inducer, which considers a battery of growth regulators like auxins such as indole-3-acetic acid, cytokinins, and others (Fernandes et al. 2012; Isaias et al. 2014; Suzuki et al. 2014; Tooker and Helms 2014). In this sense, it has been found that gall inducers can be a source of plant growth hormones, leading to gall development (Andreas et al. 2020; Tokuda et al. 2022). For example, the goldenrod gall fly, *Eurosta solidaginis* (Diptera: Tephritidae), induces several changes in the attacked tissue of their host plant, including augmented emission of terpenoids and induced β-farnesene that may be acting as a kairomone towards other herbivores (Thomas et al. 2019). In addition, it has been found that can be found in the salivary glands of these gall-inducing tephritids are capable of producing and secreting phytohormones (auxins and cytokinins) (Ponce et al. 2021).

Plant-gall inducer response has been found to have a hereditary basis, as is the case of the volatile response of the Chinese chestnut, *Castanea mollissima* (Fagaceae), when attacked by the Asian chestnut gall wasp, *Dryocosmus kuriphilus* (Hymenoptera: Cynipidae) (Table 1). In this system, under gall-inducer attack, affected plant tissue increased terpene emissions, such as nerolidol and (E, E)-α-farnesene), compounds canonically found linked to plant herbivory defenses (Wang et al. 2024; Zhu et al. 2022). By transcriptomics and phylogenetic studies, it was possible to suggest that this response was linked with the transcription of terpene synthase genes: CmAFS, liked to (E, E)-α-farnesene production, and CmNES/AFS, which is related to (E)-nerolidol biosynthesis. These two genes were significantly

Table 1 State of the art of gall VOCs and their ecological consequences

Nº	Host Plant Sp.	Family	(A) Non-galled Host Plant VOCs	(B) Gall VOCs	Proposed Semiochemical function	Gall type	Gall-Inducer sp.	Family	Superior Category	Reference
1	*Pistacia atlantica*	Anacardiaceae	Twenty mono- and 13 sesquiterpenes	VOCs + ↑ terpenes (2.5 fold)	VOCs = Gall inducer natural enemies protection (Herbivores)	Lateral bud galls	*Slavum wertheimae*	Aphididae	Hemiptera	Rostás et al. (2013)
2	*Pistacia palaestina*	Anacardiaceae	Monoterpenes and elevated sesquiterpenes	VOCs + di novo accumulation of terpenes, increased terpenes (10–60 fold).	VOCs = Gall inducer natural enemies protection (Herbivores)	Bud galls	*Baizongia pistaciae*	Aphidiiae	Hemiptera	Rand et al. (2017), Rostás et al. (2013)
3	*Schinus polygamus*	Anacardiaceae	Mono and sesquiterpenoid compounds n-heptane, n-nonane	VOCs + Terpenes: a-pinene, imonene, β-pinene and myrcene Sesquiterpenes: cis-a-bergamotene (0.8%) and g-cadinene (0.1%)[a]	VOCs = Attractors of pollinators, Plant defenses. VOCs = Ovipositional signal, Gall inducer protection against natural enemies.	Leaf gall	*Calophya* sp.	Psyllidae	Hemiptera	Damasceno et al. (2010)
4	*Ageratina adenophora*	Asteraceae	Caryophyllene, β-bisabolene, 4-thujen-2-a-yl acetate	Not reported	VOCs = Attraction of gall inducer and host plant selection.	Stem gall	*Procecidochares utilis*	Tephritidae	Diptera	Kanjana et al. (2024)

(continued)

Table 1 (continued)

Nº	Host Plant Sp.	Family	(A) Non-galled Host Plant VOCs	(B) Gall VOCs	Proposed Semiochemical function	Gall type	Gall-Inducer sp.	Family	Superior Category	Reference
5	*Silphium laciniatum*	Asteraceae	Monoterpenes: α- and β-pinene, (+)-limonene, (−)-camphene, and β-myrcene	α- and β-pinene and camphene	VOCs =Female galler host plant location VOCs = Male attraction and proxy sexual pheromone	Flowering stem gall	*Antistrophus rufus*	Cynipidae	Hymenoptera	Tooker et al. (2002, 2005)
6	*Haplopappus foliosus*	Asteraceae	Terpenes: limonene and camphene, α-thujene, p-cymene and γ-terpinene	Limonene and camphene, +, α-thujene, p-cymene and γ-terpinene -	VOCs = Attraction of galler natural enemies (Parasitoids/ Predators)	Stem gall (Apical galls)	*Haplopappusmyiia gregaria*	Cecidomyiidae	Diptera	Villagra et al. (2021)
7	*Dittrichia viscosa*	Asteraceae	Terpenes (34.6%): α-pinene, limonene and β-pinene	↑ VOC emissions. Phenol derivatives (68.7%): estragole (6 fold), Terpenes (25.1%): eucalyptol and limonene. Esters (methyl-2-propylhexanoate[a]	VOCs = Host location	Flower galls	*Myopites stylatus*	Tephritidae	Diptera	Santos et al. (2016)
8	*Baccharis spicata*	Asteraceae	(25 compounds): Alkene: 1-undecene (17.8%). Terpenes: limonene (15.3%), β-pinene (13.5%), (E)-caryophyllene (12.8%) and g-murolene (11.4%)	↑ VOCs (33 compounds): Terpenes: β-pinene (40.9%) and limonene (19.1%). In low concentration: n-Heptane, a-thujene, g-terpinene, terpinolene, a-terpineol[a], β-bourbunene, a-gurjunene, β-gurjunene and germacrene A	VOCs =Attraction of Galler Natural Enemies (Parasitoids/ Predators)	Leaf gall	*Baccharopelma spp.* (unidentified)	Psyllidae	Hemiptera	Damasceno et al. (2010)

9	*Baccharis dracunculifolia*	Asteraceae	↑ relative percentages of (E)-nerolidol and spathulenol	↑ Proportions of germacrene D, bicyclogermacrene, limonene, and b-pinene	VOCs = Attraction of galler natural enemies (Parasitoids/ Predators)	Leaf gall	*Baccharopelma dracunculifoliae*	Psyllidae	Hemiptera	Besten et al. (2014)
10	*Solidago altissima*	Asteraceae	Copaene, benzophenone and β-pinene	↑ VOCs emission	VOCs = Attraction of galler Inquilines VOCs = Modification of plant-induced responses to subsequent herbivory	Bud galls (rosette gall)	*Rhopalomyia solidaginis*	Cecidomyiidae	Diptera	Uesugi et al. (2016)
11	*Solidago altissima*	Asteraceae	GLV, ↓ concentration of terpenoids[b]	↓ - GLV, ↑ terpenoids	VOCs (β-farnesene) = kairomone, anti-herbivore defenses, Plant-plant communication	Bud galls	*Eurosta solidaginis*	Tephritidae	Diptera	Helms et al. (2014), Thomas et al. (2019)
12	*Solidago altissima*	Asteraceae	GLV and α and β-pinene, camphene, Myrcene, (Z)-3-HAC, β-phellandrene limonene, (E)-β-ocimene, linalool d, nonatriene, Methyl salicylate, β-Caryophyllene, α-Humulene, Tridecatetraene.	↓ VOCs	VOCs = Modification of plant-induced responses to subsequent herbivory	Stem gall	*Gnorimoschema gallaesolidaginis*	Gelechiidae	Lepidoptera	Tooker et al. (2008)

(continued)

Table 1 (continued)

Nº	Host Plant Sp.	Family	(A) Non-galled Host Plant VOCs	(B) Gall VOCs	Proposed Semiochemical function	Gall type	Gall-Inducer sp.	Family	Superior Category	Reference
13	*Eremanthus erythropappus*	Asteraceae	Pentacyclic triterpenes. friedelan-3-one, hentricontane, 1,19-eicosadiene, and nonscosane	Volatile sesquiterpenes, triterpenes (ɤ-taraxasterol[a], and β-amyrone), tricosane) and β-amyrone, and the alkane tricosane	VOCs = Allelopathic effect VOCs = galling insect defenses	Stem gall	*Neolasioptera* sp	Cecidomyiidae	Diptera	de Castro et al. (2022)
14	*Clusia lanceolata*	Clusiaceae	α-caryophyllene, β-caryophyllene, germacrene D, β-copaene =; ↑ flavonoids	Total phenol content higher	VOCs = anti-herbivore defenses, plant protection VOCs = Induced plant defenses	Leaf gall	*Clusiamyia nitida*	Cecidomyiidae	Diptera	Ferreira et al. (2014)
15	*Castanea mollissima*	Fagaceae	(E)-nerolidol, geranial, α-ionone (apocarotenoid), and (E, E)-α-farnesene	↑ terpenes: E)-nerolidol and (E, E)-α-farnesene (significantly induced)	VOCs = Plant defenses against herbivory	Bud galls	*Dryocosmus kuriphilus*	Cynipidae	Hymenoptera	Wang et al. (2024)
16	*Quercus robur*	Fagaceae	GLV and isoprene emissions ((E)-3-Hexenal, 6-Methyl-5-hepten-2-one, 2-Ethyl-1-hexanol, α-Pinene, β-Ocimene, Limonene, Benzaldehyde	↑ lipoxygenase volatiles: α-Pinene, Limonene, etc). ↑ total LOX	VOCs =. stress-induced emissions (LOX pathway volatiles) VOCs = indirect defenses by gall wasp parasitoids and predators	Leaf gall	*Neuroterus anthracinus*	Cynipidae	Hymenoptera	Jiang et al. (2018)

17	*Quercus robur*	Fagaceae	GLV and isoprene emissions	↑ lipoxygenase volatiles, Monoterpenes: camphene, β-myrcene, β-pinene, and linalool) and sesquiterpenes (α-Bergamotene, etc) and benzenoid (benzaldehyde, benzothiazole), ↑total LOX	VOCs = stress-induced emissions (LOX pathway volatiles) indirect defenses by gall wasp parasitoids and predators	Leaf gall	*Neuroterus albipes*	Cynipidae	Hymenoptera	Jiang et al. (2018)
18	*Quercus robur*	Fagaceae	GLV and isoprene emissions	↑ lipoxygenase volatiles, Monoterpenes (camphene, β-myrcene, β-pinene, and linalool) and sesquiterpenes (α-Bergamotene, etc) and benzenoid (benzaldehyde, benzothiazole), ↑total LOX	VOCs = stress-induced emissions (LOX pathway volatiles) indirect defenses by gall wasp parasitoids and predators	Leaf gall	*Cynips* (*C. divisa* and *C. quercusfolii*)	Cynipidae	Hymenoptera	Jiang et al. (2018)
19	*Tilia cordata*	Malvaceae	GLV	↑ Monoterpene, ↑ fatty acid-derived VOCs, isoprene, benzenoids and carotenoid breakdown products	VOCs = stress-induced emissions	Leaf gall	*Eriophyes elixis*	Eriophyidae	Trombidi-formes (Acari)	Jiang et al. (2020)
20	*Alnus glutinosa*	Fagaceae	GLV	Monoterpene emissions suppressed.	VOCs = stress-induced emissions	Leaf gall	*Eriophyes inangulis*	Eriophyidae	Trombidi-formes (Acari)	Jiang et al. (2020)

(continued)

Table 1 (continued)

N°	Host Plant Sp.	Family	(A) Non-galled Host Plant VOCs	(B) Gall VOCs	Proposed Semiochemical function	Gall type	Gall-Inducer sp.	Family	Superior Category	Reference
21	*Tibouchina pulchra*	Melastomataceae	GLV, ↑ Flavonoid	↑ Phenols	VOCs = anti-herbivore defenses, plant protection VOCs = induced defenses	Leaf gall	unidentified	Unidentified	Lepidoptera	Motta et al. (2005)
22	*Myrcia splendens*	Myrtaceae	↑ 2-E-hexenal, sesquiterpene 5, β-caryophyllene, and β-bourborene	↑ Methyl salicylate, 3-*Z*-hexen-1-ol, 2-*E*-hexen-1-ol and 1-hexanol.	VOCs = Host Location VOCs. = Attraction of galler's natural enemies (parasitoids and predators)	Leaf gall (leaf rolling gall)	*Nexothrips* sp.	Terebrantia	Thysanoptera	di Castro et al. (2018)
23	*Eucalyptus camaldulensis*	Myrtaceae	α-pinene, tricyclene, sabinene, β-myrcene, p-cymene, terpinen-4-ol, cryptone, spathulenol	p-mentha-2-4(8) diene, δ-elemene, β-elemene, E-caryophyllene and bicyclogermacrene[a]	Plant defenses mechanisms, plant-plant communication	Leaf gall	*Leptocybe invasa*	Eulophidae	Hymenoptera	Mohamed et al. (2016)
24	*Eucalyptus grandis × E. tereticornis*	Myrtaceae	Alcohol, aldehyde, benzenoid, hydrocarbon, ketone and terpenoid	↑ aldehydes, benzenoids, hydrocarbons, ketones and terpenoids, Decreased alcohols and aldehydes	VOCs. = Attraction of galler's natural enemies (parasitoids and predators)	Leaf gall	*Leptocybe invasa*	Eulophidae	Hymenoptera	Huang et al. (2022)
25	*Olea europaea*	Oleaceae	Germacrene-D, (E,E)-alpha-farnesene, and (Z,E)-alpha-farnesene, alpha-copaene, (E)-4,8-dimethylnona-1,3,7-triene, (E)-beta-guaiene and heptadecane	↑ VOCs	VOCs. = Attraction of galler's natural enemies (parasitoids and predators)	Leaf gall	*Dasineura oleae*	Cecidomyiidae	Diptera	Caselli et al. (2022)

26	*Oryza sativa*	Poaceae	S)-linalool, 4,8-dimethyl-1,3,7-nonatriene, (E) caryophyllene and (R/S)-(E)-nerolidol	VOCs compounds ratio change	VOCs = Host location, Host Recognition VOCs = Gall inducer repellency, plant induced defenses	Stem gall	*Orseolia oryzivora*	Cecidomyiidae	Diptera	Ogah et al. (2017)
27	*Salix myrsinites*	Salicaceae	GLV, isoprene	DMNT, sesquiterpenes and GLVs	VOCs = Response under elevated temperature VOCs= Herbivore enemy attractants, Plant defenses	Leaf gall	*Aculus tetanothrix*	Eriophyidae	Trombiformes (Acari)	Swanson et al. (2021)
28	*Salix fragilis and S. × rubens*	Salicaceae	GLV, At flowering (90 fold), 1,4-dimethoxybenzene (flowers)	Not reported	VOCs (flower scent) = Host location, Host Recognition	Leaf galls (bean-shaped galls)	*Pontania proxima*	Tenthredinidae	Hymenoptera	Swanson et al. (2021)
29	*Populus deltoides × P. laurifolia*	Salicaceae	GLV	(B) methyl jasmonate and GLV (heptanal, 2-ethyl hexanal). monoterpene eucalyptol and sesquiterpene β-eudesmol, and a nitrogen-containing compound (geranyl nitrile), monoterpenes (α-pinene,	VOCs = stress volatiles: anti-herbivore defenses, 3Δ -carene, limonene) and benzenoids (benzaldehyde), constitutive emissions of isoprene.	Leaf gall (petiole galls)	*Pemphigus spyrothecae*	Aphididae	Hemiptera	Ye et al. (2019)

(continued)

Table 1 (continued)

Nº	Host Plant Sp.	Family	(A) Non-galled Host Plant VOCs	(B) Gall VOCs	Proposed Semiochemical function	Gall type	Gall-Inducer sp.	Family	Superior Category	Reference
30	*Ulmus davidiana var. japonica*	Ulmaceae	cis-3-hexenyl esters, perillene and caryophyllene, (Z)-3-hexenyl butanoate, (Z)-3-hexenyl 2-methyl butanoate, (Z)-3-hexenyl butanoate, (Z)-3-hexenyl isovalerate, (Z)-3-hexenyl valerate, (Z)-3-hexenyl tiglate, (Z)-3-hexenyl caproate	VOCs but ↓, except for caryophyllene ↑	VOCs = Anti plant defenses manipulated response	Leaf gall	*Tetraneura nigriabdominalis*	Aphididae	Hemiptera	Takei et al. (2015)
31	*Vittis berlandieri x V. riparia*	Vitaceae	(39 compounds) C6-compounds, terpenes, aromatic compounds, alcohols and n-alkanes.	dodecanoic acid = to . All other (12) ↑ ↑ terpenoids (betacaryophyllene, geraniol, beta-myrcene). ↑ aromatic compounds phenylacetaldehyde, methyl salicylate and eugenol ↑ C6-compounds (E)-hex-2-enal and hexanal	Plant defenses against herbivory	Root gall	*Daktulosphaira vitifoliae*	Phylloxeridae	Hemiptera	Lawo et al. (2011)

The first column represents the case number, followed by the host plant species and its family. Column A represents VOCs from host plants without gall induction (represented by the green plant icon). Column B corresponds to the reported Gall VOCs, illustrated with the plant with red spheres (galls) icon. The proposed chemical function is presented in the next column, using the icons mentioned above for galled and non-galled plants. The following columns describe call type, Gall-inducer species, and taxonomic classification; the last column corresponds to the cited reference

↑: Increased emissions of

↓: decreased

GLV green leaves volatiles, *DMNT* (E)-4,8–dimethyl–1,3,7-nonatriene

[a]only in …(neo-synthesis)

[b]amount of VOCs similar in Tooker et al. (2008)

upregulated upon gall wasp invasion and are suggested as part of the plant's defense arsenal against several antagonists and diseases (Wang et al. 2024).

Moreover, growing evidence supports the pivotal role of symbiotic associates contributing to the synthesis of phytohormones or other effectors that are key in the alterations found in plant structures and secondary metabolites (Gätjens-Boniche 2019; Gätjens-Boniche et al. 2023; Ponce et al. 2021). This is an additional level of complexity that is well established for plant multitrophic chemical ecology. However, their effect on plant-gall VOCs phenotype and their ecological and evolutionary implications have barely been explored (Borges 2018).

From the perspective of their ecological consequences, both herbivore-induced plant volatile (HIPVs) defense and counter-defense patterns have been proposed associated with alterations of VOCs in plants affected by cecidogenesis (Lev-Yadun and Inbar 2024; Tooker et al. 2008). Indirect plant defenses are hypothesized to explain several secondary metabolite changes during cecidogenesis, including VOC alterations (dos Isaias and de Oliveira 2012; Tooker and Helms 2014). Still, while inducible plant counter-responses may operate as a defensive reaction against the early signs of gall induction, gall-inducing art may also be able to "hijack" host metabolic and physiological responses (Cook and Gullan 2008; Murakami et al. 2021; Stone and Schönrogge 2003; Tooker and De Moraes 2008). It is proposed that to achieve gall formation, cecidogenic organisms must be capable of reducing or suppressing HIPVs. In this case, this often involves the alteration of plant VOCs, influencing them for their own gain (Barônio and Oliveira 2019). As an alternative to metabolic hijacking, it is suggested that gall-induced may handle the expression of plant defensive traits, including HIPVs (Hall et al. 2017). In this chapter, I review the nature of plant volatiles related to gall induction, exploring the different sources of variability of plant VOCs involved in the interaction with cecidogenic insects (Fig. 1). I discuss how these diverse cues may play a role in establishing plant-gall inducer association and its various ecological outcomes.

2.1 *Non-galled Plants May Attract Gall Makers*

Cecidogenic arthropods may use healthy plant VOCs to achieve several ecological functions at the adult or free-living stage, such as feeding, mating, and oviposition. In this sense, their basal relationship with plant volatiles may not differ much from other free-living herbivores.

As it has been found for other herbivorous arthropods (Tooker and Hanks 2004), the volatile metabolite mixture of plants may function during host finding, attraction, and recognition of gall-forming arthropods. For instance, reproduction-related cues such as floral VOCs can be eavesdropped by gall-inducers, which can use these signals to find suitable host plants, as it has been found for leaf-galling sawflies (Hymenoptera: Tenthredinidae) attracted to flowering willows (Kehl et al. 2010). This casual exploitative interaction may be the base for further intricate associations, such as the cases where gall-inducers may also function as pollinators of their host plant or even other plant species that are not their target for oviposition, such as it has been reported for gall midges (Diptera: Cecidomyiidae), which can contribute to plant reproduction through pollination. This has been demonstrated for several other plant lineages, such as Araceae, Moraceae, and Schisandraceae, and with different degrees of intimacy and dependency with these visitors, from casual and opportunistic visitors to strict brood-site pollination (Borges 2016; Etl et al. 2022; Gan et al. 2022; Gardner et al. 2018). In addition, plant emissions from other organs (flowers, leaves, fruits, roots, etc.) can be used by gall-inducers to locate suitable hosts and as oviposition signals. This has been found for several cecidogenic insect orders, including cynipid wasps, gall midges, and micro-lepidopterans (Tooker et al. 2005).

2.2 Gall-Induced Plant VOCs Change as HIPVs

Gall-forming arthropods, such as insects and acari, interacting with host plant tissues during gall induction can qualitatively and quantitatively modify the host plant's secondary metabolite production, including VOCs (Damasceno et al. 2010; Jiang et al. 2020; Santos et al. 2016; Swanson et al. 2021; Tooker et al. 2002). Changes in plant VOCs driven by gall induction can vary the proportion of secondary volatile metabolites emitted by the plant (Huang et al. 2022). This may lead to changes that could benefit the attacked plants, such as making the attacked plant less attractive (or even repellent) to gall-inducers looking for host plants for oviposition, as it has been found in African rice gall midge, *Orseolia oryzivora* (Diptera: Cecydomiidae), this could qualify as induced plant defense as well (Ogah et al. 2017).

In several cases, gall VOC induction produces di novo volatile compounds (neosynthesis) on affected tissues that can assume novel ecological and semiochemical roles. For example, gall induction in Candeia plants, *Eremanthus erythropappus* (Asteraceae), by an undescribed *Neolasioptera* (Diptera: Cecidomyiidae) gall midge drives the production of several secondary metabolites not found on non-galled host tissues and the inhibition of others previously identified as part of the plant chemical repertoire. Among these, it is possible to find volatile sesquiterpenes proposed with allelopathic effects (de Castro et al. 2022).

3 Ecological and Developmental Context Dependency

The relevance and functions of VOCs in cecidogenesis and the interaction between plants, cecidogenic arthropods, and the associated community must be understood within the general framework of ecological chemistry and semiochemistry of plant-animal interaction and its historical (evolutionary) and contextual (ecological) dimensions (Shorthouse et al. 2005). In this sense, it has been proposed that gall-inducing arthropods can be considered a subcategory of a broader group of endophagous lifestyles, where it is also possible to find leaf-miners endophages (Giron et al. 2016). These specialized herbivores can trigger specific alterations or "manipulation" of the plant's secondary metabolisms after herbivore damage driven by factors released by the attacker, including herbivore-induced plant VOCs (HIPVs) (Klimm et al. 2020; Nyman and Julkunen-Tiitto 2000). In some studied cases, the fine-tuning of gall-induced alterations of host plant VOCs can be pretty delicate; for instance, in *Vitis* sp. gall VOCs induced by the Grape phylloxera, *Daktulosphaira vitifoliae* (Hemiptera: Phylloxeridae), not only there was a marked concentration change in released secondary metabolites such as geraniol, eugenol, vanillin and delta-cadinene from young root galls in comparison with healthy tissue, but also when these young nodisities, triggered by the second nymphal stage gall-inducer, are contrasted against older galls (mature nodosities), it was possible to detect significant differences on several VOCs like phenylmethanol, methyl salicylate, 4-ethenyl-2-methoxyphenol as well as eugenol, that were present in elevated amounts in the former (Eitle et al. 2019; Lawo et al. 2011). These ontogenetic differences in gall VOCs are represented in Fig. 1.

Regarding plant VOCs, it has been found that the effect of microorganism infections (like fungi), free-living herbivores, and endophagous herbivores and parasites (including gall-inducers), could lead to an overall reduction of constitutive VOC emissions (*e.g.*, green leaf volatiles) coupled with an increase, or even a *di novo* production, of specific volatile secondary metabolites (Table 1 and Fig. 1). This trend varies depending on each case. It is possible to find systems where gall induction and development trigger the reduction (or even the suppression) of most plant VOCs (reviewed by Borges 2018). This reduction of volatile secondary metabolites has been proposed to make plants less detectable to gall-inducer's natural enemies (*e.g.*, predators and parasitoids) (Fig. 1), disabling plant-induced defenses or making them less prone to defend themselves against subsequent herbivory (Tooker et al. 2008; Uesugi et al. 2016). For example, the induction of galls by *Gnorimoschema gallaesolidaginis* (Lepidoptera: Gelechiidae) in the goldenrod *Solidago altissima* (Asteraceae) has been demonstrated to reduce plant VOC emissions. It is proposed that it is capable of suppressing plant indirectly induced defenses that otherwise will be expressed through its fragrance, as is the case when this host plant is attacked by a generalist herbivorous insect, such as larval the tobacco budworm *Chloridea virescens* (Lepidoptera: Noctuiidae) (Tooker et al. 2008). The influence of the gall inducer on the affected plant has been proposed to prevent galler's natural enemies from detecting it. Still, it has also been found to make host plants less vulnerable to other herbivores (including budworms) (Tooker et al. 2008). In the same model plant, gall induction by *Eurosta solidaginis* (Diptera: Tephritidae) increased the emission of terpenoids in affected plants, which in turn is proposed to activate plant-plant communication defense strategies against herbivores (Helms et al. 2014; Mohamed 2016; Thomas et al. 2019). Thus, a complex context-dependent dynamic must be considered before assuming a given adaptive function of these modified signals.

Experimental evidence demonstrates that this modification reduces responsiveness toward plant signals, mediating anti-herbivory responses, such as jasmonic acid (JA) (Takei et al. 2015). Along with suppressing JA, it has also been found in several gall-inducing models that salicylic acid (SA) leads to the upregulation of the route of plant responses facing pathogen infection (Eitle et al. 2019; Erb 2018; Takei et al. 2015; Tooker et al. 2008).

On the other hand, different cases have been found where gall induction increased VOC emissions. These involve cecidogenic arthropods from varied groups and host plants from diverse families (Caselli et al. 2022; Damasceno et al. 2010; Jiang et al. 2020; Rand et al. 2017; Rostás et al. 2013; Uesugi et al. 2016). For example, in eriophyid mites (Acari, Eriophyoidea), leaf gall induction resulted in variable increased emissions of several VOCs, including fatty acid-derived volatiles, isoprene, benzenoids, and carotenoid byproducts in infection (Jiang et al. 2020). In the same way, the attack of *Daktulosphaira vitifoliae* (Hemiptera: Phylloxeridae), leading to root gall formation in grapevines, has been found to trigger an overall increase in root volatile emissions in *Vitis* sp., along with the singular production of beta-caryophyllene detected only in mature galls of the Grape phylloxera (Lawo et al. 2011). This increased emission of VOCs is linked to the gall-inducer's manipulation of the expression of several plant defense biosynthesis and signaling pathways, including the transcriptional suppression of the JA along with the expression of SA responses (Eitle et al. 2019; Nabity et al. 2013). This pattern has been proposed as a mechanism related to protecting galls and their occupants against microorganism attacks (Eitle et al. 2019). On the other hand, SA signaling is linked to acquired resistance, which is activated by SA-mediated innate responses in plant-plant communication (Ninkovic et al. 2021). Therefore, these modifications may also prime the surroundings against further gall inducer attacks.

In this sense, the ecological context may provide different adaptive values to the induced changes in host plant VOCs (Helms et al. 2019; Moreira and Abdala-Roberts 2019). For example, plant monoterpene volatiles from the compassplant, *Silphium laciniatum* (Asteraceae), have been found to contribute to host plant location for the females of the cynipid gall-inducer *Antistrophus rufus* (Hymenoptera: Cynipidae) (Tooker et al. 2005). At the same time, the emission of α- and β-pinene and camphene from flowering stem galls has been reported as a proxy for sexual pheromone in males of this gall-forming wasps (Tooker et al. 2002).

Furthermore, in a realistic scenario, changes in host plant VOCs may lead to the recruitment of diverse organisms interacting with plants, including, but not exclusively, gall inducer's natural enemies (indirect, induced plant defense response). This may be coupled with improving the conditions for completing the inducer's life cycle or adaptive/reproductive benefits (Moreira and Abdala-Roberts 2019), but not in all cases (reviewed by Borges 2018). A positive outcome can be achieved by herbivores, either by using these modified signals as cues for host location, oviposition, or mating, as illustrated above for cynipid gall-inducers and several other cases reviewed by previous authors (Barônio and Oliveira 2019). In addition, herbivore-modified plant VOCs may influence the behavior and performance of other competing or mutualistic herbivores or affect the attraction of natural enemies (*e.g.*, predators and parasitoids), interacting indirectly with the inducer through the semiochemical consequences of their influences over host plant VOCs (Davidson-Lowe and Ali 2021; El-Sayed et al. 2016; Swanson et al. 2021; Valle et al. 2023).

Looking at the evidence of other free-living and endophagous herbivores, it is possible to suggest that the fact that herbivores may contribute with signals that mediate or trigger plant's secondary metabolites change is not always related to a benefit to these inducers. Therefore, the adaptive or beneficial consequences of induced VOC changes may depend on the ecological context.

In the case of gall-forming animals, the intimacy of the relationship and the strong specificity and dependency of cecidogenic arthropods on their host plants may introduce a historical, evolutionary dimension. The modification of plant signals by the gall inducer or its symbiotic associates is often a necessary step for completing their life cycle (Raman 2005b; Shorthouse et al. 2005). In this situation, the role of this modification becomes a must for the existence of these endophagous herbivorous lineages. Furthermore, in comparison with free-living herbivores, endophagous herbivores, including gall inducers, may represent a temporally extended alteration of plant phenotype (including VOCs); this may reinforce the specialization levels and expand the community-level repercussion of these altered signals (Barônio and Oliveira 2019; Uesugi et al. 2016). Therefore, although gall induction can be considered a subclass of endophagous herbivores capable of triggering VOC responses in their host plants, it must be regarded as in the multitrophic context of plant interactions, their peculiarities that may make this kind of association unique, such as the spatiotemporal characteristics of gall induction and the cues these produce.

Plant VOCs may also work as defense responses triggered by gall inducers if they can directly influence another plant to modify its chemical phenotype toward an increased defense (Isaias et al. 2014). This has been reported in several plant-gall models where volatiles are modified after colonization, such as the case of the Chinese Chestnut and its cynipid gall inducers (Wang et al. 2024). In addition, plant-plant communication effects of galled plant VOCs may also influence other interactions, such as other herbivores. For example, in goldenrod, *Solidago altissima* (Asteraceae), gall induction by *Eurosta solidaginis* (Diptera: Tephritidae) triggers changes in host plant VOCs, corresponding to a reduction in green leaf volatiles and an increase in terpenoid emissions (Thomas et al. 2019). This modified signal can influence changes in the phytochemical emission of nearby goldenrod plants; this galled plant-plant cue is then exploited by the specialized *Uroleucon nigrotuberculatum* (Hemiptera: Aphididae) (Thomas et al. 2019).

4 Conclusions

In its 2018 review, Borges concluded the critical lack of knowledge regarding plant VOCs under cecidogenesis (Borges 2018). She rightfully stressed the need for an integrative comprehension of the roles of plant volatinome under the process of gall induction and in the multitrophic context where this occurs. In the same way,

Barônio and Oliveira also remarked that changes in plant volatile metabolites happen in a complex communicational context, where plant-plant communication and other simultaneous interactions cannot be underestimated when trying to draw a general picture of gall induction VOC roles (Barônio and Oliveira 2019). In addition to their remarks, it is fundamental to recognize plant-gall inducer as an emergent, co-constructed holobiont-type association, where plants, gall inducers, and associates are "symbiotic partners" (as in "living together") or "bionts" (Catania et al. 2017; Gehman and Harley 2019; Gilbert 2011; Margulis and Fester 1991; McFall-Ngai 2002).

These bionts display different degrees of functional integration, and their coaction triggers bidirectional influences over each other traits, including biochemistry and physiology, behavior, development and reproduction, among various dimensions of the plant-gall inducer holobiont biology (Prathapan 2024). As a result, novel dynamic phenotypes originated, and a novel emergent community is structured as "emergent properties" of this specialized association, where semiochemical signals may play fundamental roles (Chang et al. 2023). Thus, it is not only a consideration of the common characteristic of gall-inducers with general plant-herbivore chemoecological (e.g., plant defense theory) and physiological background that must be considered, but the emergent properties that include the lasting effect of gall-induction on plant phenotype and intimate associations, such as microorganism, and truly emergent (new) partners that consider natural enemies like parasitoids and predators, but also gall herbivores, inquilines and successors (Gätjens-Boniche 2019). Several of these inquilines, successors, and opportunistic herbivores cannot be found where plants do not have galls (Raman 2005b; Shorthouse et al. 2005). Thus, their existence depends on the co-niche construction of the ecosystem engineering resulting from the gall-induction process (Borges 2017). Therefore, the cues associated with this process may be paramount, for instance, in their host and oviposition location behavior. In fact, for these last categories, the evidence of the roles of plant-gall VOCs is almost nonexistent [but see (Rostás et al. 2013; Uesugi et al. 2016; Villagra et al. 2021)]. There is little evidence showing the relevance of VOCs for inquilines and gall herbivores. However, the semiochemical value of plant-gall VOCs for many other members of the gall communities still needs to be tested. An interdisciplinary approach is required to cover plant response studies and determine gall associates to fill these gaps. This combination will allow us to evaluate each partner's role and assess this ecological framework's contribution to the broad theories available for plant-animal chemical ecology.

Cecidogenesis is a widespread phenomenon found in all kinds of terrestrial ecosystems and plant lineages. Through this chapter, it was demonstrated that the volatile cues involved in the process are capable of influencing not only plant defense and gall-inducer survival and reproduction but also participating in plant-plant communication and may meditate behavioral responses of successors, inquilines, and herbivores (Table 1 and Fig. 1). It has been estimated that plant VOC changes induced by cecidogenesis may also participate in the carbon-induced terrestrial climate feedback loop as a stable source of biogenic volatile organic compounds (Noe and Niinemets 2020). Through their oxidation, these emissions lead to secondary organic aerosol formation, which participates in solar radiation absorption and cloud formation (Yu et al. 2021). It has been suggested that gall-induced VOC emissions from deciduous tree species lead to a global reduction in isoprene emissions; thus, this process may also impact Earth's global climate system (Noe and Niinemets 2020). Therefore, the study of plant-gall inducer VOCs is a relevant process with ecological and environmental implications, in need of further research, and with potential applied developments.

References

Andreas P, Kisiala A, Emery RJN, De Clerck-Floate R, Tooker JF, Price PW, Miller DG III, Chen M-S, Connor EF (2020) Cytokinins are abundant and widespread among insect species. Plants (Basel) 9:208. https://doi.org/10.3390/plants9020208

Barônio GJ, Oliveira DC (2019) Eavesdropping on gall–plant interactions: the importance of the signaling function of induced volatiles. Plant Signal Behav 14:1665454. https://doi.org/10.1080/15592324.2019.1665454

Bedetti CS, Jorge NC, Trigueiro F, Bragança GP, Modolo LV, Isaias R (2018) Detection of cytokinins and auxin in plant tissues using histochemistry and immunocytochemistry. Biotech Histochem 93:149–154. https://doi.org/10.1080/10520295.2017.1417640

Besten M, Nunes D, Granato D, Sens S, Wisniewski A Jr, Simionatto E, Riva-Scharf D (2014) Volatile components from galls induced by Baccharopelma dracunculifoliae (Hemiptera: Psyllidae) on leaves of Baccharis dracunculifolia (Asteraceae). Química Nova 38. https://doi.org/10.5935/0100-4042.20140288

Borges RM (2016) On the air: broadcasting and reception of volatile messages in brood-site pollination mutualisms. In: Blande JD, Glinwood R (eds) Deciphering chemical language of plant communication. Springer International Publishing, Cham, pp 227–255

Borges RM (2017) Co-niche construction between hosts and symbionts: ideas and evidence. J Genet 96:483–489. https://doi.org/10.1007/s12041-017-0792-9

Borges RM (2018) The galling truth: limited knowledge of gall-associated volatiles in multitrophic interactions. Front Plant Sci 9:1–8. https://doi.org/10.3389/fpls.2018.01139

Bouwmeester H, Schuurink RC, Bleeker PM, Schiestl F (2019) The role of volatiles in plant communication. Plant J 100:892–907. https://doi.org/10.1111/tpj.14496

Cardé R, Bell W (1995) Chemical ecology of insects 2. Chapman & Hall, New York

Carneiro RGS, Castro AC, Isaias RMS (2014) Unique histochemical gradients in a photosynthesis-deficient plant gall. S Afr J Bot 92:97–104. https://doi.org/10.1016/j.sajb.2014.02.011

Carneiro RGS, Isaias RMS, Moreira ASFP, Oliveira DC (2017) Reacquisition of new meristematic sites determines the development of a new organ, the Cecidomyiidae gall on Copaifera langsdorffii Desf. (Fabaceae). Front Plant Sci:8. https://doi.org/10.3389/fpls.2017.01622

Caselli A, Favaro R, Petacchi R, Angeli S (2022) Infestation of the gall midge Dasineura oleae provides first evidence of induced plant volatiles in olive leaves. Bull Entomol Res 112:481–493. https://doi.org/10.1017/S0007485321001000

Catania F, Krohs U, Chitto M, Ferro D, Ferro K, Lepennetier G, Görtz H-D, Schreiber R, Kurtz J, Gadau J (2017) The hologenome concept: we need to incorporate function. Theory Biosci 136:3–4. https://doi.org/10.1007/s12064-016-0240-z

Chang C-Y, Bajić D, Vila JCC, Estrela S, Sanchez A (2023) Emergent coexistence in multispecies microbial communities. Science 381:343–348. https://doi.org/10.1126/science.adg0727

Cook LG, Gullan PJ (2008) Insect, not plant, determines gall morphology in the Apiomorpha pharetrata species-group (Hemiptera: Coccoidea). Aust J Entomol 47:51–57. https://doi.org/10.1111/j.1440-6055.2007.00605.x

Damasceno FC, Nicolli KP, Caramão EB, Soares GLG, Zini CA (2010) Changes in the volatile organic profile of Schinus polygamus (Anacardiaceae) and Baccharis spicata (Asteraceae) induced by galling psyllids. J Braz Chem Soc 21:556–563. https://doi.org/10.1590/S0103-50532010000300023

Davidson-Lowe E, Ali JG (2021) Herbivore-induced plant volatiles mediate behavioral interactions between a leaf-chewing and a phloem-feeding herbivore. Basic Appl Ecol 53:39–48. https://doi.org/10.1016/j.baae.2021.03.005

de Castro JN, Guedes LM, Aguilera N, Becerra J, dos Isaias RMS (2022) Allelopathic potential of the extracts of non-galled stems and globoid stem galls of *Eremanthus erythropappus* (DC) McLeish (Asteraceae). Biochem Syst Ecol 100:104379. https://doi.org/10.1016/j.bse.2021.104379

de Silva AFM, Kuster VC, Rezende UC, de Oliveira DC (2019) The early developmental stages of gall-inducing insects define final gall structural and histochemical profiles: the case of Bystracoccus mataybae galls on matayba guianensis. Botany 97:427–438. https://doi.org/10.1139/cjb-2019-0017

Desnitskiy AG, Chetverikov PE, Ivanova LA, Kuzmin IV, Ozman-Sullivan SK, Sukhareva SI (2023) Molecular aspects of gall formation induced by mites and insects. Life 13:1347. https://doi.org/10.3390/life13061347

di Castro JN, Souza-Silva ÉA, Alvarenga DR, Saboia G, Soares GLG, Zini CA, Cavalleri A, Isaias RMS (2018) Structural and chemical profiles of Myrcia splendens (Myrtaceae) leaves under the influence of the galling Nexothrips sp. (Thysanoptera). Front Plant Sci 9:1521

dos Isaias RMS, de Oliveira DC (2012) Gall phenotypes – product of plant cells defensive responses to the inducers attack. In: Mérillon JMRK (ed) Plant defence: biological control, progress in biological control, vol 12. Springer, Dordrecht, pp 273–290

Dudareva N, Negre F, Nagegowda DA, Orlova I (2006) Plant volatiles: recent advances and future perspectives. Crit Rev Plant Sci 25:417–440. https://doi.org/10.1080/07352680600899973

Eitle MW, Griesser M, Vankova R, Dobrev P, Aberer S, Forneck A (2019) Grape phylloxera (*D. vitifoliae*) manipulates SA/JA concentrations and signalling pathways in root galls of *Vitis* spp. Plant Physiol Biochem 144:85–91. https://doi.org/10.1016/j.plaphy.2019.09.024

El-Sayed AM, Knight AL, Byers JA, Judd GJR, Suckling DM (2016) Caterpillar-induced plant volatiles attract conspecific adults in nature. Sci Rep 6:37555. https://doi.org/10.1038/srep37555

Erb M (2018) Volatiles as inducers and suppressors of plant defense and immunity — origins, specificity, perception and signaling. Curr Opin Plant Biol 44:117–121. https://doi.org/10.1016/j.pbi.2018.03.008

Etl F, Francke W, Schönenberger J, Dötterl S (2022) Chemical attraction of gall midge pollinators (Cecidomyiidae: Cecidomyiinae) to Anthurium acutangulum (Araceae). J Chem Ecol 48:263–269. https://doi.org/10.1007/s10886-022-01349-3

Fernandes GW, Santos JC (2014) Neotropical insect galls. Springer, Dordrecht

Fernandes W, Carneiro MAA, Isaias RMS (2012) Gall-inducing insects: from anatomy to biodiversity. In: Panizzi AR, Parra JRP (eds) Insect bioecology and nutrition for integrated pest management. CRC Press Taylor & Francis Group, New York, pp 369–395

Ferreira RO, de Carvalho Junior AR, da Silva TMG, Castro RN, da Silva TMS, de Carvalhoa MG (2014) Distribution of metabolites in galled and non-galled leaves of Clusia lanceolata and its antioxidant activity. Braz J Pharmacogn 24:617–625. https://doi.org/10.1016/j.bjp.2014.11.005

Ferreira BG, Oliveira DC, Moreira ASFP, Faria AP, Guedes LM, França MGC, Álvarez R, Isaias RMS (2018) Antioxidant metabolism in galls due to the extended phenotypes of the associated organisms. PLoS One 13:1–19. https://doi.org/10.1371/journal.pone.0205364

Gan S-R, Du W, Wang X-F (2022) Functional differentiation of floral color and scent in gall midge pollination: a study of a Schisandraceae plant. Plan Theory 11:974. https://doi.org/10.3390/plants11070974

Gardner EM, Gagné RJ, Kendra PE, Montgomery WS, Raguso RA, McNeil TT, Zerega NJC (2018) A flower in fruit's clothing: pollination of jackfruit (Artocarpus heterophyllus, Moraceae) by a new species of Gall Midge, Clinodiplosis ultracrepidata sp. nov. (Diptera: Cecidomyiidae). Int J Plant Sci 179:350–367. https://doi.org/10.1086/697115
Gätjens-Boniche O (2019) The mechanism of plant gall induction by insects: revealing clues, facts, and consequences in a cross-kingdom complex interaction. Rev Biol Trop 67:1359–1382. https://doi.org/10.15517/rbt.v67i6.33984
Gätjens-Boniche O, Jiménez-Madrigal JP, Whetten RW, Valenzuela-Diaz S, Alemán-Gutiérrez A, Hanson PE, Pinto-Tomás AA (2023) Microbiome and plant cell transformation trigger insect gall induction in cassava. Front Plant Sci 14:1237966. https://doi.org/10.3389/fpls.2023.1237966
Gehman A-LM, Harley CDG (2019) Symbiotic endolithic microbes alter host morphology and reduce host vulnerability to high environmental temperatures. Ecosphere 10:e02683. https://doi.org/10.1002/ecs2.2683
Gilbert SF (2011) Symbionts as an epigenetic source of heritable variation. In: Transformations of Lamarckism: from subtle fluids to molecular biology. MIT Press, Cambridge, MA, pp 283–293
Giron D, Huguet E, Stone GN, Body M (2016) Insect-induced effects on plants and possible effectors used by galling and leaf-mining insects to manipulate their host-plant. J Insect Physiol 84:70–89. https://doi.org/10.1016/j.jinsphys.2015.12.009
Guedes LM, Aguilera N, Torres S, Gavilán E, Rosales N (2024) Gall inducer Dasineura sp. alters the polyphenol profile and antioxidant activity of Peumus boldus stems. Tree Physiol 44:tpae024. https://doi.org/10.1093/treephys/tpae024
Hahn PG, Maron JL (2016) A framework for predicting intraspecific variation in plant defense. Trends Ecol Evol 31:646–656. https://doi.org/10.1016/j.tree.2016.05.007
Hall CR, Carroll AR, Kitching RL (2017) A meta-analysis of the effects of galling insects on host plant secondary metabolites. Arthropod Plant Interact 11:463–473. https://doi.org/10.1007/s11829-016-9486-0
Hardy NB, Cook LG (2010) Gall-induction in insects: evolutionary dead-end or speciation driver? BMC Evol Biol 10:19–21. https://doi.org/10.1186/1471-2148-10-257
Helms AM, De Moraes CM, Mescher MC, Tooker JF (2014) The volatile emission of Eurosta solidaginis primes herbivore-induced volatile production in Solidago altissima and does not directly deter insect feeding. BMC Plant Biol 14:173. https://doi.org/10.1186/1471-2229-14-173
Helms AM, Ray S, Matulis NL, Kuzemchak MC, Grisales W, Tooker JF, Ali JG (2019) Chemical cues linked to risk: cues from below-ground natural enemies enhance plant defences and influence herbivore behaviour and performance. Funct Ecol 33:798–808. https://doi.org/10.1111/1365-2435.13297
Huang Z-Y, Liu Z-J, Wang X-Y, Zhang Z-L, Lu W, Zheng X-L (2022) Electroantennographic and olfactory responses of Quadrastichus mendeli to eucalyptus volatiles induced by the gall-forming insect Leptocybe invasa. Pest Manag Sci 78:2405–2416. https://doi.org/10.1002/ps.6869
Isaias RMDS, De Oliveira DC (2012) Gall phenotypes – product of plant cells defensive responses to the inducers attack. In: Mérillon JM, Ramawat KG (eds) Plant defence: biological control. Springer, Dordrecht, pp 273–290
Isaias RM, de Oliveira DC, da Silva Carneiro RG, Kraus JE (2014) Developmental anatomy of galls in the neotropics: arthropods stimuli versus host plant constraints. In: Fernandes GW, Santos JC (eds) Neotropical insect galls. Springer, Dordrecht, pp 15–34
Jiang Y, Veromann-Jürgenson L, Ye J, Niinemets Ü (2018) Oak gall wasp infections of *Quercus robur* leaves lead to profound modifications in foliage photosynthetic and volatile emission characteristics. Plant Cell Environ 41:160–175. https://doi.org/10.1111/pce.13050
Jiang Y, Ye J, Veromann-Jürgenson L-L, Niinemets Ü (2020) Gall- and erineum-forming Eriophyes mites alter photosynthesis and volatile emissions in an infection severity-dependent manner in broad-leaved trees Alnus glutinosa and Tilia cordata. Tree Physiol 00:1–22. https://doi.org/10.1093/treephys/tpaa173

Kanjana N, Ahmed MA, Ma L, Zhang L (2024) Volatile signaling in weed plant Ageratina Adenophora: understanding the key emissions influencing Procecidochares utilis attraction to gall formation
Kehl A, Dötterl S, Aas G, Rambold G (2010) Is flower scent influencing host plant selection of leaf-galling sawflies (Hymenoptera, Tenthredinidae) on willows? Chemoecology 20:215–221. https://doi.org/10.1007/s00049-010-0050-6
Klimm FS, Weinhold A, Volf M (2020) Volatile production differs between oak leaves infested by leaf-miner Phyllonorycter harrisella (Lepidoptera: Gracillariidae) and galler Neuroterus quercusbaccarum (Hymenoptera: Cynipidae). Eur J Entomol 117:101–109. https://doi.org/10.14411/eje.2020.011
Kuster VC, Rezende UC, Cardoso JCF, dos Isaias RMS, de Oliveira DC (2019) How galling organisms manipulate the secondary metabolites in the host plant tissues? A histochemical overview in neotropical gall systems. In: Mérillon J-M, Gopal Ramawat K (eds) Co-evolution of secondary metabolites. Reference series in phytochemistry. Springer International Publishing, Berlin, pp 823–842
Lawo NC, Weingart GJF, Schuhmacher R, Forneck A (2011) The volatile metabolome of grapevine roots: first insights into the metabolic response upon phylloxera attack. Plant Physiol Biochem 49:1059–1063. https://doi.org/10.1016/j.plaphy.2011.06.008
Lev-Yadun S, Inbar M (2024) The potential of visual and olfactory signals in gall defence. Indian J Entomol 287–294. https://doi.org/10.55446/IJE.2023.871
Mani MS (1965) Ecology of plant galls, 1st edn. Springer, Dordrecht
Margulis L, Fester R (1991) Symbiosis as a source of evolutionary innovation. MIT Press, Cambridge, MA
McFall-Ngai MJ (2002) Unseen forces: the influence of bacteria on animal development. Dev Biol 242:1–14. https://doi.org/10.1006/dbio.2001.0522
Mescher MC (2012) Manipulation of plant phenotypes by insects and insect-borne pathogens. In: Hughes DP, Jacques Brodeur FT (eds) Host manipulation by parasites. Oxford University Press, Oxford, UK, pp 73–92
Mohamed, M. E. (2016) The interaction between the gall wasp Leptocybe invasa and Eucalyptus camaldulensis leaves: A study of phyto-volatile metabolites. Jour Pharm Phyto 8(4):90–98. https://doi.org/10.5897/JPP2015.0384
Moreira X, Abdala-Roberts L (2019) Specificity and context-dependency of plant–plant communication in response to insect herbivory. Curr Opin Insect Sci 32:15–21. https://doi.org/10.1016/j.cois.2018.09.003
Motta LB, Kraus JE, Salatino A, Salatino MLF (2005) Distribution of metabolites in galled and non-galled foliar tissues of *Tibouchina pulchra*. Biochem Syst Ecol 33:971–981. https://doi.org/10.1016/j.bse.2005.02.004
Murakami R, Ushima R, Sugimoto R, Tamaoki D, Karahara I, Hanba Y, Wakasugi T, Tsuchida T (2021) A new galling insect model enhances photosynthetic activity in an obligate holoparasitic plant. Sci Rep 11:13013. https://doi.org/10.1038/s41598-021-92417-3
Nabity PD, Haus MJ, Berenbaum MR, DeLucia EH (2013) Leaf-galling phylloxera on grapes reprograms host metabolism and morphology. Proc Natl Acad Sci USA 110:16663–16668. https://doi.org/10.1073/pnas.1220219110
Neilson EH, Goodger JQD, Woodrow IE, Møller BL (2013) Plant chemical defense: at what cost? Trends Plant Sci 18:250–258. https://doi.org/10.1016/j.tplants.2013.01.001
Ninkovic V, Markovic D, Rensing M (2021) Plant volatiles as cues and signals in plant communication. Plant Cell Environ 44:1030–1043. https://doi.org/10.1111/pce.13910
Noe SM, Niinemets Ü (2020) Impact of gall-forming insects on global BVOC emissions and climate: a perspective. Front For Glob Change 3. https://doi.org/10.3389/ffgc.2020.00009
Nyman T, Julkunen-Tiitto R (2000) Manipulation of the phenolic chemistry of willows by gall-inducing sawflies. Proc Natl Acad Sci USA 97:13184–13187. https://doi.org/10.1073/pnas.230294097

Ogah EO, Smart LE, Woodcock CM, Caulfield JC, Birkett MA, Pickett JA, Nwilene FE, Bruce TJ (2017) Electrophysiological and behavioral responses of female African rice gall midge, Orseolia oryzivora Harris and Gagné, to host plant volatiles. J Chem Ecol 43:13–16. https://doi.org/10.1007/s10886-016-0788-6

Ponce GE, Fuse M, Chan A, Connor EF (2021) The localization of Phytohormones within the gall-inducing insect Eurosta solidaginis (Diptera: Tephritidae). Arthropod Plant Interact 15:375–385. https://doi.org/10.1007/s11829-021-09817-5

Prathapan P (2024) Characterisation of the fig-fig wasp holobiont. Biosystems 237:105162. https://doi.org/10.1016/j.biosystems.2024.105162

Qi Y, Duan C, Ren L, Wu H (2020) Growth dynamics of galls and chemical defence response of Pinus thunbergii Parl. to the pine needle gall midge, Thecodiplosis japonensis Uchida & Inouye (Diptera: Cecidomyiidae). Sci Rep 10:12289. https://doi.org/10.1038/s41598-020-69231-4

Qian Q, Cui J, Miao Y, Xu X, Gao H, Xu H, Lu Z, Zhu P (2024) The plant volatile-sensing mechanism of insects and its utilization. Plan Theory 13:185. https://doi.org/10.3390/plants13020185

Raman A (2005a) Biology, Ecology, and Evolution of Gall-Inducing Arthropods, Vols 1 and 2

Raman A (2005b) Biology ecology and evolution of gall inducing arthropods, vol 2. Science Publishers, Enfield

Raman A (2011) Morphogenesis of insect-induced plant galls: facts and questions. Flora: morphology. Distrib Funct Ecol Plants 206:517–533

Rand K, Bar E, Ari MB, Davidovich-Rikanati R, Dudareva N, Inbar M, Lewinsohn E (2017) Differences in monoterpene biosynthesis and accumulation in Pistacia palaestina leaves and aphid-induced galls. J Chem Ecol 43:143–152. https://doi.org/10.1007/s10886-016-0817-5

Rosenkranz M, Chen Y, Zhu P, Vlot AC (2021) Volatile terpenes – mediators of plant-to-plant communication. Plant J 108:617–631. https://doi.org/10.1111/tpj.15453

Rostás M, Maag D, Ikegami M, Inbar M (2013) Gall volatiles defend aphids against a browsing mammal. BMC Evol Biol 13:1–11. https://doi.org/10.1186/1471-2148-13-193

Santos SAP, Mota L, Malheiro R, Silva F, Campos M, de Pinho PG, Pereira JA (2016) Changes in volatile compounds of Dittrichia viscosa caused by the attack of the gall-forming dipteran Myopites stylatus. Ind Crop Prod 87:71–77. https://doi.org/10.1016/j.indcrop.2016.04.002

Schoonhoven LM, Jermy T, van Loon JJA (2013) Insect-plant biology. Oxford University Press, New York

Shorthouse JD, Wool D, Raman A (2005) Gall-inducing insects - Nature's most sophisticated herbivores. Basic Appl Ecol 6:407–411. https://doi.org/10.1016/j.baae.2005.07.001

Stone GN, Schönrogge K (2003) The adaptive significance of insect gall morphology. Trends Ecol Evol 18:512–522. https://doi.org/10.1016/S0169-5347(03)00247-7

Suzuki H, Yokokura J, Ito T, Arai R, Yokoyama C, Toshima H, Nagata S, Asami T, Suzuki Y (2014) Biosynthetic pathway of the phytohormone auxin in insects and screening of its inhibitors. Insect Biochem Mol Biol 53:66–72. https://doi.org/10.1016/j.ibmb.2014.07.008

Swanson L, Li T, Rinnan R (2021) Contrasting responses of major and minor volatile compounds to warming and gall-infestation in the Arctic willow Salix myrsinites. Sci Total Environ 793:148516. https://doi.org/10.1016/j.scitotenv.2021.148516

Takei M, Yoshida S, Kawai T, Hasegawa M, Suzuki Y (2015) Adaptive significance of gall formation for a gall-inducing aphids on Japanese elm trees. J Insect Physiol 72:43–51. https://doi.org/10.1016/j.jinsphys.2014.11.006

Thomas AM, Williams RS, Swarthout RF (2019) Distribution of the specialist Aphid Uroleucon nigrotuberculatum (Homoptera: Aphididae) in response to host plant semiochemical induction by the gall fly Eurosta solidaginis (Diptera: Tephritidae). Environ Entomol 48:1138–1148. https://doi.org/10.1093/ee/nvz078

Tokuda M, Suzuki Y, Fujita S, Matsuda H, Adachi-Fukunaga S, Elsayed AK (2022) Terrestrial arthropods broadly possess endogenous phytohormones auxin and cytokinins. Sci Rep 12:4750. https://doi.org/10.1038/s41598-022-08558-6

Tooker JF, De Moraes CM (2008) Gall insects and indirect plant defenses. A case of active manipulation? Plant Signal Behav 3:503–504
Tooker JF, Hanks LM (2004) Stereochemistry of host plant monoterpenes as mate location cues for the gall wasp Antistrophus rufus. J Chem Ecol 30:473–477. https://doi.org/10.1023/B:JOEC.0000017995.83676.c9
Tooker JF, Helms AM (2014) Phytohormone dynamics associated with gall insects, and their potential role in the evolution of the gall-inducing habit. J Chem Ecol 40:742–753
Tooker JF, Koenig WA, Hanks LM (2002) Altered host plant volatiles are proxies for sex pheromones in the gall wasp *Antistrophus rufus*. Proc Natl Acad Sci 99:15486–15491. https://doi.org/10.1073/pnas.252626799
Tooker JF, Crumrin AL, Hanks LM (2005) Plant volatiles are behavioral cues for adult females of the gall wasp Antistrophus rufus. Chemoecology 15:85–88. https://doi.org/10.1007/s00049-005-0298-4
Tooker JF, Rohr JR, Abrahamson WG, De Moraes CM (2008) Gall insects can avoid and alter indirect plant defenses. New Phytol 178:657–671. https://doi.org/10.1111/j.1469-8137.2008.02392.x
Uesugi A, Morrell K, Poelman EH, Raaijmakers CE, Kessler A (2016) Modification of plant-induced responses by an insect ecosystem engineer influences the colonization behaviour of subsequent shelter-users. J Ecol 104:1096–1105. https://doi.org/10.1111/1365-2745.12587
Valle D, Mujica V, Gonzalez A (2023) Herbivore-dependent induced volatiles in pear plants cause differential attractive response by lacewing larvae. J Chem Ecol 49:262–275. https://doi.org/10.1007/s10886-023-01403-8
Villagra C, Vera W, Lenitz S, Bergmann J (2021) Differences in volatile emissions between healthy and gall-induced branches of Haplopappus foliosus (Asteraceae). Biochem Syst Ecol 98:104309. https://doi.org/10.1016/j.bse.2021.104309
Wang W, Wang M, Feng J, Zhang S, Chen Y, Zhao Y, Tian R, Zhu C, Nieuwenhuizen NJ (2024) Terpene Synthase gene family in Chinese Chestnut (Castanea mollissima BL.) Harbors two Sesquiterpene synthase genes implicated in defense against Gall Wasp Dryocosmus kuriphilus. J Agric Food Chem 72:1571–1581. https://doi.org/10.1021/acs.jafc.3c07086
War AR, Paulraj MG, Ahmad T, Buhroo AA, Hussain B, Ignacimuthu S, Sharma HC (2012) Mechanisms of plant defense against insect herbivores. Plant Signal Behav 7:1306–1320
Ye J, Jiang Y, Veromann-Jürgenson LL, Niinemets Ü (2019) Petiole gall aphid (Pemphigus spyrothecae) infestation of Populus × petrovskiana leaves alters foliage photosynthetic characteristics and leads to enhanced emissions of both constitutive and stress-induced volatiles. Trees Struct Funct 33:37–51. https://doi.org/10.1007/s00468-018-1756-2
Yu H, Holopainen JK, Kivimäenpää M, Virtanen A, Blande JD (2021) Potential of climate change and herbivory to affect the release and atmospheric reactions of BVOCs from boreal and subarctic forests. Molecules 26:2283. https://doi.org/10.3390/molecules26082283
Zhao C, Escalante LN, Chen H, Benatti TR, Qu J, Chellapilla S, Waterhouse RM, Wheeler D, Andersson MN, Bao R, Batterton M, Behura SK, Blankenburg KP, Caragea D, Carolan JC, Coyle M, El-Bouhssini M, Francisco L, Friedrich M, Gill N, Grace T, Grimmelikhuijzen CJP, Han Y, Hauser F, Herndon N, Holder M, Ioannidis P, Jackson L, Javaid M, Jhangiani SN, Johnson AJ, Kalra D, Korchina V, Kovar CL, Lara F, Lee SL, Liu X, Löfstedt C, Mata R, Mathew T, Muzny DM, Nagar S, Nazareth LV, Okwuonu G, Ongeri F, Perales L, Peterson BF, Pu LL, Robertson HM, Schemerhorn BJ, Scherer SE, Shreve JT, Simmons D, Subramanyam S, Thornton RL, Xue K, Weissenberger GM, Williams CE, Worley KC, Zhu D, Zhu Y, Harris MO, Shukle RH, Werren JH, Zdobnov EM, Chen MS, Brown SJ, Stuart JJ, Richards S (2015) A massive expansion of effector genes underlies gall-formation in the wheat pest mayetiola destructor. Curr Biol 25:613–620. https://doi.org/10.1016/j.cub.2014.12.057
Zhu X, Li L, Hsiang T, Zha Y, Zhou Z, Chen R, Wang X, Wu Q, Li J (2020) Chemical composition and attractant activity of volatiles from Rhus potaninii to the Spring Aphid Kaburagia rhusicola. Molecules 25:3412. https://doi.org/10.3390/molecules25153412

Zhu C, Wang W, Chen Y, Zhao Y, Zhang S, Shi F, Khalil-Ur-Rehman M, Nieuwenhuizen NJ (2022) Transcriptomics and antioxidant analysis of two Chinese chestnut (Castanea mollissima BL.) varieties provides new insights into the mechanisms of resistance to gall wasp Dryocosmus kuriphilus infestation. Front Plant Sci:13. https://doi.org/10.3389/fpls.2022.874434

The Physiological and Ecological Processes of Anthocyanins Accumulation and Red Gall Coloration

Denis Coelho de Oliveira, Moshe Inbar, and Simcha Lev-Yadun

1 General Aspects

Galls are neoformed plant organs that develop following the reprogramming of host plant cells and tissues induced by yet-unknown triggers originating from foreign organisms, including viruses, bacteria, and insects (Oliveira and Isaias 2010; Harris and Pitzschke 2020). The gall-inducing insects can manipulate the host plant's development, structure, and physiology for their own benefit gaining better nutrient supply and defense from biotic and abiotic factors (Oliveira et al. 2016). Among many changes, gall inducers often cause high production of reactive oxygen species (ROS) (Isaias et al. 2015; Guedes et al. 2022), triggering the gall induction, development, and shape acquisition (Oliveira et al. 2014; Price et al. 1987; Stone and Schonrogge 2003). Being static, conspicuous, long-lasting, and irreplaceable, galls (i.e., gall inducers) are vulnerable to attack by a variety of natural enemies, such as other herbivores, predators, parasitoids, and pathogens. Gall defense is usually achieved through plant-derived structures and chemical profiles. In some cases, the gall-inducing insects protect the galls themselves (Aoki and Kurosu 2010; Crespi et al. 1997; Inbar 1998).

D. C. de Oliveira (✉)
Institute of Biology, Universidade Federal de Uberlândia (UFU), Uberlândia, Minas Gerais, Brazil
e-mail: denisoliveira@ufu.br

M. Inbar
Department of Evolutionary & Environmental Biology, University of Haifa, Haifa, Israel

S. Lev-Yadun
Department of Biology & Environment, University of Haifa Oranim, Tivon, Israel

D. C. de Oliveira, R. M. dos Santos Isaias (eds.), *Plant Galls*,
https://doi.org/10.1007/978-3-031-80064-1_19

A specialized secondary metabolic pathway found in vascular land plants responds to both abiotic and biotic-induced stresses, particularly through phenylpropanoid biosynthesis, leading to the production of anthocyanins (Sharma et al. 2024) and consequent red pigmentation in many (but not all) galls. Anthocyanins are phenolic derivatives accountable for the red, blue, and purple coloration in fruits, flowers, leaves, and some vegetables (Davis et al. 2022). The fundamental structure of anthocyanins comprises a flavylium cation (c6-c3-c6), capable of bonding with various sugars, hydroxyl, or methoxyl groups, leading to the production and identification of hundreds of anthocyanin variations (He and Giusti 2010). Different sugars such as rhamnose, xylose, galactose, arabinose, and especially glucose bond to anthocyanins (Horbowicz et al. 2008) that can be mono-, di- or triglycosides (Giusti et al. 1998). The main aglycones of anthocyanins are cyanidin, peonidin, delphinidin, pelargonidin, malvidin, and petunidin (Tanaka et al. 2008). The color and stability of anthocyanins are affected by their structure, pH, temperature, light, oxygen, enzymes, and their association with acids, metals, sugars, and proteins (Francis 1989; Sharma et al. 2024). Once anthocyanins are produced under biotic and abiotic stress, their primary function is associated with scavenging free radicals (reactive oxygen species—ROS) by donating protons from a hydroxyl group, as well as supporting an unpaired electron (Miguel 2011), thereby serving a physiological role as antioxidants (Wiczkowski et al. 2013).

Reactive oxygen species, the ROS, may function as signaling molecules, influencing growth, development, and responses to stress-induced conditions in plant tissues (Mittler et al. 2006; Couee et al. 2006), phenomena also observed in galls (Isaias et al. 2015). Therefore, it stands to reason that the level of stress induced by the gall-inducing organism on plant tissues, and how the host responds to this stress, is relevant for determining various gall traits, including coloration. The conspicuous yellow, orange, or red coloration in galls (the most frequently observed colors) attributed to the accumulation of anthocyanins or carotenoids (Inbar et al. 2010) may be altered by any changes in the metabolic pathways of the host plant or within gall tissues. These changes may be induced by various factors, which may trigger the strong expression of anthocyanins as a determinant for red coloration. While red coloration can be a conservative trait in some galls (Rostás et al. 2013), in other gall systems, this red coloration can vary throughout gall development phases and among gall parts and may be influenced by levels of light exposure, overaccumulation of sugar, constraints imposed by the site of gall induction, and salivary secretions of the gall-inducing insect (Connor et al. 2012; Bomfim et al. 2019; Korgaonkar et al. 2021; Cardoso et al. 2022), without a (not tested) defensive role. In this chapter, we delve into the factors that could trigger the strong expression of anthocyanin accumulation in some gall-inducing insect systems, leading to red coloration, as well as widening their possible biological and ecological functions.

2 Exposure to High-Light Can Induce Anthocyanin Accumulation in Galls

We acknowledge the influence of light exposure on red color determination in certain galls (Wool 2004; Inbar et al. 2010; Lev-Yadun 2016; Protasov et al. 2007). In a globoid gall induced by an unidentified species of Diptera (Cecidomyiidae) on *Qualea parviflora* (Vochysiaceae), Bomfim et al. (2019) systematically quantified the levels of gall's exposure to sunlight as related to the anthocyanin concentrations and photosynthetic capacity. They found that globoid red galls exposed to high-light conditions accumulate more anthocyanins than shaded green galls. However, in the same host plant, the intralaminar gall induced by another unidentified Cecidomyiidae species does not increase the anthocyanin accumulation even under high-light exposure. The gall inducers—*Q. parviflora* systems (Fig. 1) and the fact that many galls are always green indicate that red coloration in galls exposed to high-light conditions is not a rule and appears to be species-specific and idiosyncratic. However, the

Fig. 1 (**A, B**) Galls induced by *Palaeomystella oligophaga* (Lepidoptera) on buds of *Macairea radula* (Melastomataceae), (**C, D**) Galls induced on leaves of *Qualea parviflora* on shaded (**C**) and high-light exposure (**D**)

adaptive significance of anthocyanin accumulation and red coloration as a defense against high light exposure requires further testing.

Plant tissues exposed to high-light conditions, including UV radiation and an excess of PAR, can increase the production of stressor molecules such as ROS (Schadev et al. 2021). Additionally, the process of gall induction and development, as well as the feeding activity of the gall-inducer, lead to an increase in oxidative stress, evaluated by the accumulation of ROS (Isaias et al. 2015; Kuster et al. 2019). The overexpression of ROS serves as an important trigger for anthocyanin biosynthesis and accumulation in land plants (Davis et al. 2022; Sharma et al. 2024) and, consequently, in galls as well. An overlap of these stress-induced stimuli may enhance the mechanism of antioxidant defense in both plant tissues and galls, including the biosynthesis of anthocyanins. This process may result in some galls exhibiting red coloration in an idiosyncratic manner.

Many galls induced on leaves, for example, are green due to the presence of a chlorophyllous tissue, allowing this new plant organ to photosynthesize (Shorthouse et al. 2005; Martini et al. 2020). Photosynthesis plays an important role in avoiding hypoxia and hypercarbia by maintaining redox homeostasis in gall tissues (Isaias et al. 2015; Pincebourde and Casas 2016). Therefore, it can be inferred that the production of anthocyanins in galls exposed to high-light conditions may be crucial for maintaining photosynthetic activity, even at low levels, and consequently for maintaining gall tissue homeostasis. For instance, the red galls induced on *Qualea parviflora*, which show a higher concentration of anthocyanins compared to the green ones, are more efficient in dissipating energy via NPQ (non-photochemical quenching), an important metabolic pathway to protect the photosynthetic apparatus in galls (Bomfim et al. 2019). Additionally, red galls, due to the higher accumulation of anthocyanins, can exhibit higher antioxidant activity compared to green galls, as observed in galls induced by a cynipid on *Nothofagus obliqua* (Guedes et al. 2022).

Red gall coloration has been related to high-light exposure in different gall systems. Another example is the galls induced on *Eucalyptus camaldulensis* (Myrtaceae) by the gall-inducer *Ophelimus maskelli* (Hymenoptera), which may show red coloration depending on their position on the host plant and consequently their exposure to light (Protasov et al. 2007). While the red coloration phenomenon in galls due to anthocyanin accumulation under high-light exposure is not a rule, it can confer physiological advantages in terms of antioxidant defense and protection against UV radiation, thereby safeguarding the galls' photosynthetic apparatus. The physiological advantages of the accumulation of anthocyanins leading to red phenomenon in galls should be systematically tested to better evaluate their adaptive value for the gall-inducing insects.

3 Accumulation of Sugar and Its Relationships with Gall Coloration

Galls are a true sink of photosynthates stimulated by the gall-inducing organisms, mobilizing high amounts of sugars from other plant parts (Larson and Whitham 1991; Inbar et al. 1995; Oliveira et al. 2017). Generally, the presence of cytokinins drives the establishment of source-sink relationships in plants and the priority to receive photosynthates (Sakakibara 2006; Kieber and Schaller 2014). This occurs because both cytokinins and auxins are synthesized in some of the plant meristems where cell division and elongation processes are intense (Gomes and Scortecci 2021; Wang et al. 2021). These sites have a high demand for sugar to maintain cellular machinery and cell wall growth. Over the last years, several studies reported high levels of cytokinins and auxins in the gall-inducing insets compared with the gall and non-galled tissues, evidencing the role of exogenous phytohormones in gall induction and development (Mapes and Davies 2001a, b; Straka et al. 2010; Andreas et al. 2020). Although expression of genes encoding cytokinins have not been detected in gall tissues, the elevated levels of cytokinins found in certain gall-inducing insects appear to be linked either to their dietary intake (Frébort et al. 2011) or to their own production of cytokinins (Mooi et al. 2024).

In galls, phytohormones such as auxins and cytokinins can be detected particularly in meristematic sites where cell division and elongation occur (Bedetti et al. 2014, 2018). These cellular processes result in elevated levels of complexity in galls, leading to the formation of various tissue compartments (Kurzfeld-Zexer et al. 2015; Ferreira et al. 2019). Nutritive gall tissues, for example, can store carbohydrates (i.e., glucose, fructose, and sucrose and water-soluble polysaccharides) as well as proteins and lipids (Bronner 1992; Rostás et al. 2013; Ferreira et al. 2019; Naves et al. 2021). In the outer cortex of common gall storage tissues, starch is commonly detected (Bronner 1992) and can be hydrolyzed into simple sugars, which are then transported cell-to-cell (Guo et al. 2023) to other gall compartments to supply the gall metabolism and the insect's demands (Oliveira et al. 2011). As a phenolic derivative, anthocyanins could be bonded to different sugars such as rhamnose, xylose, galactose, arabinose, and especially glucose (He and Giusti 2010). Therefore, the high sugar concentration in galls may trigger the production of secondary metabolites, such as anthocyanins, with stimulation and histolocalization occurring especially in the gall's outer cortex (Motta et al. 2005; Kuster et al. 2019). Also, during the senescence process, after the emergence of gall-inducers and inhabitants, the remobilization of nutrients to the host plant can occur. This process was proposed to lead to changes in gall coloration due to the degradation of chlorophyll (White 2010).

Globoid galls induced by *Palaeomystella olygophaga* (Lepidoptera) on buds of *Macairea radula* (Melastomataceae) exhibit a wide diversity of coloration, ranging from light green to red (Fig. 1a, b) (Rezende et al. 2021). These galls develop on *M. radula* as stem-like structures due to the shortening of internodes, accompanied by the development of leaf-like projections (Rezende et al. 2023). Found in open areas of the Brazilian Cerrado biome, these galls are exposed to high temperatures and high-light levels. Although there is no direct evidence linking high-light exposure to anthocyanin overexpression in *M. radula* galls, the accumulation of water-soluble polysaccharides in red galls compared to green ones was observed (Naves et al. 2021). One of the most significant water-soluble polysaccharides in plants are pectins, which appear to increase anthocyanin thermostability and prevent its degradation (Fernandes et al. 2020). However, the red galls on *M. radula* demonstrate lower concentrations of total soluble sugars compared to the green ones, and there is also no difference in starch concentrations (Naves et al. 2021). Similar results were observed in galls induced on ovules within floral buds of *Miconia chamissois* (Melastomataceae) by the gall-inducing *Allorhogas uberlandiensis* (Hymenoptera), which produce both green and red fruit-like galls (see Fig. 2a, b).

Connor et al. (2012) proposed that gall induction and development require the action of cytokinins and auxins produced by host tissues, as well as those secreted by certain gall-inducing insects. These phytohormones stimulate a strong sink for photosynthates in galls, and consequently, the excess sugar could be allocated for the biosynthesis of phenolic derivatives, such as anthocyanins, resulting in red coloration. This concept is known as the fabricational noise hypothesis (Connor et al. 2012). With the over-accumulation of sugars, the biosynthesis of phenolic derivatives has an important role in the carbon balance, reducing the accumulation of sugar in leaf tissues, for example (Lo Piccolo et al. 2020). As mentioned above, various sugars can be bounded to anthocyanin molecules, providing evidence of the sugar requirements for the formation of this phenolic derivative. Indeed, galls can accumulate high concentrations of both glucose and phenolic derivatives (Zoric et al. 2019; Kuster et al. 2019). However, Gerchman et al. (2013) commented that we should be careful with the above hypothesis because there is no mandatory relationship between cytokinins and anthocyanin production in plants, and that other metabolic pathways should be considered. Thus, Gerchman et al. (2013) pointed out that while the fabricational noise hypothesis proposed by Connor et al. (2012) seems to have the potential to explain some cases of red gall coloration in nature, it is not the rule.

4 The Constraints Imposed by the Galling Site Can Determine Red Gall Coloration

While gall development must occur within the structural and physiological constraints of the host plant's organs and tissues (Isaias et al. 2014), the cecidogenetic field induced by the inducer can also influence the sites adjacent to the galls (Mani 1964). The cecidogenetic field can be defined as the ability of gall-inducing insects to alter the structural and physiological pathways of host plant tissues through the application of certain salivary secretions, thereby exerting control over gall development. Once we can consider the cecidogenetic field impact on both the site of gall

Fig. 2 (**A–C**) Galls induced in the ovule of floral buds of *Miconia chamissois* (Melastomataceae), inducing green and red fruit-like gall phenotypes. (**D**) Leaf galls induced on *Manihot esculenta*, (**E**) red galls on leaves

development and on adjacent tissues, we can hypothesize that the constraints imposed on the adjacent tissues may impact gall structure and metabolism. Many galls, for example, are induced in the host epidermal cells, which can redifferentiate into a typical nutritive tissue (Oliveira and Isaias 2010; Ferreira et al. 2019). The epidermis appears to be one of the most reactive host tissues during gall induction and development, exhibiting an increase in indumentum in the form of trichomes and other emergences (Mani 1964; Ferreira et al. 2017). However, the role of trichomes in gall defense is controversial (Stone and Schonrogge 2003; Rezende et al. 2021). Trichomes and emergences in galls can accumulate both phenolic derivatives and sugars (Kuster et al. 2019; Ferreira et al. 2017), metabolites that have cross-linked metabolic pathways with anthocyanins. Even though anthocyanins can be produced at different times and sites in various plant organs and tissues, stimulated by both biotic and abiotic factors, their production appears to be more abundant in the epidermis (Davis et al. 2022; Jordheim et al. 2016; Cooney et al. 2015). Therefore, if the site of gall induction has a predisposition to produce high amounts of anthocyanins, we can infer that the gall may also produce them, and the red gall coloration may be a by-product of the host local traits, as proposed by Cardoso et al. (2022), in a host variation hypothesis.

This hypothesis was proposed based on the globoid galls induced by *P. olygophaga* on the buds of *M. radula*. The *Macairea* galls that develop as stem galls induce the shortening of internodes and the development of leaf-like structures with many emergences and trichomes (Rezende et al. 2023). These indumenta on gall structure occur because the leaves of *M. radula* have the ontogenetic capacity to develop non-glandular and glandular trichomes, as well as emergences, both on adaxial and abaxial leaf surfaces. This last structure, the emergences, is also present in the stem of this host species (Somavilla and Graciano-Ribeiro 2011) and is not a new organ specific to *Macairea* galls. Another example was found by an unidentified species of Diptera on the leaves of cassava (*Manihot esculenta*). These galls are red when the site of gall induction has a high concentration of red anthocyanins pigments (Fig. 2c, d). In these systems, the individuals are positioned side by side, experiencing the same environmental conditions. A differential accumulation of anthocyanins in cassava leaves is not a novelty, as it is a well-established qualitative and quantitative trait commonly employed for selecting resistant genotypes (Khumaida et al. 2015).

According to Cardoso et al. (2022), galls induced on redder sites have a higher probability of having red coloration than those induced on non-redder sites. Thus, the constraints imposed by host plant tissues seem to be important for red color determination, although the gall-inducing insect is considered a strong manipulator of the host plant's traits (Stone and Cook 1998; Pan et al. 2015; Hearn et al. 2019; Martini et al. 2019). Therefore, we have to be cautious because genetic variation within the gall-inducing organism can also lead to color variation in galls. Thus, despite the evidence of host constraints in gall color determination, the host variation hypothesis warrants further exploration.

5 Gall-Inducing Salivary Secretions Determine Red Coloration

The development of red galls can be related to the secretion of a specific protein into the galls by a gall-inducing aphid, resulting in the differential expression of a small number of plant genes, including those associated with anthocyanin biosynthesis. This DGC protein is expressed by the determinant gall color gene (dgc) in the aphid's salivary glands on galls induced on leaves of *Hamamelis virginiana* (Hamamelidaceae). In this system, 4% of the galls induced by the aphid *Hormaphis cornu* expressed red coloration, while 96% were green, randomly distributed on leaves. The dgc gene is a member of a diverse family of novel predicted secreted proteins characterized by a pair of widely spaced cysteine-tyrosine-cysteine (CYC) residues, named BICYCLE proteins. The BICYCLE genes are strongly expressed in the salivary glands of the gall-inducing aphid *H. cornu* generations, suggesting that they may regulate many aspects of gall development. The secretion of DGC protein into the gall tissue can result in the downregulation of several anthocyanin biosynthetic genes. However, the complete silencing of dgc in aphid salivary glands leads to the upregulation of anthocyanin genes, resulting in the accumulation of this pigment and the expression of red coloration in galls (Korgaonkar et al. 2021). The role of BICYCLE genes in the salivary glands of gall-inducing aphids in the development of various aspects of galls is still incipient. Thus, new studies manipulating this gene family in different gall systems and in transgenic plants are necessary, as well as testing whether aphids with complete dgc silencing show any adaptive significance.

6 Visual Defensive Role of Red Gall Pigmentation

Plant pigments can play an important role in animal-plant interactions, such as in pollination and seed dispersal (Chittka and Raine 2006; Schaefer and Ruxton 2011; Barônio and Oliveira 2019) and may also visually or chemically advertise danger to other organisms, a biological phenomenon called aposematism (Lev-Yadun 2001, 2016; Ruxton et al. 2004). Among the typical aposematic colors in plants, such as yellow, orange, purple, black, white, and brown, red appears to be more commonly observed (Lev-Yadun 2001, 2016). In this way, the aposematic gall hypothesis argues that visual (warning coloration) or olfactory signals may protect well-defended galls from initial attack by natural enemies (Inbar et al. 2010; Lev-Yadun and Inbar 2024).

The gall warning coloration hypothesis has scarcely been tested experimentally while controlling for other factors. Stone and Schonrogge (2003) suggested that only indirect evidence can corroborate the adaptive significance of the interaction between gall-inducing organisms and their natural enemies. In recent years, it has been shown that the chemical composition of gall tissues, originating from their secondary metabolites, indeed provides evidence of defense against natural enemies (Davidovich-Rikanati et al. 2022; Nyman and Julkunen-Titto 2000; Rand et al. 2014, 2017; Hall et al. 2017). Galls often exhibit compartmentalized tissues in which secondary metabolites, such as phenolics and tannins, are stored, functioning as defensive chemical traits that protect the gall-inducer organism (Cornell 1983; Hartley 1998; Kuster et al. 2019) but may simultaneously attract some of their natural enemies through emitted volatiles (Barônio and Oliveira 2019). However, gall-inducing insects can also control the chemical signaling emitted from galls to reduce the detection efficiency of their natural enemies (Tooker et al. 2008).

Galls induced by *Slavum wertheimae* (Hemiptera: Fordini) on shoots of *Pistacia atlântica* (Anarcadiaceae) is, so far, the only system experimentally tested considering the visual aposematic gall hypothesis. These red galls emit higher levels of volatiles compared to non-galled leaves. As a result, goats prefer to consume the leaves rather than the galls (Rostás et al. 2013). Despite biotic or abiotic factors that may induce anthocyanin accumulation and consequently red phenomenon in galls, experimental analyses that may or may not corroborate the visual aposematism hypothesis in these gall systems are necessary. We encourage conducting in situ or field experiments to test the aposematic hypothesis in insect-induced galls. These studies should also investigate the production pathways of secondary metabolites and other induced factors to explain anthocyanin accumulation and the red coloration of galls.

7 Final Remakes

Galls express a huge diversity of morphologies, structures, metabolic composition, and coloration, that have adaptive value. Here we focused on the accumulation of anthocyanins in many red galls. The accumulation of anthocyanins in galls may be determined by various factors such as high-light exposure, sugar accumulation, constraints imposed by the host plant tissues, and/or control by the salivary secretions of the gall-inducing insects. Additionally, the accumulation of metal ions has been linked to various physiological pathways in galls (Ariola et al. 2024) and to the determination of anthocyanin coloration in plants (Sharma et al. 2024), opening a new frontier for understanding the red gall phenomenon. We conclude that the red gall phenomenon appears to be physiologically idiosyncratic and dependent on the stimulation of the derivative phenolics pathway, especially anthocyanin metabolism. Once the gall-inducing organism controls the host-plant characters for its own benefit, the visual red warning signal may potentially deter natural enemies. Experimental examination of the alternative or simultaneous roles of anthocyanin accumulation in galls are still needed. We advocate for on-site experiments while considering physiological and ecological functions.

References

Andreas P, Kisiala A, Emery RJN, de Clerck-Floate RM, Tooker JF, Price PW, Miller DG III, Chen MS, Connor EF (2020) Cytokinins are abundant and widespread among insect species. Plants 9:208. https://doi.org/10.3390/plants9020208

Aoki S, Kurosu U (2010) A review of the biology of Cerataphidini (Hemiptera, Aphididae, Hormaphidinae), focusing mainly on their life cycles, gall formation, and soldiers. Psyche J Entomol 2010:380351

Ariola IA, Costa EC, Oliveira DC, Isaias RMS (2024) Soil-plant-gall relationships: from gall development to ecological patterns. Biol Rev 99:1948–1975. https://doi.org/10.1111/brv.13106

Barônio GJ, Oliveira DC (2019) Eavesdropping on gall-plant interactions: the importance of the signaling functions of induced volatiles. Plant Signal Behav 14:1665454

Bedetti CS, Modolo LV, Isaias RMS (2014) The role of phenolics in the control of auxin in galls of *Piptadenia gonoacantha* (Mart.) MacBr (Fabaceae: Mimosoideae). Biochem Syst Ecol 55:53–59

Bedetti CS, Jorge NC, Trigueiro F, Bragança GP, Modolo LV, Isaias RMS (2018) Detection of cytokinins and auxin in plant tissues using histochemistry and immunocytochemistry. Biotech Histochem 1:1–6

Bomfim PM, Cardoso JC, Rezende UC, Martini VC, Oliveira DC (2019) Red galls: the different stories of two gall types on the same host. Plant Biol 21:284–291

Bronner R (1992) The role of nutritive cells in the nutrition of cynipids and cecidomyiids. In: Shorthouse JD, Rohfritsch O (eds) Biology of insect-induced galls. Oxford University Press, New York, pp 118–140

Cardoso JCF, Gonçalves PHP, Oliveira DC, Rezende U (2022) Host plant intraspecific variation determines gall traits. Plant Biol 25:208–214

Chittka L, Raine NE (2006) Recognition of flowers by pollinators. Curr Opin Plant Biol 9:428–435

Connor EF, Bartlett L, O'Toole S, Byrd S, Biskar K, Orozco J (2012) The mechanism of gall induction makes galls red. Arthropod Plant Interact 6:489–495

Cooney LJ, Schaefer HM, Logan BA, Cox B, Gould KS (2015) Functional significance of anthocyanins in peduncles of *Sambucus nigra*. Environ Exp Bot 119:18–26

Cornell HV (1983) The secondary chemistry and complex morphology of galls formed by the Cynipinae (Hymenoptera): why and how? Am Midl Nat 110:225–234

Couee I, Sulmon C, Gouesbet G, El Amrani A (2006) Involvement of soluble sugars in reactive oxygen species balance and responses to oxidative stress in plants. J Exp Bot 57:449–459. https://doi.org/10.1093/jxb/erj027

Crespi BJ, Carmean DA, Chapman TW (1997) Ecology and evolution of galling thrips and their allies. Annu Rev Entomol 42:51–71

Davidovich-Rikanati R, Bar E, Hivert G, Huang XQ, Hoppen-Tonial C, Khankin V, Rand K, Abofreih A, Muhlemann JK, Marchese JA, Shotland Y, Dudareva N, Inbar M, Lewinsohn E (2022) Transcriptional up-regulation of host-specific terpene metabolism in aphid-induced galls of Pistacia palaestina. J Exp Bot 73:555–570

Davis KM, Landi M, van Klink JW, Schwinn KE, Brummell DA, Albert NW, Chagné D, Jibran R, Kulshrestha S, Zhou Y, Bowman J (2022) Evolution and functions of red pigmentation in land plants. Ann Bot 130:613–636

Fernandes A, Brandão E, Raposo F, Maricato É, Oliveira J, Mateus N, Coimbra MA, Freitas V (2020) Impact of grape pectic polysaccharides on anthocyanins thermostability. Carbohydr Polym 239:116240

Ferreira BG, Álvarez R, Avritzer SC, Isaias RMS (2017) Revisiting the histological patterns of storage tissues: beyond the limits of gall-inducing taxa. Botany 95:173–184

Ferreira BG, Álvarez R, Bragança GP, Alvarenga DR, Perez-Hidalgo N (2019) Feeding and other gall facets: patterns and determinants in gall structure. Bot Rev 85:78–106. https://doi.org/10.1007/s12229-019-09207-w

Francis FJ (1989) Food colorants: anthocyanins. Crit Rev Food Sci Nutr 28:273–314

Frébort I, Kowalska M, Hluska T, Frébortová J, Galuszka P (2011) Evolution of cytokinin biosynthesis and degradation. J Exp Bot 62:2431–2452

Gerchman Y, Lev-Yadun S, Inbar M (2013) Red gal pigmentation: cytokinin stimulation is not everything. Arthropod Plant Interact 7:335–337

Giusti MM, Ghanadan H, Wrolstad RE (1998) Elucidation of the structure and conformation of red radish (*Raphanus sativus*) anthocyanins using one- and two-dimensional nuclear magnetic resonance techniques. J Agric Food Chem 46:4858–4863

Gomes GLB, Scortecci K (2021) Auxin and its role in plant development: structure, signaling, regulation and response mechanisms. Plant Biol 23:894–904

Guedes LM, Torres S, Sáez-Carillo K, Becerra J, Pérez CI, Aguilera N (2022) High antioxidant activity of phenolic compounds dampens oxidative stress in *Espinosa nothofagi* galls induced on *Nothofagus obliqua* buds. Plant Sci 314:111114

Guo W-J, Pommerrenig B, Neuhaus HE, Keller I (2023) Interaction between sugar transport and plant development. J Plant Physiol 288:154073

Hall CR, Carroll AR, Kitching RL (2017) A meta-analysis of the effects of galling insects on host plant secondary metabolites. Arthropod Plant Interact 11:463–473

Harris MO, Pitzschke A (2020) Plants make galls to accommodate foreigners: some are friends, most are foes. New Phytol 225:1852

Hartley SE (1998) The chemical composition of plant galls: are levels of nutrients and secondary compounds controlled by the gall former? Oecologia 113:492–501

He J, Giusti MM (2010) Anthocyanins: natural colorants with health-promoting properties. Annu Rev Food Sci Technol 1:163–187

Hearn J, Blaxter M, Schönrogge K, Nieves-Aldrey JL, Pujade-Villar J, Huguet E, Drezen JM, Shorthouse JD, Stone GN (2019) Genomic dissection of an extended phenotype: oak galling by a cynipid gall wasp. PLoS Genet 15:e1008398

Horbowicz M, Kosson R, Grzesiuk A, Debski H (2008) Anthocyanins of fruits and vegetables – their occurrence, analysis and role in human nutrition. Veg Crops Res Bull 68:5

Inbar M (1998) Competition, territoriality and maternal defense in a gall-forming aphid. Ethol Ecol Evol 10:159–170

Inbar M, Eshel A, Wool D (1995) Interspecific competition among phloem-feeding insects mediated by induced host-plant sinks. Ecology 76:1506–1515

Inbar M, Izhaki I, Koplovich A, Lupo I, Silanikove N, Glasser T, Gerchman Y, Perevolotsky A, Lev-Yadun S (2010) Why do many galls have conspicuous colors? A new hypothesis. Arthropod Plant Interact 4:1–6

Isaias RMS, Oliveira DC, Carneiro RGS, Kraus JE (2014) Development anatomy of galls in the Neotropics: arthropods stimuli versus host plant constraints. In: Fernandes GW, Santos JC (eds) Neotropical insect galls. Springer, Dordrecht, pp 15–34

Isaias RMS, Oliveira DC, Moreira AS, Soares GL, Carneiro RG (2015) The imbalance of redox homeostasis in arthropod-induced plant galls: mechanisms of stress generation and dissipation. Biochim Biophys Acta 1850:1509–1517

Jordheim M, Calcott K, Gould KS, Davies KM, Schwinn KE, Andersen OM (2016) High concentrations of aromatic acylated anthocyanin found in cauline hairs in *Plectranthus ciliatus*. Phytochemistry 128:27–34

Khumaida N, Ardie SW, Dianasan M, Syuku M (2015) Cassava (*Manihot esculenta* Crantz.) improvement through gamma irradiation. Procedia Food Sci 3:27–34

Kieber JJ, Schaller GE (2014) Cytokinins. Arabidopsis Book 12:e0168. https://doi.org/10.1199/tab.0168

Korgaonkar A, Han C, Lemire AL, Siwanowicz I, Bennouna D, Kopec RE, Andolfatto P, Shigenobu S, Stern DL (2021) A novel family of secreted insect proteins linked to plant gall development. Curr Biol 31:1836–1849

Kurzfeld-Zexer L, Lev-Yadun S, Inbar M (2015) One aphid species induces three gall types on a single plant: comparative histology of one genotype and multiple extended phenotypes. Flora 210:19–30

Kuster VC, Rezende UC, Cardoso JCF, Isaias RMS, Oliveira DC (2019) How do galling organisms manipulate the secondary metabolites in the host plant tissues? A histochemical overview in neotropical gall systems. In: Merillon JM, Ramawat KG (eds) Co-evolution of secondary metabolites. Springer, Cham, pp 823–842

Larson KC, Whitham TG (1991) Manipulation of food resources by a galling aphid: the physiology of sink-source interactions. Oecologia 188:15–21

Lev-Yadun S (2001) Aposematic (warming) coloration associated with thorns in higher plants. J Theor Biol 210:385–388

Lev-Yadun S (2016) Defensive (anti-herbivory) coloration in land plants: anti-herbivory plant coloration and morphology. Springer, Zug

Lev-Yadun S, Inbar M (2024) The potential of visual and olfactory signals in gall defence. Indian J Entomol 86:287–294

Lo Piccolo E, Landi M, Massai R, Remorini D, Guidi L (2020) Girledinduced anthocyanin accumulation in red-leafed *Prunus cerasifera*: effect on photosynthesis, photoprotection and sugar metabolism. Plant Sci 294:110456. https://doi.org/10.1016/j.plantsci.2020.110456

Mani MS (1964) The ecology of plant galls. Dr. Junk Pub, The Hague

Mapes CC, Davies PJ (2001a) Indole-3-acetic acid and ball gall development on *Solidago altissima*. New Phytol 151:195–202. https://doi.org/10.1046/j.1469-8137.2001.00161.x

Mapes CC, Davies PJ (2001b) Cytokinins in the ball gall of *Solidago altissima* and the gall forming larvae of *Eurosta solidaginis*. New Phytol 151:203–212. https://doi.org/10.1046/j.1469-8137.2001.00158.x

Martini VC, Kuster VC, Moreira ASFP, Oliveira DC (2019) Galling insects as phenotype manipulators of cell wall composition during the development of galls induced on leaves of Aspidosperma tomentosum (Apocynaceae). S Afr J Bot 127:226–233

Martini VC, Moreira ASFP, Kuster VC, Oliveira DC (2020) Photochemical performance and source-sink relationships in galls induced by *Pseudophacopteron longicaudatum* (Hemiptera) on leaves of *Aspidosperma tomentosum* (Apocynaceae). Photosynthetica 58:827–835

Miguel MG (2011) Anthocyanins: antioxidant and/or anti-inflammatory activities. J Appl Pharm Sci 1:7–15

Mittler R, Kim Y, Song L, Coutu J, Coutu A, Ciftci-Yilmaz S, Lee H, Stevenson B, Zhu J (2006) Gain- and loss-of-function mutations in Zat10 enhance the tolerance of plants to abiotic stress. FEBS Lett 580:6537–6542. https://doi.org/10.1016/j.febslet.2006.11.002

Mooi N, Roy SW, Connor EF (2024) Cytokinin biosynthesis in Hexapoda and Insecta: a bioinformatic analysis. Arthropod Plant Interact 19:1–14. https://doi.org/10.1007/s11829-023-10022-9

Motta LB, Kraus J, Salatino A, Salatino MLF (2005) Distribution of metabolites in galled and non-galled foliar tissues of *Tibouchina pulchra*. Biochem Syst Ecol 33:971–981

Naves GFA, Kuster VK, Machado M, Dias OS, Martini VC, Oliveira DC (2021) Metabolite investments and stress levels among tissue compartments of *Palaeomystella olygophaga* (Lepidoptera) galls on *Macairea radula* (Melastomataceae). Aust J Bot 69:131–142

Nyman T, Julkunen-Titto RR (2000) Manipulation of the phenolic chemistry of willows by gall-inducing sawflies. Proc Natl Acad Sci USA 97:13184–13187

Oliveira DC, Isaias RMS (2010) Redifferentiation of leaflet tissues during midrib gall development in *Copaifera langsdorffii* (Fabaceae). S Afr J Bot 76:239–248

Oliveira DC, Carneiro RGS, Magalhães TA, Isaias RMS (2011) Cytological and histochemical gradients on two *Copaifera langsdorffii* Desf. (Fabaceae)—Cecidomyiidae gall systems. Protoplasma 248:829–837

Oliveira DC, Moreira ASFPM, Isaias RMS (2014) Functional gradients in insect gall tissues: studies on Neotropical host plants. In: Fernandes GW, Santos JC (eds) Neotropical insect galls. Springer, Dordrecht, pp 35–50

Oliveira DC, Isaias RMS, Fernandes GW, Ferreira BG, Carneiro RGS, Fuzaro L (2016) Manipulation of host plant cells and tissues by gall-inducing insects and adaptive strategies used by different feeding guilds. J Insect Physiol 84:103–113

Oliveira DC, Moreira ASFP, Isaias RMS, Martini VC, Rezende UC (2017) Sink status and photosynthetic rate of the leaflet galls induced by *Bystracoccus mataybae* (Eriococcidae) on *Matayba guianensis* (Sapindaceae). Front Plant Sci 8:1249

Pan LY, Chen WN, Chiu ST, Raman A, Chian TC, Yang MM (2015) Is a gall an extended phenotype of the inducing insect? A comparative study of selected morphological and physiological traits of leaf and stem galls on *Machilus thunbergii* (Lauraceae) induced by five species of *Daphnephila* (Diptera: Cecidomyiidae) in northeastern Taiwan. Zool Sci 32:314–321

Pincebourde S, Casas J (2016) Hypoxia and hypercarbia in endophagous insects: larval position in the plant gas exchange network is key. J Insect Physiol 84:137–153. https://doi.org/10.1016/j.jinsphys.2015.07.006

Price PW, Fernandes GW, Waring GL (1987) Adaptive nature of insect galls. Environ Entomol 16:15–24

Protasov A, La Salle J, Blumberg D, Brand D, Saphir N, Assael F, Fisher N, Mendel Z (2007) Biology, revised taxonomy and impact on host plants of *Ophelimus maskelli*, an invasive gall inducer on *Eucalyptus* spp. in the Mediterranean area. Phytoparasitica 35:50–76

Rand K, Bar E, Ben-Ari M, Lewinsohn E, Inbar M (2014) The mono -and sesquiterpene content of aphid-induced galls on *Pistacia palaestina* is not a simple reflection of their composition in intactleaves. J Chem Ecol 40:632–642

Rand K, Bar E, Ben-Ari M, Davidovich-Rikanati R, Dudareva N, Inbar M, Lewinsohn E (2017) Differences in monoterpene biosynthesis and accumulation in *Pistacia palaestina* leaves and aphid-induced galls. J Chem Ecol 43:143–152

Rezende UC, Cardoso JCF, Hanson P, Oliveira DC (2021) Gall traits and galling insect survival in a multi-enemy context. Rev Biol Trop 69:291–301

Rezende UC, Gonçalves PHP, Kuster VC, Carneiro RGS, Oliveira DC (2023) Every end is a new beginning: histological features of galls induced on *Macairea radula* (Melastomataceae) allow a post-senescence colonization. Flora 306:152369

Rostás M, Maag D, Ikegami M, Inbar M (2013) Gall volatiles defend aphids against a browsing mammal. BMC Evol Biol 13:193

Ruxton GD, Sherratt TN, Speed MP (2004) Avoiding attack: the evolutionary ecology of crypsis, warning signals and mimicry. Oxford University Press, Oxford

Sakakibara H (2006) Cytokinins: activity, biosynthesis, and translocation. Annu Rev Plant Biol 57:431–449. https://doi.org/10.1146/annurev.arplant.57.032905.105231

Schadev S, Ansari SA, Ansari MI, Fujita M, Hasanuzzaman M (2021) Abiotic stress and reactive oxygen species: generation, signaling, and defense mechanisms. Antioxidants 10:277

Schaefer HM, Ruxton GD (2011) Plant-animal communication. Oxford University Press, New York

Sharma H, Sharma P, Kumar A, Neena C, Dhatt AS (2024) Multifaceted regulation of anthocyanin biosynthesis in plants: a comprehensive review. J Plant Growth Regul 24:3048–3062

Shorthouse JD, Wool D, Raman A (2005) Gall-inducing insects – nature's most sophisticated herbivores. Basic Appl Ecol 6:407–411

Somavilla NS, Graciano-Ribeiro D (2011) Análise comparativa da anatomia foliar de Melastomataceae em ambiente de vereda e cerrado sensu stricto. Acta Bot Bras 25:764–775

Stone GN, Cook JM (1998) The structure of cynipid oak galls: patterns in the evolution of an extended phenotype. Proc R Soc Lond B 265:979–988

Stone GN, Schonrogge K (2003) The adaptive significance of insect gall morphology. Trends Ecol Evol 18:512–522

Straka JR, Hayward AR, Emery RJN (2010) Gall-inducing *Pachypsylla celtidis* (Psyllidae) infiltrate hackberry trees with high concentrations of phytohormones. J Plant Interact 5:197–203. https://doi.org/10.1080/17429145.2010.484552

Tanaka Y, Sasaki N, Ohmiya A (2008) Biosynthesis of plant pigments: anthocyanins, betalains and carotenoids. Plant J 54:733–749

Tooker JF, Rohr JR, Abrahamson WF, De Moraes CM (2008) Gall insects can avoid and alter indirect plant defenses. New Phytol 178:657–671

Wang M, Gourrierec JL, Jiao F, Demotes-Mainard S, Prez-Garcia M, Ogé L, Hamama L, Crespel L, Bertheloot J, Chen J, Grappin P, Sakr S (2021) Convergence and divergence of sugar and cytokinin signaling in plant development. Int J Mol Sci 22:1282

White TCR (2010) Why do many galls have conspicuous colors? An alternative hypothesis revisited. Arthropod Plant Interact 4:149–150

Wiczkowski W, Szawara-Nowak D, Topolska J (2013) Red cabbage anthocyanins: profile, isolation, identification, and antioxidant activity. Food Res Int 51:303–309

Wool D (2004) Galling aphids: specialization, biological complexity, and variation. Annu Rev Entomol 49:175–192

Zoric AS, Marina F, Tosevski I, Tosti T, Jovic J, Krstic O, Velijovic-Javanovic S (2019) Resource allocation in response to herbivory and gall formation in *Linaria vulgaris*. Plant Physiol Biochem 135:224–232

Natural Enemies' Influence on the Population, Structure, and Function of Insect-Induced Galls

Pedro Henrique Pereira Gonçalves, Elaine Cotrim Costa, Bruno Garcia Ferreira, and Denis Coelho de Oliveira

1 Introduction

Galls are plant structures formed due to the cellular reprogramming of the host plants, triggered by mechanical and chemical stimuli from gall-inducing organisms, resulting in cell hypertrophy and tissue hyperplasia (Oliveira et al. 2016; Harris and Pitzschke 2020). The adaptive advantages of the galling habit over the free-living habit include a greater availability of nutritional resources and protection against environmental factors and natural enemies (Stone and Schönrogge 2003). However, this relationship with natural enemies involves an evolutionary war between the different organisms associated with galls (Hori 1992; Rohfritsch 1992; Oliveira and Isaias 2010; Stone and Schönrogge 2003).

For the success of the interaction, the gall-inducing insects need to adjust their life cycles to the vegetative and/or reproductive phenological cycle of the host plants (e.g. Pfeffer et al. 2018; Costa et al. 2021a, b). From an evolutionary perspective, the life cycles of gall-inducing insects are adjusted to the host plant phenology and the occurrence of natural enemies. Consequently, gall induction and development may

P. H. P. Gonçalves (✉) · D. C. de Oliveira
Institute of Biology, Universidade Federal de Uberlândia (UFU), Uberlândia, Minas Gerais, Brazil
e-mail: denisoliveira@ufu.br

E. C. Costa
Instituto de Ciências Biológicas, Universidade Federal do Rio Grande – Campus Carreiros, Carreiros, Rio Grande, Rio Grande do Sul, Brazil

B. G. Ferreira
Universidade Federal do Rio de Janeiro, Instituto de Biologia, Departamento de Botânica, Rio de Janeiro, RJ, Brazil

D. C. de Oliveira, R. M. dos Santos Isaias (eds.), *Plant Galls*,
https://doi.org/10.1007/978-3-031-80064-1_20

be synchronized with periods when reactive plant tissues and nutritional components are most available and with strategies to avoid interactions with competing species (Weis and Abrahamson 1988; Magalhães et al. 2014; Oliveira et al. 2016; Pfeffer et al. 2018). Such synchronization can result in different life histories of the gall-inducing insects, ranging from uni- to multivoltine cycles, according to plant phenology and biotic and abiotic factors (Oliveira et al. 2016). The fundamental role of a gall is to provide protection and nutrition to the offspring of the gall inducer. However, these consistent structures are exploited by various associated organisms to ensure essential resources like nutrition and shelter for their own offspring (Stone and Schönrogge 2003; Harper et al. 2004; Redfern 2011).

Among the hypotheses of the adaptive value of gall induction, the one that presupposes protection against natural enemies is the most difficult to demonstrate, since gallers are impacted by natural enemies of different guilds, and because we cannot compare the gall inducers with their free-living ancestors (Price et al. 1986; Stone and Schönrogge 2003; Luz and Mendonça Junior 2019). Examples of such enemies are the cecidophages, kleptoparasites, inquilines, which consume the gall tissues, and the most frequent, predatory parasitoids, which completely consume the larva or nymph of the gall-inducing insect (Luz and Mendonça Junior 2019). The death of the gall inducer by one of such guilds interrupts the developmental stimuli from the galling insect (Rezende et al. 2019). Additionally, parasitoids and inquilines can modify plant tissues through direct and indirect interactions. Direct modifications by parasitoids include a slight increase in gall tissues and, in some cases, the premature senescence of galling tissues (Rezende et al. 2019; Costa et al. 2022). On the other hand, inquilines modify gall tissues to induce the formation of a new gall within the primary gall, resulting in endogalls (Fenwick 2013; Luz and Mendonça Junior 2019).

This chapter explores the intricate interplay between gall-inducing insects, host plants, and their natural enemies, especially parasitoids and inquilines. It closely examines how galls are formed through various interactions, highlighting the important role of natural enemies in controlling galling insect populations. We also investigate how parasitoids can directly and indirectly alter the behavior of these gall-inducing insects, consequently affecting the structure and functionality of galls. Inquilines further influence the structure and morphology of galls through direct interactions with plant tissues, sometimes resulting in the formation of "galls within galls", known as endogalls. In such cases, inquilines are species-specific and can only survive and induce alterations in galls formed by a specific gall-inducing organism. As we examine these multitrophic interactions, we aim to understand how these organisms fit together in the broader context of complex ecological relationships. By exploring these aspects, this chapter emphasizes the important role of interactions with natural enemies in shaping gall-inducing life cycles and in the development of gall tissues, giving hints about aspects we still need to learn in this area.

2 Multitrophic Relationship on Galls: The Role of Natural Enemies

Galls are part of a complex multitrophic network of interactions, where each element plays an important role in forming a microecosystem (Fig. 1). Abiotic factors such as water availability, soil composition, fire events, temperature, and drought can regulate the population fluctuations within the community established in this

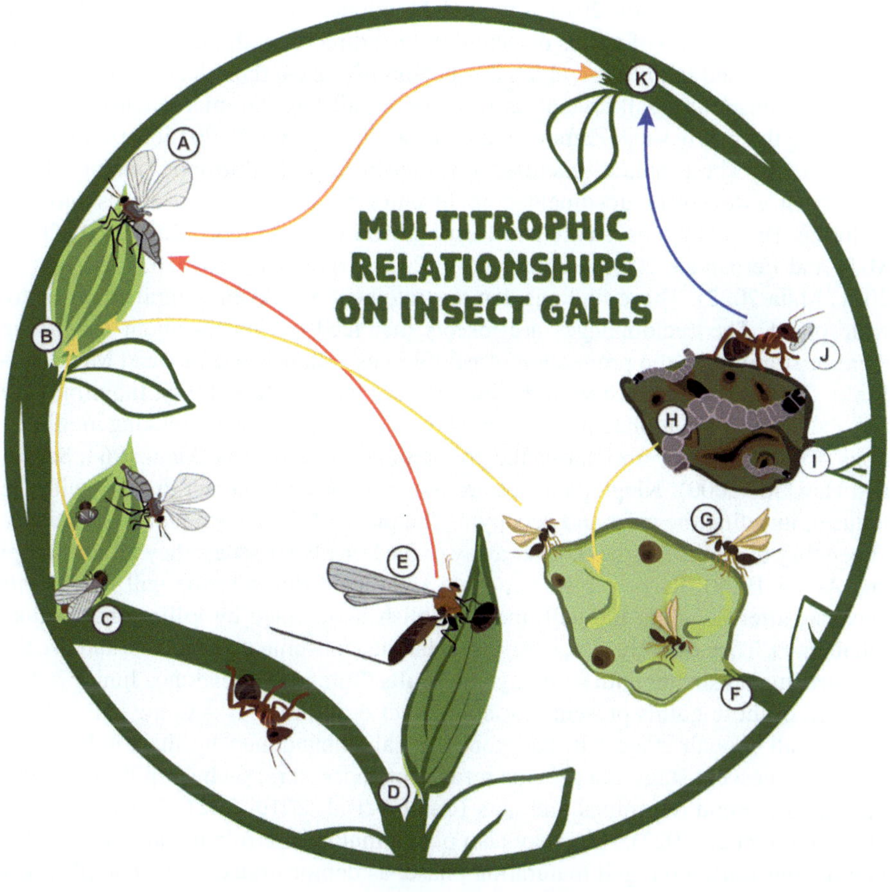

Fig. 1 Multitrophic relationships on insect galls. (A) Cecidomyiidae adult galling insect; (B) Unattacked leaf gall; (C) Adult cecidophages diptera emerged from the leaf gall with the galler; (D) Gall attacked and modified by parasitoid; (E) Adult hymenopteran parasitoid of galling insect; (F) Leaf galls modified by inquilines forming endogalls. (G) Adult inquiline hypergaller. (H) Kleptoparasite, eating gall tissues after killing the galler and finishing its developing on senescent gall tissues. (I) Gall in senescence. (J) Successor ant that uses the remaining's of the gall as nests. (K) non-galled leaf. Red arrowed line: Direct interaction with the galling insect. Yellow arrowed line: direct interaction with gall tissues. Orange Arrowed line: oviposition on available sites. Blue arrowed line: Mutualism Ant-Plant

microecosystem, thereby influencing the life cycles of the organisms involved in these interactions (Oliveira et al. 2014; Bonsignore and Bernardo 2018; Martini et al. 2021; Kuzmanich et al. 2023; Rezende et al. 2023; Zhao et al. 2023). Furthermore, the biotic elements, such as natural enemies, can also interfere with the structure of the interactions, playing crucial roles in regulating gall-inducing insect populations within ecosystems and influencing the gall structure development (Abrahamson et al. 1989; Perdikis et al. 2015).

Among natural enemies, some guilds directly interact with the gall inducers by attacking them, as the kleptoparasites, parasitoids, and predators (Abrahamson et al. 1989; Bourg and Hanson 2014; Luz and Mendonça Junior 2019; Rezende et al. 2019). Another group of guilds does not interact directly with the inducers, such as cecidophages and inquilines, which can potentially cause their death by competing pressure, consuming gall crucial tissues, and modifying the microenvironment by damaging the gall tissues (Zamora and Gómez 1993; Luz et al. 2015; Rezende et al. 2019, 2021). These natural enemies, inserted in these multitrophic relationships, have been extensively documented in inventories, observed in various biomes, including Brazilian ecosystems like Restinga and Cerrado (e.g. Maia 2001, 2022; Maia and Fernandes 2004; Campos et al. 2021; Maia and Silva 2021; Marchiori 2021; Maia 2022). These findings demonstrate how attacks of natural enemies are widespread. The cecidophages are insects that feed on galls without modifying them or stimulating the production of new tissues (Mani 1964; Luz and Mendonça Junior 2019). They coexist with the inducer and have high mobility. Inquilines are also exclusively phytophagous but coexist with the gall-inducer, inducing modifications of gall tissues or stimulating the production of new tissues (Mani 1964; Sanver and Hawkins 2000). Kleptoparasites are omnivorous and interact directly with the inducer, invading the galls and occupying the place of the gall-inducing organisms. When they do not chase the inducer away to feed on plant tissues, they kill it (Crespi and Abbot 1999). Kleptoparasites pose a significant threat to the gall inhabitants because, after invading the gall, they establish dominance by killing the original inhabitants. They directly attack the insect larvae, consume their bodies, and use the gall structure as shelter until hatching into adults (Toma and Mendonça Junior 2014).

Each of these guilds presents variable direct or indirect risks to the gall inducer during gall growth stages. In addition, the galls abandoned by their inducers or those in senescent stage can provide refuge for successors, such as spiders, insects, myriapods, pseudoscorpions and ants (Santos et al. 2019, 2021; Maia and Silva 2021; Jorge et al. 2022). Ants may also play a mutualistic role in gall multitrophic interactions during the gall maturation phase, as demonstrated in the complex networks involving plants, gall inducers, and ants, which seem to promote an indirect plant defense against other enemies (Giannetti et al. 2019; Aranda-Rickert et al. 2017). For example, nectar-like secretions of cynipid galls in South American Prosopis attract ants for cynipid progeny protection, making them analogous to extrafloral nectaries (EFNs) as a food source. However, unlike EFNs, these secretions protect the gall-inducing wasp, functioning similarly to arthropod honeydew (Aranda-Rickert et al. 2017). The benefits to all parties, including the host plant, depend on the broader ecological context and other interactions within this system.

The synchronization between the life history of this associated fauna and the phenology of the galls facilitates multitrophic interactions. Thus, by adjusting the timing of interactions, some insect guilds exhibit higher rates of attacking gall tissues and the gall-inducing insect, while others utilize the gall structures for shelter and nesting (Lima and Calado 2018; Santos et al. 2023; Rezende et al. 2023).

The gall multitrophic net is complex and context dependent. Campos et al. (2021) observed in Cerrado (Neotropical savannas) that parasitoids, particularly those belonging to Hymenoptera, were more prevalent than inquilines, predators, and successors, consistent with prior findings (Maia 2001; Maia and Fernandes 2004). These parasitoids, known for targeting plant galls, often result in high mortality rates among gall-inducing larvae (Weis and Abrahamson 1985; Price et al. 1986). Additionally, Maia and Silva (2021) also demonstrated that the parasitoids were found to be the most frequent attackers of galls in restingas, coastal ecosystems in Brazil. Among hymenopteran parasitoids, Eulophidae and Eurytomidae were predominant, as observed in other Brazilian ecosystems, as well. The arthropod orders documented in this study were also reported as part of the associated fauna in other Brazilian biomes (Maia and Silva 2021). The taxonomic knowledge of these inhabitants remains very deficient, in most regions of Brazil.

Maia (2022) registered many cecidophages, inquilines, and kleptoparasites, mostly associated with Cecidomyiidae-induced galls, in Brazilian restingas. On such ecosystems, cecidophages represented five insect orders: Coleoptera, Diptera (Cecidomyiidae, Chloropidae, Muscomorpha, Sciaridae, and Tephritidae), Hemiptera (Aphididae, and Membracidae), Lepidoptera, and Thysanoptera. Among them, lepidopterans and coleopterans were the most frequent, being recorded in 38% and 36% of the galls with cecidophagous insects. These two insect orders were also indicated as the most represented among the secondary fauna inhabiting cynipid galls (Hymenoptera). Kleptoparasites were represented by Cecidomyiidae (Diptera) and Tanaostigmatidae (Hymenoptera), which were obtained from leaf, stem, bud, and fruit galls, being more frequent in the leaf galls (69%). Cecidomyiids, in this case, were considered kleptoparasites based on all criteria except for the omnivory. The inquilines, on the other hand, were represented only by Eulophidae (Hymenoptera), and modifications in gall tissues were found in three gall morphotypes. Some authors believe that the modification of plant tissues by inquilines only occurs in systems where the gall inducer and the inquiline are phylogenetically related (Mani 1964; Van Noort et al. 2007; Luz and Mendonça Junior 2019). This criterion was not met in systems found across Restinga galls, and therefore, other four criteria were proposed: food habit—they feed on gall tissues, coexistence with the inducer, modification or production of new gall tissues, and reduced mobility (Maia 2022).

2.1 Natural Enemies' Effects on Galling Insect Populations

Parasitoids and inquilines stand out among the diverse organisms that exploit galls as resources, exerting substantial influence on the survival of gall-inducing insects (Abrahamson et al. 1989). Parasitoids, for instance, employ a unique deadly strategy: they lay their eggs on or inside other organisms, primarily insects and arachnids. These unwitting hosts become nurseries for the parasitoid offspring (Askew 1971; Price 1980). As the parasitoid larvae grow, they feed on the host tissues, ultimately leading to the host's death (Price 1980). These organisms, the most common natural enemies of gall-inducing insects in nature, also impose considerable pressure on gall-inducer populations in a top-down regulated control (Weis and Abrahamson 1985; Price and Clancy 1986; Hawkins and Gagné 1989; Hawkins et al. 1997). This guild has distinct behaviors that can be categorized as ecto and endoparasites (Godfray 1994; Hanson and Nishida 2014). The ectoparasite females inject venom into the host larva, typically paralyzing it (Askew and Shaw 1974; Hanson and Nishida 2014), whereas the endoparasite females lay eggs inside the host larva's body without inducing paralysis (Askew and Shaw 1974; Hanson and Nishida 2014). The ectoparasites primarily feed on the surface of the host larva, while the endoparasites feed within the host larva's body, temporarily delaying its development. In both scenarios, the gall inducer's body is eventually consumed (Godfray 1994; Hanson and Nishida 2014; Mayhew 2016).

By parasitizing the gall-inducing organisms, the parasitoids reduce their fitness and survival rates, thereby impacting the overall structure and development of the galls and controlling the gall-inducing population (Abrahamson et al. 1989). As an example, Kelch et al. (2016) showed a top-down effect of parasitoids on a gall population associated with *Fagus sylvatica* L. (Fagaceae) in a forest-fragmented landscape, where endoparasitoids had a considerable edge influence on *Mikiola fagi* Hartig, 1839 (Diptera: Cecidomyiidae) mortality, with a total parasitism rate of 0.39 in the forest interior versus 0.27 along the forest edge. The comparison of these mean values with the three remaining mortality causes (predation, unknown and fungus) indicated that the total parasitism rate might be the most important top-down factor on *M. fagi* and a key to explaining higher survival at the forest edge.

In a recent paper conducted in the forest steppe, it was demonstrated that *Sigmophora brevicornis* Panzer, 1804 (Hymenoptera: Eulophidae), a gregarious ectoparasitoid, exerts significant control over the population density of *Asphondylia miki* Wachtl, 1880 (Diptera: Cecidomyiidae), a gall midge species associated with the alfalfa *Medicago sativa* L. (Fabaceae) (Kaplin et al. 2024). This ectoparasitoid targets well-developed galls hosting medium- and late-instar larvae and pupae of the gall midge, thereby impacting its population dynamics, consuming over 60% of gall-inducing insect population. Additionally, other ectoparasitoids such as *Eurytoma dentata* Mayr, 1878 (Hymenoptera: Pteromalidae) parasitize the galls induced by *Asphondylia midges*, further influencing the regulation of gall midge populations. By parasitizing the larvae and pupae within the gall, this ectoparasitoid disrupts gall formation and development, potentially reducing the overall damage caused by the gall midge on the host plant. Furthermore, the role of parasitoids as natural enemies contributes to the biological control of *A. miki* populations.

The occurrence of two or more parasitoid species attacking the same host suggests their potential collaborative role in controlling the gall-inducer population, as demonstrated in 23 gall morphotypes (Maia and Silva 2021). Multiparasitism (attack of more than one parasite) can be linked with hyperparasitism (parasitoids attacking other species of parasitoids) (Maia and Monteiro 1999), a case not yet documented in the Brazilian Cerrado. The presence of multiple guilds within the same gall morphotype serves as an example of how the associated fauna can contribute to complex food webs. By collecting data on the presence of this guild and its attacks on gallers, we can infer that the importance of parasitoids in the ecology of galls is crucial. They play a central role in controlling host populations and influencing the structure and functioning of the surrounding ecological communities.

The natural enemies directly impact the population of gall inducers, although these effects can be difficult to evaluate. Inquilines, for instance, can often cause the death of the gall-inducing insect through competitive pressure, reducing the fitness of the primary gall-inducer offspring, or by consuming resources that would otherwise be available for the development of galler larvae (Maia 2022). The inquiline species *Dasineura balsamicola* Lintner, 1888 (Diptera: Cecidomyiidae) exerts significant control over the population of its host, the gall-inducing insect *Paradiplosis tumifex* Gagné, 1978 (Diptera: Cecidomyiidae) (Guay et al. 2016) on the host tree *Abies balsamea* (L.) Mill. (Pinaceae). Initially. *D. balsamicola* larvae actively seek out young galls of *P. tumifex*, positioning themselves within the developing gall tissues, leading to a close association where both larvae feed, and develop together within the gall. As summer progresses, the development of *P. tumifex* is strongly limited by that of the inquiline, allowing only *D. balsamicola* larvae to survive and exit the gall in the autumn to overwinter. Gall dissection studies have revealed varying infestation levels of inquilines by *D. balsamicola* and parasitism of *P. tumifex* by endoparasitoids, indicating the potential for complex interactions influencing the overall population dynamics of the gall-inducing insect species. Moreover, factors such as overwintering habitats, environmental temperatures, spring emergence, and egg-laying by midges (Osgood and Gagné 1978; Tabadkani et al. 2011) can significantly affect the relationship between the gall-inducing and inquilines at the population level, influencing the abundance and survival of both species within the host plant-galling system. In summary, an inquiline can play a crucial role in shaping the population dynamics and interactions within a gall-making system, impacting the survival, development, and overall success of its host.

In addition to their impacts on gall populations, inquilines can also exert significant effects that extend to the development and metabolism of the gall structure. By cohabiting within the gall, inquilines may introduce biochemical changes or alter the nutrient dynamics within the gall environment (Fenwick 2013), consequently influencing its developmental trajectory. This influence can lead to observable changes in the gall morphology, potentially impacting the overall fitness and reproductive success of the primary gall-inducing organism. Furthermore, the presence of inquilines within the gall ecosystem may introduce competition for resources or alter resource availability, thereby affecting the survival and development of the primary gall-inducing species (Baine et al. 2023). These effects highlight the intricate interplay between inquilines and primary gall inducers, emphasizing the complexity of interactions within gall-making systems, and their broader ecological implications.

2.2 *Natural Enemies' Effects on Galls Structure and Metabolism*

The presence of natural enemies, such as parasitoids and inquilines, can impact the structure and development of galls. Inquilines alter gall development by cohabiting within gall tissues, introducing biochemical changes, and modifying developmental pathways, creating new structures called endogalls. Additionally, parasitoids disrupt gall formation by modifying gall-inducing larvae behavior, indirectly influencing gall development. These interactions underscore the complex interplay between gall-inducing insects and their antagonists, re-shaping the dynamics of gall development.

2.2.1 Inquiline Effects on Gall Structure

The main function of a gall is to provide nutrition and protection to the offspring of the gall inducer (Stone and Schönrogge 2003), but some organisms, like inquilines, can impact the function of this structure. Inquilines are specialist herbivores that live inside the galls formed by other organisms (Sanver and Hawkins 2000). They feed on the gall tissues but do not directly feed on the primary gall inducer (Mani 1964). Inquilines are not only capable of exploiting the galls to obtain nutritional resources but in some cases, they also induce changes in gall structure and metabolism (Stone and Schönrogge 2003; Redfern 2011; Fenwick 2013; Ward et al. 2020).

Hawkins and Goeden (1984) reported the manipulation of gall tissues by inquilines. They described a new species of a *Tetrastichus* sp. inquiline wasp that oviposits into galls formed by the midge Atriplex sp. and modifies its tissue structure, forming endogalls. The authors suggest that the formation of endogalls is due to the mitotic activation of parenchyma cells within the gall and suggests that the sclerification of the endogall wall is not due to mechanical damage, but rather to some unknown mechanism. The formation of the new tissues by inquilines compromises the gall-inducing insect, ultimately causing its death. These results opened new avenues of research into the diversity and complexity of gall-forming insect communities. The ability of *Tetrastichus* wasp to induce the endogalls is a remarkable example of plant-insect co-evolution. The wasp has evolved a way to exploit the resources of the gall, while also killing the midge larva that created the gall in the first place by competing stress. This suggests that the wasp and the midge have been engaged in a long-term evolutionary arms race, with each species adapting to the other's strategies. Overall, Hawkins and Goeden's (1984) findings provided valuable insight base into the diversity and complexity of inquiline-gall interaction.

Inquilines that developed endogalls, also described as hypergallers (Baine et al. 2023), manipulate plant responses to control cecidogenesis at a specific level. The inquiline wasp *Periclistus* spp. (Hymenoptera: Cynipidae), that attacks the galls induced by *Diplolepis* spp. (Hymenoptera: Cynipidae) is a prime example of this strategy, as it can modify galls formed by other gall-inducing insects to create its own unique habitat (Brooks and Shorthouse 1988; Fenwick 2013). Two species of *Periclistus* sp. wasps oviposit into immature galls of *D. polita* and *D. nebulosa* on *Rosa aciculares* Lindl. (Rosaceae) leaves, respectively. The wasp eggs induce the galls to enlarge and develop Periclistus-specific tissues, creating a safe and nutritious environment for its wasp larvae. One remarkable feature of these endogalls is a second layer of sclerenchyma that develops around each inquiline chamber. This layer provides additional protection for the *Periclistus* sp. larvae and isolates them from the *Diplolepis* sp. larvae. The developmental pattern of these endogalls also differs from that of the primary gall inducer. Accordingly, the ability of *Periclistus* sp. to induce the formation of this second layer of sclerenchyma suggests the evolution of unique mechanisms for manipulating gall development (Fenwick 2013).

Maia (2001, 2024) recorded cecidomyiid galls modified by inquilines in four gall systems of Restinga environment, *Myrciamyia maricaensis* Maia, 1995 on *Myrcia ovata* Cambess (Myrtaceae); *Paulliniamyiia ampla* Maia, 2001 on *Paullinia weinmaniaefolia* Mart (Sapindaceae) (Maia 2001); *Stephomyia rotundifoliorum* Maia, 1993 on *Eugenia astringens* Cambess (Myrtaceae) and *Stephomyia mina* Maia, 1993 on *Neomitranthes obscura* (DC.) N. Silveira (Myrtaceae). at the time no analysis regarding the effects on plant tissues were made. *S. rotundifoliorum* induces cylindrical bud galls on *E. astringens* that are completely modified in size by an unknown inquiline, increasing the overall robustness of tissues (Fig. 2a–c). The *M. maricaensis* galls are attacked by one inquiline that completely modify its morphotype, creating various endogalls inside the primary gall (Fig. 2d–f). The gall tissues change drastically with the presence of the inquiline, developing hypertrophy of vascular bundles, a robust common storage tissue, and typical nutritive tissue around the endogall chamber (Fig. 2f), which is feed by them. The new tissues induced by natural enemy compressing the galling insect larvae, causing its eminent death. Additionally, the *P. weinmannifolia* galls attacked by an Eulophidae inquiline show the same pattern of formation of the endogalls (Fig. 2g–i), where the hypergallers induce the formation of the same new tissues as the ones developed on *M. maricaensis*, although these natural enemies do not compress the larval chamber that much (Fig. 2i). More recently, Maia (2024) also reported the formation of endogalls in other two gall-inducing systems that occur in the Restinga environment. Additionally, there are also galls of *S. mina* Maia, 1993 (Diptera: Cecidomyiidae) on *Neomitranthes obscura* (DC.) N.Silveira (Myrtaceae), attacked by an inquiline capable of inducing endogalls (Fig. 2j–l).

Hypergall, thus endogalls induced by hypergaller inquilines, outcomes depend on interactions between hypergall-inducing species and primary gall inducers. As

Fig. 2 Galls and graphic representations of their associated endogall's structure. (**a–c**) Unattacked bud galls and endogalls on *Eugenia astringens*. (**a**) *E. astringens* bud galls induced by *Stephomyia rotundifoliorum*. (**b**) Endogalls. (**c**) Illustration of endogalls disposition within gall tissues. (**d–f**) Formation of endogalls by unknown hypergaller inquiline sp. on galls induced by *Myrciamiia maricaensis* on *Myrcia ovata*. (**d**) Mature unmodified galls of *M. maricaensis*. (**e**) Modified *M. maricaensis* galls by unknown inquiline hypergaller sp., note the huge difference in morphotype. (**f**) Illustration of endogall anatomic structure. (**g–i**) Formation of endogalls by unknown hypergaller inquiline sp. on galls induced by *Paulliniamiia ampla* on *Paullinia weinmannifolia*. (**g**) Mature unattacked *P. ampla* gall. (**h**) Modified galls by hypergaller inquiline forming endogalls. (**i**) llustration of endogalls in *P. ampla* galls, note that the inquilines endogalls might not compress the galler's larval chamber. (**j–l**) *Neomitranthes obscura* galls and endogalls. (**j**) *Stephomyia mina* galls induced on *N. obscura* leaves. (**k**) Endogalls induced by unknown inquiline hypergaller species. (**l**) Illustration of endogalls disposition on anatomic structure. Scale bar: 5.0 mm. CST Common Storage Tissue, C Cortex, ENT Endogall Nutritive Tissue, EP Epidermis, EST Endogall Storage Tissue, GLC Galling Larval Chamber, IH Inquiline hypergaller, La Galler Larvae, ILC Inquiline Larval Chamber, NT Nutritive Tissue, MZ Mechanical Zone

shown by Baine et al. (2023), *Aciurina* sp. (Diptera: Tephritidae) galls on *Ericameria nauseosa* (Pursh) G.L. Nesom & G.I. Baird (Asteraceae) are hypergalled by *Rhopalomyia bigeloviae* Cockerell, 1889 (Diptera: Cecidomyiidae). These hypergalls were significantly larger than those without them, occurred at all seven collecting sites, and were present on 39% of collected *Aciurina* galls. Although anatomical analysis was not performed, the authors provided some insight into the overall gall structure: the hypergall formation starts shortly after the start of the development of *Aciurina* sp. galls. The single-chambered hypergalls, which protrude about 1 mm from the primary gall surface when fully developed, remain hidden under the thick white hairs of the primary gall for most of the hypergaller larvae's development. Hypergaller midges overwinter as larvae in small hypergalls that enlarge in early spring. After 6 months, the midges pupate, with adults emerging later that same month.

One major effect of hypergalls is the galling insects' death, however, the findings of Baine et al. (2023) show no correlation between the number of hypergalls and the size of the primary galler. Thus, *Rhopalomyia* does not compete for nutrition with the primary gall inducer, even when hypergalls entirely cover a primary gall. Therefore, hypergall outcomes are complex and context-dependent, in some cases the presence of inquilines/hypergallers does not cause the galling insect death. In fact, in some cases they can serve as protection for the gall-inducing insect, as the enlargement of the structure can effectively hide the inducer from other natural enemies.

A question remains: are hypergallers gall-inducing insects or only opportunistic species? The answer to that is much deeper and more complex than it appears. One of the hypotheses of the evolution of hypergallers is that these inquilines are "ex-gall-inducers" that lost their ability to induce galls on unmodified plant tissues (Mani 1964; Van Noort et al. 2007). Hypergaller cynipids, for instance, have usually lost the ability to induce galls but retained the ability to modify the gall tissue directly surrounding them into the characteristic nutritive tissue also found in the larval chambers of the gall inducer (Sanver and Hawkins 2000; László and Tóthmérész 2006). Gobbo et al. (2020) evaluated the lifestyle transition from inquiline to gall inducer by comparing the genomes of the cynipid *Synergus itoensis*, a gall inducer, with those of related inquilines. They identified genes that are likely involved in the transition from inquilinism to gall induction and found that these genes are under positive selection in the gall-inducing lineage. They also performed a selection analysis to identify these genes and found that they are enriched for functions related to plant-insect interactions. Overall, these hypergallers can be considered gall-inducing insects but only when attacking modified plant tissues.

2.2.2 Parasitoid Effects on Gall Structure

Gall development is a complex process influenced by a variety of factors, including the gall-inducing insect, the host plant, and the environment (Mani 1964; Shorthouse and Rohfritsch 1992; Isaias et al. 2014). Once the gall-inducing insect manipulates the host plant tissue to its own benefit, its feeding behavior can strongly determine the gall structure and morphology (Stone and Schönrogge 2003; Oliveira et al. 2016; Ferreira et al. 2019). However, parasitoids can alter the behavior of gall-inducing insects, depending on the specificity of the parasitoid-host-plant interaction, thereby influencing the structure and morphology of the gall (Askew 1961; Fujii et al. 2014). Some structural traits, such as the parasitoid ovipositor length, the gall size, the gall wall thickness, and the gall development stage influence the parasitism rates (Waring and Price 1989; Luz et al. 2021; Zhao et al. 2023; Fujii et al. 2014). Some parasitoids may have a more pronounced impact on gall development than others, while others may exhibit greater selectivity in the types of galls they target (Bailey et al. 2009). Generally, parasitoid attacks are associated with the premature senescence of induced galls, this can occur due to the disruption of gall development by alteration in the galling insects' feeding behavior (Rezende et al. 2019; Costa et al. 2022).

Host manipulation by insect parasitoids has primarily been studied in association with mobile hosts, with limited research on endophytic, immobile, or sessile hosts (Vinson and Iwantsch 1980; Libersat et al. 2009). Parasitoids exert profound effects on host behavior and physiology, often manipulating them to benefit their own development and reproduction (Nufio and Papaj 2001; Weinersmith 2019). For instance, parasitoid wasps like Ampulex compressa Fabricius, 1781 (Hymenoptera: Ampulicidae) inject venom into their cockroach hosts, triggering behavioral changes that render the host immobile and submissive, creating an environment conducive to the development and emergence of the parasitoid larva (Libersat et al. 2009). This manipulation of host behavior is a common strategy employed by parasitoids to ensure the success of their life cycle (Libersat et al. 2009).

The impact of parasitoids on the behavior of gall-inducing insects appears to affect the structural, cytological, and histochemical properties of gall tissues (Bronner 1992; Rezende et al. 2019). For instance, Costa et al. (2022) demonstrated that *Torymus* sp. (Hymenoptera: Torymidae) and Platygastridae (Hymenoptera) parasitoids, attacking *Lopesia* sp. (Diptera: Cecidomyiidae) galling insect larvae, disrupt their feeding activity, resulting in cytological and histochemical changes in gall tissues (Fig. 3a–d). Torymus sp., an ectoparasitoid, ultimately kills *Lopesia* sp., halting gall development (Fig. 3c). On the same note, Platygastridae an endoparasitoid also changes *Lopesia* sp. gall features (Fig. 3d). In both parasitized galls, the impairment of the feeding activity of the galling *Lopesia* sp. by parasitoids impacts the storage and nutritive cells toward cell collapse, disintegration of membrane systems, plastids, and pyknotic nuclei, leading a cell death and precocious senescence of the galls (Fig. 3c, d). Despite the difference in parasitoid attack strategies, the lipid gradient is maintained in the nutritive cells of both endo- and ectoparasitized galls, indicating the continuation of gall metabolism even after changes in *Lopesia* sp. larval feeding activity stimuli. The translocation and storage of lipids are crucial for the survival and future reproduction of the parasitoids (Mayhew 2016), as the polyembryonic development of Platygastridae larvae indirectly relies on lipids accumulated in the nutritive cells of globoid bivalve-shaped galls, which are ingested by *Lopesia* sp. larvae while alive (Hanson and Nishida 2014; Iwabuchi 2019; Costa

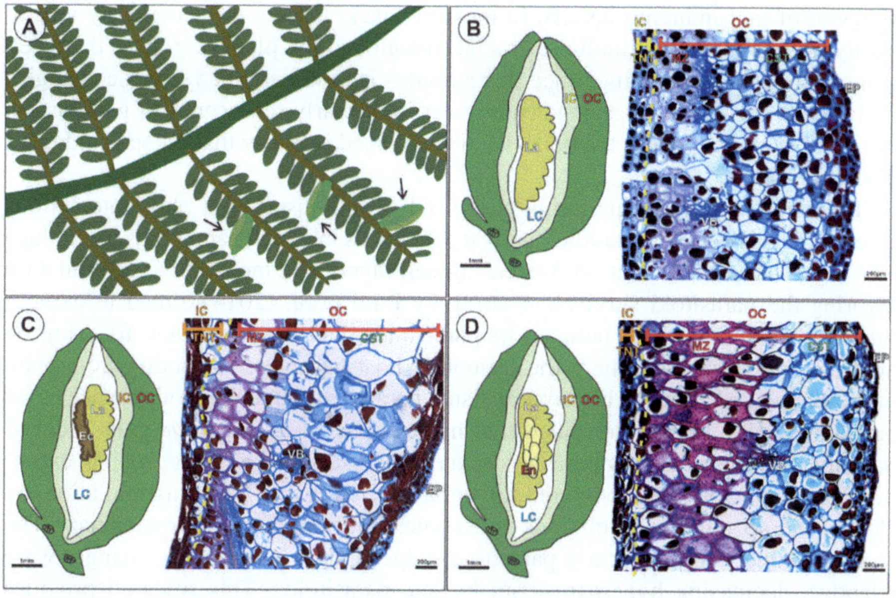

Fig. 3 Parasitoid attack effect on gall tissues. (**a**) Graphic illustration of Mimosa gemmulata galled and non-galled leaflets. (**b**) Unattacked *Lopesia* sp. gall and its anatomic structure. (**c**) *Lopesia* sp. galls attacked by a *Torymus* sp. ectoparasitoid, evidencing precocious senescence. (**d**) *Lopesia* sp. galls attacked by a Platygastridae sp. endoparasitoid, evidencing precocious senescence. CST Common Storage Tissue, Ec Ectoparasitoid, En Endoparasitoid, EP Epidermis, La Galling larvae, LC Larval Chamber, IC Inner Cortex, OC Outer Cortex, MZ Mechanical Zone, TNT Typical Nutritive Tissue, VB Vascular Bundle

et al. 2022). After the attack, host plant cell responses show disintegration of membrane systems, plastids, and nuclei. Similar histochemical profiles for lipids, reactive oxygen species (ROS), and secondary metabolites indicate comparable gall metabolism and ultimately lead to premature gall senescence. Thus, parasitoid-induced changes in host physiology can significantly affect gall development and senescence, irrespective of the parasitoid life cycle.

The galls induced by *Palaeomystella oligophaga* Becker & Adamski, 2008 (Lepidoptera: Momphidae) on *Macairea radula* (Bonpl.) DC. (Melastomataceae) showed a typical nutritive tissue by lepidopterans (Rezende et al. 2019), whose nutritive cells are filled with lipids that the larvae consume directly through its chewing feeding habit (Guedes et al. 2023). The stoppage of larval feeding and insect pupation triggered the histochemical and cytological breakdown of gall tissues, especially the nutritive tissue, initiating gall senescence. However, the attack by both cecidophages and parasitoids can stop the feeding behavior of the galling larvae, by leading them to death, which triggers the histochemical and cytological breakdown early in gall tissue (Rezende et al. 2019). These findings underscore the dependency of typical nutritive tissue development and its metabolism on uninterrupted larval feeding stimulus. The attack of parasitoids modifies the feeding behavior of gall-inducing insects, in most of cases, ceasing the feeding by venom paralysis. The feeding stimuli are crucial to maintain the plant responses that keep gall metabolism functioning, once this stimulus is modified, the response changes. In this case, this behavior change culminates in the early senescence of gall tissues. But what would happen if the behavior was changed in a way that the gall-inducing insect still feeds?

The behavioral manipulation in parasitoid-galling insect-host plant interactions is few explored yet. Some studies show mechanisms in which parasitoids can induce active gall tissue modification, increasing cell layers and metabolite accumulation favoring the parasitoid survival. A study by Fujii et al. (2014) aimed to explore interactions among a gall inducer, its parasitoids, and the host plant, focusing on how the endoparasitoid alters the immobile host herbivore to modify host plant morphology and prevent hyperparasitism, parasitoids attacking other species of parasitoids. Key findings revealed that hypertrophy of *M. pustulae* galls reduced parasitism by late attacking parasitoids, and galls attacked by *Platygaster* sp. exhibited larger thicknesses compared to unparasitized galls. This manipulation by *Platygaster* sp. culminates on thick galls, potentially avoiding hyperparasitism. The study also uncovered that early parasitoid-induced plant morphology changes were adaptive, decreasing hyperparasitism by late parasitoids. This research provides insights into the intricate interactions among gall inducers, parasitoids, and host plants, highlighting adaptive strategies employed by parasitic organisms for survival and reproductive success in their ecological niche.

Another evidence of behavioral manipulation is the occurrence of the endoparasitoid *Platygaster* sp. within galls induced by a cecidomyiid on *Piptocarpha macropoda* (DC.) Baker (Asteraceae) (Rezende et al. 2022). The tri-trophic interaction resulted in an enlargement of gall size and an increase in the number of nutritive cell layers, which accumulated lipids as energetic reserves (Rezende et al. 2022). This hyperstimulation in the plant cells is likely attributed to the heightened feeding requirements of both the cecidomyiid and the endoparasitoid. Despite the larger gall size, parasitized cecidomyiid larvae maintain comparable body sizes to unparasitized larvae, indicating that the endoparasitoid alters the gall tissue structure without augmenting the size of the galling insect. Additionally, the presence of the endoparasitoid influences the timing of gall senescence, causing parasitized galls to undergo senescence earlier than their unparasitized counterparts. In summary, the results suggest that the endoparasitoid *Platygaster* sp. has a complex and multifaceted impact on the gall-inducing cecidomyiid and consequently on the structure of galls.

3 Some Questions Remain…

Natural enemies exert a significant impact on the gall multitrophic system. Some, like parasitoids and inquilines, have a higher rate of attack, making their effects more pronounced. Parasitoids are commonly used as biological controls for free-living organisms, as their attacks ultimately lead to death, halting the production of further offspring. In insect galls, parasitoids occupy the same guild, exerting substantial effects on gall-inducing insect populations and on the structure of the multitrophic community of arthropods associated with the gall and the host plant. Although inquilines might significantly control gall-inducing insect populations, they almost usurp the role of the galler and can cause more damage to the host plant. This is because they are usually gregarious and induce hypertrophy of the attacked galls, forming endogalls.

The strategy behind the modifications of gall tissues by parasitoids is still unknown, leading to several questions and assumptions when gathering data from different authors. Is the parasitoid indeed capable of manipulating the behavior of the gall-inducing insect, and how often does the effect of parasitoid attack reflect in gall tissues? This type of interaction is difficult and rare to observe, as gall systems are closed microecosystems, and any disturbance might interfere with the behaviors of both insects, potentially causing high levels of stress and resulting in their death. Additionally, can a single species of parasitoid modify the behavior of more than one gall-inducing insect, especially when they are gregarious? Or is it more common for parasitoids to specialize in attacking species of solitary galling insects? When galling insects are gregarious, with more than one larva in the larval chamber, multiple stimuli from them are required to maintain the gall-inducing structure; therefore, parasitoid attacks in this case might reflect on the gall structure. Hence, are solitary parasitoid larvae more likely to induce differences in gall tissues than gregarious parasitoid larvae? The impact of gregarious parasitoids might be more severe than that of solitary ones, resulting in no time for the behavior manipulation of the galling insect to be reflected in gall tissues, as the consumption of the host galler might be faster in this case.

Few studies demonstrate the developmental changes caused by inquilines to form endogalls, which need to be more profoundly studied. Are their biochemical cues the same as the primary gall inducer? Since the maintenance of the functional gall structure is mostly due to feeding behavior and the compounds involved in this interaction, inquilines might utilize the same triggering compounds when they attack the gall structure. What might be the effects on gall-inducing insect offspring that are not killed by the competing pressure of endogalls? Gall formation is a stressful process that accumulates many degradative compounds such as ROS, and the accumulation of stress-related compounds could lead to a decrease in the fitness of the galling insect offspring in this case. Thus, even in cases where endogalls do not kill the galling insect, they surely affect its fitness, as the gall structure becomes a shared environment among numerous individuals. Furthermore, understanding the intricacies of these interactions could shed light on broader ecological dynamics within galling systems.

In general, studying the outcomes of specific interactions, such as the effects that natural enemies have on gall structure and metabolism, can be the first step towards something much greater. Galls are underestimated when it comes to fully describing the system and the diversity that remains unknown and might be lost due to the degradation of natural ecosystems. Therefore, by understanding the mechanisms that hold gall multitrophic systems together, we might be able to fully comprehend their importance in maintaining local and spatial diversity in endangered ecosystems across the globe. Additionally, unraveling these complexities could lead to more effective conservation strategies aimed at preserving these delicate ecosystems.

References

Abrahamson WG, Sattler JF, McCrea KD et al (1989) Variation in selection pressures on the goldenrod gall fly and the competitive interactions of its natural enemies. Oecologia 79:15–22. https://doi.org/10.1007/BF00378234

Aranda-Rickert A, Rothen C, Diez P, González AM, Marazzi B (2017) Sugary secretions of wasp galls: a want-to-be extrafloral nectar? Ann Bot 120:765–774. https://doi.org/10.1093/aob/mcx075

Askew RR (1961) The biology of the British species of the genus *Olynx* Förster (Hymenoptera: Eulophidae), with a note on seasonal color forms in the Chalcidoidea. Proc R Entomol Soc Lond A 36:103–112

Askew RR (1971) Parasitic insects. American Elsevier, New York

Askew RR, Shaw MR (1974) An account of the Chalcidoidea (Hymenoptera) parasitising leaf-mining insects of deciduous trees in Britain. Biol J Linn Soc 6(4):289–335

Bailey R, Schönrogge K, Cook JM, Melika G, Csóka G et al (2009) Host niches and defensive extended phenotypes structure parasitoid wasp communities. PLoS Biol 7(8):e1000179. https://doi.org/10.1371/journal.pbio.1000179

Baine Q, Casares EE, Carabotta EC, Martison VG, Martison EO (2023) Galls on galls: a hypergall-inducing midge and its parasitoid community. Ecology 104(5):e4018. https://doi.org/10.1002/ecy.4018

Bonsignore C, Bernardo U (2018) Effects of environmental parameters on the chestnut gall wasp and its complex of indigenous parasitoids. Sci Nat 105. https://doi.org/10.1007/s00114-018-1545-1

Bourg A, Hanson P (2014) Host specificity of gall midges (Diptera: Cecidomyiidae) on ten species of Inga (Fabaceae). In: Fernandes G, Santos J (eds) Neotropical insect galls. Springer, Dordrecht. https://doi.org/10.1007/978-94-017-8783-3_10

Bronner R (1992) The role of nutritive cells in the nutrition of cynipids and cecidomyiids. In Shorthouse JD, Rohfritsch O (eds) Biology of insect-induced galls. Oxford University Press, pp 135–157

Brooks SE, Shorthouse JD (1988) Developmental morphology of stem galls of *Diplolepis nodulosa* (Hymenoptera: Cynipidae) and those modified by the inquiline *Periclistus pirate* (Hymenoptera: Cynipidae) on *Rosa blanda* (Rosaceae). Can J Bot 76:365–381

Campos GBD, Costa EC, Santos DLS, Carvalho-Fernandes SP, Santos-Silva J (2021) Insect galls and associated fauna in two areas of Cerrado sensu stricto in the State of Bahia, Brazil. An Acad Bras Cienc 93:e20201442. https://doi.org/10.1590/0001-3765202120201442

Costa EC, Martini VC, Souza-Silva A, Lemos-Filho JP, Oliveira DC, Isaias RMS (2021a) How galling herbivores share a single superhost plant during their phenological cycle: the case of *Mimosa gemmulata* Barneby (Fabaceae). Trop Ecol 63:61–74. https://doi.org/10.1007/s42965-021-00182-1

Costa EC, Oliveira DC, Ferreira D, Isaias R (2021b) Structural and nutritional peculiarities related to lifespan differences on four *Lopesia* induced bivalve-shaped galls on the single super-host *Mimosa gemmulata*. Front Plant Sci 12. https://doi.org/10.3389/fpls.2021.660557

Costa EC, Oliveira DC, Isaias RMS (2022) Parasitoid impairment on the galling *Lopesia* sp. activity reflects on the cytological and histochemical profiles of the globoid bivalve-shaped gall on *Mimosa gemmulata*. Protoplasma 259:1585–1597. https://doi.org/10.1007/s00709-022-01756-z

Crespi B, Abbot P (1999) The behavioral ecology and evolution of kleptoparasitism in Australian gall thrips. Fla Entomol 82(2):147–164. https://doi.org/10.2307/3496568

Fenwick B (2013) Comparative morphogenesis of cynipid leaf galls induced by *Diplolepis polita* and *Diplolepis nebulosa* (Hymenoptera: Cynipidae) and modification by inquilines of the genus *Periclistus* (Hymenoptera: Cynipidae). Master's thesis

Ferreira BG, Álvarez R, Bragança GP et al (2019) Feeding and other gall facets: patterns and determinants in gall structure. Bot Rev 85:78–106. https://doi.org/10.1007/s12229-019-09207-w

Fujii T, Matsuo K, Abe Y, Yukawa J, Tokuda M (2014) An endoparasitoid avoids hyperparasitism by manipulating immobile host herbivore to modify host plant morphology. PLoS One 9(7):e102508. https://doi.org/10.1371/journal.pone.0102508

Giannetti D, Castracani C, Spotti FA, Mori A, Grasso DA (2019) Gall-colonizing ants and their role as plant defenders: from 'bad job' to 'useful service'. Insects 10:1–21

Gobbo E, Lartillot N, Hearn J, Stone GN, Abe Y, Wheat CW, Ronquist F (2020) From inquilines to gall inducers: Genomic signature of a life-style transition in *Synergus* gall wasps. Genome Biol Evol 12(11):2060–2073

Godfray HCJ (1994) Parasitoids: behavioral and evolutionary ecology. Princeton University Press Books, Princeton

Guay JF, Bulot D, Béland JM, Cloutier C (2016) Gall-maker *Paradiplosis tumifex* (Diptera: Cecidomyiidae) and its inquiline *Dasineura balsamicola* (Diptera: Cecidomyiidae): an update on epidemic episodes and seasonal ecology in Québec, Canada. Can Entomol 148(4):452–465. https://doi.org/10.4039/tce.2015.80

Guedes LM, Costa EC, Isaias RMS, Sáez-Carillo K, Aguilera N (2023) Structural patterns of Lepidoptera galls and the case of *Andescecidium parrai* (Cecidosidae) galls on *Schinus polygama* (Anacardiaceae). J Plant Res 136:715–728. https://doi.org/10.1007/s10265-023-01472-6

Hanson P, Nishida K (2014) Insect galls of Costa Rica and their parasitoids. In: Fernandes GW, Santos JC (eds) Neotropical insect galls. Springer, Germany, pp 497–517

Harper LJK, Schönrogge KY, Lim P, Francis P, Lichtenstein CP (2004) Cynipid galls: insect-induced modifications of plant development create novel plant organs. Plant Cell Environ 27(3):327–335

Harris MO, Pitzschke A (2020) Plants make galls to accommodate foreigners: some are friends, most are foes. New Phytol 225:1852–1872. https://doi.org/10.1111/nph.16340

Hawkins BA, Goeden RD (1984) Organization of a parasitoid community associated with a complex of galls on *atriplex* spp. in southern California. Ecol Enthomol Lond 9:271–292

Hawkins BA, Gagné RJ (1989) Determinants of assemblage size for the parasitoids of Cecidomyiidae (Diptera). Oecologia 81:75–88

Hawkins BA, Cornell HV, Hochberg ME (1997) Predators, parasitoids and pathogens as mortality agents in phytophagous insect populations. Ecology 78:2145–2152

Hori K (1992) Insect secretions and their effect on plant growth, with special reference to hemipterans. In: Shorthouse JD, Rohfritsch O (eds) Biology of insect-induced galls. Oxford University Press, New York, pp 157–170

Isaias RMS, Carneiro RGS, Santos JC, Oliveira DC (2014) Gall morphotypes in the Neotropics and the need to standardize them. In: Fernandes GW, Santos JC (eds) Neotropical insect galls. Springer, New York, pp 51–67

Iwabuchi K (2019) Polyembryonic insects: an extreme clonal reproductive strategy. Springer Nature Singapore Pte Ltd., Tokyo University of Agriculture and Technology Fuchu, Tokyo, p 209. https://doi.org/10.1007/978-981-15-0958-2

Jorge NC, Vasconcelos HL, Freitas MSC, Vale FHA, Isaias RMS (2022) The peculiar post-senescence in globoid stem galls triggered by fungi and arboreal ants on *Eremanthus erythropappus* (DC.) McLeisch (Asteraceae). S Afr J Bot 150:1026–1030

Kaplin V, Kosheleva O, Volodina I (2024) Relationships of the gall midge *Asphondylia miki* Wachtl (Diptera, Cecidomyiidae) with its host plant, the alfalfa *Medicago sativa*, and ectoparasitoids of its larvae and pupae (Hymenoptera, Eulophidae) in the forest-steppe of the middle volga region. Entomol Rev 103:524–533. https://doi.org/10.1134/S0013873823050032

Kelch NS, Neves FS, Fernandes GW, Wirth R (2016) Mechanisms driving galling success in a fragmented landscape: synergy of habitat and top-down factors along temperate forest edges. PLoS One 11(6):e0157448. https://doi.org/10.1371/journal.pone.0157448

Kuzmanich N, Giorgis MA, Bernaschini L, Tavella J, Salvo A (2023) Galling insect communities mediate the effects of fire on their associated parasitoid communities. Insect Conserv Divers 16:817–827

László Z, Tóthmérész B (2006) Inquiline effects on a multilocular gall community. Acta Zool Acad Sci Hung 52(4):373–383

Libersat F, Delago A, Gal R (2009) Manipulation of host behavior by parasitic insects and insect parasites. Annu Rev Entomol 54(1):189–207. https://doi.org/10.1146/annurev.ento.54.110807.090556

Lima VP, Calado D (2018) Morphological characterization of insect galls and new records of associated invertebrates in a Cerrado area in Bahia State, Brazil. Braz J Biol 78(4):636–643. https://doi.org/10.1590/1519-6984.169502

Luz FA, Mendonça Junior MS (2019) Guilds in insect galls: who is who. Fla Entomol 102(1):207–210. https://doi.org/10.1653/024.102.0133

Luz FA, Gonçalves GL, Moreira GR, Becker VO (2015) Description, molecular phylogeny, and natural history of a new kleptoparasitic species of gelechiid moth (Lepidoptera) associated with Melastomataceae galls in Brazil. J Nat Hist 49:1849–1875

Luz FA, Goetz APM, Mendonca MDS Jr (2021) What drives gallers and parasitoids interacting on a host plant? A network approach revealing morphological coupling as the main factor. Ecol Entomol 46:334–341

Magalhães TA, Oliveira DC, Suzuki AYM (2014) Patterns of cell elongation in the determination of the final shape in galls of *Baccharopelma dracunculifoliae* (Psyllidae) on *Baccharis dracunculifolia* DC (Asteraceae). Protoplasma 251:747–753. https://doi.org/10.1007/s00709-013-0574-z

Maia VC (2001) The gall midges (Diptera, Cecidomyiidae) from three restingas of Rio de Janeiro State, Brazil. Rev Bras Zool 18(2):583–629. http://dx.doi.org/10.1590/S0101-81752001000200028

Maia VC (2022) Review of the inquilinous fauna associated with insect galls in Brazilian restingas. Braz J Biol 81(4):949–968. https://doi.org/10.1590/1519-6984.234454

Maia VC (2024) Gall-inducing insects from the Maricá Environmental Protection Area (RJ, Southeastern Brazil). Rev Bras Entomol 68(2):e20230086. https://doi.org/10.1590/1806-9665-RBENT-2023-0086

Maia VC, Fernandes GW (2004) Insect galls from Serra de São José (Tiradentes, Minas Gerais, Brazil). Braz J Biol 64:423–445

Maia VC, Monteiro RF (1999) Espécies cecidógenas (Oiptera: Cecidomyiidae) e parasitóides (Hymenoptera) associadas a *Guapira opposita* (Vell.) Reitz. (Nyctaginaceae) na Restinga da Barra de Maricá, Rio de Janeiro. Rev Bras Zool 16(2):483–487

Maia VC, Silva BG (2021) Insect galls of the Brazilian Cerrado: associated fauna. Biota Neotrop 21(3):e20211202. https://doi.org/10.1590/1676-0611-BN-2021-1202

Mani MS (1964) Ecology of plant galls. Springer

Marchiori CH (2021) Mini review of parasitoids collected in the Cerrado Biome in Brazil. Int J Environ Agric Biotechnol 6:3–182

Martini V, Raymundo DP, Júnior J, Oliveira DC (2021) Bottom-up and top-down forces in plant-gall relationships: testing the hypotheses of resource concentration, associational resistance and host fitness reduction. Ecol Entomol 46:1072–1081. https://doi.org/10.1111/een.13043

Mayhew PJ (2016) Comparing parasitoid life histories. Entomol Exp Appl 159(2):147–162

Nufio CR, Papaj DR (2001) Host marking behavior in phytophagous insects and parasitoids. Entomol Exp Appl 99:273–293. https://doi.org/10.1046/j.1570-7458.2001.00827.x

Oliveira DC, Isaias RMS (2010) Redifferentiation of leaflet tissues during midrib gall development in *Copaifera langsdorffii* (Fabaceae). S Afr J Bot 76:239–248

Oliveira DC, Moreira ASFP, dos Santos Isaias RM (2014) Functional gradients in insect gall tissues: studies on neotropical host plants. In: Fernandes G, Santos J (eds) Neotropical insect galls. Springer, Dordrecht. https://doi.org/10.1007/978-94-017-8783-3_3

Oliveira DC, Isaias RMS, Fernandes GW, Ferreira BG, Carneiro RGS, Fuzaro L (2016) Manipulation of host plant cells and tissues by gall inducing insects and adaptive strategies used by different feeding guilds. J Insect Physiol 84:103–113. https://doi.org/10.1016/j.jinsphys.2015.11.012

Osgood EA, Gagné RJ (1978) Biology and taxonomy of two gall midges (Diptera: Cecidomyiidae) found in galls on balsam fir needles with description of a new species of *Paradiplosis*. Ann Entomol Soc Am 71:85–91

Perdikis D, Arvaniti K, Malliaraki S, Angelaki A (2015) The outbreak of the olive leaf gall midge populations and the importance of indigenous natural enemies in its control. In: Perdikis D, Stathas G, Papadopoulos N, Lucchi A (eds) Proceedings of the IOBC/WPRS working group "integrated protection of olive crops", Kalamata, Greece, 11–14 May 2015, pp 119–122

Pfeffer L, Rezende UC, Barônio GJ, de Oliveira DC (2018) Building two houses on a single host plant: galling insect synchronizes its life cycle with plant phenology. Oecol Aust 22:438–448. https://doi.org/10.4257/oeco.2018.2204.07

Price PW (1980) Evolutionary biology of parasites. Princeton University Press, Princeton

Price PW, Clancy KM (1986) Interactions among three trophic levels: gall size and parasitoid attack. Ecology 67(6):1593–1600. https://doi.org/10.2307/1939090

Price PW, Waring G, Fernandes G (1986) Hypotheses on the adaptive nature of galls. Proc Entomol Soc Wash 88:361–363

Redfern M (2011) Plant galls. In: The new naturalist library 117. Collins, London

Rezende UC, Custodio JCF, Kuster VC, Goncalves LA, Oliveira DC (2019) How the activity of natural enemies changes the structure and metabolism of the nutritive tissue in galls? Evidence from the *Palaeomystella oligophaga* (Lepidoptera)–*Macairea radula* (Metastomataceae) system. Protoplasma 256:669–667. https://doi.org/10.1007/s00709-018-1321-2

Rezende UC, de Castro Jorge N, Dolabela B et al (2022) Endoparasitoidism: Reflections on the structure of a cecidomyiid-gall on Piptocarpha macropoda (DC.) Baker (Asteraceae). Biol Bull Russ Acad Sci 49(Suppl 1):S198–S202. https://doi.org/10.1134/S1062359022130179

Rezende UC, Cardoso JCF, Hanson P, Oliveira DC (2021) Gall traits and galling insect survival in a multi-enemy context. Rev Biol Trop 69(1):291–301. https://doi.org/10.15517/rbt.v69i1.42826

Rezende UC, Cardoso JCF, Gonçalves PHP, Oliveira DC (2023) The multitrophic system configuration on galls of *Macairea radula* (Melastomataceae) induced by *Palaeomystella oligophaga* (Lepidoptera) depends on abiotic events. Int J Trop Insect Sci 43:2095–2104. https://doi.org/10.1007/s42690-023-01104-5

Rohfritsch O (1992) Patterns in gall development. In: Shorthouse JD, Rohfritsch O (eds) Biology of insect-induced galls. Oxford University Press, New York, pp 60–86

Santos MG, Porto GF, Lancellotti IR, Feitosa RM (2019) Ant fauna associated with *Microgramma squamulosa* (Kaulf.) de la Sota (Polypodiaceae) fern galls. Rev Bras Entomol 63:101–103

Santos JC, Maruyama PK, Fernandes GW (2021) Vacant rooms? The secondary use of stem galls by ants in *Eremanthus erythropappus* (Asteraceae). J Environ Anal Prog 6:108–112

Santos ATB, Moitinho RR, Lima VP, Calado DC (2023) From leaves to inflorescences: Gall induction of *Iatrophobia brasiliensis* Rubsaamen, 1915 on inflorescences of *Manihot caerulescens* Pohl (Euphorbiaceae) during the dry season. Rev Bras Entomol 67(4)

Sanver D, Hawkins BA (2000) Galls as habitats: the inquiline communities of insect galls. Basic Appl Ecol 1(1):3–11

Shorthouse JD, Rohfritsch O (1992) Biology of insect-induced galls. Oxford University Press, New York, p 285

Stone GN, Schönrogge K (2003) The adaptive significance of insect gall morphology. Trends Ecol Evol 18(10):512–522

Tabadkani SM, Khansefid M, Ashouri A (2011) Monogeny, a neglected mechanism of inbreeding avoidance in small populations of gall midges. Entomol Exp Appl 140:77–84

Toma TSP, Mendonça Junior MS (2014) Population ecology of galling arthropods in the Neotropics. In: Fernandes GW, Santos JC (eds) Neotropical insect galls. Springer, New York, pp 69–98

Van Noort S, Stone GN, Whitehead VB, Nieves-Aldrey JL (2007) Biology of *Rhoophilus loewi* (Hymenoptera: Cynipoidea: Cynipidae), with implications for the evolution of inquilinism in gall wasps. Biol J Linn Soc 90:153–172

Vinson SB, Iwantsch GF (1980) Host regulation by insect parasitoids. Q Rev Biol 55(2):143–165. https://doi.org/10.1086/411731

Ward AKG, Sheikh SI, Forbes AA (2020) Diversity, host ranges, and potential drivers of speciation among the inquiline enemies of oak gall wasps (Hymenoptera: Cynipidae). Insect Syst Divers 4(6):1–13. https://doi.org/10.1093/isd/ixaa017

Waring GL, Price PW (1989) Parasitoid pressure and the radiation of a gall forming group (Cecidomyiidae: *Asphondylia* spp.) on creosote bush (*Larrea tridentata*). Oecologia 79:293–299

Weinersmith KL (2019) What's gotten into you? a review of recent research on parasitoid manipulation of host behavior. Curr Opin Insect Sci 33:37–42

Weis AE, Abrahamson WG (1985) Potential Selective Pressures by Parasitoids on a Plant-Herbivore Interaction. Ecology 66(4):1261–1269. https://doi.org/10.2307/1939179

Weis AE, Abrahamson WG (1988) Potential selective pressures by parasitoids on a plant herbivore interaction. Ecology 66:1261–1269

Zamora R, Gómez JM (1993) Vertebrate herbivores as predators of insect herbivores: an asymmetrical interaction mediated by size differences. Oikos 66:223–228

Zhao Q, Zhang DK, Jiang LL, Guo J, Hu HY (2023) Globular galls on Haloxylon (Centrospermae: Chenopodiaceae) induced by gall midges, and the corresponding parasitoids in China. Appl Entomol Zool 58:345–356

Index

D. C. d. Oliveira, R. Isaias (eds.), *Plant Galls*,
https://doi.org/10.1007/978-3-031-80064-1

MIX
Papier aus verantwortungsvollen Quellen
Paper from responsible sources
FSC® C105338

If you have any concerns about our products,
you can contact us on
ProductSafety@springernature.com

In case Publisher is established outside the EU,
the EU authorized representative is:
Springer Nature Customer Service Center GmbH
Europaplatz 3, 69115 Heidelberg, Germany

Printed by Libri Plureos GmbH
in Hamburg, Germany